T0141890

Lecture Notes on Data Engineering and Communications Technologies   **165**

Series Editor

Fatos Xhafa, *Technical University of Catalonia, Barcelona, Spain*

The aim of the book series is to present cutting edge engineering approaches to data technologies and communications. It will publish latest advances on the engineering task of building and deploying distributed, scalable and reliable data infrastructures and communication systems.

The series will have a prominent applied focus on data technologies and communications with aim to promote the bridging from fundamental research on data science and networking to data engineering and communications that lead to industry products, business knowledge and standardisation.

Indexed by SCOPUS, INSPEC, EI Compendex.

All books published in the series are submitted for consideration in Web of Science.

Yap Bee Wah · Michael W. Berry ·
Azlinah Mohamed · Dhiya Al-Jumeily, OBE
Editors

Data Science and Emerging Technologies

Proceedings of DaSET 2022

 Springer

Editors
Yap Bee Wah
UNITAR Graduate School
UNITAR International University
Selangor, Malaysia

Azlinah Mohamed
Institute for Big Data Analytics and Artificial
Intelligence
Universiti Teknologi MARA (UiTM)
Shah Alam, Selangor, Malaysia

Michael W. Berry
University of Tennessee
Knoxville, TN, USA

Dhiya Al-Jumeily, OBE🆔
School of Computer Science
and Mathematics
Liverpool John Moores University
Liverpool, UK

ISSN 2367-4512   ISSN 2367-4520 (electronic)
Lecture Notes on Data Engineering and Communications Technologies
ISBN 978-981-99-0743-4   ISBN 978-981-99-0741-0 (eBook)
https://doi.org/10.1007/978-981-99-0741-0

© The Editor(s) (if applicable) and The Author(s), under exclusive license
to Springer Nature Singapore Pte Ltd. 2023, corrected publication 2023
This work is subject to copyright. All rights are solely and exclusively licensed by the Publisher, whether the whole or part of the material is concerned, specifically the rights of translation, reprinting, reuse of illustrations, recitation, broadcasting, reproduction on microfilms or in any other physical way, and transmission or information storage and retrieval, electronic adaptation, computer software, or by similar or dissimilar methodology now known or hereafter developed.
The use of general descriptive names, registered names, trademarks, service marks, etc. in this publication does not imply, even in the absence of a specific statement, that such names are exempt from the relevant protective laws and regulations and therefore free for general use.
The publisher, the authors, and the editors are safe to assume that the advice and information in this book are believed to be true and accurate at the date of publication. Neither the publisher nor the authors or the editors give a warranty, expressed or implied, with respect to the material contained herein or for any errors or omissions that may have been made. The publisher remains neutral with regard to jurisdictional claims in published maps and institutional affiliations.

This Springer imprint is published by the registered company Springer Nature Singapore Pte Ltd.
The registered company address is: 152 Beach Road, #21-01/04 Gateway East, Singapore 189721, Singapore

Preface

This volume constitutes the proceedings of the International Conference on Data Science and Emerging Technologies (DaSET 2022) held from December 20 to 21, 2022, on a virtual platform. DasET 2022 aims to provide a platform bringing together experts from academia, industries, government, and professional bodies to share recent trends in Artificial Intelligence for Data-Driven Decisions and Emerging Technologies. DaSET is committed to creating a forum that brings academic and industry practitioners to share and establish collaborations toward impactful innovative research for community development, business success, and economic prosperity. This conference is an international conference in collaboration with Liverpool John Moores University, UK. DasET 2022 was also supported by Universiti Teknologi MARA, Universiti Teknologi Malaysia, and the following international universities Institut Teknologi Sepuluh Nopember, Indonesia; Chulalongkorn University, Thailand; Charles Sturt University, Australia; Institut Teknologi Bandung, Indonesia, Prince of Songkla University, Thailand, and Data Analytics and Collaborative Computing Group, University of Macau, China. We also have strong support from the Department of Statistics (DOSM), Malaysia, Malaysia Digital Economy Corporation (MDEC), STATWORKS (M), Sdn. Bhd, Microsoft, and SIRIM Berhad Malaysia.

From a total of 82 submitted papers, 38 were selected after a rigorous review process for oral presentation, and the Best Paper Awards were given for each track. The authors and presenters for these 38 papers represented 12 different countries.

We are proud to have seven distinguished keynote speakers: Dato' Sri Dr Uzir Mahidin, Chief Statistician, Department of Statistics Malaysia; Professor Dhiya Al-Jumeily, OBE, Liverpool John Moores University, UK; Professor Jamila Mustafina, Visiting Professor, University of Anbar, Iraq; Professor Alfredo Cuzzocrea, University of Calabria, Italy; Associate Professor Dr Siva Kumar Balasundram, Universiti Putra Malaysia, Malaysia; Dr Esther Loo, Malaysia Airlines Sdn Bhd and Adjunct Professor of UNITAR, and Mr Navin Sinnathamby, Malaysia Digital Economy Corporation (MDEC). All the distinguished speakers shared various data science and emerging technologies perspectives and projects which are beneficial for academics and industry practitioners.

We would like to thank Professor Emeritus Tan Sri Dato' Sri Ir Dr Sahol Hamid Bin Abu Bakar, Vice Chancellor of UNITAR International University for his great leadership, advice, and support of local and international academic activities to foster collaborations that lead to the exchange of knowledge and skills for research with impactful outcomes for social and economic prosperity.

We also thank Series Editor, Springer, Lecture Notes on Data Engineering and Communications Technologies for the opportunity to organize this guest-edited volume. We are grateful to Mr. Aninda Bose (Senior Publishing Editor, Springer India Pvt. Ltd.), for the excellent collaboration, patience, and help during the preparation of this volume. We are confident that the volume will provide insightful information to researchers, practitioners, and graduate students in the areas of data science, artificial intelligence, and emerging technologies which are important in the digital information era.

We thank all the reviewers for their time spent reviewing the papers. Last but not least, we thank all the DaSET 2022 committees for working tirelessly to ensure a successful conference.

Organization

Patron

Sahol Hamid Bin Abu Bakar (Vice Chancellor)
UNITAR International University, Malaysia

Conference Chairs

Yap Bee Wah
UNITAR International University, Malaysia

Dhiya Al-Jumeily, OBE
Liverpool John Moores University, UK

International Advisory Committee

Mario Koeppen
Kyushu Institute of Technology, Japan

Heri Kuswanto
Institute Teknologi Sepuluh Nopember

Ku Ruhana Ku Mahmud
Universiti Utara Malaysia, Malaysia

Naomi Bt Salim
Universiti Teknologi Malaysia, Malaysia

Mohammed Bennamoun
University of Western Australia, Australia

Chidchanok Lursinsap
Chulalongkorn University, Thailand

Proceeding Editors

Yap Bee Wah
UNITAR International University, Malaysia

Michael W. Berry
University of Tennessee, USA

Azlinah Mohamed
Universiti Teknologi MARA, Malaysia

Dhiya Al-Jumeily, OBE
Liverpool John Moores University, UK

Finance and Registration Committee

Chong Kim Loy
UNITAR International University, Malaysia

Normaiza Binti Mohamad
UNITAR International University, Malaysia

Danny Ngo Lung Yao
UNITAR International University, Malaysia

Secretary

Fazida Salim	UNITAR International University, Malaysia
Lim Shu Yun	UNITAR International University, Malaysia

Technical Program Committee

Abdulaziz Al-Nahari	UNITAR International University, Malaysia
Azlin Ahmad	Universiti Teknologi MARA, Malaysia
Wasiq Khan	Liverpool John Moores University, UK
Abir Jaafar Hussain	Liverpool John Moores University, UK
Anazida Binti Zainal	Universiti Teknologi Malaysia, Malaysia
Wan Fairos Wan Yaacob	Universiti Teknologi MARA, Malaysia
Sapto Indratno	Institut Teknologi Bandung, Indonesia
Syerina Azlin Mohd Nasir	Universiti Teknologi MARA, Malaysia
Ahmad Zia Ul-Saufie	Universiti Teknologi MARA, Malaysia

Program Book Committee

Noor Lees Ismail	UNITAR International University, Malaysia
Iznora Aini Binti Zolkifly	UNITAR International University, Malaysia
Azrin Abdul Razak	UNITAR International University, Malaysia
Rohaizah Abd Latif	UNITAR International University, Malaysia
Jan Lunn	Liverpool John Moores University, UK

Logistics Committee

Norita Binti Zakaria	UNITAR International University, Malaysia
Asmaie Nor Aliuddin Ekhsan Firoz	UNITAR International University, Malaysia
Aminah Binti Abd Karim	UNITAR International University, Malaysia
Feroza Ahmad Faiz	UNITAR International University, Malaysia
Nurul Nabihah binti Zainol	UNITAR International University, Malaysia
Abdul Halim Jamaluddin	UNITAR International University, Malaysia
Haliza Binti Mohd Said	UNITAR International University, Malaysia
Rudzi Binti Munap	UNITAR International University, Malaysia
Ker Boon Chin	UNITAR International University, Malaysia
Mohd Aiman Mohd Sani	UNITAR International University, Malaysia
Ahmad Ruzaini bin Rahim	UNITAR International University, Malaysia

Harpreet Kaur A/P. Gochan Singh   UNITAR International University, Malaysia
Visnuvarthen A/L. Sakayam   UNITAR International University, Malaysia
Mohd Farizudin Mohammad Fauzi   UNITAR International University, Malaysia
Muhammad Shahrir Aizat   UNITAR International University, Malaysia
  Muhammad Shuhaili

Publicity Committee

Badrie Abdullah (Advisor)   UNITAR International University, Malaysia
Noor Azma Binti Ismail   UNITAR International University, Malaysia
Hoshang Kolivand   Liverpool John Moores University, UK
Arno T. Immelman   Prince of Songkla University, Thailand
Paridah Binti Daud   UNITAR International University, Malaysia
Wan Zakiyatussariroh Wan Husin   Universiti Teknologi MARA, Malaysia
Bander Ali Saleh   Universiti Teknologi Malaysia, Malaysia
Adatul Mukarromah   Institut Teknologi Sepuluh Nopember, Indonesia

Corporate Committee

Mohamad Shah Andrew Ibrahim   UNITAR International University, Malaysia
Stella Chua Ching Yee   UNITAR International University, Malaysia
Janette Pang Wei Fong   UNITAR International University, Malaysia
Nikki Poh Li Yi   UNITAR International University, Malaysia

International Scientific Committee

Adel Al-Jumaily   Charles Sturt University, Australia
Siddhivinayak A. Kulkarni   MIT-World Peace University, Pune, India
Simon Fong   University of Macau, China
Richard Millham   Durban University of Technology, South Africa
Layth Sliman   Efrei, Paris, France
Faiz Ahmed Mohamed Elfaki   Qatar University, Qatar

Reviewers

Abdulaziz Al-Nahari   UNITAR International University, Malaysia
Abdullah Almogahed   Universiti Utara Malaysia, Malaysia
Abdullah Alqahtani   University of Idaho, USA

Abdulrazak Yahya	University Malaysia Sarawak, Malaysia
Abir Hussain	University of Sharjah, United Arab Emirates
Achmad Choiruddin	Institut Teknologi Sepuluh Nopember, Indonesia
Adam Shariff Adli Aminuddin	Universiti Malaysia Pahang, Malaysia
Adel Al-Jumaily	Charles Sturt University, Australia
Agus Harjoko	Universitas Gadjah Mada, Indonesia
Ahmad Zia Ul-Saufie Mohamad Japeri	Universiti Teknologi MARA, Malaysia
Aigul Khanova	Kazan Federeal University, Russia
Alaa Aljanaby	Waikato Pathways College, New Zealand
Ali Gurbuz	Mississippi State University, USA
Ameer Albahem	Funnelback Core Search, Australia
Anazida Zainal	Universiti Teknologi Malaysia, Malaysia
Ashraf Osman	University Malaysia Sabah, Malaysia
Azizi Ab. Aziz	Universiti Utara Malaysia, Malaysia
Azlan Iqbal	Universiti Tenaga Nasional, Malaysia
Azlin Ahmad	Universiti Teknologi MARA, Malaysia
Azlinah Mohamed	Universiti Teknologi MARA, Malaysia
Azman Taa	Universiti Utara Malaysia, Malaysia
Bagus Jati Santoso	Institut Teknologi Sepuluh Nopember, Indonesia
Bander Al-rimy	Universiti Teknologi Malaysia, Malaysia
Bilal Khan	University of California, USA
Bong Chih How	Universiti Malaysia Sarawak, Malaysia
Chew XinYing	Universiti Sains Malaysia, Malaysia
Chidchanok Lursinsap	Chulalongkorn University, Thailand
Chin Kim On	Universiti Malaysia Sabah, Malaysia
Chong Kim Loy	UNITAR International University, Malaysia
Choo Yun Huoy	Universiti Teknikal Malaysia Melaka, Malaysia
Choong Seon Hong	Kyung Hee University, Korea (South)
Costin Badica	University of Craiova, Romania
Dedy Prastyo	Institut Teknologi Sepuluh Nopember, Indonesia
Dipti Theng	G. H. Raisoni College of Engineering, India
Dittaya Wanvarie	Chulalongkorn University, Thailand
Eng Harish Kumar	King Khalid University, Saudi Arabia
Faisal Saeed	Birmingham City University, UK
Faiz Ahmed Elfaki	Qatar University, Qatar
Fathey Mohammed	Universiti Utara Malaysia, Malaysia
Feras Zen Alden	Cihan University Erbil, Iraq
Fuad A. Ghaleb	Universiti Teknologi Malaysia, Malaysia
Gamal Abdulnaser Alkawsi	Universiti Tenaga Nasional, Malaysia
Gualberto Asencio-Cortes	Pablo de Olavide University, Spain
Hamzah Abdul Hamid	Universiti Malaysia Perlis, Malaysia

Heri Kuswanto	Institut Teknologi Sepuluh Nopember, Indonesia
Hizir Sofyan	Syiah Kuala University, Indonesia
Hojjat Adeli	The Ohio State University, USA
Hoshang Kolivand	Liverpool John Moores University, UK
Ilham Slimani	Mohammed First University, Morocco
Iznora Aini Zolkifly	UNITAR International University, Malaysia
Jamila Mustafina	Kazan Federal University, Russia
Jan Lunn	Liverpool John Moores University, UK
José Peña	Universidad Politécnica de Madrid, Spain
Karim Al-Saedi	Mustansiriyah University, Iraq
Lau Sian Lun	Sunway University, Malaysia
Layth Sliman	EFREI- Paris, France
Majed Nasser	Universiti Sains Malaysia, Malaysia
Marshima Mohd Rosli	Universiti Teknologi MARA, Malaysia
Mas Rina Mustaffa	Universiti Putra Malaysia, Malaysia
Michael Houle	National Institute of Informatics, Japan
Michael Tan Loong Peng	Universiti Teknologi Malaysia, Malaysia
Mohamed Imran Mohamed Ariff	Universiti Teknologi MARA, Malaysia
Mohammad Alaomari	Universiti Tenaga Nasional, Malaysia
Mohammed Bennamoun	The University of Western Australia, Australia
Mohd Fadzil Hassan	Universiti Teknologi PETRONAS, Malaysia
Muhammad Firdaus Mustapha	Universiti Teknologi MARA, Malaysia
Ng Kok Haur	University of Malaya, Malaysia
Nor Saradatul Akmar Zulkifli	Universiti Malaysia Pahang, Malaysia
Noraini Seman	Universiti Teknologi MARA, Malaysia
Noriko Etani	Kyoto University, Japan
Nur Iriawan	Institut Teknologi Sepuluh Nopember, Indonesia
Nur Atiqah Sia Abdullah	Universiti Teknologi MARA, Malaysia
Nurulaqilla Khamis	Universiti Teknologi Malaysia, Malaysia
Pakawan Pugsee	Chulalongkorn University, Thailand
Qais Alnuzaili	Al-Nasser University, Yemen
Raseeda Hamzah	Universiti Teknologi MARA, Malaysia
Rayner Alfred	Universiti Malaysia Sabah, Malaysia
Redhwan Al-Amri	Taylor's University, Malaysia
Retantyo Wardoyo	Universitas Gajah Mada, Indonesia
Richard Millham	Durban University of Technology, South Africa
Rizauddin Saian	Universiti Teknologi MARA, Malaysia
Rizwan Aslam Butt	NED University of Engineering and Technology, Pakistan
Rohit Gupta	Indian Institute of Technology, India
Roselina Sallehuddin	Universiti Teknologi Malaysia, Malaysia
Ruhaila Maskat	Universiti Teknologi MARA, Malaysia

Saiful Akbar — Institut Teknologi Bandung, Indonesia
Sakhinah Bakar — Universiti Kebangsaan Malaysia, Malaysia
Sharifalillah Nordin — Universiti Teknologi MARA, Malaysia
Shuzlina Abdul-Rahman — Universiti Teknologi MARA, Malaysia
Shuzlina Rahman — Universiti Teknologi MARA, Malaysia
Siddhivinayak Kulkarni — Griffith University, Australia
Siti Othman — Universiti Teknologi Malaysia, Malaysia
Siti Sakira Kamaruddin — Universiti Utara Malaysia, Malaysia
Sultan Almalki — University of Idaho, USA
Suraya Masrom — Universiti Teknologi MARA, Malaysia
Syerina Azlin Md Nasir — Universiti Teknologi MARA, Malaysia
Temitayo Fagbola — Durban University of Technology, South Africa
Thiago Silva — University of São Paulo, Brazil
Umara Urooj — Universiti Teknologi Malaysia, Malaysia
Usama Al-mulali — Multimedia University, Malaysia
Uswah Khairuddin — Universiti Teknologi Malaysia, Malaysia
Vincenzo Nicosia — University of Catania, Italy
Wahyu Wibowo — Institut Teknologi Sepuluh Nopember, Indonesia
Wan Fairos Wan Yaacob — Universiti Teknologi MARA, Malaysia
Wasiq Khan — Liverpool John Moores University, UK
Xianwei Zhou — University of Science and Technology Beijing, China

Yahia Baashar — University Sarawak Malaysia, Malaysia
Yap Bee Wah — UNITAR International University, Malaysia
Yazan AL-Khassawneh — Zarqa University, Jordan
Yuen Yee Yen — Multimedia University, Malaysia
Yuhanis Yusof — Universiti Utara Malaysia, Malaysia
Yusuke Nojima — Osaka Metropolitan University, Japan
Zalhan Mohd Zin — Universiti Kuala Lumpur Malaysia France Institute, Malaysia

Zuraida Abal Abas — Universiti Teknikal Malaysia Melaka, Malaysia

Acknowledgements

Organized by

In collaboration with

Acknowledgement

We would like to thank Professor Emeritus Tan Sri Dato' Sri Ir Dr. Sahol Hamid Bin Abu Bakar, Vice Chancellor of UNITAR International University, for his outstanding leadership, advice, and support of local and international academic activities to foster collaborations that lead to the exchange of knowledge and skills for research with impactful outcomes for social and economic prosperity. We are also grateful to Mr. Puvan, CEO of UNITAR, and the corporate staff for their ongoing assistance and support.

We are very grateful for the support from Universiti Teknologi MARA, Universiti Teknologi Malaysia, Institut Teknologi Sepuluh Nopember, Indonesia; Chulalongkorn University, Thailand; Charles Sturt University, Australia; Institut Teknologi Bandung, Indonesia, Prince of Songkla University, Thailand, and Data Analytics and Collaborative Computing Group, University of Macau, China. We also appreciate the strong support from the Department of Statistics (DOSM), Malaysia, Malaysia Digital Economy Corporation (MDEC), STATWORKS (M) Sdn Bhd, Microsoft, and SIRIM Berhad Malaysia.

We would like to thank our keynote speakers: Dato' Sri Dr. Mohd Uzir Mahidin, Chief Statistician, Department of Statistics Malaysia; Professor Dhiya Al-Jumeily, OBE, Liverpool John Moores University, UK; Professor Jamila Mustafina, Visiting Professor, University of Anbar, Iraq; Associate Professor Dr. Siva Kumar Balasundram, Universiti Putra Malaysia, Malaysia; Dr. Esther Loo, Malaysia Airlines Bhd and Adjunct Professor of UNITAR, Mr. Navin Sinnathamby, Malaysia Digital Economy Corporation(MDEC) and Ms. Rohini a/p Sooriamoorthy, STATWORKS Group. We thank all our esteemed speakers for sharing various data science and emerging technologies perspectives and projects which are beneficial for academics and industry practitioners.

We gratefully acknowledge the tremendous support from all the technical reviewers who generously sacrificed their time to review the papers. We are grateful to Mr. Aninda Bose (Executive Editor, Springer Nature) for the excellent collaboration, patience, and help during the preparation of this volume. Lastly, we thank all conference committees for their dedication and commitment to ensuring the success of DaSET2022.

Contents

Artificial Intelligence

Extractive Text Summarization Using Syntactic Sub-graph Models 3
    Yazan Alaya AL-Khassawneh, Essam Said Hanandeh,
    and Sattam Almatarneh

Analysis of Big Five Personality Factors to Determine the Appropriate
Type of Career Using the C4.5 Algorithm 18
    Rizka Shinta Wulandari, Casi Setianingsih, and Purba Daru Kusuma

Predicting Disaster Type from Social Media Imagery via Deep Neural
Networks Directed by Visual Attention 37
    Shatheesh Kumar Govindarajulu, Megan Watson, Sulaf Assi,
    Manoj Jayabalan, Panagiotis Liatsis, Jamila Mustafina,
    Normaiza Mohamad, Kdasy Al-Muni, and Dhiya Al-Jumeily, OBE

Dissemination Management for Official Statistics Using Artificial
Intelligence-Based Media Monitoring 52
    Veronica S. Jamilat, Nur Hurriyatul Huda Abdullah Sani,
    Nur Aziha Mansor Noordin, Tengku Noradilah Tengku Jalal,
    and Sasongko Yudho

Computational Vision

A Naive but Effective Post-processing Approach for Dark Channel Prior
(DCP) 67
    Danny Ngo Lung Yao, Abdullah Bade, Iznora Aini Zolkifly,
    and Paridah Daud

COVID-19 Face Mask Classification Using Deep Learning 77
    Nik Amnah Shahidah Abdul Aziz, Muhammad Firdaus Mustapha,
    and Siti Haslini Ab Hamid

Gender Classification Using CNN Transfer Learning and Fine-Tuning 92
    Muhammad Firdaus Mustapha, Nur Maisarah Mohamad,
    and Siti Haslini Ab Hamid

Multi Language Recognition Translator App Design Using Optical
Character Recognition (OCR) and Convolutional Neural Network (CNN) 103
    Mohamad Khairul Naim Zulkifli, Paridah Daud, and Normaiza Mohamad

Autonomous Driving Through Road Segmentation Based on Computer
Vision Techniques 117
  Aditi Jain, Matthew Harper, Manoj Jayabalan, Jamila Mustafina,
  Wasiq Khan, Panagiotis Liatsis, Noor Lees Ismail,
  and Dhiya Al-Jumeily, OBE

Cybersecurity

Phishing Attack Types and Mitigation: A Survey 131
  Mohammed Fahad Alghenaim, Nur Azaliah Abu Bakar,
  Fiza Abdul Rahim, Vanye Zira Vanduhe, and Gamal Alkawsi

Review of Smart Home Privacy-Protecting Strategies from a Wireless
Eavesdropping Attack 154
  Mohammad Ali Nassiri Abrishamchi and Anazida Zainal

Development of Graph-Based Knowledge on Ransomware Attacks Using
Twitter Data 168
  Abdulrahman Mohammed Aqel Assaggaf, Bander Ali Al-Rimy,
  Noor Lees Ismail, and Abdulaziz Al-Nahari

Big Data Analytics

BigMDHealth: Supporting Multidimensional Big Data Management
and Analytics over Big Healthcare Data via Effective and Efficient
Multidimensional Aggregate Queries over Key-Value Stores 187
  Alfredo Cuzzocrea

Design and Implementation of Data Warehouse Solution at Kumpulan
Wang Persaraan (KWAP) 195
  Mohamad Fairul Hussein, Paridah Daud, Omar Musa,
  Normaiza Mohamad, and Noor Lees Ismail

Consumer Behavior Prediction During Covid-19 Pandemic Conditions
Using Sentiment Analytics 209
  Saravanan Murugan, Sulaf Assi, Abbas Alatrany, Manoj Jayabalan,
  Panagiotis Liatsis, Jamila Mustafina, Abdullah Al-Hamid,
  Maitham G. Yousif, Ahmed Kaky, Danny Ngo Lung Yao,
  and Dhiya Al-Jumeily, OBE

Big Data Application on Prediction of HDD Manufacturing Process
Performance 222
  N. G. Meng Seng, Abdulaziz Al-Nahari, Noor Azma Ismail,
  and Azlin Ahmad

Visualising Economic Situation Through Malaysia Economic Recovery
Dashboard (MERD) 237
  Wan Ahmad Ridhuan Wan Jaafar, Mazliana Mustapa,
  Fatin Ezzati Mohd Aris, Noradilah Adnan, and Ahmad Najmi Ariffin

Machine/Deep Learning

Lung Nodules Classification Using Convolutional Neural Network
with Transfer Learning 253
  Abdulrazak Yahya Saleh and Ros Ameera Rosdi

Plant Growth Phase Classification Using Deep Neural Network (Case
Study of ASF in Poso District, Central Sulawesi Province) 266
  Kevin Agung Fernanda Rifki and Kartika Fithriasari

The Implementation of Genetic Algorithm-Ensemble Learning
on QSAR Study of Diacylglycerol Acyltransferase-1(DGAT1) Inhibitors
as Anti-diabetes 282
  Irfanul Arifa, Annisa Aditsania, and Isman Kurniawan

Classification of Exercise Game Data for Rehabilitation Using Machine
Learning Algorithms 293
  Zul Hilmi Abdullah, Waidah Ismail, Lailatul Qadri Zakaria,
  Shaharudin Ismail, and Azizi Abdullah

SDDLA: A New Architecture for Secured Decentralized Distributed
Learning 305
  Sufyan Almajali

Gated Memory Unit: A Novel Recurrent Neural Network Architecture
for Sequential Analysis 316
  Arav Kumar and Gabriel Nasrallah

Multi-class Classification for Breast Cancer with High Dimensional
Microarray Data Using Machine Learning Classifier 329
  Mohammad Nasir Abdullah, Bee Wah Yap,
  Nik Nur Fatin Fatihah Sapri, and Wan Fairos Wan Yaacob

Machine Learning Techniques for Predicting Risks of Late Delivery 343
  Ravikanth Lolla, Matthew Harper, Jan Lunn, Jamila Mustafina,
  Jolnar Assi, Chong Kim Loy, and Dhiya Al-Jumeily, OBE

Quora Insincere Questions Classification Using Attention Based Model 357
Snigdha Chakraborty, Megan Wilson, Sulaf Assi, Abdullah Al-Hamid,
Maitham Alamran, Abdulaziz Al-Nahari, Jamila Mustafina, Jan Lunn,
and Dhiya Al-Jumeily, OBE

Suicide Ideation Detection: A Comparative Study of Sequential
and Transformer Hybrid Algorithms 373
Aniket Verma, Matthew Harper, Sulaf Assi, Abdullah Al-Hamid,
Maitham G. Yousif, Jamila Mustafina, Noor Azma Ismail,
and Dhiya Al-Jumeily, OBE

Well Log Data Preparation and Effective Utilization of Drilling Parameters
Using Data Science Based Approaches 388
Rahul Talreja, Thomas Coombs, Sulaf Assi, Noor Azma Ismail,
Manoj Jayabalan, Panagiotis Liatsis, Mohamed Mahyoub,
Abdullah Al-Hamid, and Hoshang Kolivand

Deep Learning-Based Approach for Classifying the Severity **of Metal
Corrosion Using Sem Images** 403
Saranga Veeramangal Hebbar, Basheera M. Mahmmod,
Iznora Aini Zolkifly, Abdulaziz Al-Nahari, and Sadiq H. Abdulhussain

Insurance Risk Prediction Using Machine Learning 419
Rahul Sahai, Ali Al-Ataby, Sulaf Assi, Manoj Jayabalan,
Panagiotis Liatsis, Chong Kim Loy, Abdullah Al-Hamid,
Sahar Al-Sudani, Maitham Alamran, and Hoshang Kolivand

Loan Default Forecasting Using StackNet 434
Saket Satpute, Manoj Jayabalan, Hoshang Kolivand, Jolnar Assi,
Omar A. Aldhaibani, Panagiotis Liatsis, Paridah Daud, Ali Al-Ataby,
Wasiq Khan, Ahmed Kaky, Sahar Al-Sudani, and Mohamed Mahyoub

Statistical Learning

Neural Network Autoregressive Model for Forecasting Malaysia Under-5
Mortality 451
Wan Zakiyatussariroh Wan Husin, Aina Nafisya Suhaimi,
Nur Shuhaila Meor Zambri, Muhammad Azri Aminudin,
and Nor Azima Ismail

Robustness of Support Vector Regression and Random Forest Models:
A Simulation Study 465
Supriadi Hia, Heri Kuswanto, and Dedy Dwi Prastyo

The Impact of Restricting Community Activities on COVID-19
Transmission: A Case Study in Sumatra Island, Indonesia 480
  Abdullah Sonhaji, Sapto W. Indratno, Kurnia Novita Sari,
  Adi Pancoro, Ernawati Arifin Giri-Rachman, Udjianna S. Pasaribu,
  and Susi Setiyowati

Predicting Internet Usage for Digital Finance Services: Multitarget
Classification Using Vector Generalized Additive Model with SMOTE-NC 494
  Wahyu Wibowo, Amri Muhaimin, and Shuzlina Abdul-Rahman

Text Mining and Classification

Identifying Topic Modeling Technique in Evaluating Textual Datasets 507
  Nik Siti Madihah Nik Mangsor, Syerina Azlin Md Nasir,
  Shuzlina Abdul-Rahman, and Zurina Ismail

Y-X-Y Encoding for Identifying Types of Sentence Similarity 522
  Thanaporn Jinnovart and Chidchanok Lursinsap

Evaluation of Extractive and Abstract Methods in Text Summarization 535
  Ranjita Kumari Biswal Lenka, Thomas Coombs, Sulaf Assi,
  Manoj Jayabalan, Jamila Mustafina, Panagiotis Liatsis,
  Abdullah Al-Hamid, Sahar Al-Sudani, Noor Lees Ismail,
  and Dhiya Al-Jumeily, OBE

Correction to: Well Log Data Preparation and Effective Utilization
of Drilling Parameters Using Data Science Based Approaches C1
  Rahul Talreja, Thomas Coombs, Sulaf Assi, Noor Azma Ismail,
  Manoj Jayabalan, Panagiotis Liatsis, Mohamed Mahyoub,
  Abdullah Al-Hamid, and Hoshang Kolivand

Author Index 547

About the Editors

Professor Yap Bee Wah is currently Director of Research and Consultancy Centre of UNITAR International University, Malaysia. She was formerly Faculty Member of the Centre of Statistical and Decision Science Studies at Faculty of Computer and Mathematical Sciences (FSKM), Universiti Teknologi MARA. She was also Head of Advanced Analytics Engineering Centre (AAEC), a research center of excellence in Faculty of Computer and Mathematical Sciences (2016–2020). In February 2021, AAEC became a Centre of Excellence in UiTM with the name Institute of Big Data Analytics and Artificial Intelligence (IBDAAI). She has supervised 15 Ph.D. students. She is Active Researcher and has published papers in ISI journals such as Expert Systems with Applications Journal of Statistical Computation and Simulation, Communications in Statistics Computation and Simulation and Journal of Clinical and Translational Endocrinology, and also in Scopus-indexed journals. She was Conference Chair of the International Conference on Soft Computing in Data Science (2015–2019 and 2021). She was also one of the editors of the SCDS conference proceedings published in Springer CCIS series. She was Guest Editor of Applied Soft Computing (Q1) journal and Pertanika Journal of Social Science and Humanities Special Issue (2016). She is also one of the editors of the book titled "Supervised and Unsupervised Learning for Data Science" published by Springer Nature Switzerland AG 2020. This book is in collaboration with Prof. Michael W. Berry, University of Tennessee, USA, and Prof. Azlinah Mohamed, Universiti Teknologi MARA.

Professor Michael W. Berry is Co-author and Editor of fifteen books covering topics in scientific computing, information retrieval, text/data mining, and data science. His most recent book entitled "Supervised and Unsupervised Learning for Data Science" was published in 2019 by Springer International Publishing. He is Co-editor of the Soft Computing in Data Science volumes from 2015 to 2019 published by Springer and is Co-author of popular books published by Society for Industrial and Applied Mathematics (SIAM): Understanding Search Engines: Mathematical Modeling and Text Retrieval, Second Edition, and Computational Information Retrieval. He has published over 115 refereed journal and conference publications. He has organized numerous workshops on text mining and was Conference Co-chair of the 2003 SIAM Third International Conference on Data Mining in San Francisco, CA. He was also Program Co-chair of the 2004 SIAM Fourth International Conference on Data Mining in Orlando, Florida, and is currently Honorary Co-chair of the International Conference on Soft Computing in Data Science (SCDS) series (2015–present). He is Member of SIAM, ACM, MAA, ASEE, and the IEEE Computer Society and is on the editorial board of Foundations of Data Science (AIMS) and the SIAM Journal on Matrix Analysis and Applications (SIAM). Professor Berry is also Certified Program Evaluator for the Computing Accreditation Commission (CAC) of the Accreditation Board for Engineering and Technology, Inc. (ABET).

Professor Dr. Azlinah Mohamed holds the title of Full Professor at the Institute for Big Data Analytics and Artificial Intelligence (IBDAAI), Universiti of Teknologi MARA (UiTM), Shah Alam, Malaysia. She has a strong managerial background and a series of industrial linkages. She is also one of the editors of the Soft Computing in Data Science, SCDS (2015–2019 and 2021) conference proceedings published in Springer CCIS series. She is also one of the editors of the book titled "Supervised and Unsupervised Learning for Data Science" published by Springer Nature Switzerland AG 2020. This book is in collaboration with Prof. Michael W. Berry, University of Tennessee, USA, and Prof. Yap Bee Wah, Universiti Teknologi MARA. Her current research interests are in the areas of big data, soft computing, artificial intelligence, and web-based decision support systems using intelligent agents in electronic government applications. She has good strategic appreciation and vision with a proven track record in supporting business and industry needs and highly focused with a consistent track record of successful and relevant academic programs with time and budget. Her research is well communicated in a series of conferences, journals, and high-impact journals indexed in ISI or Scopus.

Dr. Dhiya Al-Jumeily, OBE is Professor of Artificial Intelligence and President of eSystems Engineering Society. He has extensive research interests covering a wide variety of interdisciplinary perspectives concerning the theory and practice of applied artificial intelligence in medicine, human biology, environment, intelligent community, and health care. He has published well over 300 peer-reviewed scientific international publications, 17 books, and 17 chapters in multidisciplinary research areas including machine learning, neural networks, signal prediction, telecommunication fraud detection, AI-based clinical decision-making, medical knowledge engineering, human–machine interaction, intelligent medical information systems, sensors and robotics, and wearable and intelligent devices and instruments. But his current research passion is decision support systems for self-management of health and medicine. Dhiya has successfully supervised over 20 Ph.D. students' studies and has been External Examiner to various UK and international universities for undergraduate programs, postgraduate programs, and research degrees. He has been actively involved as Member of editorial board and review committee for a number of peer-reviewed international journals and acts as Program Committee Member or as General Chair for a number of international conferences. Dhiya is also successful Entrepreneur. He is Head of enterprise for the Faculty of Engineering and Technology. He has been awarded various commercial and research grants, nationally and internationally, over £7.5M from Overseas Research and Educational Partners, UK, through British Council and directly from industry with portfolio of various Knowledge Transfer Programs between academia and industry. He has a large number of international contacts and leads or participates in several international committees in his research fields. Dhiya has one patent and coordinated over ten projects at national and international levels.

Artificial Intelligence

Extractive Text Summarization Using Syntactic Sub-graph Models

Yazan Alaya AL-Khassawneh[1]([✉]), Essam Said Hanandeh[2], and Sattam Almatarneh[1]

[1] Data Science and Artificial Intelligence Department, Zarqa University, Zarqa, Jordan
{ykhassawneh,salmatarneh}@zu.edu.jo
[2] Computer Information Systems Department, Zarqa University, Zarqa, Jordan
hanandeh@zu.edu.jo

Abstract. Summarization systems are needed to handle the vast amounts of text data being collected by machine-controlled systems. These systems help users to get a better understanding of the main concepts of the paper. This type of summary is produced by people who produce extract summaries. It involves gathering significant excerpts from a text and not creating a new summary. The graph-based approaches are very popular in terms of text summarisation. This paper presents an example of generating single-document summaries using Dijkstra algorithm and Rectangles. The second phase is to construct sub-graphs and then bit vector representation. The final section is to retrieve the derived sentences.

Keywords: Extractive summarization · Graph-based summarization · Feature extraction · Dijkstra algorithm

1 Introduction

The emergence of the Internet, social networks, forums, information technologies are unfolding in an extremely rotating manner and are leading to increasingly problematic data interacting with one another in order to understand, create, store, develop and store it. The entire document still needs to be read in full to determine whether the knowledge it contains is relevant or not, however, this becomes a slow and overwhelming activity. But what if the information can be summarized in some way to produce keywords that will ease the re-leader's time and energy in making decisions? Hence, automatic text summaries are the answer to the Current Disadvantage (ATS). Let's clarify text summaries first: What is a text summary (TS)? A thematic TS could be a text that contains the most urgently needed information from one or many texts in a very simplified form.

TS: identification of the most relevant text; Interpretation of the data and receipt of a schema with the recorded information. The aim of ATS is to reduce the amount of text and at the same time to protect the first document as well as possible so that the browser can interpret the information read more quickly. The ATS has improved in quality due to the need to analyze large amounts of important information, for example: making summaries of books, comics, reviews, news, scientific articles, internet, social networks, and others. Any kind of text information is summarized.

© The Author(s), under exclusive license to Springer Nature Singapore Pte Ltd. 2023
Y. B. Wah et al. (Eds.): DaSET 2022, LNDECT 165, pp. 3–17, 2023.
https://doi.org/10.1007/978-981-99-0741-0_1

The overgrowth of data has forced researchers to look for other ways to get text summaries, and even its scope is fixed, they need tools that allow the content to be summarized with illustrated text. The ATS can be used in an extremely single document or in a multi-document, relying on the required specifications.

Since the increasing growth of net Technology, the quantity of electronic documents on the web has augmented exponentially. Today, folks will acquire info from numerous sources. Despite the progress achieved within the document account field, the matter hasn't been resolved fully. Concerning the machine-controlled document summarization, it seeks finding strategies for detective work the foremost important information existing during a text [1]. In addition, it's necessary to spot bound options through process data. That's required for reducing the amount of your time needed for accessing information [2]. Within the light-weight of that, people's interests in researching regarding automated text summarization systems has been increasing [3, 4].

Machine-controlled text counting technologies are being explored by mortals for more power through new or improved strategies [5]. However, in all of the studies that have been carried out in the area of document summary, the need for adaptation and creativity has not diminished [1]. Computerized document summarization could be a prominent subtopic of the language process (NLP) with the aim of presenting long text records in a compact and understandable manner [6].

Document summarization methods are generally divided into 2 types: extractive and abstract. The summary of the document consists of three phases: text description, punctuation and sentence composition. Theoretical document accounting strategies describe the key element of documents using language production methods and therefore re-express them to build a schema [7–9]. They are even divided into single and multiple documents according to the number of documents to be combined.

2 Related Work

Many authors have planned analysis work on ETS based on completely different techniques or ways, that helped in generating summaries of text document(s).

Anthers in [10] focused on supervised learning that analyzes the characteristics, dataset, and strategies used to select sentences. As long as there are two functions, the syntactic and the linguistic use in ETS. The range of grammatical features is generated on the basis of a data record. From this function; A relevant and informative schema (semantics) feature was created that is used in word2vec and Glove when embedding. Once each feature works together, it works based on the LSTMNN model. The corpus data set of the 41st International ACM Conference SIGIR 2018 was used, which consists of measurements that are summarized according to the ROUGE metric.

In Ref. [11] researchers mentioned many criteria for ETS from the multiple document. It is only based on two ETS criteria: reduction of redundancy and content coverage and, on the other hand, coherence and connectivity considered individually. These criteria are applied in the generic text account for multiple document extraction. DUC data sets and a mixture of objective functions are used. Here analysis of the document according to ROUGE metrics. The result was generated from the reduction in redundancy, relevance and content coverage in terms of execution time and average ROUGE.

Authors in Ref. [12] implements the TextRank algorithm and extends the preprocessing and postprocessing part. The data (records) are coded and various methods such as tfidf and Word2vec are taken into account. In addition, the implementation will be tested with Malaysian content to prove domain independence of the algorithm.

Topic-based approaches can be found in [13–15]. They are based on the distribution of words through documents, from which topics can be derived that later form summaries. [16] is building a small global network to aggregate biomedical articles.

The Automatic Text Summarization (ATS) [17] concludes that they are the ones who select the most notable sentences of the documents for their subsequent concatenation to form a summary. [18] mentions that the most successful systems use EATS approaches as they cut and merge parts of the text to create a reduced version, [19] therefore notes that EATS results in summaries with information available from the original text without changes. [20] suggest that a typical EATS consists of 2 phases, the first consisting of preprocesses.

In Ref. [21] the researchers suggested a semantic method for multi-document EATS that uses statistical methods based on machine learning based on graphs.

Ref. [22] propose a single sentence extraction model that models the relationship between sentences, [23] determine the extraction of subsets based on tree decisions based on a neural model, [24] indicate that it is to RTAE from oriented news By a hybrid algorithm between semantic analysis and random fields, coherent and detailed information.

Sentence scoring and summary collection is the most popular approach used with automatic extractive summary. Sentence scoring is implemented in most of the approaches used today. The evaluation methods are listed as word evaluation, sentence evaluation and graph evaluation [25, 26]. In word classification methods, classification is made taking into account the meaning of sentences, including the occurrence of a word in the text [27, 28] with terms such as proper names, places and objects that are assumed to be a determinant are rated higher [7, 29]. The formal properties of words (bold, italic, underlined) are taken into account in text evaluation procedures [30].

In addition, sentences that begin with phrases such as "short", "finite", and "as a result" in the text are called character phrases, and the sentences that precede such statements are called significant sentences [27]. The evaluation is also based on the title of the text to be summarized. It is assumed that sentences containing the words mentioned in the title are added to the abstract and their level of significance expanded accordingly [31].

The techniques of sentencing regularly think about the traits of sentences, giving more precedence to sentences of more size [32, 33]. Points are carried out to sentences through figuring out the area of the sentence and whether or not or now no longer they have got numerical values [28, 30, 34].

The authors of Ref. [35] explained the implementation and assessment of extractive summarization approach as a method of assisting novices with analyzing challenges. Graph-primarily based totally representations are regularly utilized in textual content evaluation strategies due to the fact they have got very green solutions.

In Ref. [36], the researchers added TextRank, which gives graph-primarily based totally precis illustration the usage of textual content intersections. Likewise, LexRank become carried out in Ref. [37] the usage of a proprietary centrality-primarily based totally algorithm, one of the node centrality techniques.

Both the TexRank and LexRank techniques have been inspired through the PageRank [38] technique, a textual content summarization approach that become added to acquire valuable sentences in a textual content the usage of shared know-how among the phrase and sentence sets [39].

The researchers of Ref. [40] identified the graphic documents within their study victimization link generation for automatic document summarization. They established the structure by unveiling the relationships of the text in the documents and analyzing the summaries by contrastive them with those made by human hand. In Ref. [41], a graph-based approach was planned to make sure linguistics consistency, with nodes pertaining to the terms of the documents and also the edges representing the semantic relationships between the nodes. In general, the graph diameter calculation is enforced for all the nodes in the graph, and the shortest and longest methods are outlined because the poorest and best bonds. Though graph structures and documents are delineated in Ref. [42], nodes and edges have been generatedwere supported native similarities. Stochastic process was wont to offer a listing of the relevant documents.

A summary technique for the medical specialty sector was bestowed in referee [43]. A graph supported ideas and relationships with a semi-dictionary - based mostly frame-work was generated employing a method referred to as Unified Medical Language, so a PageRank formula was used. In Ref. [44], the authors recommended a brand new graph based on improved random walking.

In Ref. [45], the researchers adopted a graph-based approach for carrying out an extractive summarization. The latter researchers proposed a new method for summa-rization. This method is based on hybrid modelling graph. They recommended deliv-ering an innovative hybrid similarity function (H). The latter function hybridises 4 dis-tinct measures of similarity. These measures are cosine, Jaccard, word alignment and window-based similarity. The method employed a trainable summarizer and takes into consideration several features. The impact of those features on the task of summarization has been examined.

In Ref. [46] the researcher adopted a graph-based approach for carrying out Semantic Role Labelling (SRL) to create a Graph Semantic Model (GSM) that would be used to find a summary of a single document. The experimental findings based on the proposed system indicate that the Triangle-Subgraph using SRL (Triangles SRL) is the best illus-tration for a comprehensive summary. The results show that the best average precision, recall and F-Measure are produced by our proposed method.

In Ref. [45], the researchers proposed a single document summarization technology based on hybrid graph, four different similarity measures, including cosine similarity, Jaccard similarity, word alignment similarity and window-based similarity measure, were combined to create a hybrid similarity function to calculate the weight of the graph. The experimental results showed that specific combinations of features could give higher efficiency. It also showed that some features have more effect than others on the summary creation.

Researchers in Ref. [47] proposed a graphics-based system that is used to abstract Arabic texts. Therefore, although the structure of morphological Arabic languages is more complex, which is very difficult for noun extraction, the summary method is used. Then the document is converted into graphics in which the vertex sets from the graph are represented. Modified PageRank algorithms are used; is the initialized area of each vertex due to the large number of nouns available in sentences. The nouns used in certain documents generate more information. The starting range of the generated sentences counts the number of nouns. Based on the number of nouns, the final summary of the given documents is more relevant and with more information. The performance development of the EASC Corpus data set is used and only carried out in 10,000 iterations.

Researchers in Ref. [48] suggested a group-based text summary. In ETS, linguistic or statistical techniques are used to extract key sentences from the document and concatenate those sentences for a summary in the group-based and in the semantic link network. The top sections of the group-based summary are selected and concatenated in order to produce a relevant summary. Four types of semantic links are used to generate groups, e.g. B. cause and effect links, similar links, sequential links, and part of a link. ACL Anthology, which contains a total of 173 articles from ACL2014 conferences.

3   Overview of Approach

In this section, we will describe the steps we took to generate the summary based on the sub-graph based method. The main steps are:

i.   Data pre-processing.
ii. Features Selection and Scoring.
iii. Syntactic graph representation based on similarity measure.

    1. Sentence similarity measures.
    2. Syntactic sub-graph construction.
    3. Sentences bit-vector scoring.
    4. Calculate sentence score based on bit-vector value.
    5. Extractive summary generation based on similarity measure.

iv. Evaluation and Results.

3.1   Data Pre-processing

The first step in summarizing text is data preprocessing. In our study, this step consists of 3 sub-steps: text segmentation, stop word removal and word derivation. The first step in text segmentation breaks down the text document into sentences. We used the technique. Stop word removal to remove nonsensical words. We also use the derivative algorithm to remove any affixes (prefixes or suffixes) present in a word to create the root word. In this step, we extract the important words that are present in the document and discard the rest of the existing words, which could seriously affect the similarity between the documents.

3.2 Feature Extraction

Regarding the document that is textual, it's represented by the following set $D = (S_1, S_2, \ldots, S_k)$. S_i refers to the sentence which is found within a specific document D. Then, the extraction of features is implemented to the content that is textual. The primary useful sentence along with the structure of words shall be determined. Each document shall feature structures. Such structures include: proper-noun instances, thematic-word, sentence similarities, term weights, numerical data, sentence positions, sentence lengths, and title words. Information about such structures are listed below:

1. Title words: Regarding the higher scores, they shall be assigned to sentences which involve words that have been obtained from titles. The meaning of the contents shall get conveyed in the title words. This is determined as it's presented below:

$$F1 = \frac{Words_{Title} \ Count \ in \ sentence \ S}{Word_{Count} \ in \ Title} \tag{1}$$

2. Sentence lengths: Sentences that are shortened –e.g. author lines or date- shall be removed. Regarding the normalised length of each sentence, it is assessed. Regarding the normalised length of each sentence, it is determined as it's displayed below:

$$F2 = \frac{Words_{Count} \ in \ Sentence \ S}{Words_{count} \ in \ longest \ sentence} \tag{2}$$

3. Sentence positions: Regarding the higher scores, they are assigned to the sentences that occur further ahead in their paragraph. Regarding each paragraph that has n sentences, the score of each sentence is determined as displayed:

$$F3 = \begin{cases} 1 & for \ first \ sentence \\ \frac{n-i}{n}, where \ i = 1, 2, 3, \ldots, n \ for \ other \ sentences \end{cases} \tag{3}$$

4. Numerical data: Regarding each sentence that shows numerical terms that replicates main statistical figures within the text, they shall go through the process of summarizing. As for the numerical score of each sentence, it is determined as it's listed below:

$$F4 = \frac{Numerical\_dat_{Count} \ in \ sentence \ S}{Sentence_{Length}(S)} \tag{4}$$

5. Thematic words: The quantity of domain-specific or thematic words showing maximum-possible relativeness and exist in a sentence shall be divided by the maximum quantity that exist in the sentences

$$F5 = \frac{Thematic_{word_{Count}} \ in \ sentence \ S}{Max(Thematic\_word_{Count})} \tag{5}$$

6. Sentence-to-sentence similarities: The methods of token-matching are employed for measuring the extent of similarity between each sentence and with other ones. A matrix [N][N] is set up. N refers to the overall amount of the sentences that have

been found. It also refers to the components that are diagonal and fixed at 0. The assessment of sentences shouldn't be conducted in comparison to themselves. The similarity score of each sentence shall be assessed as below:

$$F6 = \frac{\sum [(S_i, S_j)]}{MAX [(S_i, S_j)]}, \ where \ i = 1 \ to \ N \ and \ j = 1 \ to \ N \tag{6}$$

3.3 Sentence Similarity

Edges are the relations between sentences. These relations are calculated based on the similarity between sentences. These relations assigned by comparing the value of the similarity with a predefined threshold. If the value of similarity is less than the threshold, then there is no edge, otherwise there is an edge between sentences.

Cosine similarity measurement was used to find the relation between the sentences. Cosine similarity is a quantification of similarity concerning two vectors of an inner product space that gauges the cosine of the angle between them. Cosine similarity is extensively used method to discover the similarity among two texts. The Cosine similarity among two text (t_1, t_2) is

$$SIM (t_1, t_2) = \frac{\sum_{i=1}^{n} t_{1i} t_{2i}}{\sqrt{\sum t_{1i}^2} \times \sqrt{\sum t_{2i}^2}} \tag{7}$$

3.4 Document as a Graph

In the graph-based approach, each document is represented by a graph. Document is models as Graph where term represented by vertices and relation between terms is represented by edges.

$$G = (Vertex, EdgeRelation)$$

There are generally five different types of vertices in the Graph representation.

$$Vertex = (F, S, P, D, C)$$

where

$$F - Featureterm, S - Sentence, P - Paragraph, D - Document, C - Concept$$

$$EdgeRelation = (Syntax, Statistical, Semantic)$$

The edge relation found between two feature terms may vary based on the context of the graph.

1. Occurrence of words in a document's sentence, paragraph, or section
2. Common words found within a document's sentence, paragraph, or section

3. Co-occurrence found within the fixed window of n words
4. Semantic relation or words that have similar meaning, are spelled the same way but have varying meanings, or opposite word

The first step of creating graph for the text document is to assign the nodes. As mentioned earlier, the nodes can be one of five choices, either Feature-term, Sentence, Paragraph, Document or Concept.

Nodes in our work will be the sentences (every sentence in the text represents a node in the graph), and the edges between nodes are the similarity between these sentences.

Based on the similarity prices, we are able to build the graph for the text document per [49], we have a tendency to set a grading threshold (β) for sentence similarity at 0.5. In alternative that means if the similarity value is a smaller amount than 0.5, then there's no relation between sentences, that means, there is no edge between nodes.

3.5   D. Bit-Vector Scoring

A bit vector (bit array) can be used to map some domain (typically a range of integers) to values within the set (0, 1). One can interpret the values as dark/light, valid/invalid, locked/unlocked, absent/present, etc. Since there are only two values possible, it is possible to store them in one bit. Like other arrays, one can manage the access to a single bit by implementing an index to the array. If we assume its size (or length) to be equal to n bits, one can use the array to specify the domain's subset (e.g. (0, 1, 2, ..., n $-$ 1)), where a 1-bit represents the presence and a 0-bit represents the absence of a number within the set. This particular set data structure utilises about n/w words of space. Here, w represents the number of bits for every machine word. It is mainly irrelevant whether the least significant bit or the most significant bit (of the word) is indicative of the smallest-index number. However, the former is often more preferred (on little-endian machines).

Next step is to construct the sub-graph. In this study we used two different sub-graphs to find which one is the best and can create better summary. These sub-graphs are Rectangles and Shortest path using Dijkstra's algorithm.

First step is to create the adjacency matrix. Adjacency matrix (connectivity matrix; reachability matrix) A matrix used as a method of representing an adjacency structure, which in flip represents a graph. If A is the adjacency matrix similar to a given graph G, then aij = 1 if there may be an part from vertex i to vertex j in G; in any other case aij = 0. If G is a directed graph then aij = 1 if there may be an part directed from vertex i to vertex j; in any other case aij = 0. If the vertices of the graph are numbered 1,2,...m, the adjacency matrix is of a kind m $\times$ m. If A $\times$ A $\times$ ... $\times$ A (p terms, p $\leftarrow$ m) is evaluated, the nonzero entries suggest the ones vertices which are joined with the aid of using a course of duration p; certainly the cost of the (i,j)th access of Ap offers the variety of paths of duration p from the vertex i to vertex j. By inspecting the set of such matrices, p = 1, 2, ..., m $-$ 1 it is able to be decided whether or not vertices are connected.

The first sub-graph we will use is the Rectangle sub-graph. After creating the adjacency matrix, we can build list of triangles representing the text. To find the Rectangles in the graph the following is the algorithm used to construct them.

Algorithm 1 Rectangle Graph Construction

```
Input: N*N adjacency matrix , (A (I,J))
Output: Array of rectangles
Start
Rectangles_Array=[],
For each edge in the matrix A (I,J), namely XY, find all edges start with Y {
XY  ∧  YZ ──▶ XZ  }
If XZ ε A(I,J) {
For each edge in the matrix A (I,J), namely XZ, find all edges start with Z {
XZ  ∧  ZW──▶ XW  }
If XW ε A(I,J) {
Add the rectangle of edges (X,Y,Z,W) to Rectangles_Array[]}
}
Stop
```

The Dijkstra algorithm can be used to identify the shortest distance (or the lowest cost/ least effort) between a start node and another node found within the graph. The idea behind the algorithm is for it to be able to continuously compute the shortest distance from a starting point and for it to be able to exclude longer distances when creating an update.

Algorithm 2 Graph Construction using Dijkstra's algorithm

```
Function Dijkstra(Graph, source):
        Create vertex set Q
        For each vertex v in Graph:       // Initialization
                dist[v] ← INFINITY         // Unknown distance from source to
v
                prev[v] ← UNDEFINED        // Previous node in optimal path
        from source
                Add v to Q                 // All nodes initially in Q (unvisited nodes)
        dist[source] ← 0                   // Distance from source to source
        While Q is not empty:
                u ← vertex in Q with max dist[u]   // Node with the highest
similarity will be selected first
                Remove u from Q
                For each neighbour v of u:   // where v is still in Q.
                        alt ← dist[u] + length(u, v)
                        if alt > dist[v]:      // A highest similar path to v has been
found
                                dist[v] ← alt
                                prev[v] ← u
                Return dist[], prev[]
```

3.6 Sentence Scoring

The measures of sentence similarity play a significant role in carrying out the text-related research. They a significant role in carrying out the process of text summarization. In simple words, a measure represents how alike several items are when after comparing them with one another. An accurate measure for measuring sentence similarity shall improve the computer-human interaction. In the present study, we aimed to measure the sentence similarity. We aimed to measure it based on the semantic roles labelling.

We aimed to measure it based on the semantic similarity as it it's presented through equation number 7. The sentences scores that are within the range of (0–1) are defined as shown below based on the semantic similarity combined with the features chosen for each sentence.

$$Score(S_i) = Bit_{Vectore}(S_i) * \sum_{k=1}^{m} Score(F_k(S_i)) \qquad (8)$$

where $Score(S_i)$ is the score of sentence S, $Bit_{Vectore}(S_i)$ is the bit-vector valueof sentence S and $Score(F_k(S_i))$ is the score of feature K and m is the number of features used to score the sentences.

3.7   Summary Generation

After sentence scores were obtained, every sentence within the document was allotted by the sentence score value. Solely sentences with sub-graph structures are chosen for thought because these are related to no lower than 2 others. Every and each sentence is ranked in falling order per its score. Those evaluated with high scores are free for document report according to compression ratio. AN extraction, or compression rate of on the point of 20% of the core matter content has been incontestible to be as instructive of the contents as the document's complete text [50]. Within the final stage, the summarizing sentences are organized within the order of their abstract occurrences as found in the initial text.

4   Experimental Results

To experiment the implementation of the proposed graph based approach for single-document extractive summarization, we used DUC 2002 document sets (DUC, 2002) - the standard corpus widely employed in text summarization research, which contains documents along with their human model summaries. At first, we perform pre-processing on the sets of documents. This step involves sentence splitting, tokenization, stop words elimination and word stemming. Once the documents are pre-processed, we apply four different similarity measures to compute similarity between sentences. Then we represented text as graph, after that we created the sub-graph. We rank the sentences based on six features with their probabilities combined with the Bit-Vector for each sentence. Finally, the top ranked sentences were selected as a summary of the main text based on the compression rate. We employ three evaluation measures – Mean coverage score (Recall), Precision, and F-measure for the evaluation of our proposed approach. This metric assesses the quality of system summary by comparing it with human model summaries and other benchmark summarization systems.

The proposed approach was evaluated in the context of the Single Document Extractive Abstract Task using 103 articles/datasets provided by Document Understanding Evaluations 2002 (DUC, 2002). For each data set, our approach generates a summary with a compression rate of 20%.

To compare the performance of our proposed approach, we setup different comparison models, which are as follows: we compared the results for each sub-graph by comparing the best results of each single feature. Then we compared the results between the different sub-graphs to find which sub-graph can produce a better summary. After that we compared these results with different benchmark summarizers; [33, 51–53], Microsoft Word 2007 summarizer, Copernic summarizer, Best automatic summarization system in DUC 2002, Worst automatic systems in DUC 2002, and the average of human model summaries (Models).

This study focused on finding the effect of features and similarity measurement to create the Syntactic Sub-Graph to produce the summary. Based on the generalization of the obtained results, the performance of the proposed model, it is clear that using Dijkstra sub-graph, has the best value. The best value is related to Sentence to Sentence (S2S) feature, which is 49.615% and 52.307% similar to human generated summaries using ROUGE-1 for Rectangles and Dijkstra sub-graphs respectively. The best value is related to Sentence to Sentence (S2S) feature, which is 47.389% and 49.334% similar to human generated summaries using ROUGE-L for Rectangles and Dijkstra sub-graphs respectively.

For comparative evaluation, Tables 1 and 2 show the mean coverage score (recall), average precision and average F-measure obtained on DUC 2002 dataset for the proposed approach with different benchmark summarizers, using ROUGE-1 and ROUGE-L respectively.

Table 1. Comparison of Single Extractive Document Summarization using ROUGE-1 result at the 95% confidence interval for Proposed Syntactic Sub-graph Model

Method	Precision	Recall	F-Measure
H2:H1	0.61656	0.61642	0.61627
MS-Word	0.47705	0.40325	0.42888
Copernic	0.46144	0.41969	0.43611
Best-System	0.50244	0.40259	0.43642
Worst-System	0.06705	0.68331	0.1209
GSM	0.49094	0.43565	0.45542
Fuzzy	0.49769	0.45706	0.47181
BiDETS	0.52987	0.4561	0.48503
Swarm model	0.47741	0.43028	0.44669
Fuzzy swarm	0.49126	0.43622	0.45524
Rectangles	0.49722	0.49954	0.49823
Dijkstra	0.51788	0.52045	**0.52307**

Table 2. Comparison of Single Extractive Document Summarization using ROUGE-L result at the 95% confidence interval for Proposed Syntactic Sub-graph Model

Method	Precision	Recall	F-Measure
H2:H1	0.584	0.58389	0.58374
MS-Word	0.44709	0.36368	0.39263
Copernic	0.29031	0.25986	0.27177
Best-System	0.46677	0.37233	0.40416
Worst-System	0.66374	0.06536	0.11781
BiDETS	0.48783	0.42	0.44652
Swarm model	0.44143	0.39674	0.41221
Fuzzy swarm	0.45355	0.40144	0. 41937
Rectangles	0.47364	0.47511	0.47389
Dijkstra	0.49235	0.49365	**0.49334**

5 Concluding Remark

Automatic text summarisation is a top level view of a foundation text by a machine to point out the foremost important data in a very shorter version of the most text whereas still protective its major linguistics content and helps the user to apace recognise large amounts of information. The bulk of existing automatic text summarisation approaches extracts the most significant information from supply documents. Conventionally, automatic text summarisation systems mine the sentences from the source documents betting on their importance to the documents. The summarisation systems measure the weights of numerous extraction options for each text part then the weights of sentence are joined because the general weight of the text element. Finally, the sentences with the very best weight are going to be extracted. This study has shown that one graph will be reborn into another graph, with considerably smaller number of edges by count the amount of triangles. The utilization of Adjacency Matrix Representation is easy and this feature has contributed towards shorter execution times. The experimental results based on the syntactic sub-graph model show that the Dijkstra sub-graph is the best representation to create a better summary. The experimental results based on the sub-graph model show that we could improve the quality of the summary.

References

1. Ermakova, L., Cossu, J.V., Mothe, J.: A survey on evaluation of summarization methods. Inf. Process. Manag. **56**(5), 1794–1814 (2019). https://doi.org/10.1016/j.ipm.2019.04.001
2. Hark, C., Seyyarer, A., Uçkan, T., Karci, A.: Doğal dil isleme yaklasimlari ile yapisal olmayan dökümanlarin benzerliği. In: International Artificial Intelligence and Data Processing Symposium, IDAP 2017, pp. 1–6 (2017)
3. Yao, J., Wan, X., Xiao, J.: Recent advances in document summarization. Knowl. Inf. Syst. **53**(2), 297–336 (2017). https://doi.org/10.1007/s10115-017-1042-4

4. Chitturi, A.K.: Survey on abstractive text summarization using various approaches. Int. J. Adv. Trends Comput. Sci. Eng. **8**, 2956–2964 (2019). https://doi.org/10.30534/ijatcse/2019/45862019

5. Hark, C., Uçkan, T., Seyyarer, A., Karci, A.: Metin Özetleme _ İçin Çizge Tabanlı Bir Öneri. In: International Artificial Intelligence and Data Processing Symposium, IDAP 2018 (2018)

6. Joshi, A., Fidalgo, E., Alegre, E., Fernández-Robles, L.: SummCoder: an unsupervised framework for extractive text summarization based on deep auto-encoders. Expert Syst. Appl. **129**, 200–215 (2019). https://doi.org/10.1016/j.eswa.2019.03.045

7. Gambhir, M., Gupta, V.: Recent automatic text summarization techniques: a survey. Artif. Intell. Rev. **47**(1), 1–66 (2016). https://doi.org/10.1007/s10462-016-9475-9

8. Saroo Raj, R.B., Singh, G., Balaji, S., Ajit Baskar, K.H.: A model to predict loan defaulters using machine learning. Int. J. Emerg. Technol. Eng. Res. **6**(10) (2018)

9. Tan, J., Wan, X., Xiao, J.: Abstractive document summarization with a graph-based attentional neural model. In: Proceedings of the 55th Annual Meeting of the Association for Computational Linguistics, pp. 1171–1181 (2017)

10. Begum, M., Sezer, E.A., Akcayol, M.A.: Candidate sentence selection for extractive text summarization. Inf. Process. Manag. **57**(6), 102359 (2020). https://doi.org/10.1016/j.ipm.2020.102359

11. Sanchez-Gomez, J.M., Vega-Rodríguez, M.A., Perez, C.J.: Experimental analysis of multiple criteria for extractive multi-document text summarization. Expert Syst. Appl. **140**, 112904 (2020). https://doi.org/10.1016/j.eswa.2019.112904

12. Manju, K., Peter David, S., Mary, S.I.: A framework for generating extractive summary from multiple Malayalam documents (2021)

13. Issam, K.A.R., Patel, S., Subalalitha, C.N.: Topic modeling based extractive text summarization. Proc. Int. J. Innov. Technol. Explor. Eng. (IJITEE) **9**(6) (2020). ISSN 2278–3075

14. Gialitsis, N., Pittaras, N., Stamatopoulos, P.: A topic-based sentence representation for extractive text summarization. In: Proceedings of the Multiling 2019 Workshop, co-located with the RANLP 2019 Conference, vol. 1, pp. 26–34 (2019)

15. Hafeez, R., Khan, S., Abbas, M.A., Maqbool, F.: Topic based summarization of multiple documents using semantic analysis and clustering. In: 2018 15th International Conference on Smart Cities: Improving Quality of Life Using ICT and IoT (HONET-ICT), vol. 1, p. 18 (2018)

16. Moradi, M.: Small-world networks for summarization of biomedical articles. In: Section for Artificial Intelligence and Decision Support, Medical University of Vienna, Austria (2019)

17. El-Kassas, W.S., Salama, C.R., Rafea, A.A., Monhamed, H.K.: Automatic text summarization: a comprehensive survey. Expert Syst. Appl. **165**, 113679 (2020). https://doi.org/10.1016/j.eswa.2020.113679

18. Rush, M.A., Chopra, S., Weston, J.: A neural attention model for sentence summarization. In: Proceeding of the 2015 Conference on Empirical Methods in Natural Language Processing, Lisbon, September 2015, pp. 379–389 (2015). https://doi.org/10.18653/v1/D15-1044

19. Castañeda, N.H., Hernández, R.A.G., Ledeneva, Y., Castañeda, A.H.: Evolutionary automatic text summarization using cluster validation indexes. Hernández Castañeda **24**, 583–595 (2020). https://doi.org/10.13053/cys-24-2-3392

20. Quillo-Espino, J., Romero-Gonzalez, R.-M.: Where are the automatic text summaries located in the 2021? A review. Int. J. Adv. Res. Comput. Commun. Eng. **10**, 11–16 (2021). https://doi.org/10.17148/IJARCCE.2021.10402

21. Bidoki, M., Monsavi, M.R., Fakhramahgmad, M.: A semantic approach to extractive multi-document summarization: applying sentence expansion for tuning of conceptual densities. Inf. Process. Manag. **57**, 102341 (2020). https://doi.org/10.1016/j.ipm.2020.102341

22. Zhong, M., Liu, P., Chen, Y., Wang, D., Qiu, X., Huang, X.: Extractive summarization as text matching. In: Proceedings of the 58th Annual Meeting of the Association for Computational Linguistics, July 2020, pp. 6197–6208 (2020). https://doi.org/10.18653/v1/2020.acl-main.552

23. Zhou, Q., Wei, F., Zhou, M.: At which level should we extract? And empirical analysis on extractive document summarization In: Proceedings of the 28th International Conference on Computational Linguistics, Barcelona, December 2020, pp. 5617–5628 (2020). https://doi.org/10.18653/v1/2020.coling-main.492

24. Muneera, M.N., Sriramya, P.: Extractive text summarization for social news using hybrid techniques in opinion mining. Int. J. Eng. Adv. Technol. **9**, 2109–2115 (2020). https://doi.org/10.35940/ijeat.B3356.02932

25. Chelliah, B.J., Lathia, D., Yadav, S., Trivedi, M., Soni, S.S.: Sentiment analysis of Twitter data using CNN. Int. J. Emerg. Technol. Eng. Res. (IJETER) **6**(4) (2018)

26. Ferreira, R., et al.: Assessing sentence scoring techniques for extractive text summarization (2013)

27. Luhn, H.P.: The automatic creation of literature abstracts. IBM J. Res. Dev. **2**(2), 159–165 (1958)

28. Shardan, R., Kulkarni, U.: Implementation and evaluation of evolutionary connectionist approaches to automated text summarization (2010)

29. Nasr Azadani, M., Ghadiri, N., Davoodijam, E.: Graph-based biomedical text summarization: an itemset mining and sentence clustering approach. J. Biomed. Inform. **84**, 42–58 (2018)

30. Student, P.G., Coe, D.M.: A comparative study of Hindi text summarization techniques: genetic algorithm and neural network (2015)

31. Gupta, V.: Hybrid algorithm for multilingual summarization of Hindi and Punjabi documents. In: Prasath, R., Kathirvalavakumar, T. (eds.) MIKE 2013. LNCS (LNAI), vol. 8284, pp. 717–727. Springer, Cham (2013). https://doi.org/10.1007/978-3-319-03844-5_70

32. Gupta, V., Lehal, G.S.: A survey of text summarization extractive techniques. J. Emerg. Technol. Web Intell. **2**(3), 258–268 (2010)

33. Abuobieda, A., Salim, N., Albaham, A.T., Osman, A.H., Kumar, Y.J.: Text summarization features selection method using pseudo genetic-based model. In: Proceedings of the 2012 International Conference on Information Retrieval & Knowledge Management, CAMP 2012, pp. 193–197 (2012)

34. Fattah, M.A., Ren, F.: GA, MR, FFNN, PNN and GMM based models for automatic text summarization. Comput. Speech Lang. **23**(1), 126–144 (2009)

35. Nandhini, K., Balasundaram, S.R.: Improving readability through extractive summarization for learners with reading difficulties. Egypt Inform. J. **14**(3), 195–204 (2013)

36. Mihalcea, R., Tarau, P.: A language independent algorithm for single and multiple document summarization. In: Proceedings of the IJCNLP 2005, 2nd International Joint Conference on Natural Language Processing, pp. 19–24 (2005)

37. Erkan, G., Radev, D.R.: Lexrank: graph-based lexical centrality as salience in text summarization. J. Artif. Intell. Res. **22**, 457–479 (2004)

38. Brin, S., Page, L.: The anatomy of a large-scale hypertextual web search engine. Comput. Netw. ISDN Syst. **30**(1–7), 107–117 (1998)

39. Parveen, D., Ramsl, H.-M., Strube, M.: Topical coherence for graph-based extractive summarization. In: Proceedings of the 2015 Conference on Empirical Methods in Natural Language Processing, pp. 1949–1954 (2015)

40. Salton, G., Singhal, A., Mitra, M., Buckley, C.: Automatic text structuring and summarization. Inf. Process. Manag. **33**(2), 193–207 (1997)

41. Medelyan, O.: Computing lexical chains with graph clustering. In: Proceedings of the ACL 2007 Student Research Workshop, pp. 85–90 (2007)

42. Chen, Y.-N., Huang, Y., Yeh, C.-F., Lee, L.-S.: Spoken lecture summarization by random walk over a graph constructed with automatically extracted key terms. In: Twelfth Annual Conference of the International Speech Communication Association (2011)

43. Plaza, L., Stevenson, M., Díaz, A.: Resolving ambiguity in biomedical text to improve summarization. Inf. Process. Manag. **48**(4), 755–766 (2012)

44. Xiong, S., Ji, D.: Query-focused multi-document summarization using hypergraph-based ranking. Inf. Process. Manag. **52**(4), 670–681 (2016)

45. AL-Khassawneh, Y.A., Salim, N., Jarrah, M.: Improving triangle-graph based text summarization using hybrid similarity function. Indian J. Sci. Technol. **10**(8), 1–15 (2017). https://doi.org/10.17485/ijst/2017/v10i8/108907

46. AL-Khassawneh, Y.A.: The use of semantic role labelling with triangle-graph based text summarization. Int. J. Emerg. Trends Eng. Res. **8**(4), 1162–1169 (2020). https://doi.org/10.30534/ijeter/2020/34842020

47. Elbarougy, R., Behery, G., El Khatib, A.: Extractive Arabic text summarization using modified PageRank algorithm. Egypt. Inform. J. **21**(2), 73–81 (2020). https://doi.org/10.1016/j.eij.2019.11.001

48. Cao, M.Y., Hai, Z.G.: Grouping sentences as a better language unit for extractive text summarization. Futur. Gener. Comput. Syst. **109**, 331–359 (2020). https://doi.org/10.1016/j.future.2020.03.046

49. Mihalcea, R., Tarau, P.: A language independent algorithm for single and multiple document summarization (2005)

50. Morris, A.H., Kasper, G.M., Adams, D.A.: The effects and limitations of automated text condensing on reading comprehension performance. Inf. Syst. Res. **3**(1), 17–35 (1992)

51. Binwahlan, M.S., Salim, N., Suanmali, L.: Fuzzy Swarm Based Text Summarization, vol. 1 (2009)

52. Binwahlan, M.S., Salim, N., Suanmali, L.: Swarm based text summarization. In: International Association of Paper Presented at the Computer Science and Information Technology-Spring Conference, IACSITSC 2009, pp. 145–150 (2009)

53. Suanmali, L., Salim, N., Binwahlan, M.S.: Fuzzy logic based method for improving text summarization. arXiv preprint arXiv:0906.4690 (2009)

Analysis of Big Five Personality Factors to Determine the Appropriate Type of Career Using the C4.5 Algorithm

Rizka Shinta Wulandari, Casi Setianingsih[✉], and Purba Daru Kusuma

School of Electrical Engineering, Telkom University, Bandung, Indonesia
setiacasie@telkomuniversity.ac.id

Abstract. An individual must do work to meet his daily needs. According to revisesociology.com, which analyzed data from the Annual Survey for Hours and Earnings (ASHE) in the UK from 1997 to 2014, a worker would spend 92,120 h in a lifetime if assuming a full adult working life from ages 18–67. So that a person's life will be spent much work. Furthermore, an article by Upskilled Australia mentions that many factors affect a person's job satisfaction; it could be his job, benefits, or team camaraderie. However, a study shows that personality compatibility with work is the main factor that has a critical role. The more united one's character with his career, the more productive and optimistic one's work performance will be, therefore the selection of a job that suits the individual's personality becomes very essential. In this study, a system has been created that can help job seekers find out what type of work matches the person's personality so that each worker can work productively and achieves. The personality factors used are the Big Five, namely Extraversion, Agreeableness, Conscientiousness, Emotional Stability, and Intellect. This personality factor is mapped using C4.5 to get a job that matches your personality. The work fields based on the 20 Job Families are taken from the O*NET database. The system runs according to its purpose with accuracy of 99.99%. For further research, the development will be combined with other personality tests, such as the Myers-Briggs Type Indicator so that personality assessment has a broader scope.

Keywords: C4.5 · Big five · Personality · Jobs

1 Introduction

Work is something that everyone must do because we have to earn a living to continue to live life. Of course, not only that, according to Barry Schwartz, workers who are satisfied with their work find the work challenging, stimulating, engaging, and meaningful. Even some workers consider their work to be an essential thing. Seeing the current conditions, many workers who carry out their work an important thing. Seeing the current conditions, many workers are forced to carry out their work because they don't want to do it but keep doing it because later they will be given an incentive. This will result in less than

© The Author(s), under exclusive license to Springer Nature Singapore Pte Ltd. 2023
Y. B. Wah et al. (Eds.): DaSET 2022, LNDECT 165, pp. 18–36, 2023.
https://doi.org/10.1007/978-981-99-0741-0_2

optimal and unproductive work results. This will affect the performance of the place where the worker works, which needs to follow the vision to be achieved.

Personality is one aspect of psychology and an essential factor in influencing a worker's performance. Personality has been extensively researched in the field of psychology. Several theories and tests developed to measure this concept have been proposed, revised, and rejected. Nevertheless, personality tests are widely used by job seekers and recruiters as a parameter to get a job that matches workers and companies.

Seeing these conditions, the authors are compelled to conduct a study of personality tests based on the Big Five Personality inventory. The Big Five Personality has been recognized as the best and most comprehensive framework for measuring human personality based on the consensus of researchers and psychologists worldwide (Digman, 1999; Corr & Matthews, 2009). In this study, the author wants to create a system that can help job seekers to find suitable jobs according to the Big Five Personality factor with the C4.5 Algorithm as decision making.

In this study, the novelty offered is to map the results of the Big Five Personality Factors test into groups of types of work using the C4.5 Algorithm to get jobs that match the user's personality. The work field is based on the O*NET 20 Work Family database. Considering the suitability of personality with work is the main factor that is essential in increasing worker productivity.

2 Related Work

Personality theory in personality psychology is currently widely applied to determine a person's interests and talents, a person's mental health level, and also to define a person's career steps. Modern personality psychology uses the scientific inference method to test theories. There are several ways to measure personality, namely self-report tests, Q-sort tests, ratings and ratings by others, biological measurements, behavioral observations, interviews, expressive behavior, document analysis, projective tests, demographics and lifestyles, and online internet analysis of social media and Big Data [1].

Big Five factors affect specific job performance based on studies showing that personality aspects are valid predictors in measuring academic or work performance [2]. An individual's career success can be attributed to Extraversion, Emotional Stability, and Agreeableness. Individuals with high Extraversion, high Emotional Stability, and low Agreeableness experience a higher level of career satisfaction than those with low Extraversion, low Emotional Stability, and high Agreeableness. Extraversion positively correlates with career success, such as salary, promotion, and career satisfaction [3].

In a study conducted by Mar'i et al. the author made a professional recommendation system based on the Big Five personality traits using the Fuzzy Inference System (FIS) method. The five Big Five Personality Factors are used as inputs for this system. The data type of each big five personality factor value is in the form of numerical data. The result is that the system has an accuracy of 63% [4].

In the study to determine the hospital readmission rate of diabetes patients by Tamin et al. the authors conducted five experiments with different data samples. This study implements the C4.5 Algorithm to make the decision tree because the C4.5 Algorithm can overcome missing values and pruning. The highest accuracy is found in the 4th data sample with numeric and nominal attribute types, which produce an accuracy value of 74.5% [5].

In another study by Hssina et al. the authors conducted a comparative analysis of the ID3 and C4.5 decision tree algorithms. This research focuses on the critical elements of their decision tree construction from a data set. The results of this study confirm that C4.5 is the most powerful algorithm [6].

3 Research Method

3.1 Personality Psychology

Personality psychology is a sub-section of psychology that studies the human personality, distinguishing one human from another. The purpose of personality psychology is to build a personality theory with scientific, systematic, testable, comprehensive, and practical observations, for example, on job interests, health, and so on [7].

3.2 Big Five

The Big Five is a scientific classification of personality traits, a system that maps corresponding personality traits based on people's assessment of each other into the big five factors. The personality traits classified into five factors were derived from statistical studies of responses to personality items [3]. The five factors of these personality traits are Extraversion (E), Agreeableness (A), Conscientiousness (C), Emotional Stability (ES), and Intellect (I). Extraversion includes being sociable, talkative, assertive, and active. Agreeableness includes being cooperative, flexible, tolerant, and forgiving. Conscientiousness includes being reliable, striving for achievement, hardworking, diligent, and orderly. Emotional stability includes being calm, confident, and resilient. Finally, the intellect is curious and broad-minded, intelligent, and cultured [8].

An inventory that is often used for research in personality is the IPIP (International Personality Item Pool). IPIP is a collaboration of scientists in building measurements for personality. IPIP is in the public domain, meaning anyone can access personality scale items without the built-in ones and without having to ask for permission or pay. For example, one of the scales on the IPIP that is often used to measure the Big Five is the IPIP-BFM-50 (IPIP Big Five Factor Marker) which Goldberg built (2019) [9].

3.3 C4.5 Algorithm

C4.5 is the development of the ID3 Algorithm built by J. Ross Quinlan [10]. The actions carried out include C4.5 which can overcome continuous and discrete attributes, can overcome the problem of missing attribute values and can do pruning after the decision tree is made. The following are the C4.5 steps in making a decision tree [11]:

1. Selecting an attribute as the root node
2. Create a branch for each node value
3. Split cases in branches

4. Repeat the process for each branch until all cases on the branch have the same class

Select an attribute as the root, and it is based on the highest gain ratio value of the existing attributes. The concept of gain ratio is the development of information gain. The gain ratio can be used to reduce the effect of bias resulting from information gain [12]. The gain ratio adjusts the information gain for each attribute allowing the breadth and uniformity of attribute values [13]. The gain ratio formula, namely:

$$GainRatio(S, A) = \frac{Gain(S, A)}{SplitInfo(S, A)} \tag{1}$$

S = case set
A = attribute
Gain(S, A) = information gain on attribute A
SplitInfo(S, A) = split information on attribute A

Split information is the divisor in the gain ratio formula. The higher the split information value, the lower the gain ratio value. The number of values owned by categorical attributes and how uniformly these values are distributed will affect the value of split information [13]. The following is the split information formula:

$$SplitInfo(S, A) = -\sum_{i=1}^{n} \frac{S_i}{S} log_2 \frac{S_i}{S} \tag{2}$$

S = case set
A = attribute
Si = number of samples for attribute A

To calculate the gain value of the attribute, Eq. (3) is used as follows [11]:

$$Gain(S, A) = Entropy(S) - \sum_{i=1}^{n} \frac{|S_i|}{S} * Entropy(S_i) \tag{3}$$

S = case set
A = attribute
n = number of partition attributes A
|Si| = number of cases on partition i
|S| = number of cases in S

To calculate entropy, use the formula:

$$Entropy(S) = \sum_{i=1}^{n} -p_i * log_2 p_i \tag{4}$$

S = case set
A = attribute
n = number of partitions S

pi = the proportion of Si to S

The advantages of using C4.5 include the results from decision tree analysis that are easy to understand and build. In addition, they do not require much data compared to other classification algorithms, model results are easy to understand, use statistical techniques that can be validated, and computation time is faster than different classification Algorithms [14].

4 System Design and Testing

4.1 System Overview

In general, the system that will be made describes that the user will take the Big Five personality test with the IPIP-BFM-50 item, which has been adapted into Indonesian. The Big Five personality test will produce the user's personality traits in five factors: Extraversion, Agreeableness, Conscientiousness, Emotional Stability, and Intellect.

The results of the personality test scores will be mapped using the C4.5 method to get job recommendations that follow the user's personality. Job recommendations are divided into 20 job families taken from the O*NET database.

4.2 Dataset

The Open-Source Psychometrics Project obtained the dataset used in the C4.5 model development process to map the Big Five personality factors with job recommendations. The Open-Source Psychometrics Project has been educating the public about the numerous types of personality tests, their uses, and their meaning since late 2011. In addition, this website also develops data for research and development [15].

The dataset's attributes are the five Big Five personality factors: Extraversion, Agreeableness, Conscientiousness, Emotional Stability, and Intellect. The dataset has 2038407 rows.

4.3 Big Five Personality Test Scoring

Table 1 represents the items from the IPIP-BFM-50 along with the weights of each item indicating the value of each Big Five factor.

In each item, there are five responses, namely "Very Inaccurate", "Moderately Inaccurate", "Neither Inaccurate nor Accurate", "Moderately Accurate", and "Very Accurate". For items with a rating marked (+), the response "Very Inaccurate" is worth 1, "Moderately Inaccurate" is worth 2, "Neither Inaccurate nor Accurate" is worth 3, "Moderately Accurate" is worth 4", and "Very Accurate" is 5.

As for items with a rating marked (−) the opposite applies. Responses "Very Inaccurate" scored 5, "Moderately Inaccurate" scored 4, "Neither Inaccurate nor Accurate" scored 3, "Moderately Accurate" scored 2, and "Very Accurate" scored 1. After the user has filled in all items, add up all the scores. To get the total value of each Big Five factor.

Table 1. IPIP-BFM-50 Indonesian adaptation

No	Original item	Indonesian adaptation	Scoring*
1	Am the life of the party	Menghidupkan suasana dalam suatu acara	1+
2	Feel little concern for others	Tidak terlalu memedulikan orang lain	2−
3	Am always prepared	Selalu mempersiapkan segala hal	3+
4	Get stressed out easily	Mudah merasa tertekan	4−
5	Have a rich vocabulary	Menguasai banyak kosakata	5+
6	Don't talk a lot	Tidak banyak berbicara	1−
7	Am interested in people	Peduli dengan orang lain	2+
8	Leave my belongings around	Meninggalkan barang pribadi di sembarang tempat	3−
9	Am relaxed most of the time	Merasa tenang hampir setiap saat	4+
10	Have difficulty understanding abstract ideas	Kesulitan memahami ide yang bersifat abstrak	5−
11	Feel comfortable around people	Merasa nyaman berada di sekitar orang lain	1+
12	Insult people	Bersikap kasar pada orang lain	2−
13	Pay attention to details	Memperhatikan hal-hal secara rinci	3+
14	Worry about things	Mudah khawatir	4−
15	Have a vivid imagination	Memiliki imajinasi yang sangat kuat	5+
16	Keep in the background	Lebih suka bekerja di belakang layar	1−
17	Sympathize with other's feelings	Bersimpati dengan perasaan orang lain	2+
18	Make a mess of things	Mengacaukan banyak hal	3−
19	Seldom feel blue	Jarang merasa sedih	4+
20	Am not interested in abstract ideas	Tidak tertarik dengan ide-ide abstrak	5−
21	Start conversations	Memulai suatu percakaran	1+
22	Am not interested in other people's problems	Tidak tertarik dengan masalah orang lain	2−
23	Get chores done right away	Segera mengerjakan tugas yang diberikan	3+

(continued)

Table 1. (*continued*)

No	Original item	Indonesian adaptation	Scoring*
24	Am easily disturbed	Mudah merasa terganggu	4-
25	Have excellent ideas	Memiliki ide-ide yang cemerlang	5+
26	Have little to say	Sedikit berkata	1-
27	Have a soft heart	Lemah lembut	2+
28	Often forget to put things back in their proper place	Sering lupa meletakkan barang kembali pada tempatnya	3-
29	Get upset easily	Mudah merasa kesal	4-
30	Do not have a good imagination	Tidak memiliki imajinasi yang baik	5-
31	Talk to a lot of different people at parties	Berinteraksi dengan banyak orang dalam suatu acara	1+
32	Am not really interested in others	Tidak terlalu tertarik dengan kondisi orang lain	2-
33	Like order	Menyukai keteraturan	3+
34	Change my mood a lot	Memiliki perasaan yang berubah – ubah	4-
35	Am quick to understand things	Cepat dalam memahami sesuatu	5+
36	Don't like to draw attention to myself	Tidak suka menjadi pusat perhatian	1-
37	Take time out for others	Meluangkan waktu untuk orang lain	2+
38	Shirk my duties	Mengabaikan tugas-tugas saya	3-
39	Have frequent mood swings	Memiliki suasana hati yang sering cepat berubah	4-
40	Use difficult words	Menggunakan istilah-istilah yang sulit	5+
41	Don't mind being the center of attention	Tidak keberatan menjadi pusat perhatian	1+
42	Feel others' emotions	Memahami perasaan orang lain	2+
43	Follow a schedule	Melakukan aktivitas sesuai jadwal atau agenda	3+
44	Get irritated easily	Mudah merasa jengkel	4-
45	Spend time reflecting on things	Meluangkan waktu untuk merefleksikan berbagai hal	5+

(*continued*)

Table 1. (*continued*)

No	Original item	Indonesian adaptation	Scoring*
46	Am quiet around strangers	Tidak banyak berbicara pada orang yang tidak dikenal	1−
47	Make people feel at ease	Membuat orang lain merasa nyaman	2+
48	Am exacting in my work	Telaten dalam mengerjakan tugas	3+
49	Often feel blue	Sering merasa sedih	4−
50	Am full of ideas	Memiliki banyak ide	5+

*1 Extraversion, 2 Agreeableness, 3 Conscientiousness, 4 Emotional Stability, 5 Intellect. (+ or −) indicates the direction of scoring.

4.4 Job Family and Rules

Table 2 is a job family that will be the output of a job recommendation system based on the Big Five personality test. There are 20 categories of job families taken from the O*NET database [16]. O*NET (Occupational Information Network) is a free database containing job definitions to help job seekers, students, or talents to understand today's world of work.

Table 2. Job family

No	Job family	Code
1	Computer and Mathematical	A
2	Architecture and Engineering	B
3	Food Preparation and Serving Related	C
4	Arts, Design, Entertainment, Sports, and Media	D
5	Health Care Support	E
6	Building and Grounds Cleaning and Maintenance	F
7	Construction and Extraction	G
8	Production	H
9	Installation, Maintenance, and Repair	J
10	Management	K
11	Education, Training, and Library	L
12	Sales and Related	M
13	Farming, Fishing, and Forestry	O

(*continued*)

Table 2. (*continued*)

No	Job family	Code
14	Transportation and Material Moving	P
15	Life, Physical, and Social Science	Q
16	Business and Financial Operations	R
17	Legal	S
18	Health Care Practitioners and Technical	T
19	Community and Social Services	U
20	Protective Service	V

Table 3. Job families

No	Job examples	Code
1	Actuaries, Data Scientists, Biostatisticians	A
2	Aerospace Engineers, Cartographers, Landscape Architects	B
3	Baristas, Bartenders, Chefs, and Head Cooks	C
4	Art Directors, Athletes, Disc Jockeys	D
5	Dental Assistants, Endoscopy Technicians, Medical Transcriptionists	E
6	Pest Control Workers, Tree Trimmers and Pruners, Grounds Maintenance Workers	F
7	Painters, Carpenters, Electricians, Operating Engineers	G
8	Bakers, Gem and Diamond Workers, Machinists	H
9	Aircraft Mechanics and Service Technicians, Commercial Divers, Geothermal Technicians	J
10	Clinical Research Coordinators, Investment Fund Managers, Security Managers	K
11	Curators, Tutors, Teachers, Archivists	L
12	Model, Sales Engineers, Travel Agents	M
13	Agricultural Inspectors, Animal Breeders, Forest, and Conservation Workers	O
14	Flight Attendants, Aircraft Cargo Handling Supervisors, Commercial Pilots	P
15	Economists, Psychologists, Animal Scientists	Q
16	Accountants and Auditors, Fundraisers, Management Analysts	R
17	Lawyers, Judges, Arbitrators, Mediators, and Conciliators	S
18	Genetic Counselors, Psychiatrists, Art Therapists	T
19	Counselors, Community Health Workers, Child, Family, and School Social Workers	U

(*continued*)

Table 3. (*continued*)

No	Job examples	Code
20	Animal Control Workers, Firefighters, Private Detectives, and Investigators	V

Table 3 is an example of work from each job family.

Table 4 shows the scores range for the Big Five personality test results and their labels.

Table 4. Big five personality test score range

Factor	Score	Label
Extraversion	0–16	Low
	17–32	Moderate
	33–50	High
Agreeableness	0–16	Low
	17–32	Moderate
	33–50	High
Conscientiousness	0–16	Low
	17–32	Moderate
	33–50	High
Emotional stability	0–16	Low
	17–32	Moderate
	33–50	High
Intellect/imagination	0–16	Low
	17–32	Moderate
	33–50	High

Table 5 shows the rules used to map the five Big Five personality factors with job recommendations. These rules are based on previous research by F. Mar'i, W. F. Mahmudy, and C. Yusainy (2019) [4]. The total rules are 89 rows.

Table 5. Rules

No	Big five personality factors					Job family
	E	A	C	ES	I	
1	Low	Moderate	Moderate	Low	High	A
2	Low	Moderate	Moderate	Moderate	High	A
3	Low	Moderate	Moderate	High	High	A
4	Moderate	Moderate	Moderate	Low	High	A
5	Moderate	Moderate	Moderate	Moderate	High	A
6	Moderate	Moderate	Moderate	High	High	A
7	High	Moderate	Moderate	Low	High	A
8	High	Moderate	Moderate	Moderate	High	A
9	High	Moderate	Moderate	High	High	A
10	Low	Moderate	High	Moderate	High	B
11	Moderate	Moderate	High	Moderate	High	B
12	Low	High	Moderate	Moderate	Low	C
13	Low	High	Moderate	Moderate	Moderate	C
14	Low	High	Moderate	Moderate	High	C
15	Moderate	High	Moderate	Moderate	Low	C
16	Moderate	High	Moderate	Moderate	Moderate	C
17	Moderate	High	Moderate	Moderate	High	C
18	High	High	Moderate	Moderate	Low	C
19	High	High	Moderate	Moderate	Moderate	C
20	High	High	Moderate	Moderate	High	D
21	Low	Low	High	High	High	D
22	Low	Moderate	High	High	High	D
23	Low	High	High	High	High	D
24	Moderate	Low	High	High	High	D
25	Moderate	Moderate	High	High	High	D
26	Moderate	High	High	High	High	D
27	Moderate	Low	High	High	High	D
28	Moderate	Moderate	High	High	High	D
29	Moderate	High	High	High	High	D
30	Low	Moderate	High	High	Low	E
31	Low	Moderate	High	High	Moderate	E

(continued)

Table 5. (*continued*)

No	Big five personality factors					Job family
	E	A	C	ES	I	
32	Low	Moderate	High	High	High	E
33	Moderate	Moderate	High	High	Low	E
34	Moderate	Moderate	High	High	Moderate	E
35	Moderate	Moderate	High	High	High	E
36	High	Moderate	High	High	Low	E
37	High	Moderate	High	High	Moderate	E
38	High	Moderate	High	High	High	E
39	Low	Moderate	High	Moderate	Low	F
40	Low	Moderate	High	Moderate	Moderate	F
41	Low	Moderate	High	Moderate	High	F
42	Moderate	Moderate	High	Moderate	Low	F
43	Moderate	Moderate	High	Moderate	Moderate	F
44	Moderate	Moderate	High	Moderate	High	F
45	High	Moderate	High	Moderate	Low	F
46	High	Moderate	High	Moderate	Moderate	F
47	High	Moderate	High	Moderate	High	F
48	Low	Moderate	High	Moderate	Low	G
49	Low	Moderate	High	Moderate	Moderate	G
50	Low	Moderate	High	Moderate	High	G
51	Moderate	Moderate	High	Moderate	Low	G
52	Moderate	Moderate	High	Moderate	Moderate	G
53	Moderate	Moderate	High	Moderate	High	G
54	High	Moderate	High	Moderate	Low	G
55	High	Moderate	High	Moderate	Moderate	G
56	High	Moderate	High	Moderate	High	G
57	Low	High	High	Moderate	Low	G
58	Low	High	High	Moderate	Moderate	G
59	Low	High	High	Moderate	High	G
60	High	High	High	Moderate	Low	G
61	High	High	High	Moderate	Moderate	G

(*continued*)

Table 5. (*continued*)

No	Big five personality factors					Job family
	E	A	C	ES	I	
62	High	High	High	Moderate	High	G
63	Low	High	High	Low	Low	H
64	Low	High	High	Low	Moderate	H
65	Low	High	High	Low	High	H
66	Low	High	High	High	Low	J
67	Low	High	High	High	Moderate	J
68	Low	High	Moderate	High	Low	J
69	Low	High	Moderate	High	High	J
70	Moderate	Moderate	High	Low	Low	K
71	Moderate	Moderate	High	Low	Moderate	K
72	Moderate	Moderate	High	Low	High	K
73	Moderate	Moderate	Moderate	Moderate	Moderate	L
74	Moderate	Moderate	Low	Moderate	Moderate	L
75	Moderate	Moderate	Moderate	High	Low	L
76	Moderate	Moderate	Moderate	High	Moderate	L
77	Moderate	High	Moderate	High	Moderate	M
78	Moderate	High	High	High	Moderate	M
79	Low	Low	High	Moderate	Moderate	O
80	Low	Low	High	Moderate	High	O
81	Low	Low	High	High	Moderate	O
82	Low	Moderate	Moderate	Moderate	Low	P
83	Low	Moderate	Moderate	Moderate	Moderate	P
84	Moderate	High	High	High	High	Q
85	Moderate	High	High	High	Moderate	R
86	Low	High	High	High	Moderate	S
87	Low	High	High	High	Low	T
88	Moderate	Moderate	Moderate	High	Low	U
89	Moderate	Moderate	Moderate	High	Moderate	V

4.5   C4.5 Algorithm

The dataset has attributes in the form of five Big Five personality factors. However, the total value of each factor has yet to be discovered. So, from each data, the total value of each factor must be calculated first. The assessment of each question of the Big Five personality test is in Table 1. Then the results of these calculations are used to determine work recommendations with the C4.5 method (Fig. 1).

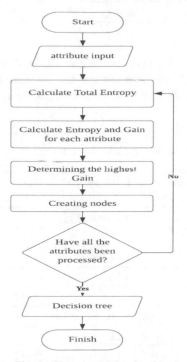

Fig. 1. Flowchart C4.5 algorithm

The initial preparation of each attribute is tested by finding the gain ratio value using the formula [17]. Then determine which attribute has the highest gain ratio value. The attribute will be the first node (root). Next, select the branch similarly by looking at the highest gain ratio of each partition result.

The following is an example of a C4.5 algorithm calculation case. For simplicity, the first 5 data sets are taken, and only two labels are taken, namely "A" and "L". The dataset is shown in Table 6.

Table 6. Sample case dataset

No	E	ES	A	C	I	job_family
1	30	24	31	32	33	A
2	29	26	28	28	31	L
3	26	27	32	27	31	L
4	35	23	30	32	36	A

(*continued*)

Table 6. (*continued*)

No	E	ES	A	C	I	job_family
5	30	22	28	29	32	L

The first step is determining the attribute that will be used as the root. The attribute that will be used as the root is the attribute that has the highest gain ratio value. Table 7 is the result of calculating the gain ratio for each attribute using the formula described above.

Table 7. Result of gain ratio calculation

Node	Attribute	Value	Number of cases	A	L	Entropy	Gain	Split info	Gain ratio
1	Total		5	2	3	0,067361111			
	E						00.32	0,05	00.45
		≤30	4	1	3	0,05625			
		>30	1	1	0	0			
	ES						00.42	0,067361111	00.43
		≤24	3	2	1	0,063888889			
		>24	2	0	2	0			
	A						00.42	0,067361111	00.43
		≤28	2	0	2	0			
		>28	3	2	1	0,063888889			
	C						0,067361111	0,067361111	01.00
		≤29	3	0	3	0			
		>29	2	2	0	0			
	I						0,067361111	0,067361111	01.00
		32	3	0	3	0			
		>32	2	2	0	0			

The highest gain ratio value is attributed C at threshold 29 with a value of 1 and attribute I at threshold 32 with the same gain ratio value as attribute C. However, the selected attribute is the attribute with the highest gain ratio value first from the left dataset attribute. Therefore, what will be the root node is attribute C (Table 8).

The second step is to create a branch for each value. For example, it is known that attribute C at threshold 29 has branch values ≤29 and >29 (see Fig. 2).

The third step is to split the cases into branches. The number of cases of attribute C ≤ 29 is 0 for label "A" and 3 for label "L". Therefore every case with a value of

Table 8. Attribute C at threshold 29

Attribute	Value	Number of cases	A	L
C				
	≤29	3	0	3
	>29	2	2	0

Fig. 2. Branch division for each value

C ≤ 29, the result is "L". Likewise, the number of cases for attribute C > 29 is 0 for the "L" label and 2 for the "A" label. Then every case with a value of C > 29, the result is "A". So the resulting C4.5 decision tree is as follows (see Fig. 3):

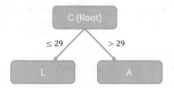

Fig. 3. Division of cases in branches

5   Testing and Result

Table 9 is a data partition test scenario. The test was carried out four times with several different partitions.

Table 9. Data partition testing scenario

Testing	Data training	Data testing	Data training size	Data testing size
1	60%	40%	1223044	815363
2	70%	30%	1426884	611523
3	80%	20%	1630725	407682
4	90%	10%	1834566	203841

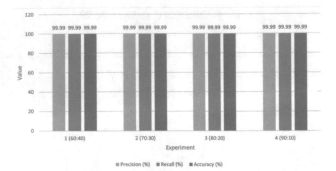

Fig. 4. Data partition testing result

Figure 4 is the result of testing the data partition on the model in the job recommendation system based on the Big Five Personality Test. Accuracy is the number of correct predictions by the classification model divided by the total number of predictions made. Precision is when the classification model predicts positively how often it is true. Finally, the recall represents how many were correctly identified by the classification model out of all possible positives.

The data partition test results show that the significant difference in the data partition does not cause a significant difference in model performance.

Table 10 is the result of testing the system on 30 respondents.

Table 10. System test results to respondents

No	Big five personality factors					Job family	Rule	Conclusion
	E	ES	A	C	I			
1	20	35	36	34	28	R	85	Valid
2	28	34	33	39	37	Q	84	Valid
3	35	23	45	35	31	G	61	Valid
4	35	26	33	42	30	G	61	Valid
5	32	28	32	33	35	B	11	Valid
6	20	34	33	40	37	Q	84	Valid
7	27	34	34	35	32	R	85	Valid
8	33	29	37	35	28	G	85	Valid
9	26	29	39	30	27	C	16	Valid
10	26	21	40	27	33	C	17	Valid
11	24	23	42	30	48	C	17	Valid
12	31	35	32	36	32	E	34	Valid

(continued)

Table 10. (*continued*)

No	Big five personality factors					Job family	Rule	Conclusion
	E	ES	A	C	I			
13	21	20	42	30	33	C	17	Valid
14	20	28	39	32	35	C	17	Valid
15	27	35	27	29	26	V	89	Valid
16	24	27	35	31	39	C	17	Valid
17	33	20	33	36	33	G	62	Valid
18	37	29	38	39	37	G	62	Valid
19	27	34	40	38	36	Q	84	Valid
20	27	35	27	29	26	V	89	Valid
21	45	21	34	33	0	G	60	Valid
22	40	24	38	36	38	G	62	Valid
23	27	27	35	32	39	C	17	Valid
24	37	29	40	43	34	G	62	Valid
25	37	29	37	33	29	G	61	Valid
26	37	25	43	36	33	G	62	Valid
27	34	28	33	30	27	C	19	Valid
28	33	20	35	34	29	G	61	Valid
29	33	24	39	36	34	G	62	Valid
30	29	30	23	36	22	F	43	Valid

System accuracy testing is carried out to determine whether this work recommendation system has an accuracy value that is quite feasible. System accuracy testing is done by looking at the results of work recommendations from the system and then comparing them with the rules from previous studies by F. Mar'i, W. F. Mahmudy, and C. Yusainy (2019) [4]. Then the accuracy value is calculated using the following formula:

$$Accuracy = \frac{Total\ Accurate\ Data}{Total\ Data} \times 100\% \tag{5}$$

$$Accuracy = \frac{30}{30} \times 100\% = 100\%$$

Based on the results of the calculations above, the system has an accuracy value of 100%, so this system can work well following the purpose of this research, namely, to implement the Big Five personality test to find out which jobs are suitable for an individual's personality.

6 Conclusion

Based on the tests in this research, mapping the Big Five personality factors with job recommendations using the C4.5 method has an accuracy performance of 99.99%. To create an effective decision tree with C4.5, it is determined by what questions to ask and when. The gain ratio and entropy can evaluate the exact question. The results of system testing with thirty tested data obtained an accuracy rate of 100% to determine the job family.

References

1. Friedman, H.S., Schustack, M.W.: Personality: Classic Theories and Modern Research. Pearson Education, Inc., New York City (2016)
2. Ziegler, M., Bensch, D., Maaß, U., Schult, V., Vogel, M., Bühner, M.: Big five facets as predictor of job training performance: the role of specific job demands. Learn. Individ. Differ. **29**, 1–7 (2014)
3. Seibert, S.E., Kraimer, L.M.: The five-factor model of personality and career success. J. Vocat. Behav. **58**, 1–21 (2001)
4. Mar'i, F., Mahmudy, W.F., Yusainy, C.: Sistem Rekomendasi Profesi Berdasarkan Dimensi Big Five Personality Menggunakan Fuzzy Inference System Tsukamoto. Jurnal Teknologi Informasi dan Ilmu Komputer (JTIIK) **6**, 456–466 (2019)
5. Tamin, F., Iswari, N.: Implementation of C4.5 algorithm to determine hospital readmission rate of diabetes patient. In: 2017 4th International Conference on New Media Studies (CONMEDIA) (2017). https://doi.org/10.1109/conmedia.2017.8266024
6. Hssina, B., Merbouha, A., Ezzikouri, H., Erritali, M.: A comparative study of decision tree ID3 and C4.5. Int. J. Adv. Comput. Sci. Appl. **4**(2) (2014). https://doi.org/10.14569/specialis sue.2014.040203
7. Cervone, D., Pervin, A.L.: Personality: Theory and Research. Wiley, Hoboken (2013)
8. Sackett, R.P., Walmsley, T.P.: Which personality attributes are most important in the workplace? Perspect. Psychol. Sci. **9**, 538–551 (2014)
9. Akhtar, H., Saifuddin, A.: Indonesian adaptation and psychometric properties evaluation of the big five personality inventory: IPIP-BFM-50. Jurnal Psikologi **46**, 32–44 (2019)
10. Larose, D.T., Larose, C.D.: Discovering Knowledge in Data. Wiley, Hoboken (2014)
11. Mardi, Y.: Data mining: Klasifikasi Menggunakan Algoritma C4.5. Jurnal Edik Informatika **2**, 213–219 (2017)
12. Quinlan, J.R.: C4.5: Programs for Machine Learning. Morgan Kaufmann Publishers, San Mateo (1993)
13. Bramer, M.: Principles of Data Mining. Springer, London (2020)
14. Rahim, R., et al.: C4.5 classification data mining for inventory control. Int. J. Eng. Technol. **7**, 68–72 (2018)
15. Take a personality test - open source psychometrics project. Take a personality test - Open Source Psychometrics Project. https://openpsychometrics.org/. Accessed 24 Aug 2021
16. Build your future WITH O*net online. O*NET OnLine. https://www.onetonline.org/. Accessed 24 Aug 2021
17. Jadhav, D., Bakade, T., Deshpande, S., Pethe, O., Kale, S., Madane, Y.: Big five personality prediction using machine learning algorithms. Math. Stat. Eng. Appl. **71**(3), 1128–1133 (2022)

Predicting Disaster Type from Social Media Imagery via Deep Neural Networks Directed by Visual Attention

Shatheesh Kumar Govindarajulu[1], Megan Watson[2], Sulaf Assi[2(✉)],
Manoj Jayabalan[1], Panagiotis Liatsis[3], Jamila Mustafina[4], Normaiza Mohamad[5],
Kdasy Al-Muni[6], and Dhiya Al-Jumeily OBE[1]

[1] Faculty of Engineering and Technology, Liverpool John Moores University, Liverpool L3 3AF, UK

[2] School of Pharmaceutical and Biomolecular Science, Liverpool John Moores University, Liverpool L3 3AF, UK
s.assi@ljmu.ac.uk

[3] Department of Electrical Engineering and Computer Science, Khalifa University, Abu Dhabi, UAE

[4] Kazan Federal University, Kazan, Russia

[5] Faculty of Business and Technology, UNITAR International University, Petaling Jaya, Malaysia

[6] Saudi Ministry of Health, Najran, Saudi Arabia

Abstract. Social media has become the primary source for the public for seeking news and updates in crisis such as disasters. However, the information sought from social media in disasters is usually in the form posts (images or texts) with unorganized content that often contains duplicate, feeds, inappropriate and irrelevant posts. Processing these posts and generating meaningful information out of them is a challenge. This research proposed deep neural network-based design driven by visual-attention mechanism for classifying disaster types from social media imagery. Deep neural networks were applied to raw datasets consisting of 71K images obtained from actual disasters and were split into training validation and test sets. Three approaches were applied including 'Base Model', 'Bottleneck Attention Module' and 'Focus Attention Module'. The Base Model showed the highest accuracy, but the Focus Attention Module learnt faster than models and enabled to cut down the training time. The research enhanced disaster management capabilities of government, first responders, non-governmental organizations and other relevant aid agencies.

Keywords: Social media · Disasters · Image classification · Deep neural networks · Bottleneck attention module · Focus attention module

1 Introduction

1.1 Background

Social media impact on today's lifestyle is profound and continues to influence our culture, habits and way of thinking. Social media impacts everyday life not only for

© The Author(s), under exclusive license to Springer Nature Singapore Pte Ltd. 2023
Y. B. Wah et al. (Eds.): DaSET 2022, LNDECT 165, pp. 37–51, 2023.
https://doi.org/10.1007/978-981-99-0741-0_3

individuals but also for governments, corporates and small enterprises [1]. Social media has become the primary source of news for the public, even the mainstream news outlets and media leverage power of social media to drive traffic to their digital presence [2]. It is not surprising to realize during disasters, social media has become one of the primary news sources through which public gather information.

A disaster causes serious disruptions to societal function, burdens government agencies to manage the crisis and response [3]. A big part of crisis management is to continuously process news feeds, public enquiries, social media posts, and act accordingly. Many of the government agencies' standard operating procedures for crisis response and management are built on traditional sources of information like news, public enquiry calls but not necessarily for handling millions of social media posts that surface in minimal time [4, 5]. Social media posts are not selected and organized content, so processing actionable information can be a problem. On top of it, one needs to account for the duplicate feeds, irrelevant, and inappropriate posts that accompany social media timeline, especially during crisis [5].

During an evolving crisis, the complexity of deriving meaningful information from social media posts in a time critical manner is challenging. It gets even harder, when the social media posts content is not just text but also accompanies with or exclusively rich content such as photos, videos, etc. There are mechanisms to process text feeds or posts but to caption a photo or a video and derive real-time meaningful information and possibly insights for crisis management falls on setting up a very complex system that should also be accurate and optimally operable [6, 7].

The focus of this research is to propose a deep neural network solution aided by a visual attention mechanism to classify disaster images posted on social media. This solution potentially can come in aid of government agencies, non-government organizations, etc. to improve effectiveness of crisis management and help tailor response to the impacted community.

Deep convolutional neural networks based on AlexNet, VGGNet, Inception Network, MobileNets, etc. perform very well to classify images but it processes a lot more details and noise when an entire image is consumed and thus possibly result in inaccurate information. Visual attention mechanism can overcome this deficiency by focusing on only salient aspects of an image to classify information and thus can perform better [8].

The necessity to pair the visual attention mechanism to a deep neural network, is to improve prediction accuracy of a disaster type from social media imagery content in the event of a crisis developing. Adding a visual attention component to an existing image classifying deep neural network morphs its behavior such as human attention. Xu et al. 2015 demonstrated similar ability of neural networks to caption images better with visual attention [9]. There is more evidence of better classification performance with visual attention through the works of An et al. 2021 [10] in classifying the medical images and Haut et al. 2019 [11] in Hyperspectral image classification.

While the objective of this research is limited to disaster classification from image, the proposed solution can be potentially scaled for other needs including assessing the severity of damage, predicting optimal mitigation and humanitarian response, etc....

1.2   Rationale

Damage in disasters is enormous and not often it is possible to identify the nature of the disasters, causes and degree of harm. Evidence collected requires time in terms of recovery, transport and analysis and has many issues related to sample integrity and continuity. Images are usually quick, easy to take and taken non-destructively and that speeds up the investigation process. Imagery content coming from a disaster event can provide many levels of information for the first responders, government decisions makers and other relevant organizations to respond and control an evolving crisis. The biggest challenge in developing artificial intelligence models for disaster response is that there is scarce amount of curated content from disaster events. Li et al. 2019 deployed a model pre-trained on the different domain source to identify if a damage occurred or not due to a disaster event [12]. Imran et al. 2020 developed an automatic image processing system to assess damage severity from imagery [13]. This system was deployed and activated during Hurricane Dorian and ran for 13 days with help of a volunteer response organization. This research builds on the previous models by combining visual attention mechanism to improve performance. Specifically, this research used attention models that allowed neural network to focus on each piece of evidence in a disaster image until the whole pieces are classified. While the focus is on the disaster type classification for this research, it could potentially yield a framework that can be extended in other areas of crisis management and beyond.

2   Disaster Types Prediction Models

2.1   Research Methodology

The disaster image dataset primarily contains of data collected during 2017 disasters and were housed in opensource databases – Multimodal Crisis Dataset, Artificial Intelligence for Digital Response, Damage Assessment Dataset, and Damage Multimodal Dataset. Alam et al. 2021, as part of their research effort, sourced information from the above databases and taxonomically organized approximately 71K images [14]. Alam et al. 2021, research included manual effort to annotate tweets and associated images from major 2017 natural disasters that included wildfires, earthquakes, floods, and hurricanes.

These disaster images are labelled and catalogued in a file that is in a tab-separated values (TSV) fileformat and segmented for training and testing. Any missing data are handled accordingly and only, vital data will be used for model building. All the paths of files are disciped in Table 1. Listed below are the data artifacts.

2.2   Data Selection

The raw dataset consists of ~71K images from actual disasters, and image metadata. The data provided was split into training, validation, and test files at source. The ratio of training to validation sets was 3:1. Images associated in the non-training datasets were consistently found to be corrupt and missing EXIF (Exchangeable Image File) data causing the Image APIs to fail during modelling. Corrupt images were removed

Table 1. Data folders

Data/	Root directory
aidr_disaster_types/	Disaster Images by Type from AIDR system
aidr_info/	Disaster Images by Information from AIDR system
damage_image_dataset/	Disaster Images from Damage Assessment Dataset
crisismmd/	Disaster Images from Multimodal Crisis Dataset
multimodal-deep-learning-disaster-response/	Disaster Images from Damage Multimodal Dataset

Table 2. Available disaster data

Disaster types	Informativeness	Humanitarian categories	Damage severity assessment
Earthquake	Informative	Affected, Injured, Dead people	Little; No Damage
Fire	Not informative	Infrastructure; Utility damage	Mild Damage
Flood		Not Humanitarian	Severe Damage
Hurricane		Rescue Volunteering; Donation effort	
Landslide			
Other			

and in addition, some of the images were also missing metadata, thus resulting in a much smaller dataset of ~41K images for modelling.

A good portion (~40%) of ~41K images were uninformative and unrelated to disasters. Thus, were excluded from the dataset, retaining only a relevant portion of training samples required for non-disaster classification. Image classes 'Fire', 'Landslide' and 'Other Disaster' were excluded from dataset due to lower number of samples as shown on Table 2. The image metadata includes name of the disaster event (event_name), image identifier (image_id), type of disaster (disaster_types), and if the image content is useful or not (informative). Each sample record also includes information about damage severity (damage_severity), humanitarian aid needed (humanitarian), and image file path (image_path).

2.3 Data Transformation

The deep neural network architecture solutioned for this research includes MobileNets V3. MobileNets V3 includes input pre-processing layers for rescaling inputs to float tensors of image pixels with values in the $[-1, 1]$ range. This implies no image normalization is needed to feed into the model. Further with MobileNets V3, the input comprises of three input channels – Height, Width and RGB layers with the default shape of (224, 224, 3). This requires the input images are reshaped to (224, 224, 3) specification.

To bind a specific data record to an image on the file system, the image path (image path) is transformed with the physical image location on the file system for retrieval and processing during modelling. Approximately, 9200 images were synthetically added to sample set through Image Augmentation by introducing gaussian blur, horizontal image flipping, and vertical image flipping of randomly selected images from the original data source.

Finally, the deep neural network for this research was built with TensorFlow, an open-source machine learning library. The input feed into these networks are multidimensional arrays called Tensors and are driven through descriptive pipelines called Tensor Datasets. Specifically for this research solution, the input data to the neural network comprises of a crisis image, it's label as one hot encoded vector along with its numerical class weight. Batches of this input data are transformed, pre-processed, and created as Tensor Dataset and fed to the neural network.

2.4 Attention Model

This research focuses on developing deep neural network with attention model to optimize the performance of the underlying convolutional network in image classifications. Two types of attention-based mechanisms are utilized to explore the performance.

Bottleneck Attention Module (BAM)
For the first of two approaches, the deep learning model for the disaster classification will utilize dual attention mechanism, in the form of channel and spatial attentions to extract salient features. The continuing emphasis with the solution is that attention mechanism would also be light weight and adds little to no overhead to network performance. Bottleneck Attention Module (BAM) was developed by Park et al. 2020 as a self-sufficient and light weight component that can be embedded within an existing deep learning network to enhance performance [15]. BAM extracts channel (what?) and spatial (where?) features that can enhance the salient features and improve the quality of classification.

Focus Attention Module (FAM)
The second approach in building the attention-based model was inspired by the work of Xu et al. 2015 [2]. Xu et al. 2015 prescribed Recurrent Neural Network (RNN) based soft attention to caption the contents of an image [9]. For this research involving crisis images, the image classification network is based on a deep Convolutional Neural Network (CNN). The task of the attention module is to generate focus on the salient parts of the image and direct the outcome based on a focal point.

2.5   Experiments

In total, there were five noteworthy experimental designs explored and with iterative customizations to optimize the individual models as explained in Table 3.

Table 3. Model experiments

Experiments	Key components
MobileNet V3 Base Model	MobileNet V3 Large + No Attention Module
With Channel Attention	MobileNet V3 + Channel Attention
With Spatial Attention	MobileNet V3 + Spatial Attention
Mixed Attention	MobileNet V3 + Channel Attention + Spatial Attention
With Focus Attention Model	MobileNet V3 + Focus Attention

2.6   Tools and Environment

Sufficient resource capacity around hardware and software are required to run the machine learning workload to develop the model(s). Listed below in Table 4, are resources specification used for realizing the research design and model evaluations.

Table 4. Software and libraries

Jupyter lab	Keras	Scikit image
Anaconda	Scikit Learn	Pillow – Python Imaging Library
Python	Pandas	OS
Tensorflow	Numpy	CUDA and cuDNN
Tensorflow Hub	Seaborn	OpenCV
Tensorflow Addons	Matplotlib	Pool, Netron

Machine Learning Infrastructure

- Cloud Provider – Google Cloud Platform (GCP)/Google Colab
- Virtual Machine (VM) Image Name - tf2-cpu.2-4.m65
- Operating System Environment - Debian GNU/Linux 10 (buster) (GNU/Linux 4.19.0-16-cloud-amd64 x86_64\n)
- CPU Platform - Intel Broadwell
- GPU - NVIDIA Corporation GK210GL [Tesla K80] Tesla V100-SXM-16GB
- Storage – Cloud Storage

3 Results and Discussion

The research involved developing experimental models, with extensive amount of model training, continuously refining the model performance and evaluating results. While this effort may not require setting up an end-to-end machine learning pipeline infrastructure, but sufficient resource capacity around hardware and software are required to run the machine learning workload to develop the model(s). The results below showed findings for the three approaches i.e. Base Model, BAM and FAM.

3.1 Base Model Results

The base model ran for 26 epochs and took 57 min. This model produced top two results based on 'Precision' at 0.8500, 'Accuracy' at 0.9152 and 'Recall' at 0.8023. The best loss value achieved at 0.8635 (Fig. 1).

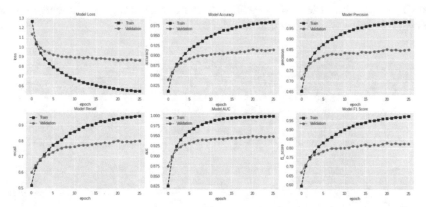

Fig. 1. Base model learning curves

3.2   Channel Attention Model Results

The Channel Attention Model ran for 14 epochs and took 27 min. This model had the lowest 'Precision' at 0.4778, lowest 'Accuracy' at 0.7290 and the highest 'Recall' at 0.9517. The best loss value achieved at 0.9394 (Fig. 2).

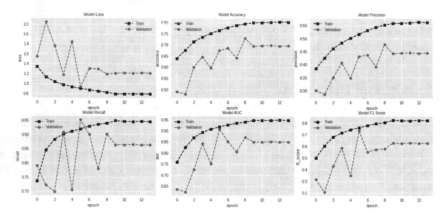

Fig. 2.   Channel attention model learning curves

3.3   Spatial Attention Model Results

The Spatial Attention Model ran for 23 epochs and took 44 min. This model produced following scores with 'Precision' at 0.5227, 'Accuracy' at 0.7705 and 'Recall' at 0.9515. The best loss value achieved at 0.8269 (Fig. 3).

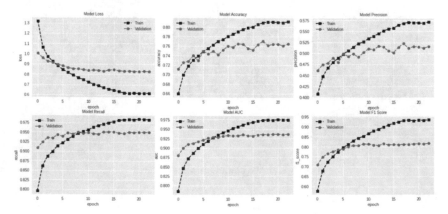

Fig. 3.   Spatial attention model learning curves.

3.4 Mixed Attention Model Results

The Mixed Attention Model ran for 19 epochs and took 37 min. This model produced following scores with 'Precision' at 0.5900, 'Accuracy' at 0.8205 and 'Recall' at 0.9265. The best loss value achieved at 0.8374 (Fig. 4).

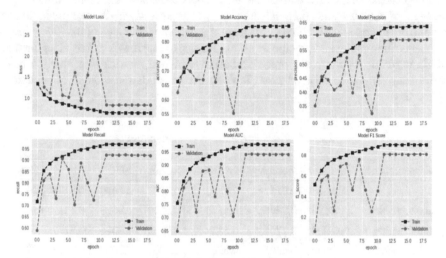

Fig. 4. Mixed attention model learning curves.

3.5 Focus Attention Model Results

The Focus Attention model ran for 13 epochs and took 24 min. This model produced top two results based on 'Precision' at 0.8602, 'Accuracy' at 0.9132 and 'Recall' at 0.7857. The best loss value reached at 0.8027 (Fig. 5).

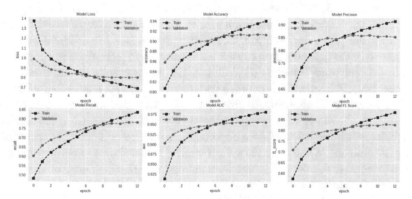

Fig. 5. Focus attention model learning curves.

3.6   Results Analysis

The foundation of this result analysis will be based on how accurate and precise the models are. The accuracy measures the distance of prediction to the ground truth and while precision measures the quality of the prediction. It is important from a disaster identification process perspective that the model predicts more relevant results to determine the nature of response and communication. Any inconsistency in disaster prediction can prove costly in crisis. An incorrect broadcast of emergency message or disaster response can be more damaging and precipitate a bad situation to one that is worse. Keeping these criteria in mind, accuracy and precision becomes pivotal metrics in evaluation of the experiments conducted. Listed below (Table 5) are the validation scores from individual experiments.

Table 5. Model experiment validation results

Experiment models	Loss	Accuracy	Precision	Recall	AUC	F1	Epochs
Base	0.8635	0.9152	0.8500	0.8023	0.9489	0.8279	26
Channel attention	0.9394	0.7290	0.4778	0.9517	0.9075	0.7312	14
Spatial attention	0.8269	0.7705	0.5227	0.9515	0.9370	0.8173	23
Mixed attention	0.8374	0.8205	0.5900	0.9265	0.9409	0.8108	19
Focus attention	0.8027	0.9132	0.8602	0.7857	0.9544	0.8279	13

The first experiment called 'Base Model' involved 'MobileNet V3 Large' with no attention module and produced top two results (Fig. 1) with lowest number of model parameters to train. This experiment ran for 26 epochs for a total time of 51 min, the longest training time. The training automatically 35 stopped early with the precision score plateauing. Based on the AUC score, it had the second-best label classification behind 'Focus Attention Module (FAM)' model (Experiment – 5).

Experiments (2, 3, & 4) based on the 'Bottleneck Attention Module (BAM)' yielded lowest validation accuracy and precision results. The validation results were fluctuating (wavy) for better part of training stages, possibly due to lack of sufficient information being learned or the data shuffling caused fewer samples to learn from in comparison to validation set. The models belonging to these experiments eventually stabilised and had the highest 'Recall' scores. This coupled with relatively lowest precision and accuracy, one can say that the 'Bottleneck Attention Module (BAM)' algorithm predicts most of the relevant results but also included irrelevant ones. It's obvious from the results that spatial features (where?) played more important role than the channel features (what?) in learning the problem. When paired together (what and where?), the experiment improved its performance, but the overall scores were not up to mark still. These outcomes led to the next set of experiments with 'Focus Attention Module (FAM)' algorithm.

'Focus Attention Module (FAM)' experiments yielded best of validation scores at 0.9132 for accuracy, and 0.8602 for precision. Classification loss is crucial in measuring the prediction deviation from the ground truth and the model with lowest loss value was

'Focus Attention Model' at 0.8027 and the 'Channel Attention Model' was relatively worst at 0.9394. The reason for high loss value is due to over confidence in model's prediction when it failed.

The AUC score for this experiment model was the best at 0.9544, meaning it had the best disaster classification ability in comparison to its peers in the experiments. The 'Focus Attention Module (FAM)' experiments also took least number of epochs and time to train.

Overall, 'Focus Attention Module (FAM)' model performed the best and at the same time sped up the training by 51% when compared to its nearest rival 'Base Model'. This model added very little parameter overhead on top of the 'Base Model' that it was built on. This indicates 'Focus Attention Module (FAM)' algorithm effectively used the model parameters better than other models and was able to direct right attention on discriminatory features for disaster classifications.

3.7 Results Implications

With this research and continued analysis, evolved better interpretations of results and key findings. One of the paradoxical situations was high loss scores despite the model performing better in disaster image classifications with very high accuracy, precision, and AUC scores. When the model misclassifies, it obviously is unable to distinguish the disaster features well enough to classify correctly.

'Focus Attention Module (FAM)' model misclassified some images which were indeed indistinguishable to human eyes as well. It is understandable from these outcomes why the model could not distinguish them either and it also developed a high confidence in its label prediction that threw the validation loss by a good margin.

The 'Bottleneck Attention Module (BAM)' models underachieved in disaster image classifications and the overall performance metrics were not up to mark. The key reason is for its underwhelming performance is due to the deep placement of this BAM module in the neural network. In convolutional neural networks, the deeper layers have already learned and digested the spatial and channel information within feature extracts. The pixel level information such as location is not discernible separately. As a result, BAM algorithm that relies on spatial and channel data didn't benefit. Placing BAM module in early CNN layers will improve performance, but the research design relied on 'MobileNet V3 Large' out of box and that limited the position within the neural network for integration with BAM module for attention.

'Focus Attention Module (FAM)' neural network has two layers of decisions. The first decision layer learns about the focus region by contextualizing the extracted feature for classification and directs second decision layer to classify the disaster images based on the attention probability. This algorithm was more effective in generating attention and was 51% quicker to learn than its nearest rival ('Base Model') with only fewer parameter overhead.

Based on the validation metrics, 'Base Model' and the visual attention based on 'Focus Attention Module (FAM)' model performed equally well but the later model learned faster and significantly cutdown the training time.

3.8 Discussion

The aim of this research is to propose a deep learning neural network-based design driven by visual attention mechanism to classify disaster types from social media imagery.The research explored two visual attention-based techniques and four related experiments. It became evident as the research progressed that a lightweight neural network is essential for success of operationalizing disaster classification process. 'MobileNet V3' from Google fit the profile [16]. It has optimized architecture for mobile device CPUs that comes with a lightweight decoder module called 'Lite Reduced Atrous Spatial Pyramid Pooling' (LR-ASSP) and the state of the art (SOTA) performance in classifications, detection, and segmentation. In this research, 'MobileNet V3' is the underpinning for the deep neural network design for disaster image classification.

The performance of the visual attention-based models using 'Bottleneck Attention Module (BAM)' relative to other experiments were underwhelming. It was discovered during research analysis that the performance was due to lack of sufficient spatial and channel data that neural network was providing to the 'Bottleneck Attention Module (BAM)' module. In the deeper stages of the neural networks, the spatial and channel information are already learned and those datapoints are digested within the feature extracts but are inaccessible and unavailable as discrete datapoints. 'Bottleneck Attention Module (BAM)' algorithm developed by Park et al. 2020 relies on capturing the spatial and channel information to drive attention focus [8]. Further research is needed to explore, if this algorithm will fare better for disaster image classification if the 'Bottleneck Attention Module (BAM)' module is included in the earlier layers of the neural network.

The second visual attention-based approach taken was with 'Focus Attention Module (FAM)'. The later experiments based on 'Focus Attention Module (FAM)' performed well with the accuracy at 0.9132 and AUC score at 0.9544 but its precision (0.8602) and recall (0.7857) numbers could do better. Essentially, these numbers are not much different from performance of 'Base Model' that didn't have attention module. In fact, the F1 Scores (0.8279) are same between them. 'Focus Attention Module (FAM)' learned faster and significantly sped up the training by 51% than the 'Base Model' but based on disaster image classification metrics alone both the model variants were about the same.

'Focus Attention Module (FAM)' model's classification performance metrics accomplished may rise the question about the need for vision-based attention mechanism for disaster image classification. Contrary to the outcomes of this research, there are numerous references to attention guided networks performing well in other fields of interests. In this use case, we cannot ignore the fact disaster imagery comes with lots of clutter, irrelevant noises, and similarities in aftermath imagery of destructions. It is evident from the validation loss that the performance of 'Focus Attention Module (FAM)' based deep neural network was impacted by these factors. These factors emphasize the need for exploring further and improving the current attention driven neural network design established within this research.

There are opportunities here to improve quality (precision) of the classification by reducing false positives and false alarms (negatives). The future research could build on

'Focus Attention Module (FAM)' baseline design and incorporate a multi-headed attention neural network. These attention-heads have individual tasks where one attention-head is focussed on recognizing the noise and clutter to avoid, while the second attention-head is tasked with spatially identifying the disaster specific regions and dynamically developing disaster image mask filters. The third attention-head uses these masks to filter the portion of the original image and direct the neural network classification based on the filtered image.

This is akin to a design proposed by Guan et al. 2020 for Thorax Disease Classification [16]. The results in the research paper by Guan et al. 2020 are interesting and reemphasizes the original premise for this research that a vision driven attention models can aid better in disaster image classification [16]. Understandably, the future solution is not just limited to this recommendation but can build on this research for end-to-end crisis management pipeline.

4 Conclusions

This research proposed a deep learning neural network aided by visual attention mechanism to classify disaster types from social media imagery. Based on the qualitative analysis of model metrics for this research solution, it could be concluded that the model learns significantly faster, lightweight and performs well in the disaster image classification. One of the objectives for this research was that visual attention enhances image classification performance of conventional convolutional neural network. In this regard, the research explored attentions mechanisms with multiple experiments based on spatial, channel and feature contexts.

From a data perspective, there was abundance of disaster images in social media and the rest in the unknown ether. There was very limited set of curated content of disaster images though, which narrowed this research's boundaries. However, considering limited research on this topic, real-world solutions utilizing artificial intelligence in disaster classifications, and of course, the quality of the model evaluation metrics, this research merits consideration.

Emergency management organization could utilize this research solution as a foundational AI model to build improvements on and potentially engineer new frameworks that can be extended in other areas of crisis management and beyond. The solution presented here fills a gap in crisis management capabilities by utilizing social media imagery to identify disaster types. When this capability paired with aerial and satellite images can potentially provide holistic disaster assessments for aid agencies in decisions, directing resources, broadcasting formal alerts, etc. It is also evident from this research, there is systemic gap in collecting, organizing, and maintaining disaster imagery by aid agencies. There needs to be organized efforts by aid agencies to establish open standards and protocols for disaster information exchange.

There are other practical implications of this research, while it is primarily focused on the disaster classification from social media imagery, it also paves way for new ideas in the sphere of crisis management. The current solution can be extended to the following but not limited to expanding capabilities to identifying multiple disasters, assessing humanitarian aid, resource planning and generating captions from disasters.

If aid agencies harness the power of social media and artificial intelligence, then there are limitless opportunities to improve crisis management, reimagine crisis response and better community wellbeing and experience during a crisis. This research lends a thought to aid agencies, and potentially opens new avenues to revise established emergency response methods and protocols. Further research and study are needed to build on the research solution provided here.

The model produced high accuracy and AUC scores at 0.9132 and 0.9544 respectively. It can further improve in performance. The causes impacting the model performance were clutter, irrelevant noises, and similarities in disaster imagery. New research should prioritize eliminating or decreasing the impact of these causes. In this regard, one of the considerations should be to explore multi-head attention strategies with each attention module specialized to generate task specific focus and perform collectively disaster classification. In theory, this type of neural network design potentially gives more levers to optimize and extract better classification performance.

Convolutional Neural Networks (CNN) models are trained to be domain specific and can lean to be computationally heavy and demanding. 'Vision Transformer (ViT)' developed by Dosovitskiy et al. 2021 is a next generation of vision models from 'Google Research' that can be domain agnostic and computationally efficient [17]. 'Vision Transformer (ViT)' splits an image into fixed size patches, each patch is identified with a sequential spatial position, and fed serially into a transformer encoder for image classification. 'Vision Transformer (ViT)' is a multi-head self-attention vision model that comes with a higher precision rate (based on publicly available reports and feedback) and is an intriguing option to explore further. It can potentially improve on the best precision score (0.8602) accomplished with this research solution. These recommendations and beyond are worth pursuing in the future research.

References

1. Swart, J.: Tactics of news literacy: how young people access, evaluate, and engage with news on social media. New Media Soc. 14614448211011447 (2021)
2. Naeem, S.B., Bhatti, R., Khan, A.: An exploration of how fake news is taking over social media and putting public health at risk. Health Info. Libr. J. 38(2), 143–149 (2021)
3. Alexander, D.E.: On the meaning of impact in disaster risk reduction. Int. J. Disaster Risk Sci. 1–6 (2022)
4. Liu, J., Chen, Y., Chen, Y.: Emergency and disaster management-crowd evacuation research. J. Ind. Inf. Integr. 21, 100191 (2021)
5. Phengsuwan, J., et al.: Use of social media data in disaster management: a survey. Future Internet 13(2), 46 (2021)
6. Alalawneh, A.A., Al-Omar, S.Y.S., Alkhatib, S.: The complexity of interaction between social media platforms and organizational performance. J. Open Innov.: Technol. Market Complex. 8(4), 169 (2022)
7. Martín-Rojas, R., García-Morales, V.J., Garrido-Moreno, A., Salmador-Sánchez, M.P.: Social media use and the challenge of complexity: evidence from the technology sector. J. Bus. Res. 129, 621–640 (2021)
8. Yeh, C.H., Lin, M.H., Chang, P.C., Kang, L.W.: Enhanced visual attention-guided deep neural networks for image classification. IEEE Access 8, 163447–163457 (2020)

9. Xu, K., et al.: Show, attend and tell: neural image caption generation with visual attention. In: International Conference on Machine Learning, pp. 2048–2057. PMLR (2015)
10. An, F., Li, X., Ma, X.: Medical image classification algorithm based on visual attention mechanism-MCNN. Oxidative Med. Cellular Longevity **2021** (2021)
11. Haut, J.M., Paoletti, M.E., Plaza, J., Plaza, A., Li, J.: Visual attention-driven hyperspectral image classification. IEEE Trans. Geosci. Remote Sens. **57**(10), 8065–8080 (2019)
12. Li, X., Yan, D., Wang, K., Weng, B., Qin, T., Liu, S.: Flood risk assessment of global watersheds based on multiple machine learning models. Water **11**(8), 1654 (2019)
13. Imran, M., Alam, F., Qazi, U., Peterson, S., Ofli, F.: Rapid damage assessment using social media images by combining human and machine intelligence. arXiv preprint arXiv:2004.06675 (2020)
14. Alam, F., Qazi, U., Imran, M., Ofli, F.: HumAID: human-annotated disaster incidents data from Twitter with deep learning benchmarks. In: ICWSM, pp. 933–942 (2021)
15. Park, J., Woo, S., Lee, J.Y., Kweon, I.S.: A simple and light-weight attention module for convolutional neural networks. Int. J. Comput. Vis. **128**(4), 783–798 (2020)
16. Guan, Q., Huang, Y., Zhong, Z., Zheng, Z., Zheng, L., Yang, Y.: Thorax disease classification with attention guided convolutional neural network. Pattern Recogn. Lett. **131**, 38–45 (2020)
17. Dosovitskiy, A., et al.: An image is worth 16×16 words: transformers for image recognition at scale. arXiv preprint arXiv:2010.11929 (2020)

Dissemination Management for Official Statistics Using Artificial Intelligence-Based Media Monitoring

Veronica S. Jamilat[1][(✉)], Nur Hurriyatul Huda Abdullah Sani[1],
Nur Azila Mansor Noordin[1], Tengku Noradilah Tengku Jalal[1], and Sasongko Yudho[2]

[1] Department of Statistics Malaysia, Core Team Analitik Data Raya, Block C6, Complex C,
Putrajaya, Malaysia
veronica@dosm.gov.my
[2] eBdesk Malaysia Sdn Bhd, 3A-2 TH Uptown 3, Jalan SS 21/39, Damansara Utama,
47400 Petaling Jaya, Selangor, Malaysia

Abstract. Online media has been essential to our everyday lives as it spreads crucial information to the masses. Unlike traditional media, online news and social media let people freely communicate in two ways, usually by using reactions and comments, which creates public opinion and engagement on news or topics. Public opinion and engagement are beneficial in creating awareness of important news or issues. However, it will also lead to misuse, misinterpretation, and spreading of false information. For that reason, organisations often use media monitoring as a tool to gain insights about products or services. Nowadays, media intelligence, an artificial intelligence-based media monitoring, has been widely used across organisations to transform massive information from various online media platforms into useful insights. Department of Statistics Malaysia (DOSM), a national statistics provider, uses media intelligence to prevent misuse and misinterpretation of official statistics. It allows DOSM to identify the degree of happiness and centrality among the public on official statistics. This paper presents the application of using Public Maturity Assessment on Official Statistics (PMAOS), a system curated by Intelligence Media Analysis (IMA) software. PMAOS uses Natural Language Processing (NLP), syntactic and semantic techniques to understand the structure of a text, and SentiWordNet Lexicon as a base for sentiment classification. Regular monitoring of official statistics by DOSM in social media and online news found that the degree of centrality at the right source for official statistics is high. The evidence shows no or least occurrence of misuse or misinterpretation of official statistics as it is centralised across government agencies.

Keywords: Artificial intelligence · Media intelligence · Sentiment analysis · Official statistics · PMAOS

1 Introduction

The Department of Statistics Malaysia (DOSM) is a government organisation entrusted with producing official statistics that the government mainly uses as evidence-based

© The Author(s), under exclusive license to Springer Nature Singapore Pte Ltd. 2023
Y. B. Wah et al. (Eds.): DaSET 2022, LNDECT 165, pp. 52–64, 2023.
https://doi.org/10.1007/978-981-99-0741-0_4

decision-making in formulating, evaluating, and reviewing the national development policy. In practice, national statistics offices worldwide disseminate official statistics based on Special Data Dissemination Standard (SDDS). SDDS prescribes best practices that can be observed or monitored by the users of statistics. These practices are referred to as monitorable elements [1, pp. 1]. Following SDDS, DOSM disseminates official statistics through its portal and social media platforms.

Online media has been essential to our everyday lives as it spreads crucial information to the masses. However, the media's role does not just end there. The media strengthens our society as opinions are later formed based on the daily news. Therefore, getting a strong grip on how media intelligence works will prepare DOSM to prevent misuse, misinterpretation, and spreading of false information.

Media intelligence is a process of analysing public opinion through online media platforms for organisations to measure and manage products and services performance across the broader spectrum of dissemination. Findings and understanding the trends will give insights to strengthen the organisation's dissemination strategy. Media intelligence performs data science and data mining steps and procedures to process large amounts of data without neglecting the velocity, veracity, variety, and value of the data. Thus, utilising big data enables DOSM to produce valuable insights into understanding the public's opinion, which requires complicated tasks. In this regard, Natural Language Processing (NLP) enhances how the machine is trained to understand the human language.

Sentiment analysis is a sub-field of NLP that tries to identify and extract opinions from a given text. Sentiment analysis is also called opinion mining. It is the field of study that analyses people's opinions, sentiments, evaluations, appraisals, attitudes, and emotions towards entities such as products, services, organisations, individuals, issues, events, topics, and their attributes [2]. It is one of the most active research areas in natural language processing and is also widely studied in data mining, web mining, and text mining. The first step in setting up sentiment analysis is to break down the text-based formats of the news title and the news article to identify which parts of the sentences are the statements and what sentences bring positive or negative sentiments [3].

The growing importance of sentiment analysis coincides with the growth of social media such as reviews, forum discussions, blogs, micro-blogs, Twitter, and social networks. Rather than only relying on the data itself, sentiment and public opinion sometimes might influence the decisions made. It is beneficial in social media monitoring as it allows us to gain an overview of more comprehensive public opinion behind specific topics.

With soaring social media activity, emotions are viewed as a valuable commodity from a business perspective. By carefully evaluating people's opinions and sentiments, organisations can rationally find out what people think about the products and services, then incorporate feedback to prevent misinterpretation.

In official statistics dissemination, misinterpretation may lead to misleading statistics. Misleading statistics refer to the intentional or accidental misuse of numerical data. The result provides misleading information that creates an inaccurate description of the topic or issues. Misuse of statistics is common in news, media, politics, and others [4].

Each day, DOSM are bombarded by numbers in the media. Many of the facts and figures quoted in the news, such as unemployment, inflation, divorce rate, etc. However, analysing how newspapers or the public can spin the same facts differently is important. It may cause by several factors such as misunderstanding the data, using incomparable definitions, and deliberately misinterpreting the information. As a result, this analysis helps DOSM improve the dissemination of the official statistics.

2   Study Background

2.1   Motivation

Online media, especially social media, have effectively changed how people access information. Unfortunately, not all information shared on social media platforms is from reliable sources. False information, such as fake news, appears on the social media news feed with or without our consent. As DOSM is a government agency entrusted to provide and disseminate official statistics to the public, the challenge is to prevent official statistics from becoming false information by misinterpretation or misuse of these statistics.

DOSM took a preventive measure through its dissemination management for official statistics using media intelligence to ensure accurate official statistics reach the public. This paper presents how DOSM conducted media monitoring and ontology analysis using Public Maturity Assessment on Official Statistics, known as PMAOS, a system curated by Intelligence Media Analysis (IMA) software. PMAOS is an innovative approach that utilises the capability of Artificial Intelligence (AI) for efficiently monitoring public opinion on official statistics and the degree of centrality at the right source. PMAOS are media intelligence platforms that analyse Malaysians' responses to the official statistics published by DOSM on online platforms. Through these platforms, DOSM can assess the public's satisfaction based on the indicators provided on PMAOS. Sample Heading (Third Level). Only two levels of headings should be numbered. Lower level headings remain unnumbered; they are formatted as run-in headings.

2.2   Literature Review

Today, people rely on social media such as Facebook, Twitter, blog, and forum, and they usually have accounts with multiple social media services. There can be a situation where decisions have been made based on evidence from data but had to be reconsidered due to public opinion and sentiment. Paul Hoffman in Parul Pandey [5] once said, "If you want to understand people, especially your customers, then you have to be able to possess a strong capability to analyse text."

Dissemination of official statistics is the stage of statistical processing in which data collected and compiled by statistical offices are made available to the public. Data dissemination defined as the activity of making official statistics, statistical analyses, statistical services, and metadata accessible to users [6, pp. 301]. The dissemination and use of statistics facilitate government and the public in informed decision-making, which should be seen as the main goal of national statistical systems.

AI holds great potential for statistical organisations. Compiling statistics can be made more efficient by automating specific processes or assisting people in performing the processes. It also enables statistical organisations to use new data types such as imagery and social media.

Broadly, AI is defined as the theory and development of computer systems that perform tasks typically requiring human intelligence, such as visual perception, speech recognition, decision-making, and translation between languages [7].

More importantly, AI enables machines and systems to analyse their surroundings and make autonomous decisions to achieve specific goals. AI is a strategic priority and a key driver of economic development. It has the potential to solve a wide range of problems, including disease treatment and reducing the environmental impact of agriculture [8].

3 Methodology

Data mining is the process of detecting patterns in raw and extensive data extractions from various data sources. Text mining is part of data mining which primarily an AI technology that utilises NLP. In the case of extracting media data from various media platforms, without NLP, it would not be able to understand the meaning of the news, discussions, captions, or any postings. Data mining alone will solely focus on the structure of the media data rather than what the postings mean.

3.1 Media Intelligence

Media intelligence provides various analysis with media content as the main data source. PMAOS utilizes web crawling to gather countless online news sources in Malaysia using Python script. Web crawling, commonly known as spider, spiderbot, or crawler, is the process of storing online contents by indexing URLs or HTMLs into a database or files.

Web crawling typically works by extracting web pages by a search engine which later indexes the downloaded pages so that users can systematically retrieve the data in the future [9]. This method is necessary to extract large quantities of data from any website [10]. Figure 1 below briefly shows the flow of the web crawling process for any website.

Most media content is unstructured and in text-based formats. Thus, data processing is more complex than datasets in structured formats. Establishing media intelligence allows DOSM to oversee various indicators on any issues, such as the news exposure of a subject matter, identifying the individuals who are controlling the narrative, the media share, trending keywords or phrases, and the sentiments from the public. The AI technology can also conduct ontology-based analysis that constructs the relationships between various individuals or entities by finding similar distinctions that bound them together. Therefore, NLP is significant in media intelligence to make it easier for DOSM to analyse data as it uses human language to understand the data and provide better insight.

NLP works by synthesising or reading data the way humans understand languages, be it English or Malay. Simulating the way humans understand any language requires a multitude of steps for a computer as NLP is highly ambiguous. Thus, through AI, it can be disambiguated to improve the clarity of data. The way this disambiguating process

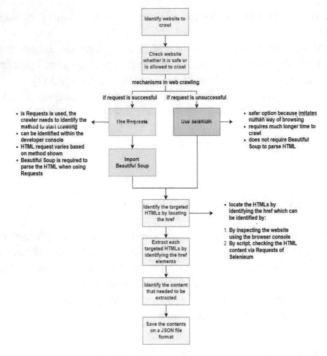

Fig. 1. Web crawling process

works is by disintegrating long texts into several language analysis stages based on the text's syntax, semantics and pragmatics. In order for media intelligence to be established according to the NLP technology, news sources will need to be collected to complete the corpus creation.

Corpus primitively is defined as a collection of authentic text organised into datasets that is readable for machine learning. Hence, all of the online news sources collected which typically are in unstructured formats will then be needed to be transformed into machine-readable texts [11].

Sentiment analysis requires the corpus to be annotated. The corpus is a collection of texts in a structured format. By adding value to the corpus, it will allow the machine to identify words more effectively for the sentiment analysis to take place [11]. The following part will explain the process of corpus annotation from the collection of texts captured through web crawling to support the sentiment analysis of PMAOS. The process from the news articles to form a sentiment analysis is shown in Fig. 2.

3.2 The Steps

The tokenization step in Fig. 2 is the stage that breaks down texts into short sentences and words called tokens. Tokens are cleaned by removing unnecessary words and punctuations to ensure the corpus annotation process is more efficient.

Figure 3 shows the different stages of NLP conducted in PMAOS. Several layers support the NLP technology: the morphological layer, lexical layer, syntactic layer, semantic

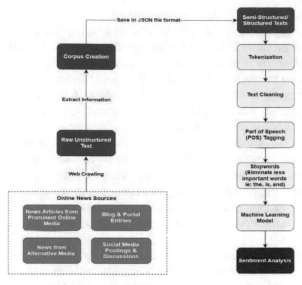

Fig. 2. Simplified flow for sentiment analysis

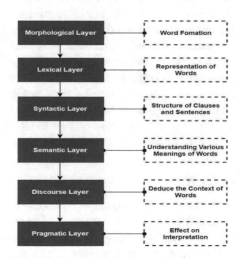

Fig. 3. Layers of Natural Language Processing

layer, discourse layer, and pragmatic layer [12]. The morphological layer processes the text by breaking down the structure of words into smaller units such as stem words, root words, prefixes, suffixes, and others. Meanwhile, the lexical layer finds the relationship between morphemes such as suffixes or prefixes and extracts the word into its root form. Under the lexical stage, the Part of Speech (POS) tagging is also conducted. It categorizes the text or corpus into its grammatical function, whether the word is an adverb, noun, verb, or others. These POS classifications will also be done according to the language's dictionary, whether in English or Bahasa Melayu.

Next, the syntactic layer focuses on the structure of the sentences from the corpus. From the POS tagging, the syntactic layer will analyse the tags based on the structure by assessing and parsing the sentence with the correct grammar at the sentence level. Furthermore, under the syntactic layer, the machine will assign each tokenized word structure based on the parse tree, which is the most widely used syntactic structure. Using a parse tree to present the syntax structure will help the machine understand the syntax better and allow multiple passes without the need to re-parse the input as it is an in-memory structure that matches the grammatical functions.

The semantic layer then analyses the meaning behind sentences or phrases and whether the meaning makes sense. The semantic layer also analyses whether certain group words can be treated as a single word, such as referring to a person's name, organisation, category, or object. Through this process, the machine can detect several notable parts of the crawled news articles, such as the date it was published, the headline, the news lead, the news content, the media organisation or author, and statements found on the news article.

Subsequently, the discourse interprets the meaning of the sentence or word by considering the previous word or sentence. The discourse layer focuses on the language by relating it to the social context. Hence, through the discourse layer, the machine can identify whether the sentence brings positive or negative sentiments based on the effect of each word or sentence. This process is essential as each word carries different sentiment. However, the following word or sentence might change the meaning and concludes an opposite sentiment as compared to the individual tokenized words.

Lastly, pragmatics analysis intertwines with the discourse layer in terms of studying the context of the text. However, the pragmatic layer is focused more on the effects of the context of the text. In other words, pragmatics in NLP studies the utterances of words, phrases, and sentences. Under the pragmatics layer, the machine will assess all the news articles and detect the relationship or association of individuals that appear on specific issues. Figure 4 shows a detailed rundown of the NLP process.

3.3 Media Monitoring Analysis

PMAOS allowed DOSM to manage dissemination by monitoring media share, news exposure, top influencer, top person, sentiment analysis, and dependency analysis. Media share is calculated based on the number of news of a media being exposed to a specific topic. The higher the news exposure is, the bigger the media share compared to others. News exposure is the publicity of news received. Typically, if the public anticipates the topic or news, its publicity is high and remains relevant for a period of time. Top influencer is calculated based on the number of statements of influencers exposed. The more statements exposed, the higher the chance of somebody being able to influence the perception of a specific topic. The individual who gives the most statements leads to being the one controlling the narration in the media. Top person is calculated based on the number of news a person is mentioned in the content. It can also represent popularity. The more an individual's name is mentioned in the media, the more popular they are. Sentiment Analysis is a process of massive exploration of information stored on the web to perceive and categorise the views expressed that transform people's opinions towards a particular topic into positive, negative, or neutral. Dependency Analysis

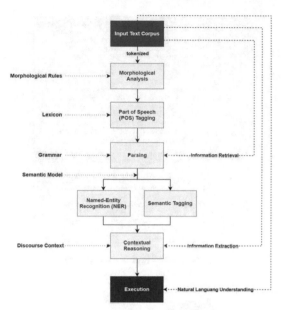

Fig. 4. Process of NLP

provides exposure comparison of selected topics to understand better how one topic is compared with others. Ontology analysis constructs the relationships between various entities by finding similar distinctions that bound them together.

3.4 Use Case

This paper uses Gross Domestic Products (GDP), one of the economic indicators, as a use case to demonstrate how DOSM conducted AI-based media monitoring. The monitoring approach focuses on the dependency between GDP and two other economic indicators, Unemployment and Inflation. The study period is from 1st January to 31st July 2022. Topics and keywords used for each indicator are shown in Fig. 5.

4   Results

Figure 6 shows the media share on the economy GDP topic, with Bernama, the Malaysian national news agency, contributing 22.39% of the total media share. The rest of the contributors are from reliable news agencies, indicating that GDP dissemination is highly centralised among reliable sources, preventing GDP misinterpretation and spreading false news.

As shown in Fig. 7, higher news exposure in February caused the GDP release on the economy bounced back in the final stretch of 2021. Meanwhile, July recorded the highest news exposure. July gives the GDP 329, followed by June with 321 news exposure. More exposure in June and July is because the public is anticipating the economic performance for the second quarter of 2022 during worldwide economic uncertainty.

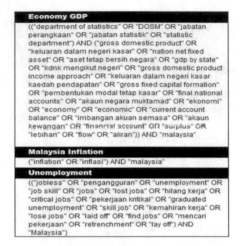

Fig. 5. List of topics & keywords

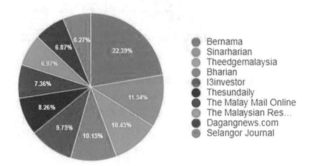

Fig. 6. Percentage of media share for economy GDP

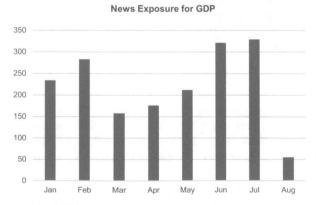

Fig. 7. Numbers of news exposure for economy GDP

As for the social media platform, Twitter users give the highest exposure in June 2022 on GDP and more exposure from April 2022 to July 2022 (Fig. 8). In addition, Fig. 8 shows no significant changes in exposure from Facebook users between the months.

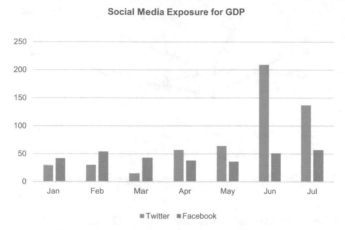

Fig. 8. Active accounts exposing economy GDP

Figure 9 shows that online news and Twitter have more positive sentiments than Facebook. The highest positive sentiments are from Twitter users.

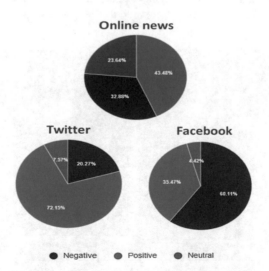

Fig. 9. Sentiment Analysis on Online News, Twitter, and Facebook for economy GDP

Figure 10 shows the sentiment on online news by month series from January to July 2022. The month of February showed the highest positive sentiment from the media, where most media framed articles on the recovery of Malaysia's economy following the release of positive GDP in the fourth quarter of 2021.

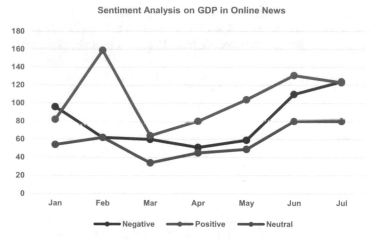

Fig. 10. Trend of sentiment on economy GDP in online news

Based on the ontology analysis results in Fig. 11, the connecting lines between clusters show significant dependency between these three topics. Closer distance and a big group of news framed unemployment and Malaysia inflation in the same articles shows that the degree of centrality between both topics is high. In contrast, the degree of centrality of GDP towards unemployment and inflation is relatively small. The public seems to anticipate more on unemployment and inflation topics because of the higher release frequency, which is monthly compared to the quarterly release on GDP.

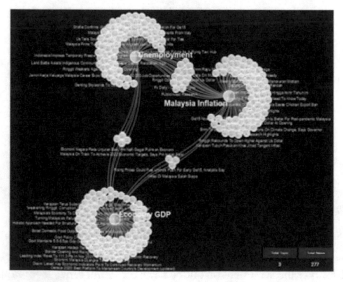

Fig. 11. News ontology analysis

Influencer ontology analysis in Fig. 12 shows that the distance between influencers of respective topics is approximately equally distanced. The influencers across these topics are mostly ministers and government officials.

Ontology analysis results suggest that the centrality at the right source for official statistics is high and centralised across government agencies. The results show no or least occurrence of misuse or misinterpretation of official statistics.

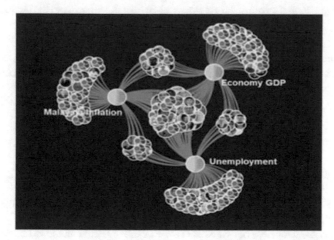

Fig. 12. Influencer ontology analysis

5 Discussion and Conclusion

Media monitoring is not new in the world of marketing. It has been used for decades and continues evolving in parallel with the explosion of online media usage. Media monitoring has also increasingly been utilised across Government institutions for various needs, especially on public opinion as input for policy-making. Not only is DOSM responsible for providing time based official statistics on SDDS, but also to ensure accurate information reaches the public. For that reason, DOSM utilises an Artificial Intelligence-based media monitoring, PMOAS. Through PMOAS, DOSM obtained insights on the official statistics from the media and public. This tool allows DOSM to identify the degree of happiness and centrality in official statistics.

AI is undoubtedly a powerful tool when transforming massive data into insights. As a national statistics provider, the application of AI is vast. In a broader context, it can be extended into the Generic Statistical Business Process Model, which comprises comprehensive business processes needed to produce official statistics. For example, AI-based Optical Character Recognition (OCR) software can convert an image of archived official statistics records or publications into an accessible electronic version with text. Another AI application in improving surveys and census implementation is enhancing security features that protect responses by monitoring and refining security measures to enable all legitimate responses while maintaining security.

The full potential of AI in national statistics offices is yet to be unlocked. To date, the international statistical community is actively engaging in experimental projects on producing official statistics using unconventional data, which will change the landscape of official statistics production worldwide.

In conclusion, official statistics dissemination management using AI-based media monitoring practiced in DOSM demonstrates a preventive measure against misuse and misinterpretation of official statistics among the public. The practice aligns with the international statistical community's effort to utilise AI that contributes better data for better lives.

Acknowledgment. The authors would like to thank the Department of Statistics Malaysia, Mrs. Jamaliah Jaafar, and Dr. Mohamad Shukor Talib for the input and effortless guidance in completing this paper.

References

1. The special data dissemination standard: guide for subscribers and users. International Monetary Fund, Washington, D.C. (2013)
2. Bing, L.: Sentiment Analysis and Opinion Mining (Synthesis Lectures on Human Language Technologies). University of Illinois, Chicago (2012)
3. Baccianella, S., Esuli, A., Se, F.: SENTIWORDNET 3.0: an enhanced lexical resource for sentiment analysis and opinion mining. In: Proceedings of the Seventh International Conference on Language Resources and Evaluation (LREC 2010). European Language Resources Association (ELRA), Valletta (2010)
4. Misleading Statistics Examples – Discover the Potential for Misuse of Statistics & Data in the Digital Age. https://www.datapine.com/blog. Accessed 2 Aug 2022
5. Sentiment Analysis is difficult, but AI may have an answer. Towards Data Science. https://towardsdatascience.com/. Accessed 2 Aug 2022
6. Guidance on Modernizing Statistical Legislation, ECE, U., Geneva (2018)
7. Knowles, E.: Artificial Intelligence. The Oxford Dictionary of Phrase and Fable. Oxford University Press Inc., New York (2006)
8. Artificial intelligence in EU enterprises. https://ec.europa.eu/eurostat/. Accessed 2 Aug 2022
9. Nemeslaki, A., Pocsarovszky, K.: Web crawler research methodology. In: 22nd European Regional Conference of the International Telecommunications Society (ITS2011), Budapest (2011)
10. Bird, S., Klein, E., Loper, E.: Natural Language Processing with Python: Analyzing Text with the Natural Language Toolkit, 1st edn. O'Reilly Media Inc., California (2009)
11. Olive, J., Christianson, C., McCary, J.: Handbook of Natural Language Processing and Machine Translation, 1st edn. Springer, New York (2011)
12. Hamouda, A., Rohaim, M.: Reviews classification using SentiWordNet lexicon. J. Comput. Sci. Inf. Technol. (OJCSIT) (2011)

Computational Vision

A Naive but Effective Post-processing Approach for Dark Channel Prior (DCP)

Danny Ngo Lung Yao[1](✉), Abdullah Bade[2], Iznora Aini Zolkifly[1], and Paridah Daud[1]

[1] Faculty of Business and Technology, UNITAR International University, 47301 Petaling Jaya, Selangor, Malaysia
`danny.ngo@unitar.my`
[2] Faculty of Science and Natural Resources, University Malaysia Sabah, Jalan UMS, 88400 Kota Kinabalu, Sabah, Malaysia

Abstract. Dark Channel Prior (DCP) is originally introduced to remove the haze effects from a digital image. Though the effectiveness of the DCP approach on haze removal, the DCP approach often leads the recovered image to become darker even though the haze effects had been removed. The images with the same or similar levels of color mean are likely without the problem of color shifts. Therefore, we have created a color mean adjustment method to adjust for the color shift by balancing the means of the color channels. CLAHE is employed in this study to boost the image's contrast, while the color mean adjustment method is utilized to smooth its final appearance. Throughout the experiments like visual comparison analysis, Peak Signal-to-Noise Ratio (PSNR), and Universal Quality Index (UQI), our proposed method proved to be a highly effective post-processing approach for the DCP approach as it suppresses more image noises, enhance image quality, and, more importantly, allows the DCP approach to be used on underwater images. Besides, our proposed method also resolves the dark look issue of the DCP approach.

Keywords: Color mean adjustment · CLAHE · DCP

1 Introduction

Dark Channel Prior (DCP) [4] was first suggested for haze reduction, stating that most non-sky areas of outdoor haze images will consist of many dark pixels. The introduction of DCP eliminated haze effects from hazy images in a very simple and efficient fashion. However, the idea of DCP also causes the reconstructed image to have an inherently gloomy appearance. Numerous studies have been undertaken to enhance the DCP approach, particularly the dark look issue after the haze removal procedure. For example, Multi-scale Retinex (MSR) algorithm [5] and Bilateral Filter in Local Contrast Correction (BFLCC) method [9] have been suggested as post-processing stages for the DCP approach depending on the fraction of sky areas presented in the hazy images [8]. Recently, Color Channel Compensation (3C) method [3] was presented for opponent color space as the pre-processing stage for most image improvement techniques, which

© The Author(s), under exclusive license to Springer Nature Singapore Pte Ltd. 2023
Y. B. Wah et al. (Eds.): DaSET 2022, LNDECT 165, pp. 67–76, 2023.
https://doi.org/10.1007/978-981-99-0741-0_5

demonstrates the possibility that the dark appearance problem of the DCP approach may be resolved by color compensation. The energy of light is lost in the process of transmitting where the amount lost varies with the environment and eventually resulted in the loss of color intensity. Recompensing the color loss will help to obtain back the loss amount to the image which eventually helps to obtain a natural image brightness and color balance.

(a)   (b)

(c)   (d)

Fig. 1. IUWB. (a) input (b) DCP approach (c) Underwater image (d) DCP approach.

The notion of color compensation was first introduced in the Underwater White Balance (UWB) method [2] while attempting to prevent the loss of color channels caused by light absorption. The existence of color cast is caused by the color imbalance between color channels which is related to the loss of color intensity. UWB method recompenses the loss of blue and red channels based on the green channel which is said to be well preserved through the observation of underwater images. The color compensation in the UWB method reduced the color cast greatly and shows the potential to recover the color balance. Later, the Improved UWB (IUWB) [7] technique offered a mean adjustment method for the blue and red channels based on the green channel, while simultaneously enhancing the UWB method using a standard deviation ratio. Figure 1 shows the IUWB where the underwater effects are removed from the images, but the color distortion appeared in image (d) by surprise. Even if the performance of the mean adjustment is inconsistent as the color distortion is often occurred unexpectedly, the potential for making the image's color more realistic is promising as the color cast was removed in an effective manner.

To resolve the inconsistent performance of the IUWB technique which often resulted in producing unexpected color distortion, Generalized Color Compensation (GCC) [11] enhanced the mean adjustment method and offered a color mean adjustment method, which adjusts the mean of the RGB channels according to the maximum channel. This is because adjusting the mean of the blue and red channels is underestimated the loss in the green channel which possibly leads to color imbalance. The color mean adjustment method produces more uniform results and can eliminate various types of color casts regardless of whether they are bluish, greenish, or yellowish. In addition, an Underwater Image Enhancement Framework (UIEF) [12] was developed to integrate the GCC with another color balancing approach, the Statistical Gray World Algorithm (SGWA), to enhance the underwater images and remove the color cast. Evaluating the efficiency of the color mean adjustment method, this study proposed the color mean adjustment method as a post-processing approach to address the dark appearance of the image obtained after the DCP approach.

2 Dark Channel Prior

The Dark Channel Prior (DCP) approach [4] eliminates the haze effects based on the previous knowledge that most non-sky regions in outdoor haze images consist of dark pixels. Based on the DCP approach, the dark channel is determined as the minimum intensity among color channels c at the local patch $\Omega(x)$ using Eq. (1). Later, the transmission t is calculated using Eq. (2), and the guided filter is used to improve it. The ambient light A is produced by picking the brightest pixel among the dark channels. Equation (3) yields the recovered image, t_0 is the lowest limit of the transmission t and J is the haze-free image. Figure 2 shows the output of the DCP approach. Though the haze effects are greatly reduced, the image's appearance becomes gloomy.

$$J^{dark}(x) = \min_{y \in \Omega(x)} \left(\min_c \left(J^c(y) \right) \right) \tag{1}$$

$$t(x) = 1 - \min_{y \in \Omega(x)} \left(\min_c \left(\frac{J^c(y)}{A^c} \right) \right) \tag{2}$$

$$J(x) = \frac{I(x) - A}{\max(t(x), t) = t_0} + A \tag{3}$$

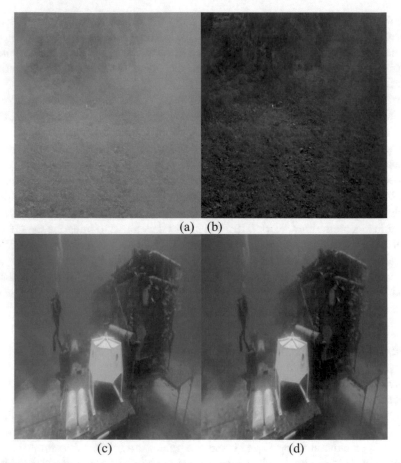

(a)   (b)

(c)   (d)

Fig. 2. DCP approach. (a) Haze images (b) DCP approach (c) Underwater image (d) DCP approach.

3 Proposed Method

The color mean adjustment method can shift the color mean and remove the color cast without producing any further distortion on the image. Therefore, we offered post-processing using the color mean adjustment method to resolve the DCP's dark appearance problem. Figure 3 shows the process of the proposed method where the input image is processed by the DCP approach and then post-processing with color mean adjustment and CLAHE.

Our proposed method will remove the haze effects from the input image by implementing the DCP approach. After that, the color mean adjustment method will adjust the mean of RGB channels as Eqs. (4), (5), and (6) based on the maximum channel I_{max} where I_c is the intensity of the color channel and $\overline{I_c}$ is the mean of the color channel. The mean of color channels with a slight difference will help to shift the color and make the image appeared natural. However, color mean adjustment method will smooth the obtained image where the image contrast will be reduced at a great level. Hence, the image contrast will be enhanced through CLAHE after the color mean adjustment method.

$$I_R^* = I_R - \overline{I_R} + \overline{I_{max}} \tag{4}$$

$$I_G^* = I_G - \overline{I_G} + \overline{I_{max}} \tag{5}$$

$$I_B^* = I_B - \overline{I_B} + \overline{I_{max}} \tag{6}$$

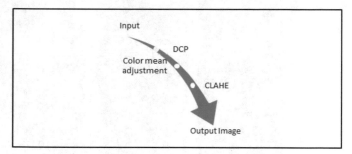

Fig. 3. Proposed method

4 Experiments

Several experiments like visual comparison analysis, Peak Signal-to-Noise (PSNR), and Universal Quality Index (UQI) [10] are used to evaluate the performance of the suggested technique. The example images are from the O-HAZE [1] and UIEBD [6] databases. The visual comparison analysis will compare images based on the presence of a dark appearance.

The result of the visual comparison analysis of the hazy images is shown in Table 1. The DCP approach has created images, like outdoor (image 1) and outdoor (image 2) with a gloomy look. The DCP approach with color mean adjustment method reduces the image's darkness while maintaining its smoothness. Meanwhile, the DCP approach with CLAHE simply lessens the appearance of darkness even though the image contrast had been enhanced. Our suggested solution which post-processed the DCP approach with color mean adjustment and CLAHE had eliminated the darkness and enhances contrast. The findings of visual comparison analysis on underwater images are shown in Table 2. The DCP approach demonstrates the ability to eliminate the haze effects from underwater images, but not the underwater look caused by light absorption. This is due to the DCP approach only take account the scattering effects as the light absorption effects are insignificant in the atmospheric environment. While light absorption effects getting greater due to the properties of the water medium, the DCP approach is unable to remove those effects.

Table 1. Visual comparison analysis on haze images.

Methods	Output		
	Outdoor (Image 1)	*Outdoor (Image 2)*	*Outdoor (Image 3)*
Input			
DCP			
DCP with color mean adjustment method			
DCP with CLAHE			
DCP with color mean adjustment method and CLAHE			

Table 2. Visual comparison analysis on underwater images.

Methods	Output		
	Underwater (Image 1)	*Underwater (Image 2)*	*Underwater (Image 3)*
Input			
DCP			
DCP with color mean adjustment method			
DCP with CLAHE			
DCP with color mean adjustment method and CLAHE			

Concerning Table 2, color mean adjustment method assists the DCP approach in reducing underwater effects. This resolves the limitation of the DCP approaches while dealing with the underwater images, but the images smoothen more after the color mean adjustment. Meanwhile, the image contrast is greatly improved after the DCP approach is post-processed by CLAHE though the underwater effects remain. Our suggested approach which post-processing the DCP approach with the color mean adjustment method and CLAHE demonstrated that the underwater effects were eliminated and the image contrast is enhanced while the haze effects had been removed from the input images.

Table 3. UQI.

| | UQI | | | |
	DCP	DCP with color mean adjustment method	DCP with CLAHE	DCP with color mean adjustment method and CLAHE
Outdoor (Image 1)	0.64	0.89	0.84	0.92
Outdoor (Image 2)	0.78	0.96	0.91	0.92
Outdoor (Image 3)	0.84	0.92	0.89	0.89
Underwater (Image 1)	0.48	0.62	0.71	0.75
Underwater (Image 2)	0.61	0.86	0.77	0.88
Underwater (Image 3)	0.53	0.87	0.74	0.92

The UQI test examines the image quality generated, with the image quality being better if the UQI result is closer to 1. Table 3 displays the UQI test results. According to Table 3, color mean adjustment method and CLAHE alone may assist the DCP in creating higher-quality images. However, our suggested technique, DCP with color mean adjustment method and CLAHE, yielded greater UQI values, particularly for underwater images.

The PSNR test is used to evaluate the image noise contained in the output image where the image noises are fewer if the PSNR value obtained is higher. Table 4 shows the results of the PSNR test. According to Table 4, DCP with color mean adjustment method and DCP with CLAHE produced fewer noises than the DCP. Our proposed method which was the DCP with color mean adjustment method and CLAHE performed better in suppressing the image noises, especially for the underwater images. Figure 4 shows the line chart of PSNR test. The result clearly indicates the proposed method had performed better for underwater images.

Table 4. PSNR

	PSNR			
	DCP	DCP with color mean adjustment method	DCP with CLAHE	DCP with color mean adjustment method and CLAHE
Outdoor (Image 1)	12.58	16.78	15.22	17.40
Outdoor (Image 2)	17.86	24.11	18.64	17.68
Outdoor (Image 3)	16.13	18.31	16.42	16.20
Underwater (Image 1)	10.49	11.92	12.89	13.52
Underwater (Image 2)	11.36	15.52	13.42	15.57
Underwater (Image 3)	14.03	17.97	16.90	19.86

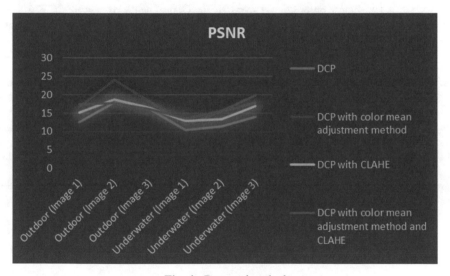

Fig. 4. Proposed method

5   Conclusion

In a nutshell, our suggested solution resolves the DCP's gloomy issue while also migrating the DCP toward underwater image improvement. Based on the PSNR test, the color mean adjustment method and CLAHE post-processing techniques for the DCP scheme demonstrate strength in suppressing image noises, while the UQI test reveals the suggested method acquired improved image quality. The proposed method is suitable for the images that undergo color shifting or having color cast. Though the effectiveness of color mean adjustment method is observed, the increased image smoothness needs to be resolved in the future such as through image sharpening.

References

1. Ancuti, C.O., Ancuti, C., Timofte, R., De Vleeschouwer, C.: O-HAZE: a dehazing benchmark with real hazy and haze-free outdoor images. In: 2018 IEEE/CVF Conference on Computer Vision and Pattern Recognition Workshops (CVPRW), pp. 867–8678 (2018). https://doi.org/10.1109/CVPRW.2018.00119
2. Ancuti, C.O., Ancuti, C., De Vleeschouwer, C., Bekaert, P.: Color balance and fusion for underwater image enhancement. IEEE Trans. Image Process. 27(1), 379–393 (2018)
3. Ancuti, C.O., Ancuti, C., De Vleeschouwer, C., Sbert, M.: Color channel compensation (3C): a fundamental pre-processing step for image enhancement. IEEE Trans. Image Process. 29, 2653–2665 (2020)
4. He, K., Sun, J., Tang, X.: Single image haze removal using dark channel prior. IEEE Trans. Pattern Anal. Mach. Intell. 33(12), 2341–2353 (2011). https://doi.org/10.1109/TPAMI.2010.168
5. Johson, D., Rahman, Z., Woodell, G.: A multiscale retinex for bridging the gap between color images and the human observation of scenes. IEEE Trans. Images Process. 6, 965–976 (1997)
6. Li, C., et al.: An underwater image enhancement benchmark dataset and beyond. IEEE Trans. Image Process. 29, 4376–4389 (2020). https://doi.org/10.1109/TIP.2019.2955241
7. Moon, S.W., Lee, H.S., Eom, I.K.: Improvement of underwater colour correction using standard deviation ratio. Electron. Lett. 56(20), 1051–1054 (2020)
8. Pei, S., Lee, T.: Effective image haze removal using dark channel prior and post-processing. In: 2012 IEEE International Symposium on Circuits and Systems (ISCAS), pp. 2777–2780 (2012). https://doi.org/10.1109/ISCAS.2012.6271886
9. Schettini, R., Gasparini, F., Corchs, S., Marini, F.: Contrast Image Correction Method. J. Electron. Imaging 19(2), 023005 (2010)
10. Wang, Z., Bovik, A.C.: A universal image quality index. IEEE Signal Process. Lett. 9(3), 81–84 (2002)
11. Yao, D.N.L., Bade, A., Waheed, Z.: Recompense the color loss for underwater image using generalized color compensation (GCC) technique. In: 14th Seminar on Science & Technology, pp. 96–99 (2021)
12. Yao, D.N.L., Bade, A., Waheed, Z.: Underwater image enhancement framework using the synthesis of colour compensation and balance methods. Ph.D. thesis, University Malaysia Sabah (2022)

COVID-19 Face Mask Classification Using Deep Learning

Nik Amnah Shahidah Abdul Aziz[1], Muhammad Firdaus Mustapha[2]([✉]),
and Siti Haslini Ab Hamid[3]

[1] Faculty of Computer and Mathematical Sciences, Universiti Teknologi MARA Cawangan
Kelantan Kampus Kota Bharu, Lembah Sireh, 15050 Kota Bharu, Kelantan, Malaysia
[2] Faculty of Computer and Mathematical Sciences, Universiti Teknologi MARA Cawangan
Kelantan, Bukit Ilmu, 18500 Machang, Kelantan, Malaysia
mdfirdaus@uitm.edu.my
[3] Department of Information Technology, FH Training Center, 16800 Pasir Puteh, Kelantan,
Malaysia

Abstract. The COVID-19 pandemic has triggered a global health disaster because
its virus is spread mainly through minute respiratory droplets from coughing,
sneezing, or prolonged close contact between individuals. Consequently, World
Health Organization (WHO) urged wearing face masks in public places such as
schools, train stations, hospitals, etc., as a precaution against COVID-19. However,
it takes work to monitor people in these places manually. Therefore, an automated
facial mask detection system is essential for such enforcement. Nevertheless, face
detection systems confront issues, such as the use of accessories that obscure the
face region, for example, face masks. Even existing detection systems that depend
on facial features struggle to obtain good accuracy. Recent advancements in object
detection, based on deep learning (DL) models, have shown good performance
in identifying objects in images. This work proposed a DL-based approach to
develop a face mask detector model to categorize masked and unmasked faces in
images and real-time streaming video. The model is trained and evaluated on two
different datasets, which are synthetic and real masked face datasets. Experiments
on these two datasets showed that the performance accuracy rate of this model is
99% and 89%, respectively.

Keywords: Face detection · Object detection · COVID-19 · Deep learning

1 Introduction

The ongoing COVID-19 outbreak has necessitated people wearing face masks as a
precaution to safeguard against contracting the virus and reduce viral transmission. The
World Health Organization (WHO) considers face mask-wearing the best way to combat
this pandemic [1–6]. Before the pandemic, only a small percentage of people used face
masks to protect against environmental pollution [5, 7]. Many healthcare workers use
face masks during routine operations. However, keeping track of large groups of people
gets more complicated.

© The Author(s), under exclusive license to Springer Nature Singapore Pte Ltd. 2023
Y. B. Wah et al. (Eds.): DaSET 2022, LNDECT 165, pp. 77–91, 2023.
https://doi.org/10.1007/978-981-99-0741-0_6

Furthermore, face detection has recently become more noticeable, and its uses in the industry have significantly progressed [8]. Nevertheless, there are still challenges with recent advancements in face detection, including variations in illumination, posture, facial expressions, and the existence of occlusions (facial occlusion) [8–14]. As mentioned in previous research, quality issues like occlusions, facial expressions, different angles of faces, and small face sizes can reduce the overall recognition accuracy [15]. Facial occlusion is typical in real-world applications, yet occlusion-related problems have gotten comparatively little attention [9, 10]. Facial occlusion occurs when items like sunglasses, face masks, scarves, and others cover certain parts of the face. Moreover, facial occlusion makes detecting faces more complicated, impacting the detection algorithms' efficiency [8]. Even existing models based on facial features find it difficult to reach good accuracy. Most importantly, the model's performance drops significantly when tested on real-world occluded faces [16].

Also, to overcome occlusion-related problems, deep learning (DL) is a popular method among many researchers. The evolution of DL, mainly convolutional neural networks (CNN), is a powerful classification method used for image recognition, object detection, image classification, and face recognition. Working with CNN obtains better performance detecting individual faces with a face mask in a particular area and robustly performs training [17]. CNN has detected significant characteristics without human supervision, making it the most used [18]. CNN can also train datasets containing large amounts of data or images, saving time and human resources [2]. Additionally, CNN architectures are an essential component of the object detector (e.g., AlexNet, VGGNet, ResNet, etc.). These networks will extract features from input images [19], which helps reduce the amount of redundant data in datasets.

In addition, several methods related to CNN have been used in previous studies to improve facial occlusion problems. Xia et al. [20] used CNN with multi-task learning to improve automated teller machine (ATM) monitoring functionalities. This study obtained 100% accuracy on the AR face database [21] with the most basic facial occlusion and 97.24% accuracy on a newly created Deep-FO dataset with over 50000 color images. According to Annagrebah et al. [22], not all proposed methods to detect and recognize faces are the best ways to solve occlusion problems. For example, the Haar-cascades-based feature performs well in detecting faces and bodies in images, but it is challenging to recognize complicated facial poses. Bhuiyan et al. [23] built a system to classify people wearing face masks using YOLOv3, and their model reached 96% detection accuracy. Though the datasets collected are not that varied, it gives good accuracy in testing when using some real-world data. Gathani and Shah [24] used CNN as part of a learning-based system to classify people, whether they were wearing face masks or not. This model obtained 68.72% precision detection accuracy, and 85.82% mean average precision. Loey et al. [25] proposed pre-trained CNN models (e.g., ResNet50 and YOLOv2) to detect individual faces with medical masks. This study obtained 81% accuracy in medical mask detection. Nonetheless, there are limitations identified in previous studies, as shown in Table 1. Table 2 shows an overview of face mask detection and classification models to identify individual faces with and without face masks.

Moreover, surgical mask datasets on Kaggle and PyImageSearch datasets are easily accessible. Some research, such as [31, 32], used masked faces (MAFA) datasets gathered

Table 1. Limitations identified in previous studies.

References	Limitations
[16, 23]	Trained with limited datasets
[24–28]	Not evaluated in real-world environments
[3, 25, 27, 29]	Trained using synthetic datasets
[4, 26]	Focused on face masks only
[16]	Focused on face masks and sunglasses only
[25]	Video footage does not identify facial masks

Table 2. An overview of face mask detection and classification.

References	Methods	Libraries	Accuracy (%)
[4]	CNN/ MobileNetV2	OpenCV/ TensorFlow/ Keras	–
[23]	CNN/ YOLOv3	LabelIMG	96.00
[24]	CNN	TensorFlow	68.72 (masked) 98.61 (non-masked)
[25]	ResNet50/ YOLOv2	TensorFlow/ MATLAB	81.00
[28]	CNN/ VGG19	OpenCV/ TensorFlow/ Keras	98.70
[30]	Faster-RCNN/ ResNet50	Google Colab	81.00

from the Internet, comprising 30,811 images and 35,806 mask faces with a minimum size of 32×32 pixels. These images come with various occlusion degrees, mask types, and orientations. Ge et al. [31] presented a model and dataset to locate regular and masked faces in natural environments. This study proposed a CNN named LLE-CNNs, which consists of three elements (proposal, embedding, and verification). This study obtained an average precision (AP) of 76.4% in MAFA datasets. Ai et al. [32] proposed a DL-based method to identify face mask-wearing in crowded areas to reduce COVID-19 spread. This study chose 15,000 images from MAFA datasets, including 7500 images with and without masks, and acquired 91.93% detection accuracy. Table 3 shows datasets used in previous studies.

Therefore, this work proposed a DL-based approach to develop a face mask detector model to categorize masked and unmasked faces in images and real-time streaming video. Besides, this work proposed a lightweight DL-based method, MobileNetV2, suitable for mobile devices or other devices with low computational power. Lastly, this work proposed two datasets: synthetic masked face and MAFA.

The remaining work structure is as follows: Sect. 2 presents the materials and methods of the model. Section 3 provides the implementation results and discussion of the model. Lastly, Sect. 4 concludes this work.

Table 3. Summary of existing datasets.

References	Datasets	Total images	Descriptions
[15, 33]	WIDER Face	32203	Contains several face images with a wide range of scale, illumination, posture, facial expression, & occlusion
[31, 32]	Masked Faces (MAFA)	30811	Contains real masked face datasets
	Masked Face-Net	133783	
	Bing	4039	
[20]	Deep FO	50000	Contains images with different head poses, occlusions, & without occlusions
	AR Face	4000	Contains many occluded faces (sunglasses & scarf)
[28]	Face Mask 12K Images	11792	Contains real masked face datasets
[34]	Labeled Faces in the Wild (LFW)	5749	Face images are retrieved from the Internet
	CASIA-WebFace	10000	
[35]	Face Disguise	2000	The datasets were gathered from 8 diverse backgrounds & 10 different occlusions
	Specs of Face (SoF)	42592	Contains images with partial occlusion & illumination conditions
[36]	Synthetic masked faces	1376	PyImageSearch reader Bhandary [37] created the dataset

2   Materials and Methods

Two experiments are conducted to demonstrate the performance of the model in detecting face mask-wearing in images and real-time streaming video. Both experiments used synthetic masked face datasets and MAFA datasets to identify whether or not a person is wearing a face mask.

2.1   Dataset Collection

Experiment 1 used synthetic masked face datasets created by PyImageSearch reader Bhandary [37], and Experiment 2 used MAFA datasets [31] representing real-world masked faces. For MAFA datasets, this work selected 2713 out of 30811 images to reduce computational cost. There are two parts to the dataset: training and testing. For training purposes, 75% of the images of each class were used, and the remaining 25% were used for testing purposes. The datasets are divided into "with mask" and "without mask". Table 4 indicates the datasets used in the experiment. Figure 1 shows sample images of synthetic masked face datasets, and Fig. 2 shows sample images of MAFA datasets.

Table 4. Datasets used in the experiment.

Experiments	References	Datasets	Type of images		
			Masked	Unmasked	Total
Experiment 1	Bhandary [37]	Training data	690	686	1376
		Testing data	320	319	639
Experiment 2	Ge et al. [31]	Training data	1358	1355	2713
		Testing data	679	673	1352

Fig. 1. Sample images of synthetic masked datasets, from Bhandary [37].

Fig. 2. Sample images of MAFA datasets, from Ge et al. [31].

2.2 Hardware and Software Requirements

Experiments are conducted on a personal computer (PC) with the following setup specifications: Intel (R) Core (TM) i5-8300H CPU @ 2.30 GHz, 4 GB RAM, 64-bit operating system, x64-based processor, Windows 10, and NVIDIA GeForce GTX 1050. To build the model, this experiment used OpenCV 4.5.5.64, Keras 2.6.0, and TensorFlow 2.6.2 libraries.

2.3 Proposed System Architecture

Training a face mask detector comprises two stages: training and deployment, as illustrated in Fig. 3. Each stage has sub-steps. The training stage consists of loading the proposed datasets from disk, training the model using Keras and TensorFlow libraries on the proposed datasets to automatically detect whether or not a person is wearing a face mask, and then serializing the model to disk. Once the model is trained, the next stage is loading the model from the disk. Following, detect faces in images and real-time streaming video with OpenCV. Then, extract each face's regions of interest (ROI). Lastly, categorize each face as "with mask" or "without mask".

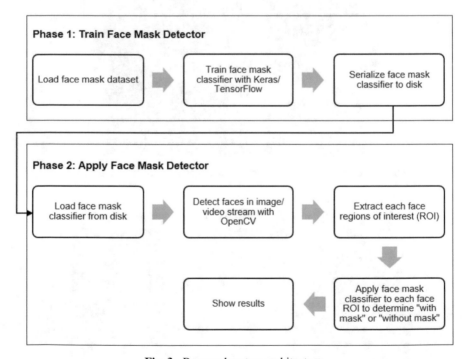

Fig. 3. Proposed system architecture.

Data Pre-processing. Resizing the input image to 224 * 224 pixels, and converting it to a single channel, then the current model's size can be reduced to maintain consistency.

Data Augmentation. Modes provide various transformations, including horizontal and vertical flips, translations, changes in scale, rotations, and shearing. Table 5 shows augmentation parameters used to produce additional images. Figure 4 depicts some of the dataset transformations.

Table 5. Augmentation parameters.

Parameters	rotation_range	zoom_range	shear_range	horizontal_flip
Details	20	0.15	0.15	True

Fig. 4. Data augmentation (Rotated 20°).

CNN Architecture. MobileNetV2 [38] expands on the ideas of MobileNetV1 using an inverted residual structure with a linear bottleneck to reduce computations and improve accuracy. Figure 5 shows the structure of MobileNetV2. Besides, MobileNetV2 is a fundamental feature extractor for object detection tasks, taking images as input and generating feature maps for each input image. Furthermore, MobileNetV2 outperforms MobileNetV1 with comparable model size and computational cost. Moreover, MobileNetV2 architecture has a 7 × 7 max pooling layer, a flattening layer, a hidden layer with a Rectified Linear Unit (ReLU) activation function, a 0.5 dropout value, two neurons in the output layer, and a SoftMax activation function. Therefore, this work used MobileNetV2 architecture, a lightweight classifier that requires less computational power and is suitable for mobile devices or other devices. Table 6 shows the hyperparameters used to train the model.

2.4 Evaluation Metrics

Several evaluation metrics, i.e., accuracy, precision, recall, and f1-score, are used to evaluate the performance of the proposed CNN model. These metrics are derived from four variables: true positive (TP), true negative (TN), false positive (FP), and false negative (FN). To calculate accuracy, precision, recall, and f1-score, use Eqs. 1, 2, 3, and 4.

$$Precision = \frac{TP}{TP + FP} \tag{1}$$

$$Recall = \frac{TP}{TP + FN} \tag{2}$$

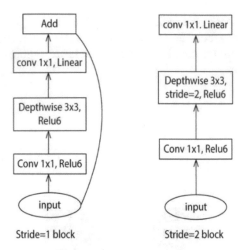

Fig. 5. MobileNetV2 [39].

Table 6. Hyperparameters for training model.

Parameters	Details
Learning rate	0.0004 (1e−4)
Epochs	20
Batch size	32
Optimizer	Adam
Loss function	Binary cross-entropy

$$Accuracy = \frac{TP + TN}{TP + TN + FP + FN} \qquad (3)$$

$$f1 - score = 2 \times \frac{precision \times recall}{precision + recall} \qquad (4)$$

3 Results and Discussion

This section discusses the results of two experiments. Experiment 1 used synthetic masked face datasets, while Experiment 2 used MAFA datasets representing real-world masked faces.

3.1 Experiment 1

Table 7 shows the final results of experiment 1. Figure 6 displays training loss and accuracy.

Table 7. Experiment 1 classification report.

	Precision	Recall	F1-score	Support
With_mask	0.99	0.99	0.99	138
Without_mask	0.99	0.99	0.99	138
Accuracy			**0.99**	276
Macro avg	0.99	0.99	0.99	276
Weighted avg	0.99	0.99	0.99	276

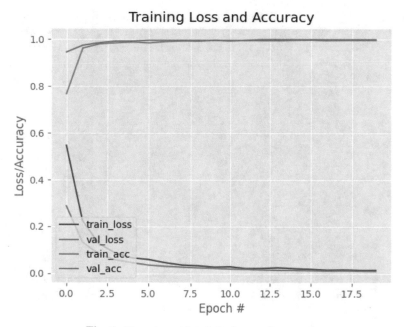

Fig. 6. Experiment 1 training loss and accuracy.

3.2   Experiment 2

Table 8 shows the final results of experiment 2. Figure 7 displays training loss and accuracy. Based on experimental results, the detection performance decreased due to train on real-world occluded images. There are some limitations to MAFA datasets. The image resolutions in MAFA datasets are in various forms. Then, low image quality and lighting conditions also affect detection performance. Lastly, the model could be "confused" with images of people not wearing face masks, such as bandanas covering their mouths and clothes wrapped around their faces. Table 9 shows a summary of two experimental results.

Furthermore, the model can detect whether or not a person is wearing a face mask in images and real-time streaming video. As displayed in Fig. 8, the green box indicates

Table 8. Experiment 2 classification report.

	Precision	Recall	F1-score	Support
With_mask	0.88	0.90	0.89	272
Without_mask	0.90	0.88	0.89	271
Accuracy			**0.89**	543
Macro avg	0.89	0.89	0.89	543
Weighted avg	0.89	0.89	0.89	543

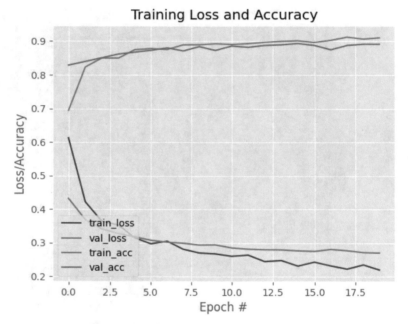

Fig. 7. Experiment 2 training loss and accuracy.

Table 9. Summary of experimental results.

	Experiment 1	Experiment 2
Datasets	Synthetic masked faces	MAFA
Accuracy (%)	99.00	89.00

the presence of the face mask, whereas the red box demonstrates the absence of the face mask. The results are shown in the following screenshots (see Fig. 8):

In this experiment, there were two different lighting conditions. As shown in Fig. 9, the first image (a) had 100% accuracy with normal lighting, but the second image (b)

Fig. 8. Detect faces in images.

had 83.93% accuracy with low-light conditions. Thus, detection performance degraded in the real-time streaming video due to lighting conditions. Below are some screenshots of the results (see Fig. 9):

(a)   (b)

Fig. 9. Detect faces in real-time video streams.

Table 10 shows the performance comparison of proposed datasets with existing models. Based on this table, the proposed model obtained a low-performance accuracy rate compared to Ai et al. [32] in MAFA datasets due to small datasets. The proposed model used 2713 images, while Ai et al. [32] used 15000 images.

Table 10. Comparison of proposed datasets with existing models.

Datasets	References	Methods	Results			
			Precision (%)	Recall (%)	F1-score (%)	Accuracy (%)
Synthetic masked faces	[36]	MobileNetV2	99.00	99.00	99.00	99.00
	Proposed	MobileNetV2	**99.00**	**99.00**	**99.00**	**99.00**
MAFA	[31]	LLE-CNNs	76.40	–	–	–
	[32]	Improved ResNetV2	97.00	97.00	97.00	91.93
	Proposed	MobileNetV2	**89.00**	**89.00**	**89.00**	**89.00**

4 Conclusion

This work has developed a face mask detector model to categorize masked and unmasked faces in images and real-time streaming video. Experimental results on synthetic and real masked face datasets demonstrated that the performance accuracy rate of this model is 99% and 89%, respectively. Nonetheless, current datasets have some drawbacks. There is an unbalanced ratio between synthetic masked face datasets and MAFA datasets. Then, current datasets have a variety of image resolutions. Lastly, poor image quality and lighting conditions reduce detection performance.

In the future, this work can be extended to improve the existing algorithm to achieve better accuracy in detecting faces under partial occlusion. Then, this work plans to retrain the model with more perfect pre-processed data, for instance, histogram equalization for contrast adjustments or conducting some data augmentation. Besides, this work plans to balance the ratios of these two proposed datasets and employ other datasets that contain a wider range of mask types with high image resolutions. Moreover, this work plans to train more training data with high-performance computing hardware. Lastly, this work can be used as a digital detector at hospitals, train stations, schools, and other indoor places.

Acknowledgement. The Fundamental Research Grant Scheme (FRGS) of the Ministry of Education (MOE) supported this work with grant number 600-IRMI/FRGS 5/3 (234/2019).

References

1. Kocacinar, B., Tas, B., Akbulut, F.P., Catal, C., Mishra, D.: A real-time CNN-based lightweight mobile masked face recognition system. IEEE Access **10**, 63496–63507 (2022). https://doi.org/10.1109/access.2022.3182055
2. Tembhare, P.U., Sonekar, N., Rohankar, S., Chandankhede, A., Kothekar, S.: Face mask detection system using deep learning. Int. J. Creat. Res. Thoughts **9**(5), 152–155 (2021)
3. Hussain, S., et al.: IoT and deep learning based approach for rapid screening and face mask detection for infection spread control of COVID-19. Appl. Sci. **11**(3495), 1–27 (2021). https://doi.org/10.3390/app11083495
4. Vinitha, V., Velantina, V.: Covid-19 facemask detection with deep learning and computer vision. Int. Res. J. Eng. Technol. **7**(8), 3127–3132 (2020)
5. Sabir, M.F.S., et al.: An automated real-time face mask detection system using transfer learning with faster-rcnn in the era of the COVID-19 pandemic. Comput. Mater. Contin. **71**(2), 4151–4166 (2022). https://doi.org/10.32604/cmc.2022.017865
6. Jignesh Chowdary, G., Punn, N.S., Sonbhadra, S.K., Agarwal, S.: Face mask detection using transfer learning of inceptionV3. In: Bellatreche, L., Goyal, V., Fujita, H., Mondal, A., Reddy, P.K. (eds.) BDA 2020. LNCS, vol. 12581, pp. 81–90. Springer, Cham (2020). https://doi.org/10.1007/978-3-030-66665-1_6
7. Harriat Christa, G., Jesica, J., Anisha, K., Sagayam, K. M.: CNN-based mask detection system using OpenCV and MobileNetV2. In: 2021 3rd International Conference on Signal Processing and Communication, pp. 115–119 (2021). https://doi.org/10.1109/ICSPC51351.2021.9451688
8. Bade, A., Sivaraja, T.: Enhanced AdaBoost haar cascade classifier model to detect partially occluded faces in digital images. ASM Sci. J. **13** (2020). https://doi.org/10.32802/asmscj.2020.sm26(5.12)
9. Min, R., Hadid, A., Dugelay, J.-L.: Efficient detection of occlusion prior to robust face recognition. Sci. World J. **2014**, 1–10 (2014). https://doi.org/10.1155/2014/519158
10. Zeng, D., Veldhuis, R., Spreeuwers, L.: A survey of face recognition techniques under occlusion. IET Biometrics **10**(6), 581–606 (2021). https://doi.org/10.1049/bme2.12029
11. Hemathilaka, S., Aponso, A.: A comprehensive study on occlusion invariant face recognition under face mask occlusions. Mach. Learn. Appl. An Int. J. **8**(4), 1 (2021). https://doi.org/10.5121/mlaij.2021.8401
12. Min, R., Hadid, A., Dugelay, J.-L.: Improving the recognition of faces occluded by facial accessories. In: 2011 IEEE International Conference on Automation Face Gesture Recognition, pp. 442–447 (2011). https://doi.org/10.1109/FG.2011.5771439
13. Akhtar, Z., Rattani, A.: A face in any form: new challenges and opportunities for face recognition technology. Comput. (Long. Beach. Calif.) **50**(4), 80–90 (2017). https://doi.org/10.1109/MC.2017.119
14. Park, S., Lee, H., Yoo, J., Kim, G., Kim, S.: Partially occluded facial image retrieval based on a similarity measurement. Math. Probl. Eng. **2015**, 1–11 (2015). https://doi.org/10.1155/2015/217568
15. Strueva, A.Y., Ivanova, E.V.: Student attendance control system with face recognition based on neural network. In: 2021 International Russian Automation Conference, pp. 929–933 (2021). https://doi.org/10.1109/RusAutoCon52004.2021.9537386
16. Erakin, M.E., Demir, U., Ekenel, H.K.: On recognizing occluded faces in the wild. In: 2021 International Conference of the Biometrics Special Interest Group, pp. 1–5 (2021). https://doi.org/10.1109/BIOSIG52210.2021.9548293
17. Sarker, I.H.: Deep learning: a comprehensive overview on techniques, taxonomy, applications and research directions. SN Comput. Sci. **2**(6), 1–20 (2021). https://doi.org/10.1007/s42979-021-00815-1

18. Alzubaidi, L., et al.: Review of deep learning: concepts, CNN architectures, challenges, applications, future directions. J. Big Data **8**(1), 1–74 (2021)
19. Zaidi, S.S.A., Ansari, M.S., Aslam, A., Kanwal, N., Asghar, M., Lee, B.: A survey of modern deep learning based object detection models. Digit. Signal Process. **126**, 103514 (2022). https://doi.org/10.1016/j.dsp.2022.103514
20. Xia, Y., Zhang, B., Coenen, F.: Face occlusion detection based on multi-task convolution neural network. In: 2015 12th International Conference on Fuzzy Systems and Knowledge Discovery, pp. 375–379 (2015). https://doi.org/10.1109/FSKD.2015.7381971
21. Martinez, A., Benavente, R.: The AR face database. Computer Vision Central Technical Report (1998)
22. Annagrebah, S., Maizate, P.A., Hassouni, P.L.: Real-time face recognition based on deep neural network methods to solve occlusion problems. In: 2019 Third International Conference on Intelligent Computing in Data Sciences, pp. 1–4 (2019). https://doi.org/10.1109/ICDS47 004.2019.8942385
23. Bhuiyan, M.R., Khushbu, S.A., Islam, M.S.: A deep learning based assistive system to classify COVID-19 face mask for human safety with YOLOv3. In: 2020 11th International Conference on Computing, Communication and Networking Technologies (2020). https://doi.org/ 10.1109/ICCCNT49239.2020.9225384
24. Gathani, J., Shah, K.: Detecting masked faces using region-based convolutional neural network. In: 2020 IEEE 15th International Conference on Industrial and Information Systems, pp. 156–161 (2020). https://doi.org/10.1109/ICIIS51140.2020.9342737
25. Loey, M., Manogaran, G., Taha, M.H.N., Khalifa, N.E.M.: Fighting against COVID-19: a novel deep learning model based on YOLO-v2 with ResNet-50 for medical face mask detection. Sustain. Cities Soc. **65**, 102600 (2021). https://doi.org/10.1016/j.scs.2020.102600
26. Ibitoye, O.: A brief review of convolutional neural network techniques for masked face recognition. In: 2021 IEEE Concurrent Processes Architectures and Embedded Systems Virtual Conference, pp. 1–4 (2021). https://doi.org/10.1109/COPA51043.2021.9541448
27. Mbunge, E., Simelane, S., Fashoto, S.G., Akinnuwesi, B., Metfula, A.S.: Application of deep learning and machine learning models to detect COVID-19 face masks - a review. Sustain. Oper. Comput. **2**, 235–245 (2021). https://doi.org/10.1016/j.susoc.2021.08.001
28. Zhang, E.: A real-time deep transfer learning model for facial mask detection. In: 2021 Integrated Communications Navigation and Surveillance Conference, pp. 1–7 (2021). https:// doi.org/10.1109/ICNS52807.2021.9441582
29. Nithyashree, V., Roopashree, S., Duvvuri, A., Vanishree, L., Madival, D.A., Vidyashree, G.: A solution to COVID-19: detection and recognition of faces with mask. In: 2021 International Conference on Intelligent Technologies, pp. 1–6 (2021). https://doi.org/10.1109/CONIT5 1480.2021.9498426
30. Ejaz, S.M., Islam, R.M.: Masked face recognition using convolutional neural network. In: 2019 International Conference on Sustainable Technology for Industry 4.0, pp. 1–6 (2019). https://doi.org/10.1109/STI47673.2019.9068044
31. Ge, S., Li, J., Ye, Q., Luo, Z.: Detecting masked faces in the wild with LLE-CNNs. In: 2017 IEEE Conference on Computer Vision and Pattern Recognition, pp. 426–434 (2017). https:// doi.org/10.1109/CVPR.2017.53
32. Ai, M.A.S., et al.: Real-time facemask detection for preventing COVID-19 spread using transfer learning based deep neural network. Electronics **11**(14), 2250 (2022). https://doi.org/ 10.3390/electronics11142250
33. Zhu, R., Yin, K., Xiong, H., Tang, H., Yin, G.: Masked face detection algorithm in the dense crowd based on federated learning. Wirel. Commun. Mob. Comput. **2021**, 1–8 (2021). https:// doi.org/10.1155/2021/8586016
34. Ku, H., Dong, W.: Face recognition based on MTCNN and convolutional neural network. Front. Signal Process. **4**(1), 37–42 (2020)

35. Shinwari, A.R., Ayoubi, M.: A comparative study of face recognition algorithms under occlusion. Kardan J. Eng. Technol. **2**(1), 86–96 (2020). https://doi.org/10.31841/KJET.202 1.15
36. Rahmani, M.K.I., Taranum, F., Nikhat, R., Farooqi, M.R., Khan, M.A.: automatic real-time medical mask detection using deep learning to fight COVID-19. Comput. Syst. Sci. Eng. **42**(3), 1181–1198 (2022). https://doi.org/10.32604/csse.2022.022014
37. Bhandary, P.: Mask Classifier (2020). https://github.com/prajnasb/observations.git. Accessed 02 June 2022
38. Sandler, M. Howard, A., Zhu, M., Zhmoginov, A., Liang-Chieh, C.: MobileNetV2: inverted residuals and linear bottlenecks. In: 2018 IEEE/CVF Conference on Computer Vision and Pattern Recognition, pp. 4510–4520 (2018). https://doi.org/10.1109/CVPR.2018.00474
39. Karim Sujon, M.R., Hossain, M.R., Al Amin, M.J., Bepery, C., Rahman, M.M.: Real-time face mask detection for COVID-19 prevention. In: 2022 IEEE 12th Annual Computing and Communication Workshop and Conference, pp. 0341–0346 (2022). https://doi.org/10.1109/ CCWC54503.2022.9720764

Gender Classification Using CNN Transfer Learning and Fine-Tuning

Muhammad Firdaus Mustapha[1], Nur Maisarah Mohamad[1](✉),
and Siti Haslini Ab Hamid[2]

[1] Faculty of Computer and Mathematical Sciences, Universiti Teknologi MARA Cawangan Kelantan, Bukit Ilmu, 18500 Machang, Kelantan, Malaysia
mdfirdaus@uitm.edu.my, 2020273478@student.uitm.edu.my
[2] Department of Information Technology, FH Training Center, 16800 Pasir Puteh, Kelantan, Malaysia
cthaslini@gmail.com

Abstract. The important task of analyzing facial images is performed by a soft biometric application known as gender classification. Convolutional Neural Network (CNN) is currently one of the deep learning techniques used to classify gender to address a variety of issues, including intelligent advertising, tourism, surveillance systems, and other fields. However, CNN requires a lot of power to process information quickly and accurately. Thankfully, transfer learning has been developed to solve this issue. The aim of this study is to evaluate the accuracy of gender classification using transfer learning techniques compared to methods that train data from scratch. The MobileNet and MobileNetv2 models are used in this study because they are among the fastest and the models with the least amount of power consumption. The transfer learning process will then be used to refine these two models by evaluating their performance and processing two different types of FaceARG datasets that have been divided into two different face skin color (Bright and Dark). According to the experimental findings, the accuracy of MobileNetv2 increased to 92% for Bright face skin color datasets and 89% for Dark face skin color datasets. This paper demonstrates that CPU usage is still relevant if the model performs transfer learning before the classification process.

Keywords: Gender classification · Transfer learning · Fine-tuning · Central processing unit · MobileNet · MobileNetV2

1 Introduction

Recent advances in deep learning (DL) have been facilitated by high-performance computing, accessibility, and availability of labeled data [1]. DL outperforms traditional computer vision methods in classification tasks and is one of the most important tools for facial image analysis [2, 3]. Generally, DL will show high accuracy when it has a large number of train data sets. However, with the variety of DL approaches nowadays, it is able to provide high accuracy even when dealing with a small number of datasets [4].

© The Author(s), under exclusive license to Springer Nature Singapore Pte Ltd. 2023
Y. B. Wah et al. (Eds.): DaSET 2022, LNDECT 165, pp. 92–102, 2023.
https://doi.org/10.1007/978-981-99-0741-0_7

Researchers actively pursued several methods to improve the accuracy of face images before the advent of DL, including the use of artificial image features and machine learning. The techniques used to solve face image problems include Long Short Term Memory (LSTM) [5], Recurrent Neural Network (RNN) [6], Generative Adversarial Network (GAN) [7], Auto Encoder [8], Convolutional Neural Network (CNN) [9], and many others. CNN is a popular method among researchers for dealing with soft biometrics issues in facial images.

DL-related soft biometrics research has added a new area of expertise to address issues with the human face. Recent research [10] has focused on the automatic extraction of individual characteristics such as gender [11, 12], age [13, 14], ethnicity [15], etc. Accurately identifying a human face using soft biometric data is difficult. However, it shows that the presence of soft biometrics can improve the overall performance of the system in scenarios where hard biometric features such as iris or face can degrade [16].

The main contribution of this study is to use a fine-tuning methodology to improve the accuracy of pretrained CNN models. The methodology is based on a balanced FaceARG dataset of face images isolated specifically for the task of gender classification in various skin colors (Bright and Dark). In this study, DL is applied using transfer learning paradigm that reuses trained networks and transfers its knowledge to a new classification tasks with partial or complete adaptation to available data. The resulting CNN model significantly outperforms existing baseline methods as tested on the balanced FaceARG dataset.

The following section provides a summary of previous studies. In Sect. 3, the proposed research methodology is covered. Section 4 describes in detail the experiment and its findings. Section 5 wraps up the paper and future research.

2 Related Works

Soft biometrics, particularly gender classification, can be useful in a variety of computer vision applications, but it has received less attention than the more well-known recognition and identification issues.

Furthermore, the study of gender classification requires a face detection process first, hence Prihodava and Jech in [17] propose a Unmanned Aerial Vehicle (UAV) method to classify areas outside the gender face that are difficult to detect. In [18], they used the CNN method by increasing the layers and processed the image using the trained ResNet face detection dlib package that detected 68 face images in frontal images and achieved gender classification accuracy of 81.9% (male) and 80.5% (female) by using the LFW dataset [19]. Palani et al. [20] also used ResNet face detection trained using the dlib package to detect faces in face mask datasets. The authors developed an algorithm using the concept of deep transfer learning and fine-tuning using the MobileNetv2 architecture to classify faces with face masks and without face masks and achieved 98% accuracy on both classifications.

A transfer learning method was used in [21] to classify gender on the VGG16 architecture and achieved 91.44% accuracy on the LFW-Gender dataset. Jiang [22] also used transfer learning using a pretrained architecture and the highest classification accuracy can reach up to 95.10%. A study conducted by Janahiraman and Subramaniam [23] on several pre-trained models and found that MobileNet has the highest misclassification with a classification rate of 49%. Sanjana et al. [24] created a new model of transfer learning with fine-tuning methods. Fine-tuning is achieved by freezing the base layer and inserting two Fully Connected (FC) heads with Rectified Linear Unit (ReLU) and sigmoid activation function. The pretrained MobileNetv2 model produced 85% accuracy in this study. Furthermore, Patel and Chaware used MobileNetv2 in [25]. They achieved 81% accuracy by adapting and adding new layers to the pretrained CNN model.

Based on these findings, the proposed study generates gender classifiers that use transfer learning techniques to improve the accuracy of pretrained CNN models created using datasets with different face skin color (Bright and Dark).

3   Methodology

The pretrained MobileNet and MobileNetv2 are CNN models that use ImageNet as weights. The baseline model is a CNN model that fully utilizes the ImageNet pretraining weights. While transfer learning is applied by refining the pretrained CNN model and adding a new FC head based on the proposed gender classification task. The entire gender classification procedure is divided into two parts (baseline and fine-tuning), each with two phases consisting of a specific process whose results are then passed on to the next step.

Figure 1 (Training Phase) and Fig. 2 (Classification Phase) represents the flow of the entire process as a procedure and displays a flow chart of the different phases and processes in each phase.

3.1   Training Phase

This phase includes two training processes, namely pretrained CNN model training algorithms (MobileNet and MobileNetv2) for the baseline and transfer learning with fine tuning.

A custom data set needs to be loaded to train the model, both for baseline and for fine-tuning. The first step involves using one-hot encoding to label images that belong to binary classes. The model is then trained using these images and their accompanying labels. The validation phase begins when the model starts to learn features and distinguish between male and female images.

The fine-tuning and baseline procedures are distinguished by the way the CNN model is loaded during the training process. Before continuing the process, the fine-tuning method uses ImageNet as its weights. During fine-tuning, different sets of unlabeled images were fed into the model and after prediction, the original and predicted labels were compared. This allows evaluation of model predictions made during the training phase before the classification phase. The baseline process, on the other hand, completely not include any weights and relies solely on data from the existing dataset. After training, the model saves all data into a model file, specifically used to analyze the test dataset for gender classification.

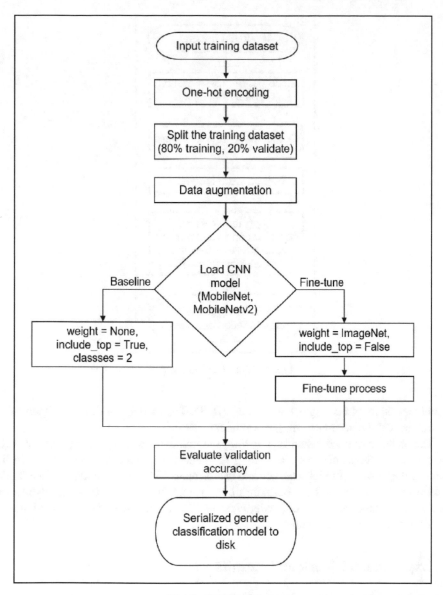

Fig. 1. Training phase.

3.2 Classification Phase

This is the deployment phase, where the trained model is used to examine the input images of the test dataset. Detection can only be made in single face images before pre-processing. Grayscale image in preprocessing stage and downsized (224 × 224 pixels). Next, the resized frame is checked. The feature extraction function then extracts the features needed for analysis, such as eyes, nose, eyebrows and jawline, and draws the regions of interest needed to make predictions. The OpenCV deep learning face detector

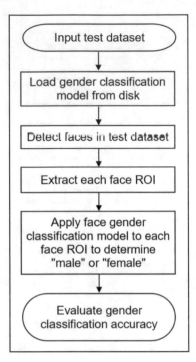

Fig. 2. Classification phase.

model is used to extract regions of interest (ROI). The model will then use OpenCV to extract the 128-D face embedding from the face ROI.

The ROI are marked instantly, allowing the model to make specific predictions for these areas, at the same time eliminating the need to calculate irrelevant areas in the frame and improving model accuracy and performance. The entire analysis of the model is reflected in the output by a bounding box around the detected face that shows the prediction of the input dataset. Gender classification confidence rates are also listed in addition to the list of predictions.

4 Experimental Work and Results

The model for this project has been trained and tested on Windows 10 laptop with 8GB of RAM and Central Processing Unit (CPU) Intel Core i5 chip - 7th Generation. Python 3.9 implement in this project.

4.1 Dataset and Hyperparameter Setting

The dataset used in this experiment is FaceARG dataset [26] which is publicly available online. The dataset collected over 175 000 facial images from the Internet and used four independent human subjects to label the images with racial information. FaceARG is the largest face database available annotated with age, race, accessory and gender information.

In the experiment, gender distribution is considered because the purpose of the research is gender classification. Since FaceARG is an "in-the-wild" face dataset, this dataset is cleaned of any accessories and this research selects a dataset of face images that are only from the front view as the proposed research dataset. As a result, the table below shows a distribution dataset divided into two categories: Bright for Asian-faced people and Dark for Afro-American-faced people. Only 6252 face images were selected from the total FaceARG and divided equally into two categories: Bright (3126) and Dark (3126). Each category is divided into two genders, 1563 (female) and 1562 (male). Table 1 and Table 2 show part of the dataset divided into two racial groups. The hyperparameters used in this project are shown in Table 3.

Table 1. Dataset separation of Asian faces.

Bright (Asian)				
Category	Total images (T)	Training images (0.8 * 0.8 * T)	Validation images (0.2 * 0.8 * T)	Test images (0.2 * T)
Female	1563	1000	250	313
		1250		
Male	1563	1000	250	313
		1250		

Table 2. Dataset separation of Afro-American faces.

Dark (Afro-American)				
Category	Total images (T)	Training images (0.8 * 0.8 * T)	Validation images (0.2 * 0.8 * T)	Test images (0.2 * T)
Female	1563	1000	250	313
		1250		
Male	1563	1000	250	313
		1250		

4.2 Result and Discussion

Table 4 and Table 5 summarizes the results of the gender classification experiment in terms of accuracy and time for the training dataset (2 500 images) and the test dataset

Table 3. Hyperparameters for gender classification.

Hyperparameter	Setting
Input size	224, 224
Initial learning rate	0.0001
Number of epochs	20
Batch size	32
Optimizer	Adam

(626 images) for both skin colors (Bright and Dark). Figure 3 and Fig. 4 shows the graph of the training loss and accuracy of baseline and fine-tune model. Meanwhile, MobileNetv2 face the overfitting problem when dealing with 20 number of epochs. The success of using transfer learning can be seen in the result of fine-tuning test accuracy through MobileNetv2 model which obtain 92% for Bright skin color and 89% for Dark skin color which only 50% on the baseline result. Furthermore, the test accuracy of MobileNet after fine tuning increase from 70% to 87% in Bright skin color dataset and 76% to 92% for Dark skin color dataset.

Table 4. Result of gender classification for bright skin color.

CNN model	Baseline		Fine-tune	
	MobileNet	MobileNetv2	MobileNet	MobileNetv2
Training acc.	85%	50%	94%	98%
Test acc.	**70%**	**50%**	**87%**	**92%**
Training time	23 m 59 s	23 m 21 s	22 m 21 s	22 m 12 s
Test time	1 m 9 s	1 m 18 s	1 m 28 s	1 m 14 s

Table 5. Result of gender classification for dark skin color.

CNN model	Baseline		Fine-tune	
	MobileNet	MobileNetv2	MobileNet	MobileNetv2
Training acc.	78%	50%	98%	96%
Test acc.	**76%**	**50%**	**92%**	**89%**
Training time	15 m 42 s	26 m 16 s	22 m 22 s	28 m 56 s
Test time	1 m 16 s	1 m 48 s	1 m 13 s	1 m 21 s

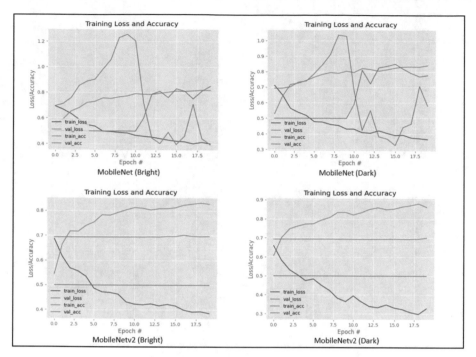

Fig. 3. Training loss and accuracy of baseline model.

The probability of obtaining high accuracy increases with decreasing training loss. It is easy to overfit binary classification. All the predictions of the overfitted model fall into one class. Therefore, only 50% of the predictions are accurate. The other classes are incorrect. As a result, the experimental results show that MobileNetv2 only achieves 50% accuracy for the base model. In other words, if the mobilenetv2 baseline model is trained in many epochs, it can achieve high accuracy without overfitting. Ultimately, compared to the baseline for both types of models, the training accuracy on the fine-tuning model is more consistent and quickly reaches high accuracy. The baseline model takes more time to stabilize than the refined model.

Additionally, the outcomes of this study have been contrasted with those of a prior study [24, 25] that applied a different kind of dataset to the fine-tuning method while using the same pre-training model, MobileNetv2. The MobileNetv2 model is superior to the MobileNet model because it achieves higher accuracy than MobileNet and is an upgraded model of the MobileNet model. Table 6 compares previous and proposed work on the MobileNetv2 model with fine-tuning method. According to the comparison, the classification accuracy using the proposed MobileNetv2 fine-tuning model achieves the highest accuracy for face with bright skin color.

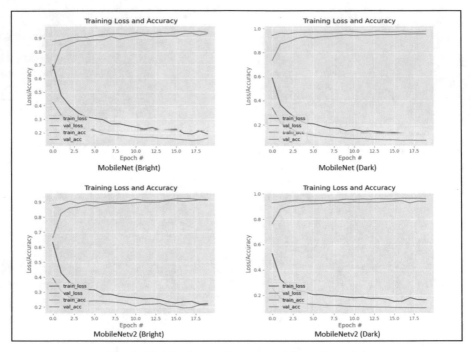

Fig. 4. Training loss and accuracy of fine-tune model.

Table 6. Result comparison with previous and proposed work.

Experiment	Image dataset	Accuracy
Sanjana et al. [24]	Messidor-II and Aptos-2019	85%
Patel and Chaware [25]	Aptos-2019	81%
Proposed Fine-Tuning (Dark)	FaceARG	89%
Proposed Fine-Tuning (Bright)	**FaceARG**	**92%**

5   Conclusion and Future Work

Experimental results show that gender classification using transfer learning can achieve competitive accuracy among CNN models that have been trained using a Central Processing Unit (CPU). Transfer learning can improve the accuracy of gender classification models even with a small number of epochs. A successful gender classification approach will improve the performance of facial recognition applications used in intelligent advertising, tourism, surveillance systems and other fields. Further studies will continue with the same research but compared with various fine-tuning approaches from previous studies using the MobileNetv2 model.

Acknowledgement. The Fundamental Research Grant Scheme (FRGS) of the Ministry of Education (MOE) supported this research with grant number 600-IRMI/FRGS 5/3 (234/2019).

References

1. Xie, Y., Ding, L., Zhou, A., Chen, G.: An optimized face recognition for edge computing, pp. 2019–2022 (2019)
2. Poojary, R., Pai, A.: Comparative study of model optimization techniques in fine-tuned CNN models. In: 2019 International Conference on Electrical and Computing Technologies and Applications, ICECTA 2019, pp. 1–4 (2019). https://doi.org/10.1109/ICECTA48151.2019. 8959681
3. Khan, W., Crockett, K., O'Shea, J., Hussain, A., Khan, B.M.: Deception in the eyes of deceiver: a computer vision and machine learning based automated deception detection. Expert Syst. Appl. **169**(November 2020), 114341 (2021). https://doi.org/10.1016/j.eswa.2020.114341
4. Wu, X., Sahoo, D., Hoi, S.C.H.: Neurocomputing recent advances in deep learning for object detection. Neurocomputing **396**, 39–64 (2020). https://doi.org/10.1016/j.neucom.2020. 01.085
5. Liu, S., Long, Y., Xu, G., Yang, L., Xu, S., Yao, X.: An optimized Capsule-LSTM model for facial expression recognition with video sequences (2021)
6. Ackerson, J.M., Dave, R., Seliya, N.: Applications of recurrent neural network for biometric authentication & anomaly detection (2021)
7. Li, T.W., Lee, G.C.: Performance analysis of fine-tune transferred deep learning. In: Proceedings of the 3rd IEEE Eurasia Conference on IOT, Communication and Engineering 2021, ECICE 2021, pp. 315–319 (2021). https://doi.org/10.1109/ECICE52819.2021.9645649
8. Kantarcı, A.: Thermal to visible face recognition using deep autoencoders (2020)
9. Kumar, R., Ashish, Jadaun, M. S., Sinha, J., Astya, R.: Transfer learning based facial recognition system. In: 2021 3rd International Conference on Advances in Computing, Communication Control and Networking, ICAC3N 2021, pp. 1646–1649 (2021). https://doi.org/10. 1109/ICAC3N53548.2021.9725614
10. Ramos-Muguerza, E., Docio-Fernandez, L., Alba-Castro, J.L.: From hard to soft biometrics through DNN transfer learning. In: 2018 IEEE 9th International Conference on Biometrics Theory, Application System, BTAS 2018 (2018). https://doi.org/10.1109/BTAS.2018. 8698589
11. Lee, B., Gilani, S.Z., Hassan, G.M., Mian, A.: Facial gender classification - analysis using convolutional neural networks. In: 2019 Digital Image Computing: Techniques and Applications, DICTA 2019 (2019). https://doi.org/10.1109/DICTA47822.2019.8946109
12. Mohamad, N.M., Haslini, S., Hamid, A.: Improving gender classification based on skin color using CNN transfer learning, pp. 186–191 (2022)
13. Zaghbani, S., Boujneh, N., Bouhlel, M.S.: Age estimation using deep learning. Comput. Electr. Eng. **68**(October 2017), 337–347 (2018). https://doi.org/10.1016/j.compeleceng.2018.04.012
14. Mustapha, M.F., Mohamad, N.M., Osman, G., Hamid, S.H.A.: Age group classification using Convolutional Neural Network (CNN). J. Phys. Conf. Ser. **2084**(1) (2021). https://doi.org/10. 1088/1742-6596/2084/1/012028
15. Loo, E.K., Lim, T.S., Ong, L.Y., Lim, C.H.: The influence of ethnicity in facial gender estimation. In: Proceedings - 2018 IEEE 14th International Colloquium on Signal Processing & Its Applications, CSPA 2018, no. March, pp. 187–192 (2018). https://doi.org/10.1109/CSPA. 2018.8368710

16. Hassan, B., Izquierdo, E., Piatrik, T.: Soft biometrics: a survey: benchmark analysis, open challenges and recommendations. Multimed. Tools Appl. (2021). https://doi.org/10.1007/s11 042-021-10622-8
17. Prihodova, K., Jech, J.: Gender recognition using thermal images from UAV. In: 2021 International Conference on Information and Digital Technologies, IDT 2021, pp. 83–88 (2021). https://doi.org/10.1109/IDT52577.2021.9497627
18. Haseena, S., Bharathi, S., Padmapriya, I., Lekhaa, R.: Deep learning based approach for gender classification. In: Proceedings of the 2nd International Conference on Electronics, Communication and Aerospace Technology, ICECA 2018, no. Iceca, pp. 1396–1399 (2018). https://doi.org/10.1109/ICECA.2018.8474919
19. Knoche, M., Hormann, S., Rigoll, G.: Cross-Quality LFW: A database for analyzing cross-resolution image face recognition in unconstrained environments. In: Proceedings - 2021 16th IEEE International Conference on Automatic Face and Gesture Recognition, FG 2021 (2021). https://doi.org/10.1109/FG52635.2021.9666960
20. Palani, S.S., Dev, M., Mogili, G., Relan, D., Dey, R.: Face mask detector using deep transfer learning and fine-tuning. In: Proceedings of the 2021 8th International Conference on Computing for Sustainable Global Development, INDIACom 2021, pp. 695–698 (2021). https://doi.org/10.1109/INDIACom51348.2021.00123
21. Mittal, S.: Gender recognition from facial images using convolutional neural network, pp. 347–352 (2019)
22. Jiang, Z.: Face gender classification based on convolutional neural networks. In: Proceedings - 2020 International Conference on Computer Information and Big Data Applications, CIBDA 2020, pp. 120–123 (2020). https://doi.org/10.1109/CIBDA50819.2020.00035
23. Janahiraman, T.V., Subramaniam, P.: Gender classification based on asian faces using deep learning. In: 2019 IEEE 9th International Conference on System Engineering and Technology, no. October, pp. 84–89 (2019). https://doi.org/10.1109/ICSEngT.2019.8906399
24. Sanjana, S., Shadin, N.S., Farzana, M.: Automated diabetic retinopathy detection using transfer learning models. In: 2021 5th International Conference on Electrical Engineering and Information Communication Technology, ICEEICT 2021, pp. 1–6 (2021). https://doi.org/10.1109/ICEEICT53905.2021.9667793
25. Patel, R., Chaware, A.: Transfer learning with fine-tuned MobileNetV2 for diabetic retinopathy. In: 2020 International Conference for Emerging Technology, INCET 2020, pp. 2020–2023 (2020). https://doi.org/10.1109/INCET49848.2020.9154014
26. Darabant, A.S., Borza, D., Danescu, R.: Recognizing human races through machine learning—a multi-network, multi-features study. Mathematics **9**(2), 1–19 (2021). https://doi.org/10.3390/math9020195

Multi Language Recognition Translator App Design Using Optical Character Recognition (OCR) and Convolutional Neural Network (CNN)

Mohamad Khairul Naim Zulkifli, Paridah Daud(✉), and Normaiza Mohamad

UNITAR International University, Petaling Jaya, Selangor, Malaysia
paridah69@unitar.my

Abstract. A mobile translator is a Phone's app that lets user to translate between languages. In this paper, we proposed multilanguage recognition translator (MLRT) mobile app to help the education system, especially for students who are new to Arabic, Malay, and English to learn the languages and translated to other languages. OCR Methodology has been chosen for this project because it is the most appropriate methodology to develop a mobile application. Data acquisition, pre-processing, segmentation, feature extraction, classification, and post-processing are the six phases for OCR methodology. The Convolutional Neural Network (CNN) algorithm is used by deep learning to identify objects in image and Optical Character Recognition (OCR) is used for feature extraction to process Arabic words and translate them into Malay. A system architecture has been created to provide an overview of how the application will run and the functionality and the framework of output to show the application works.

Keywords: Mobile App · CNN algorithm · Optical Character Recognition

1 Introduction

Every country has its own language, which everyone speaks. Arabic is currently the world's sixth most spoken language. Arabic is taught in 1187 schools in Malaysia, and more than 200 million people speak Arabic as their first language [6]. This language has a wide range of words [2]. With this many words, you need to know how to translate each one correctly. But the Arabic language has been forgotten for a long time [4]. There are probably a lot of people who don't know or don't know how to read Arabic, English, or Malay words. Vocabulary is the most important thing in learning a language. The more Arabic, English, and Malay words they know, the more they can speak, write, read, and listen. People who do not know or have difficulty reading Arabic words are unable to trans-late them. Then it will be hard to know what every Arabic, English, and Malay word means. People can't talk to each other in Arabic, English, or Malay if they don't have enough words in their vocabulary. Using old-school methods makes it hard for teachers or tutors to teach Arabic to students who don't know much of the language. Without modern

© The Author(s), under exclusive license to Springer Nature Singapore Pte Ltd. 2023
Y. B. Wah et al. (Eds.): DaSET 2022, LNDECT 165, pp. 103–116, 2023.
https://doi.org/10.1007/978-981-99-0741-0_8

tools, it would take a long time to learn how to write every word in Arabic, English, and Malay. We are lucky to live in a time with so many technologies. A mobile application platform, MLRT can translate writing in Arabic, English, and Malay. Arabic recognition systems can automatically read or search text that is written or typed in Arabic [3]. The recognizers were changed so that they could handle continuous handwriting. This made it possible to pull out words and whole lines from handwritten pages and recognize them [10].

The Arabic alphabet has 29 characters, written and read from right to left [1]. Words are written with horizontal lines. Furthermore, depending on its position within a word, each Arabic character has two to four different forms like initial, isolated, medial, and final. Sixteen Arabic characters have a single, double, triple, or zigzag called Hamza in Arabic. These diacritics are used to differentiate between characters with related main parts. In this context, there are many groups of characters with the same body but differ by the number of dots, the position of the dots, and whether they have a dot or a zigzag. An Arabic paragraphs or word formed by connecting the characters. These six letters (ا, و, ذ, د, ر, ز) cannot be joined to any other letter that comes after one of them. As a result, an Arabic character can be related with either the right or left side, both sides, or neither. For English and Malay word, it contains twenty-six characters. Each letter has an upper- and lower-case form. Below are the current characters that has been using from now on. The exact shape of printed letters depends on the typeface (and font), and the shape of handwritten letters, especially cursive, can be different from the typical printed form. Each character has its own size and font. Font is a specific size, weight, and style of typeface. Each typeface is a type-matched set, and each glyph has a part called a "sort". A typeface is a group of fonts that all have the same design.

Image processing is the process of doing things to an image to make it better or get useful information from it. Some of the most important uses of image processing in computer vision are remote sensing, face detection, fingerprint detection, character recognition, and microscope imaging. A picture is made up of smaller pictures, which are often called regions or areas of interest. Images are often made up of groups of objects, each of which makes up a region. Text recognition is part of image processing. The goal of text recognition is to turn a paper-based text document into a digital format that word processor software can work with. This system will be used to translate Arabic characters and words into Malay. The system must be able to figure out what data or Arabic characters are being entered by converting them to a single character. Text recognition is done in steps called preprocessing, segmentation, feature extraction, classification, and postprocessing. Text recognition can be done in two different ways. The first of these is online and has to do with the recognition-sensitive device. It uses the digitized trace of the pen to figure out what the symbol is [11]. The second method, called "offline", focuses on recognizing handwriting in the form of an image [8]. In this case, only the finished character or word is available.

Google ML Kit, the Google Translation app, and Tesseract OCR will be used in text recognition field. Google ML Kit is a powerful and easy-to-use software kit (SDK) for mobile developers who want to use Google's machine learning and expertise. The Google Translate app has many features and works with more than 100 languages. Tesseract OCR is an open-source text recognition (OCR) engine that can directly or through an API take printed text from photos. It also works with lots of languages. There are two common ways to tell who someone is. For instance, there are both supervised and unsupervised systems. Convolutional Neural Network (CNN), SVM, and KNN are some of the best-known supervised methods, while fuzzy logic is one of the best-known unsupervised methods. There are also 4 types of AOCR, such as Optical Character Recognition (OCR), Optical Word Recognition (OWR), Intelligent Character Recognition (ICR), and Intelligent Word Recognition (IWR). OCR is a way to turn scanned images into text that can be edited [4]. Text preprocessing and segmentation algorithms determine how well OCR can predict results. OWR is the same as OCR in that it only works on one word at a time. ICR is mostly used for handwritten text or text that repeats itself. ICR processes one character at a time using machine learning, while IWR processes one word at a time [16].

CNN are used by deep learning to identify objects in images. The CNN is a feed forward network typically analyses visual representations by analyzing grid data such as topology [17]. CNN is also known as "ConvNet". Each image in CNN is represented in form of pixel value arrays. Operation Convolution forms base of the CNN. The input layer can accept image pixels in the form of arrays as input. Hidden layers perform function extraction by executing calculations and manipulations. Figure 1 shows several hidden layers, including a convolution layer, ReLu layer, pooling layer, and fully connected layer. These hidden layers are performing features extraction from the image [15]. Finally, there is the fully connected layer, which aims to connect every neuron in the previous layer to every neuron in the subsequent layer. CNN was designed to map and process images as input data to output variables.

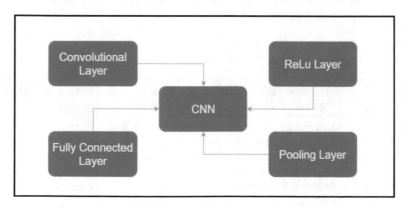

Fig. 1. CNN layer

Because of their high classification rates, Support Vector Machines (SVM) have become one of the most popular classification systems in data mining and pattern recognition applications. SVM has been successfully applied by researchers in a wide range of modern learning applications, including Optical Character Recognition (OCR). Bioinformatics, document analysis, and image classification are all examples of bioinformatics. SVM is commonly used with kernels such as linear, polynomial, RBF, and sigmoid. A multiclass SVM classification has been used in the proposed system [7] with different kernels of linear, polynomial, RBF, and sigmoid has been used. It has an extremely high recognition accuracy. The final step is recognition, which involves the SVM matching the selected class with the ASCII character and figuring the desired word in the Arabic lexicon.

K-Nearest Neighbor (KNN) algorithm is a supervised machine learning algorithm. The KNN-based algorithm can be used to solve both classification and regression problem statements. The symbol 'K' represents the number of nearest neighbours to a new unknown variable that must be classified or predicted. The hyperparameter for KNN is denoted by the symbol 'K'. The value of K must be fine-tuned for proper classification or prediction. The goal of the KNN algorithm is to find all the closest neighbours to a new unknown data point to determine what class it belongs to. It's a method based on distance. KNN calculates the distance between all points near the unknown data and selects the ones with the shortest distances to it. As a result, it's also known as a distance-based algorithm. Previously, KNN could be used to solve both classification and regression predictive problems. However, in the industry, it is more commonly used in classification problems. To evaluate any technique, generally KNN been look at 3 important aspect [13] such as ease to interpret output, calculation time and predictive power.

1.1 Comparison of Techniques

Table 1 shows the comparison between three techniques which are Convolutional Neural Network (CNN), Support Vector Machine (SVM) and K-Nearest Neighbor (KNN). The table shows that CNN is more suitable when working with images especially for recognizing characters [19].

Table 1. Comparison between techniques

Features	Convolutional Neural Network (CNN)	Support Vector Machine (SVM)	K-Nearest Neighbor (KNN)
Methods	Take in an input image, assign importance to various aspects in the image then be able to differentiate one from the others	Call the kernel trick to transform your data and then based on these transformations it finds an optimal boundary between the possible output	In classification and regression, the input consists of the k closest training examples in a data set
Position Accuracy	Higher	Moderate	Lower
Stability	Depend on	Stable	Stable

(*continued*)

Table 1. (*continued*)

Features	Convolutional Neural Network (CNN)	Support Vector Machine (SVM)	K-Nearest Neighbor (KNN)
Hardware Support	Supported on desktop and mobile	Only supported on desktop	Supported on desktop and mobile
Advantages	- Automatic detect the important features without human supervision	- Effective in high dimensional spaces	- Simple to implement - Effective if the training dataset is large
Disadvantages	- Large training dataset needed - Don't encode the position and orientation of object	- Can't perform well with large dataset	- Accuracy depends on the quality of data - Required high memory

1.2   Justification on the Chosen Technique and Features

Abhishek, Yamuna and Anjali [18], have proposed a multilangual system that can adapt the OCR Telugu scripts from the Latin-based OCR systems. For that, the authors used deep learning to help this mapping of different languages especially with the re-strictions of the Teluge script. Moreover, Mushtaq, F., et al. [19] have proposed a dataset for handwritten Urdu characters and a CNN module that can recognise the Urdu character based on what have been learnt from the created dataset. Kataria and Jethva, [20] have addressed the adoption of Latin-based OCR system to Deva-nagari script that is used for Sanskrit language. The authors used a combination of deep learning algorithms to help the recognition of the Devanagari letters from San-skrit documents.

As a result, to develop a mobile application that can translate Arabic characters into Malay, translating an Arabic characters or word, it must first be recognized using Optical Character Recognition (OCR) and Convolutional neural network (CNN). This is due to the highest accuracy that can be obtained when using CNN. The CNN have proven so effective that they are a good method of transition to any prediction problem that includes image as input data [5].

The method used in CNN is to take in an input image, assign importance to various aspects of the image, and then distinguish one from the others. Furthermore, the highest accuracy using CNN will be a great result if this technique is used for this project, but the stability of CNN is dependent on the problem domain itself, where this technique will be applied. If use the CNN technique on desktop and mobile, it will be easier to use at times when do not want to use it on desktop. This is also appropriate for the project because it detects important features automatically without human supervision, so it does not rely on a human. Optical Character Recognition (OCR) is used for feature extraction, which may be applied in the CNN that used in this research. Other features that can be utilized but not in this project include Optical Word Recognition (OWR), Intelligent Character Recognition (ICR), and Intelligent Word Recognition (IWR).

2 Methodology

The methodology used for this project is illustrated in this chapter because methodology is significant since it provides as a guideline for developing the project and solving problems using a specific technique. The OCR approach was chosen for this project, and each aspect of the OCR methodology will be detailed in this chapter and will be implemented using CNN. A system architecture is also created to explain the flow of how the application operates once it has been developed. Finally, the hardware and software requirements are listed to ensure that the application will go successfully.

2.1 Optical Character Recognition Methodology

Since the rise of digital computers in the field of image processing and pattern recognition, machine simulation of human functions has been an extremely challenging research field. The OCR is a technology that involves capturing text character by character, analyzing it, and then converting the character image into character codes. Because it automates the processing of large amounts of paper, transfers data into machines, and provides a web interface to paper documents, the OCR method is the best method for scanning typed pages of text. OCR is classified into two different categories which is offline recognition and online recognition. Offline recognition refers to systems where the input is either an image or a digitized version of the document, whereas online recognition refers to systems in which the successive points are expressed as a function of the time and the order of strokes is also available. One of the most important steps of offline character recognition system is skew detection and correction which has to be used in scanned documents as a pre-processing stage in almost all document analysis and recognition system [14]. This project made use of an offline recognition technique. OCR methodologies are classified based on two main criteria: the data acquisition process, which can be online or offline, and the text type, which can be machine-printed or handwritten. In overall, the OCR problem has six major stages, (see Fig. 2). There are several methods exist for extracting features and training of OCR systems, each with its own set of advantages and disadvantages.

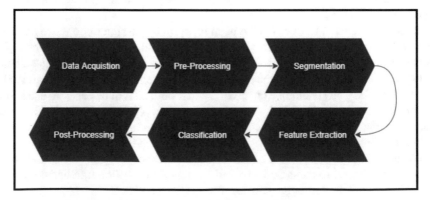

Fig. 2. Stages of OCR methodology

The stage of the OCR methodology will be represented here. The OCR approach varies greatly depending on the nature of the OCR system and the methodology is adopted. Preprocessing is the first stage of OCR, followed by segmentation, feature extraction, classification, and postprocessing. Some stages are combined or omitted in some methods, while others use a feedback mechanism to update the output of each stage.

Data Acquisition
The advancement of automatic character recognition systems has been divided into two categories based on the method of data acquisition, which are online character recognition systems and offline character recognition systems. Character recognition offline captures statistics from documents using optical scanners or cameras, whereas character recognition online uses digitize to directly capture writing based on the order of strokes, speed, pen up and pen down information. In this project, an offline recognition system is presented.

Pre-processing
The Arabic word or writing will be converted into binary images during the pre-processing stage. Binarization is the first step in most image analysis systems, and it refers to the process of converting a grayscale image into a binary image. A stage of image enhancement is required. The scheme used for image binarization, and enhancement is described in and consists of five distinct steps: preprocessing with a lowpass Wiener filter, rough estimation of foreground regions using Niblack's approach, background surface circulation by interpolating neighboring background intensities, thresholding by merging the calculated background surface with the original image, and finally a post-processing step that improves the quality of the image. Figure 3 shows the original image and Fig. 4 shows the binarized image in phase of pre-processing.

Fig. 3. The original image   **Fig. 4.** The binarized image

Segmentation
Going to follow binarization of the Arabic word or writing, a top-down segmentation approach is used. The first lines of the documents are spotted, followed by word extraction and character segmentation. There are three distinct steps in the text line segmentation methodology. The first step is to extract the binary image's connected components and estimate the average character height. A block-based Hough transform is used in the second step to detect potential text lines, and a third step is used to correct any splitting,

recognize text lines that the previous step did not reveal, and finally, separate vertically linked characters and designate them to text lines.

Once the text lines have already been located, segmentation-based projection profiles are used to detect the words. Then, to separate them into letters, use the following method. The general concept is it possible to find segmentation paths that connect the feature points on the word's skeleton and its background. First, the average height of the characters will be determined. The nodes of a word are detected, and the following steps are implemented to all connected components with a height to width ratio less than or equal to 0.5, to separate them into letters.

Step 1: Calculate the skeleton and background of the character component.
Step 2: Divide the skeleton into four sections: the top segment, the bottom segment, the stroke segment, and the hole segment.
Step 3: Discover the skeleton's feature points. The following are the various types of feature points: Fork point: The point on a segment where there are more than two connected branches. End point is a point on a segment with only one neighboring pixel. A corner-point is a point on a segment where the curvature of the segment suddenly and unexpectedly changes.
Step 4: In this step, all potential segmentation paths are built. Each segmentation path should begin with a feature point on the stroke segment as well as conclude with a feature point on the bottom segment. As a result, if there is no single feature point on the top segment, a vertical path is constructed beginning at this key point on the top segment and ending at the bottom segments. In addition, if no feature point on the bottom segment fits a feature point on the stroke segment, a vertical path is assembled beginning with this feature point on the stroke segment and ending with this feature point on the bottom segment. To assemble potential segmentation paths from bottom segment to top segment, a similar process is used in upward search.
Step 5: The best segmentation paths are chosen after locating all possible segmentation paths. To accomplish this, it is necessary to consider only segmentation paths that result in characters with width limits, beginning at the beginning of the component or from last segmentation path that was selected. The best segmentation path is chosen from among these.

This process is repeated till its character component cannot be segmented into the other letters and if there are no other possible segmentation paths. Once the characters within in the word have been located, all character components of words with width less than one will be calculated and merged with the nearest character in terms of integrating pieces of a broken character.

Feature Extraction
One of the most important roles in a recognition system is feature extraction, also known as image representation. The most basic method is to feed grayscale or binary images to a recognizer. All binary character images will be normalized while keeping the original aspect ratio in mind. To extract Arabic characters, two types of features will be used. The first set of characteristics is based on zones. The image is split into horizontal and vertical zones, and the density of each Arabic character pixel is calculated for each zone.

The area formed by the projections of the upper and lower, as well as the left and right character profiles, is calculated in the second type of feature.

First, the character image's center mass is located. Upper and lower profiles are calculated for each image column by calculating the distance between the horizontal line and the closest pixel to the top or bottom boundary of the character image. This results in two zones, upper and lower, which are then separated into vertical blocks. We compute the area of the upper/lower character profiles for each block formed. Likewise, features based on the left and right character profiles will be extracted. In the case of zone-based features, the character image is split into five horizontal and five vertical zones, yielding a total of 25 features.

In the circumstance of features based on character projection profiles, the image is divided into ten vertical zones, resulting in a total of 20 features. Equally, the image is segmented into ten horizontal zones, resulting in a total of 20 features based on the left/right projection profile. The feature extraction model, which employs 65 features, was created by combining features based on zones and features based on character profile projections. Each character is thus represented by a 65-dimensional feature. Figure 5 below shows the feature extraction that showing different part of words.

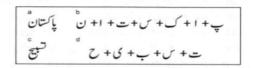

Fig. 5. Showing different parts of words

Classification

The classification stage is the decision-making part of the recognition system [9]. The effectiveness of the features determines the performance of a classifier. At this point, every word and Arabic character is converted into a text file. Characters are extracted, and each one is represented by a feature vector. The characters will then be classified using the newly created database. The neural network (CNN) was used to solve this classification problem. The third objective will be identified at this stage, which is to evaluate the accuracy of translated words in Arabic.

A neural network is a computing architecture that consists of massively parallel interconnections of adaptive 'neural' processors. Because of its parallel nature, it can achieve high accuracy at a faster rate than traditional techniques. Because of its adaptability, it can quickly adapt to changes in the data and learn about the characteristics of the input signals. A neural network is made up of many nodes, each of which feeds information to another node in the network, and the final decision is based on the complex interaction of all nodes.

Post-processing

The final stage of OCR systems is post-processing. At this point, each word or writing image of an Arabic character prints the corresponding recognized characters in structured text form. It also takes care of few things such as knowledge of the grammar for allowing great accuracy [12].

3 System Architecture

System architecture defines the structure of a software system. This is generally a collection of diagrams that represent services, components, layers, and interactions. Camera lens, image processing, mobile application, OCR software, database, user, and UI/UX framework are the major application components in this project. These components must be able to communicate with one another for the application to operate efficiently. The Arabic font, which is less well-known, is without a doubt one of the most difficult areas of pattern recognition rather than English and Malay font. Several pattern recognitions approaches, including neural networks, will be used for offline handwriting recognition. Some reading systems recognize strokes and attempt to recognize Arabic, English and Malay characters, groups of Arabic, English and Malay characters, or entire Arabic, English and Malay words.

Neural networks are made up of simple elements that work in parallel. These elements got their information from biological nervous systems. The network function is largely determined by the connections between elements, as it is in nature. By adjusting the values of connections between elements, this project will train a neural network to perform a specific function. The project made use of neural networks to adjust, or train, so that a specific input results in a specific target output. In such a case, the neural network is adjusted in OCR methodology based on a comparison of the output in the database and the target, till the network output fits the target.

In this research, many such input and target pairs are typically used to train a network. This paper proposed using OCR to process the image and a neural network to train the input and target Arabic, English and Malay writings. Following the capture of the image of handwriting. Figure 6 below shows the system architecture. This application was designed to solve the issue of a user who is unable to understand Arabic, English and Malay letters, or words. The user just needs to take a picture of handwriting Arabic word or letter as an input for this application, as shown in Fig. 6. Then, they need to drag and adjust the text that want to be translated. This function allows the user to get the text that want to be OCR or translated.

The images that have already passed this phase will be identified as final images and processed by OCR. OCR recognizes the text on a photograph and converts it to editable text. The system will use a space separated identifier to separate this material. Then each word will be compared to each word in the database's title field. If there is a match, the system will take the word and translate it. The system will compare every word in Arabic in the database to be translated when translating from Arabic, English and Malay to Arabic, English and Malay. If a match is found, the system will utilize that word as the output for the user. The comparison process was carried out using a system query. After then, the user sees the Arabic, English or Malay word or letter as a result.

The following is an explanation on how to recognize Arabic handwriting. Tesseract library will initially stabilize the final picture orientation in the on PhotoTaken activity. The width and height of the final image are then converted to ARGB 8888 before being processed by the Tesseract function. Tesseract performed the detection by invoking the baseApi.getUTF8Text () method. Tesseract interprets all Arabic words. Tesseract will then replace the detection result in the trained data with an Arabic word. For the database, server database design is required for database installation and updates. This server database was created with MySQL. An admin painstakingly entered every detail Arabic and Malay word into the database. When the application is installed, the system will download all data from the server in JSON format and save it in the database. This operation was completed via a web service that was installed on the server.

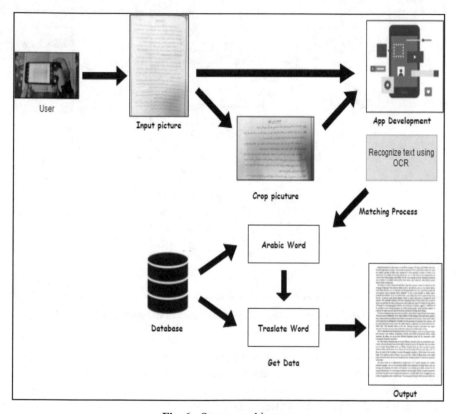

Fig. 6. System architecture

4 The Proposed System Design

A flowchart was used to illustrate each component of the application. Figure 7, shows the functional software flowchart is where all functionality and framework output are designed. The main objective is to have a detailed graph of how the application works.

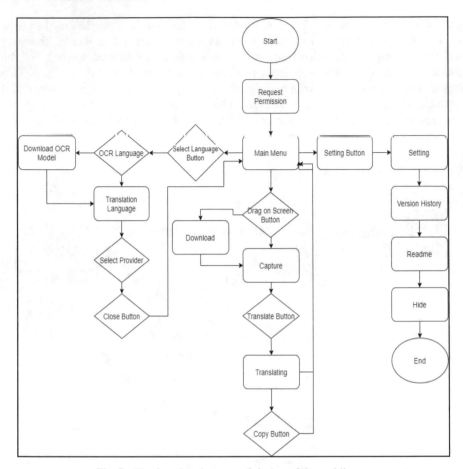

Fig. 7. The functional proposed design of the mobile app

A snapshot sample of the proposed interface for the mobile application is shown in Fig. 8. In this interface, the user can drag on their phone to create a translation area for the interface's drag on screen functionality. The user can edit the translation section to select the word or piece of writing that they want to translate. The user must then enter the translate button to translate. This will proceed to the translate interface. Google handles the translation by default. Additionally, the user has the option of copying the text of the translation or the writing that is being translated.

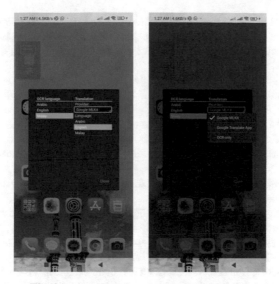

Fig. 8. A sample of source language interface

5 Conclusion

Image processing is the best way to recognize Arabic handwriting because it can help turn every Arabic character into a digital form. This information can then be used to sort data and translate it. Comparing several image processing algorithms, such as convolutional neural networks (CNN), support vector machines (SVM), and the K-Nearest Neighbors (KNN) and found that CNN was the best choice for the purpose of this research. This research aims to propose a suitable design ta make an app that uses the CNN algorithm and optical character recognition to process Arabic words along with Malay and English languages for translation. This app's extraction of Ara-bic text or the recognition of Arabic handwriting from Arabic documents is done using Optical Character Recognition (OCR) system. The proposed Mobile app can help users to translate and learn Arabic words in an efficient way.

References

1. Abubaker, A.A., Lu, J.: The optimum font size and type for students aged 9–12 reading Arabic characters on screen: a case study. J. Phys. Conf. Ser. **364**(1) (2012). https://doi.org/10.1088/1742-6596/364/1/012115
2. Akouaydi, H., Abdelhedi, S., Njah, S., Zaied, M., Alimi, A.M.: Decision trees based on perceptual codes for on-line Arabic character recognition. In: 1st IEEE International Workshop on Arabic Script Analysis and Recognition, ASAR 2017, pp. 153–157 (2017). https://doi.org/10.1109/ASAR.2017.8067778
3. Alsaeedi, A., Al Mutawa, H., Natheer, S., Al Subhi, W., Snoussi, S., Omri, K.: Arabic words recognition using CNN and TNN on a Smartphone. In: 2nd IEEE International Workshop on Arabic and Derived Script Analysis and Recognition, ASAR 2018, 57–61 (2018). https://doi.org/10.1109/ASAR.2018.8480267

4. Althobaiti, H., Lu, C.: A survey on Arabic optical character recognition and an isolated handwritten Arabic character recognition algorithm using encoded freeman chain code. In: 2017 51st Annual Conference on Information Sciences and Systems, CISS 2017, pp. 1–6 (2017). https://doi.org/10.1109/CISS.2017.7926062

5. Alwaqfi, Y.M., Mohamad, M.: A review of Arabic optical character recognition techniques & performance. Int. J. Eng. Trends Technol. 1, 44–51 (2020). https://doi.org/10.14445/223 15381/CATI1P208

6. Azeem, S.A., Ahmed, H.: Recognition of segmented online Arabic handwritten characters of the ADAB database. In: Proceedings - 10th International Conference on Machine Learning and Applications, ICMLA 2011, vol. 1, pp. 204–207 (2011). https://doi.org/10.1109/ICMLA. 2011.120

7. Chang, C.C., Lin, C.J.: LIBSVM: a library for support vector machines. ACM T. Intell. Syst. Technol. 2(3), 1–27 (2011)

8. Guo, G., Li, S., Chan, K.: Character recognition by support vector machines. In: IEEE International Conference an Automatic Face and Gesture Recognition, pp. 196–201 (2000)

9. Goswami, R., Sharma, O.P.: A review on character recognition techniques. Int. J. Comput. Appl. 83(7), 18–23 (2013). https://doi.org/10.5120/14460-2737

10. Hamdani, M., Mousa, A.E.D., Ney, H.: Open vocabulary arabic handwriting recognition using morphological decomposition. In: Proceedings of the International Conference on Document Analysis and Recognition, ICDAR, pp. 280–284 (2013). https://doi.org/10.1109/ICDAR.201 3.63

11. Papageorgiou, C.: A trainable system for object detection. Int. J. Comput. Vision 18, 1 (2000)

12. Pashte, P., Samir Kerawdekar, P.B.: A review on OCR methodology Rajendra Mane College of engineering and technology, Ambav, Ratnagiri, India. 5(02), 1049–1050 (2017)

13. Gadekar, R.R1, Bhosale, R.S2: K-Nearest neighbor classification over encrypted relational data. 2(4), 704–708 (2016). IJARIIE-ISSN(O)-2395-4396

14. Sahu, N., Sonkusare, M.: A study on optical character recognition techniques. Int. J. Comput. Sci. Inf. Technol. Control Eng. 4(1), 01–15 (2017). https://doi.org/10.5121/ijcsitce.2017.4101

15. Savita, A., Amit, C., Anand, N., Saurabh, S., Byungun, Y.: Improved handwritten digit recognition using convolutional neural network (CNN). Sensors (2020). MDPI

16. Tappert, C., Suen, C., Wakahara, T.: The state of the art in online handwriting recognition. IEEE Trans. Pattern Anal. Mach. Intell. 12, 8 (1990)

17. Simard, P.Y., Steinkraus, D., Platt, J.C.: Best practices for convolutional neural network applied to visual document analysis. In: Sevent International Conference on Document Analysis and Recognition. IEEE (2003)

18. Abhishek, B.V.S., Yamuna, K., Anjali, T.: Multilingual translational optical character recognition system for printed Telugu text. In: 2021 12th International Conference on Computing Communication and Networking Technologies (ICCCNT), pp. 1–5. IEEE (2021)

19. Mushtaq, F., Misgar, M.M., Kumar, M., Khurana, S.S.: UrduDeepNet: offline handwritten Urdu character recognition using deep neural network. Neural Comput. Appl. 33(22), 15229–15252 (2021). https://doi.org/10.1007/s00521-021-06144-x

20. Kataria, B., Jethva, D.H.B.: CNN-bidirectional LSTM based optical character recognition of sanskrit manuscripts: a comprehensive systematic literature review. Int. J. Sci. Res. Comput. Sci. Eng. Inf. Technol. (IJSRCSEIT) (2019). ISSN, 2456–3307

Autonomous Driving Through Road Segmentation Based on Computer Vision Techniques

Aditi Jain[1], Matthew Harper[1], Manoj Jayabalan[1], Jamila Mustafina[2], Wasiq Khan[1], Panagiotis Liatsis[3], Noor Lees Ismail[4], and Dhiya Al-Jumeily OBE[1(✉)]

[1] Faculty of Engineering and Technology, Liverpool John Moores University, Liverpool L3 3AF, UK
d.aljumeily@ljmu.ac.uk
[2] Kazan Federal University, Kazan, Russia
[3] Department of Electrical Engineering and Computer Science, Khalifa University, Abu Dhabi, UAE
[4] UNITAR International University, Petaling Jaya, Malaysia

Abstract. Autonomous Driving refers to self-driving vehicles without the need of intervention from the human driver. Safety enhancement, energy optimization, comfort, maintenance and cost are the key benefits of Autonomous Driving. Other benefits include - productivity, reduced congestion/traffic, prevention of car crashes, reducing carbon footprint and ease of parking in congested cities as driverless vehicles could drop passengers off and move on. Autonomous Driving can be achieved via processing visual images/videos at runtime and then converting them to vehicle control signals. This study will help in detecting objects on roads (such as moving vehicles, pedestrians, other static objects on road and road segmentation) using deep convolutional neural network (CNN), which in turn help in aiding driverless future. Semantic segmentation can help recognize objects and their location. Semantic segmentation refers to labelling images with pixel-by-pixel classification that in turn helps to perceive the surrounding environment. Human driver also perceives the driving environment in a similar manner. The purpose of this study is to do an overview and check the feasibility of semantic segmentation using deep learning algorithms in the field of Autonomous Driving specifically in road segmentation task. This study will also be comparative study on CNNs models in terms of accuracy, precision, mean IOU and processing time. The scope of this study will include examination and comparison of two popular algorithms - Fully Convolutional Network (FCNs) and Semantic Segmentation model (SegNet). In detail, this study will mainly focus on road segmentation task in Autonomous Driving with the help of CNN models and will conclude which model is best suited for road segmentation task under different weather conditions, in terms of precision, accuracy, mean IoU and processing time.

Keywords: Suicide prevention · Deep convolutional neural network · Semantic segmentation · Fully convolutional network · Semantic segmentation task · Text classification

© The Author(s), under exclusive license to Springer Nature Singapore Pte Ltd. 2023
Y. B. Wah et al. (Eds.): DaSET 2022, LNDECT 165, pp. 117–128, 2023.
https://doi.org/10.1007/978-981-99-0741-0_9

1 Introduction

Deep learning is rapidly emerging and popular technology in computer science which contributes to various fields such as computer vision [1], speech recognition, natural language processing [2], machine translation, image processing [3], video processing, audio processing [4], and data analytics. Computer Vision helps computer to inculcate a high-level understanding of digital images or videos. It is the process of analyzing, extracting, and comprehending information from a picture or a succession of images. The process of capturing, evaluating, understanding digital images, and extracting high-dimensional features from the actual world to produce symbolic information that may be turned into decisions, are all examples of computer vision tasks. It essentially works like human visual system to understand and automate tasks such as scene reconstruction, semantic segmentation [5], object detection, object recognition, motion estimation, event detection and video tracking etc. Image segmentation which is a sub domain of computer vision techniques attracted numerous works and applications in real world, medical image diagnostics, aerial picture segmentation, and autonomous vehicles are just a few examples.

Vision-based autonomous driving is one such field which is benefitting from the deep learning techniques. Most machine learning, computer vision, robotics, and transportation conferences and journals are seeing an increase in the quantity of scholarly articles on this topic [5]. Several automakers are already delivering vehicles with powerful computer vision technologies for autonomous lane following, assisted parking, and collision detection, among other applications. Constructors are working on and de-redesigning prototypes with levels 4 and 5 autonomy in the meantime. As the vast majority of automotive accidents today are caused by human mistake, 2 the development of autonomous vehicles has the potential to reduce traffic congestion, fuel consumption, and crashes, as well as boost personal mobility and save lives [6]. Estimating the amount of free road surface is critical for fully autonomous driving. Aside from avoiding obstacles, road detection help with path planning and decision-making, particularly in instances where lane markers are not visible or present. The subject of road detection has been studied for many years, and the preceding literature has a wide range of methodologies. Despite producing cutting-edge outcomes, camera-based techniques are heavily influenced by ambient illumination. As a result, their performance is likely to suffer significantly at night or when exposed to light circumstances that differ from those experienced during training [7].

The purpose of this study is to find out best model that can be used in road segmentation task which will in turn help in environment perception and vehicle-to-vehicle communication in autonomous driving. Road segmentation is the process of identifying pixels that belong on the road and detecting multiple objects on the road. The car's direction and control may then be precisely automated, which is very dependent on the extracted road. In this paper, two state-of-the-art algorithms for road segmentation task are compared: the Fully Convolutional Network (FCN) and the Semantic Segmentation Model (SegNet), both of which can yield equivalent results while taking less time to process.

2   Background

Recently, several semantic segmentation solutions based on Deep Convolutional Neural Networks (DCNN) have been proposed [8]. The amazing result gained by AlexNet [9] and GoogleNet [10] for image classification has increased the use of DCNN based computer vision applications, including semantic segmentation. Each convolutional layer in DCNN gathers features from the input images and learns abstract information. The more information a network learns, the more generalised it becomes. DL architectures, such as SegNet [1], GoogleNet [10], and VGGNet [11] deepen the network to achieve improved model accuracy and segmentation performance. As per the study conducted by Junaid, deeper networks need large amount of data to get trained, requiring a lot more data to function well. Making the networks deeper in instances where data is scarce leads to over-fitting, preventing the goal of learning generic features from being met [12]. Various end-to-end multi-task deep learning networks based on deep convolutional neural networks and long short-term memory recurrent neural networks (CNN-LSTM) are designed and compared, with the goal of obtaining not only visual spatial 17 information but also dynamic temporal information in driving scenarios, and improving steering angle and speed predictions. To increase the knowledge of driving scenarios, two supplementary tasks based on semantic segmentation and object detection are presented. Experiments are carried out on the publicly available Udacity dataset as well as a newly gathered Guangzhou Automotive Cooperate dataset. The findings reveal that the suggested network architecture is capable of properly predicting steering angles and vehicle speed. In addition, the impact of multi-auxiliary jobs on network performance is investigated using a visualisation method that displays the network's key map. Finally, the suggested network architecture has been thoroughly tested on the Grand Theft Auto V (GTAV) autonomous driving simulation platform on an experimental road with an average takeover rate of two times per 10 km. [13] As per Zohourian, there are currently two major methods for training CNN-based image processing systems. In terms of the input data model, the two methodologies differ. One method relies on a patch-wise image analysis, which involves the extraction and categorization of rectangular patches of a fixed size for each image. The other is based on full picture resolution, which analyses every pixels in an image at its original size. The network size was increased to achieve recent gains in both CNN-based approaches, Deeper networks, on the other hand, have high processing costs, making them unsuitable for embedded devices in driver assistance systems [14]. Fernandes in his paper suggested that the quality of the perception challenge determines how well a self-driving system performs. Increased availability of 3D scanners such as LIDAR has resulted from advancements in sensor technology, allowing for a more accurate representation of the vehicle's surroundings, resulting in safer systems. Since early 2010, the rapid development of self-driving systems and the resulting rise in research studies has resulted in a massive increase in the quantity and uniqueness of object detection technologies [15]. Two of the major difficulties for autonomous vehicles are obstacle detection and target tracking. Fang and Cai offer a new approach based on computer vision to enable target 18 tracking and real-time obstacle identification. Obstacle detection is handled by the ResNet-18 deep learning neural network, while real-time target tracking is handled by the Yolo-v3 deep learning neural network. The self-driving car manoeuvres to avoid obstacles and to follow camera-tracked objectives. As a result, using the PID algorithm to adjust the steering and movement of the autonomous vehicle during movement will aid the suggested vehicle in achieving stable and precise tracking [16].

3 Methodology

3.1 Dataset

The Cityscapes Dataset [17] is an extensive image dataset based on semantic understanding of urban street scenes. It consists of semantic-based, instance-based, and dense pixel annotations for 30 different types of images, divided into eight 22 categories: flat surfaces, humans, cars, constructions, nature, objects, sky, and void. This data has a variety of images captured in 50 cities of several months (spring, summer, fall), daytimes, and good weather conditions. It contains 5000 images of fine annotations and 20000 images of coarse annotations. Initially the data was recorded in the form of videos and the images were manually picked to have features like varying background, varying scene layout, and large number of dynamic objects. Academic and non-academic institutions can use this Cityscapes Dataset for non-commercial activities such as academic research, teaching, scientific publications, or personal experimentation. To acquire the dataset, cityscapes-dataset website provides option to download the images on the form of zip after login. Also, there is a provision to submit our results and compare with existing benchmarks. Figure 1 shows the steps followed in the methodology.

3.2 Exploratory Data Analysis and Segmentation

Exploratory Data Analysis (EDA), also known as Data Exploration, is a step in the Data Analysis Process that employs a variety of approaches to better comprehend the dataset in question.

For the task of image segmentation, a label (or class) for each input image is needed to identify which pixel belongs to which object. Hence, for each pixel in the image a class is assigned. Segmentation is the term for this process. A segmentation model provides a lot more information about an image.

3.3 Data Preparation

In this study, this data will be used to build and train deep neural networks for the task of semantic understanding of urban scenes at pixel-level, panoptic semantic labelling and instance-level. This dataset consists of 2975 training images, 500 validation images and 1525 testing images. The input image will be downsized, and the number of major classes is reduced from the default to 19 major classes. The two techniques will be trained in 150 epochs, and the models will be compared. Only road labels are used to train the model in this study, which focuses on road segmentation and other labels such as building, vegetation, wall, fence, sidewalk, traffic light, truck, traffic sign, sky, terrain, train, pole, person, car, rider, bus, bicycle, motorcycle, are not considered. Hence, the labels except the road labels will be dropped at this stage.

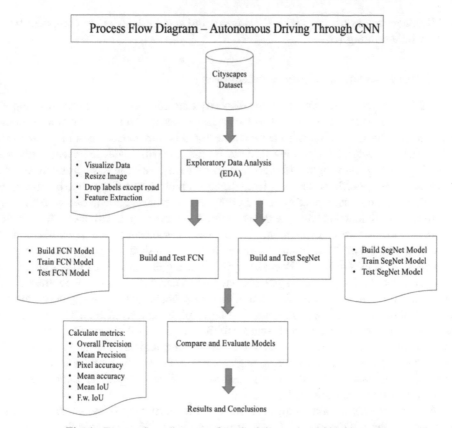

Fig. 1. Process flow diagram of methodology adopted in this study.

3.4 Data Pre-processing

The quality of data has a significant impact on model performance, thus pre-processing is a vital step in ensuring that our model receives the data in the correct format. Many experiments with a variety of pre-processing approaches has been done and found that the following procedures produced the best results. Cityscapes dataset consists of 2975 training images, 500 validation images and 1525 testing images. The images are processed in a series of steps before being utilized in training and evaluating the models.

- Images and segmentation maps are resized to 784 × 368, using bilinear and nearest neighbor interpolation respectively.
- 255, resulting in pixel values in the range [0, 1], divide the values of each image pixel.
- Training images and their associated segmentation maps are flipped horizontally with probability 0.5.
- Training images are randomly brightened or darkened by multiplying them with a random variable of uniform distribution U (0.5, 1.5).
- 50% of all training images are augmented through a Gaussian blur with standard deviation $\sigma \sim$ U (0, 5), to synthetically mimic the motion blur.

- All images are normalized with RGB mean and standard deviation, computed from the whole training set.

3.5 Class Balancing and Data Augmentation

One of the problems with datasets is the disparity between the identified classes: important classes like pedestrians and cyclists are under-represented pixel-for-pixel when compared to the sky or buildings. To compensate for this imbalance, CNN loss functions are frequently weighted according to class frequency, with under-represented classes receiving the highest weights. However, class disparity isn't the only problem. Finely segmenting an object that is far away from the ego vehicle does not appear to be a requirement for an autonomous pilot system. In this case, a crude segmentation or bounding box may be sufficient, but having access to a fine semantic segmentation of nearby items can help with free space estimation. To improve the robustness and generalisation of the end-to-end network for improved prediction, several data balancing and data augmentation strategies are used. Because varied lighting circumstances might affect the performance of an image-based system's end-to-end network, the picture contrast, saturation, and brightness are altered with a particular probability to produce more training samples of diverse lighting conditions. Furthermore, images from the side cameras are used to produce simulated failure instances for training using a data synthesis technique. The typical driving process produces too much steering angle data around zero, according to the distribution of 38 steering angles; hence, the steering angles near zero are randomly eliminated to avoid biassing toward travelling straight. For data balancing, random horizontal flips are used.

A common way of improving the results of image training is by deforming, cropping, or brightening the training inputs in random ways. With Random Scale, training the model with lower resolution images than the original resolution creates a possibility to make a crop and ultimately increase the number of training images. The Cityscapes dataset images have a resolution 2048×1024. The models were trained on 784×368 crops. Color Space Transformations Image data is encoded into three stacked matrices, each of size height $\times$ width. These matrices represent pixel values for an individual RGB color value. The effectiveness of color space transformations is intuitive to conceptualize. The color space transformations done during the training were performed as decreasing/increasing the brightness or the contrast of random crops in the images.

3.6 Feature Extraction

Feature extraction means extracting a vector (or a set of vectors) that describe an image, in order to use it in a Machine Learning system. For example, Convolutional Neural Networks learn to extract a vector summary of an image, which is later used by a classifier to predict the category of the image. To some extent, a segmentation method can be thought as a feature extractor, if you want to use as features the class of each pixel. Feature extraction is an intermediate step in computer vision, which will produce features, and later be applied in the decision making step to accomplish any task, such as segmentation, or object recognition. Segmentation is a process in image processing, which is to divide the image into different parts based on user defined criteria. And the criteria include calculating features and making decision based on the features.

3.7 Machine Learning Models and Evaluation

Two machine learning models will be developed using Python and compared for their ability to distinguish relevant driving environments. These models are a Fully Convolutional Neural Network and a Semantic Segmentation model. The models were evaluated the models in several metrics such as Overall precision, Mean precision, Mean accuracy, Mean Intersection over Union (IoU), Pixel accuracy, Frequency Weighted IoU and Processing time.

4 Results and Discussion

In the section, results of both the models will be discussed and analyzed in terms of qualitative and quantitative results obtained by them. Both the model's output will be compared and contrasted against each other, and conclusions will be derived for this 45 study. Qualitative results will be analyzed visually from the segmentation outputs and Quantitative results will be analyzed and compared numerically based on the various evaluation metrics for the models. Results from both the models will also be concluded and discussed and conclusions to further improve the models based on experiments done as part of this study will also be derived.

4.1 Qualitative Analysis of Results

This section contains example input image and its output from different models for a qualitative comparison. Figure 2 and Fig. 3 contains semantic segmentation results from both models, using the Cityscapes validation dataset.

 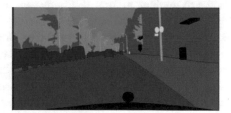

Fig. 2. Input image versus label image.

SegNet produce remarkable segmentation output e.g., detected small objects (person, and sidewalk), and less misclassification. FCN produces the worst segmentation, and it is not suitable for applications that require high performance. As seen in the Fig. 3, the SegNet approach is robust in complicated scenarios and capable of distinguishing objects of various sizes (e.g., different sizes of pedestrians and shadows). The experiment demonstrates that SegNet is a well-balanced technique that produces satisfactory results in terms of segmentation results whereas FCN failed to distinguish smaller objects, as shown in Fig. 2, leading to misclassification of segmentation results.

Fig. 3. FCN output (left) versus SegNet Output.

4.2 Quantitative Analysis of Results

In this section, both models - FCN and SegNet will be compared quantitatively in terms of various evaluation metrics and demonstrate the advantages and disadvantages of using these models in segmentation task in autonomous driving scenarios. The test result on the Cityscapes benchmark shows that our solution achieves feasible results in terms of accuracy and processing time. The image-based road segmentation, on the other hand, is affected by lighting conditions. False positives and negatives can be caused by shadows, blurring, and unclear textures. A fusion of various sensors, including camera, LiDAR, and radar, will be used in future studies to improve road segmentation and object detection. In terms of overall precision, mean precision, mean IoU, and processing time, SegNet outperforms the competition. SegNet is ready to go on a self-driving car and deliver acceptable results. FCN had the worst performance, and its processing time was the same as SegNet's. Because the trade-off between performance and processing time is modest, SegNet is a good model to use for self-driving cars. SegNet produces excellent segmentation results, such as detecting small objects (people and sidewalks) and reducing misclassification. FCN produces the poorest segmentation and is therefore unsuitable for high-performance applications. As the trade-off between performance and processing time is modest and acceptable, it can be concluded that SegNet is a good model to implement for self-driving cars scenarios.

Table 1 shows evaluation metrics results summarized for both the models. SegNet has outperformed most of the metrics as compared to FCN. Processing time for SegNet is also low and its feasible for high performance applications like autonomous driving. Overall Precision and Mean precision shows significant improvement in terms of numbers for SegNet. Mean accuracy for both the models is comparable and similar, pixel accuracy on the other hand shows slight improvement for SegNet. These results also provides evidence that the predictability of the location of particular items in a road scene might aid semantic segmentation, as well as promising results for the use of cartographic information as an image in CNNs. We expect that this series of papers will encourage the public release of real-world datasets with synchronous semantic segmentation labels and exact cartography data. At the centimeter level, the maps would need to be extremely precise. This type of image, known as a High Definition (HD) map, is already popular for autonomous driving. The only information supplied 48 by the map in this chapter is road boundaries and intersections. HD maps have more information, which could help increase segmentation accuracy even further, notably for road lines IoU. It would be possible to rasterize these HD maps in any format to use them as proposed in this study. The second contribution of this paper is presented in this section, which focuses on improving the model performance of real-time networks for semantic segmentation tasks. This strategy significantly enhances the performance of smaller, low-performing network designs while requiring no more space or processing resources. The proposed strategy for improving model performance for moderate to high performing networks on the supplied cityscapes dataset is to use a larger dataset.

Table 1. Evaluation metrics results for the FCN and SegNet

Metric	FCN	SegNet
Overall precision (%)	87.7	91.2
Mean precision (%)	53.2	63.4
Mean accuracy (%)	73.1	75.4
Mean intersection over union (m IoU) (%)	65.8	57.9
Pixel accuracy (%)	90.3	94.7
Frequency weighted IoU (%)	83.2	78.4
Processing time (seconds)	0.059	0.51

4.3 Comparison of Other Models on Cityscapes Dataset

Table 2 shows a comparison of the comprehensive performances of several models based on previous studies conducted on Cityscapes Dataset (Miao and Zhang, 2021). Our technique produces competitive outcomes judged by the mentioned criteria, even though experiments are conducted at varied resolutions. Mean IoU, which is the mean intersection over union (mIoU) is a class-averaged ratio of the intersection of the pixel-wise classification results with the ground truth to their union, is used to quantify the accuracy of the prediction outcomes along with varied image resolutions. Compared with other models, including PSPNet, and DeepLab, our model is considerably small. Compared with lightweight models, including ENet and ESPNet, our model exceed them in mIoU with equivalent speed. Compared with ICNet and Fast-SCNN, which all process 2048 × 1024 inputs as well, Fast-SCNN achieves the same accuracy as ICNet, but is more lightweight. SegNet with 640 × 360 resolution achieves less mIoU than our model with equivalent speed. BiSeNet and DFANet achieves equivalent mIoU, however BiSeNet achieves better frames per second as compared to the later. Only PSPNet gives more mIoU than our models, all the other models underperforms as compared to our models. Also, our models achieve quite good accuracy in less processing time which makes them suitable for self-driving scenarios. As large processing time leads to delay in decisions in autonomous driving, processing time metrics is of high importance along with accuracy of segmentation results.

5 Conclusion

This research examines the state-of-the-art in semantic segmentation to determine which approaches are appropriate for the task of road segmentation in self-driving scenarios.

Table 2. Comparative results of different models on cityscapes dataset

Model	Resolution	mIoU (%)	Speed (FPS)
PSPNet	713 × 713	81.2	0.78
DeepLab	1024 × 512	63.1	0.25
ENet	640 × 360	57	135.4
ICNet	2048 × 1024	69.5	30.3
ESPNet	1024 × 512	60.3	113
ERFNet	1024 × 512	68	41.7
Fast-SCNN	2048 × 1024	68	123.5
DFANet B	1024 × 1024	67.1	120
BiSeMNet	1536 × 768	68.4	105.8
SegNet	640 × 360	57	16.7

The two ways to background concepts and ideas were outlined in this paper. Furthermore, the distinctions between the studied approaches and traditional methods are listed in this work. This research uses the Cityscape dataset to conduct an experiment to determine the optimum performance approach and the trade-off between processing time and performance. The findings of the experiment reveal that SegNet is a well-balanced technique that produces acceptable results while requiring minimal processing time. Using this method, it was shown that SegNet-based semantic segmentation, which is designed to detect general images, can be used to segment road areas effectively. Application of SegNet in autonomous driving demonstrates that it can meet the demands of accuracy, efficiency, and visibility in a variety of situations. Furthermore, the performance of SegNet meets the real-time requirements of autonomous driving systems. According to the findings of this study, a modified system can increase the accuracy of discriminating between road and no-road scenarios. The Cityscapes dataset of 500 images was used to assess precision. As a result, extensive investigation of deep learning methods provides the opportunity for more in-depth understanding and future research to improve current approaches.

The major contribution of this study is SegNet and FCN model implementation that utilises data augmentation and feature representations suitable for semantic segmentation. The network builds upon Cityscapes dataset but differs in the data augmentations techniques. It is also trained with a smaller batch size. The data augmentations used in the developed network include affine transformations and Random Erasing. When developing the network, some general parameter settings for the augmentations as well as an additional erasing strategy were also studied and evaluated.

Due to the amount of work and limited time, it was not possible to realize more experiments. Executing other experiments will be part of a future work with the goal to improve performance of the model on the task of semantic segmentation. This topic is very much promising in automotive industry. The semantic segmentation has proven to help the autonomous vehicles to understand the context of the environment around the vehicle. Even though the number of annotated instances in Cityscapes dataset is significant and sufficient for deep architecture to learn and predict attributes correctly, increasing the number of instances will surely help in reaching better benchmarking values. Moreover, we can focus on more types of objects or more attributes rather than just the roads. The datasets consist of a real-world data that is why the images are noisy which makes the segmentation even more difficult. Having a model that gives a reasonable segmentation result on this real-world image would significantly help to the progress of self-driving vehicles. In addition to this, future work can consider distorting the Cityscapes dataset images to create more training data. Both approaches could possibly significantly improve the segmentation results for autonomous driving scenario. From the automotive point of view, it can result in a good performance and could create a very promising model that might significantly support a series development.

References

1. Badrinarayanan, V., Kendall, A., Cipolla, R.: SegNet: a deep convolutional encoder decoder architecture for image segmentation. IEEE Trans. Pattern Anal. Mach. Intell. **3912**, 2481–2495 (2017). CDC (2022) Facts about suicide, CDC.gov. https://www.cdc.gov/suicide/facts/index.html. Accessed 14 Oct 2022
2. Zhang, Z., Ohya, J.: Movement control with vehicle-to-vehicle communication by using end-to-end deep learning for autonomous driving. In: ICPRAM 2021 - Proceedings of the 10th International Conference on Pattern Recognition Applications and Methods, ICPRAM, pp. 377–385 (2021)
3. Farsiu, S., Robinson, M.D., Elad, M., Milanfar, P.: Fast and robust multiframe super resolution. IEEE Trans. Image Process. **1310**, 1327–1344 (2004). http://ieeexplore.ieee.org/document/1331445/
4. Pham, T., Lee, Y.-S., Mathulaprangsan, S., Wang, J.-C.: Source separation using dictionary learning and deep recurrent neural network with locality preserving constraint. In: 2017 IEEE International Conference on Multimedia and Expo (ICME), pp. 151–156. IEEE (2017). http://ieeexplore.ieee.org/document/8019516/
5. Zhao, H., Shi, J., Qi, X., Wang, X., Jia, J.: Pyramid scene parsing network. In: Proceedings - 30th IEEE Conference on Computer Vision and Pattern Recognition, CVPR 2017 (2017). https://github.com/hszhao/PSPNet
6. Zablocki, É., Ben-Younes, H., Pérez, P., Cord, M.: Explainability of vision-based autonomous driving systems: review and challenges (2021). http://arxiv.org/abs/2101.05307
7. Caltagirone, L., Scheidegger, S., Svensson, L., Wahde, M.: Fast LIDAR-based road detection using fully convolutional neural networks. In: IEEE Intelligent Vehicles Symposium, Proceedings (2017)
8. Chen, L.C., Papandreou, G., Kokkinos, I., Murphy, K., Yuille, A.L.: DeepLab: semantic image segmentation with deep convolutional nets, atrous convolution, and fully connected CRFs. IEEE Trans. Pattern Anal. Mach. Intell. **404**, 834–848 (2018)
9. Zhang, Y., Gao, J., Zhou, H.: Breeds classification with deep convolutional neural network. PervasiveHealth: Pervasive Comput. Technol. Healthc. 145–151 (2020)
10. Szegedy, C., et al.: Going deeper with convolutions. In: Proceedings of the IEEE Computer Society Conference on Computer Vision and Pattern Recognition, 07–12 June 2015, pp. 1–9 (2015)
11. Simonyan, K., Zisserman, A.: Very deep convolutional networks for large-scale image recognition. In: 3rd International Conference on Learning Representations, ICLR 2015 - Conference Track Proceedings, pp. 1–14 (2015)

12. Junaid, M., et al.: Multi-feature view-based shallow convolutional neural network for road segmentation. IEEE Access **8**, 36612–36623 (2020). https://ieeexplore.ieee.org/document/8988193/

13. Wang, D., Wen, J., Wang, Y., Huang, X., Pei, F.: End-to-end self-driving using deep neural networks with multi-auxiliary tasks. Autom. Innov. **2**(2), 127–136 (2019). https://doi.org/10.1007/s42154-019-00057-1

14. Zohourian, F., Antic, B., Siegemund, J., Meuter, M., Pauli, J.: Superpixel-based road segmentation for real-time systems using CNN. In: VISIGRAPP 2018 - Proceedings of the 13th International Joint Conference on Computer Vision, Imaging and Computer Graphics Theory and Applications, 5 September 2018, pp. 257–265 (2018)

15. Fernandes, D., et al.: Point-cloud based 3D object detection and classification methods for self-driving applications: a survey and taxonomy. Inf. Fusion **68**, 161–191 (2021). https://linkinghub.elsevier.com/retrieve/pii/S1566253520304097. Accessed 23 Oct 2021

16. Fang, R., Cai, C.: Computer vision based obstacle detection and target tracking for autonomous vehicles. In: MATEC Web of Conferences, vol. 336, p. 07004 (2021). https://www.matec-conferences.org/10.1051/matecconf/202133607004

17. Cordts, M., et al.: The cityscapes dataset for semantic urban scene understanding. In: Proceedings of the IEEE Computer Society Conference on Computer Vision and Pattern Recognition, 2016-Decem, pp. 3213–3223 (2016)

Cybersecurity

Phishing Attack Types and Mitigation: A Survey

Mohammed Fahad Alghenaim[1]([⊠]), Nur Azaliah Abu Bakar[1,2,3],
Fiza Abdul Rahim[1,2,3], Vanye Zira Vanduhe[2], and Gamal Alkawsi[3]

[1] Advanced Informatics Department Razak Faculty of Technology and Informatics, Universiti Teknologi Malaysia, 54100 Kuala Lumpur, Malaysia
aalghenaim@graduate.utm.my
[2] Istanbul Ticaret University and Istanbul Aydin University Istanbul, Istanbul, Turkey
[3] The Energy University (UNITEN), Kajang, Selangor, Malaysia

Abstract. The proliferation of the internet and computing devices has drawn much attention during the Covid-19 pandemic stay home and work, and this has led the organization to adapt to staying home. Also, to let the organization work due to the infrastructure for working on proxy during the pandemic. The alarming rate of cyber-attacks, which through this study infer that phishing is one of the most effective and efficient ways for cyber-attack success. In this light, this study aims to study phishing attacks and mitigation methods in play, notwithstanding analysing performance metrics of the current mitigation performance metrics. Results indicate that business enterprises and educational institutions are the most hit using email (social engineering) and web app phishing attacks. The most effective mitigation methods are training/awareness campaigns on social engineering and using artificial intelligence/machine learning (AI/ML). To gain zero or 100% phishing mitigation, AI/ML need to be applied in large scale to measure its efficiency in phishing mitigation.

Keywords: Social engineering · Phishing · Mitigation · Artificial intelligence · Machine learning

1 Introduction

Phishing is considered a contemporary issue in the ICT; phishing is one of the deadliest cyber security threats to internet users [58]. These threats are still growing at an exponential rate. As internet users are getting aware of the new phishing methods, at the same time, attackers are also creating new effective methods to hack on a large scale, making the internet promiscuous. Numerous working phishing concepts are in place today; new innovative methods are also in play. The most common is where a website replica is made with exact features seemingly impossible to notice deference between the original website and the clone or, in other words, replica or cloaking of sites/web applications [20, 43]. Due to this, internet users are generally fooled to think of fake or cloned websites, so they tend to input their personal information, which could be a huge threat [12, 13]. It is paramount to note the intensity of phishing; [14] and [54], stipulates

© The Author(s), under exclusive license to Springer Nature Singapore Pte Ltd. 2023
Y. B. Wah et al. (Eds.): DaSET 2022, LNDECT 165, pp. 131–153, 2023.
https://doi.org/10.1007/978-981-99-0741-0_10

that malicious software is installed through phishing to mine cryptocurrency without the computer owner's noticing.

Phishing attacks are based on obtaining desired information from users quickly and easily with the help of misdirection, panic, curiosity, or excitement [4]. Phishing generally uses social engineering fundamentals to lure internet users of their personal information, which is used to hack websites that could harm individuals or corporations [2, 6, 16, 20, 21, 52]. Attackers use a couple or numerous social engineering methods to get to internet users. One of these methods is to log in details to bank accounts or other sites; the information could, of course, damage the peace, finances, health, security, and other factors that could harm human life in general [19, 36, 38, 58, 64]. There are kinds of literature that have put together several phishing methods and systems to end or mitigate the protracted phishing on today's internet. These kinds of literature provide us with the current phishing methods and treats, current methods used for phishing mitigation, phishing penetration methods, phishing performance metrics (datasets and accuracy), and phishing prediction [37, 43, 57, 59, 64].

This study is aimed at putting together studies on the phishing mentioned above areas, as mentioned aiming to identify efficient and effective mitigation methods by answer compounding questions based on; (1) the internet application that is phished the most, also considering factors such as; (2) phishing targeted area; (3) performance metrics; (4) methods of mitigation; (5) penetration methods; (6) analysis on current phishing prediction mechanism, for five years (between 2017 to 2022).

This paper is presented in two sections: the first present a literature review discussing phishing and types of phishing techniques. Secondly, we present a systematic review of research journals and conference papers proceedings for the last five years ranging from 2018 to 2022. These journals were extracted from six primary databases: WoS, Scopus, ERIC, PubMed, IEEExplore, and JSTOR.

2   Related Studies

There are numerous types of phishing in play today; phishing entails the phisher tricking the victim by adopting technical and social approaches; hence, getting acquainted with these threats and their approaches is mandatory [58]. A phisher intends to obtain information that will aid the hacker by appearing to the unsuspecting owner as a trustworthy entity [29] juxtapose that the goal of phishing is to fool a victim by giving the attacker sensitive information, such as usernames, passwords, and credit and debit card information. There are many technical approaches to phishing; however, the most adopted technical approaches are listed and described in the following subsections [31, 54, 59].

3   Types of Phishing Email Attacks

3.1   Social Engineering

Social engineering is one of the most devastating phishing attack methods. It involves phishers playing into the victim's confidence, feelings such as compassion or anxiety, the desire to assist, and ignorance of how to accomplish their goal [16, 42, 52]. Social engineering is focused on diverting the victim from making reasonable decisions, causing the victim to make irrational decisions [35]. Fear, envy, intrigue, rage, friendship, loyalty, ambition, selflessness, a sense of obligation, and supremacy are such emotions [2, 15, 21]. By taking advantage of these emotions, the attacker causes the victim to reveal his/her personal information and assets. Such actions derive from the defensive predisposition leading one to take precautionary, urgent action.

The Method of convincing anyone to share their knowledge is known as social engineering. Social engineers rely on people's ignorance of the consequence of sharing their valuable information and their lack of knowledge for securing their systems and IT infrastructure from security attacks [28, 35, 56]. The employees of an organization may carry out these attacks through a third-party agency, violating the organization's rules for financial gain or revenge [18]. The attacker uses different tactics to gather sensitive information about the victims; this is a social engineering attack method [48], which is classified as [2, 4, 9, 16, 24, 29]:

- Human-Based Social Engineering–In which an illegitimate person can access an organization's IT infrastructure and share the information.
- Computer-Based Social Engineering–It is carried out using spam mail, attracting the user with an offer as a gift or as "you won the lottery."
- Mobile-Based Social Engineering–A service provider could acquire access to a victim's phone using a rogue program.

This phishing attack is performed using a website, email, voice or text messages, social media, etc. People usually believe that social engineering is easy to detect, but researchers have stated that they respond poorly when detecting deception and lies [6, 31, 47, 77]. Kevin Mitnick's notorious attacks demonstrated how damaging advanced social engineering attacks could be for information protection efforts in corporations and public agencies [16, 36, 42, 77, 78]. This implies that the social engineering technique can effectively persuade victims to perform notorious activities [56].

Deceptive phishing: This technique involves supplying clients with malicious links via emails and redirecting them to malicious websites where they are likely to enter sensitive information [2, 11, 33]. [9] thoroughly overviews a deceptive phishing attack and different anti-phishing techniques. They present the different methods used by phishers and the advantages and disadvantages of the different countermeasures [9].

a) Disguised Email [3, 4, 21, 36, 39, 43, 69]:
- Portal site information mail phishing attack: In recent times, personal information is being stolen continuously, and account theft and sales are occurring as a result.
- Secure email phishing attack is widely used in government and financial institutions. An attacker sends a disguised secure email. The body of the secure email can only be read after entering a specific password. However, phishing emails automatically connect to phishing sites without entering passwords when the user presses the Mail View button, induces the user to enter their account information, and provides attachments when acquiring account information.
b) E-mail body vulnerability: An attacker could insert a vulnerability into the body of the email, causing the user to connect to a phishing site simply by reading the email.

Traditional methods require mail users to click on links, read attached files, and so on. When using the email body vulnerability, a hidden malicious script is executed immediately when the email is read and connects to a phishing site. When the mail is read, it is automatically connected to the phishing site, and the attacker induces the user to enter their account information. In the case of an attack using mail body Web vulnerabilities, webmail or a mail client program supporting HTML is used.

c) Attached file camouflage: An attacker may send a phishing email disguised as an attached file; when a user clicks an attachment link, they are connected to a phishing site created by the attacker and presented with a screen prompting them to enter their account information. In general, it can be assumed that attached files are downloaded directly from the email body. However, in the case of large, attached files, they can be downloaded from a separate site, reducing the user's suspicion.

d) Malicious code hidden email: The attacker sends a compressed file, Hangul Word Processing (HWP) document file, MS Word file, etc. that hides malicious code disguised as the content of interest to the recipient or catches the recipient's attention with the subject, mail body, and attached file name. For example, they could write, please check the attached file in the body of the mail to induce the recipient to read the attached file.

e) Squatting attack: A Squatting attack is a denial-of-service (DoS) attack where a program interferes with another program through shared synchronization objects. Several attack derivatives exist for different scenarios, such as a typo-squatting attack, skill-squatting attack, and voice-squatting attack.

f) Clone phishing: Duplicating already sent emails and attaching a malicious link to them can allow for a successful attack on an unsuspecting user. Ahmad Alamgir [46] proposed a new method where websites use One Time Password and User-machine Identification system to combat phishing attacks. Webservers will send a one-time password to a user by SMS or email and create an encrypted token for the device after the user inputs the password [11, 13, 20, 41, 44, 46].

3.2   Mobile Phishing Attack

Graphical user interface (GUI) [7–10, 24]: The Android GUI framework is famous for multi-interactive activities. The GUI is what the user can see and interact with. The Android GUI provides a variety of pre-built components, such as structured layout objects (e.g., Linear Layout) and components (e.g., Button and Edit Text). These elements allow developers to build the graphical user interfaces for the app. The layout structure uses a GUI hierarchy to follow UI design principles. The Android GUI framework is a reusable and extensible set of components with well-defined interfaces that can be specialized. However, the security of the Android GUI framework remains an important yet under-scrutinized topic [28, 66]. The Android GUI framework is a reusable and extensible set of components with well-defined interfaces that can be specialized [7]. However, the security of the Android GUI framework remains an important yet under-scrutinized topic.

Similarity attacks (spoofing attacks) analyze the GUI code of the original app and partially modify the GUI code. Attackers then add logic code to manipulate the original app logic [8]. For example, attackers can crack payment apps to bypass the payment functionality.

Window overlay attacks render a window on top of the mobile screen, either partially (e.g., Toast and Dialog) or completely (e.g., similar UI pages), overlapping the original app window [29]. For example, attackers choose a particular time to render the phishing UI pages by monitoring the occurrence of the original app's login activity. This attack usually leverages the flaws of the design mechanism in mobile OS e.g., using as the code below, to get "top activity" before Android 5.1.

ActivityManager#getRun-ningTasks()

Task hijacking attacks trick the system into modifying the app navigation behaviors or the tasks (back stacks) in the system [35, 38, 60]. For example, the back button is popular with users because it allows users to navigate back through the history of activities. However, attackers may abuse the back button to mislead the user into a phishing activity (e.g., misusing "task affinity"). In short, attackers try to modify the tasks and back stack to execute phishing attacks.

3.3 Smishing

Smishing attacks use short message services or SMS, commonly known as text messages. This attack has become increasingly popular since people are more likely to trust a message that comes in through a messaging app on their phone than a message delivered via email [41, 70, 94]. Although many victims don't equate phishing scams with personal text messages, the truth is that it is easier for threat actors to find your phone number than your email [48, 59]. There is a finite number of options with phone numbers – in the U.S, a phone number is ten digits [41, 70, 93, 94]. Compare this to an email address, which is not limited by size, although there is a reasonable number of expected characters. Emails can include numbers, letters, and symbols –!, #, and %, for example. It is much easier to string together ten random digits to reach a victim than to connect to a person via an email address [4].

The hacker can simply send messages to any combination of digits the same length as a phone number, and they can try any combination of digits with no harm or foul. Gartner reports that users read 98% of text messages and respond to 45%. This makes text very logical for hackers to use as an attack vector, especially when Gartner reported that only 6% of emails receive responses.

3.4 Text Phishing

Hackers might try to accomplish many different things with a text message. This includes stealing personal details from you by posing as a representative from your bank. They could try to get you to click on a link in the text message to connect to your bank's webpage and verify a recent suspicious charge. They may ask you to call their customer service number, conveniently included within the text message, to talk to them about a recent suspicious charge or a compromised account [13, 19, 26, 41, 43, 44].

Hackers also attempt to use sympathetic measures to gather sensitive information. An example includes messages regarding hurricane relief where the threat actor asks you for a charitable donation. The hacker asks the victim to click the included link and enter your credit card information, address, and, often, the victim's insurance number [61]. Once the hacker obtains the victim's credit card number, the criminal can even charge your credit card monthly to avoid alarming the victim [72].

3.5   Access Point Phishing (Rogue Access Point)

Wi-Fi phishing: also called Evil Twin, is regarded as Wi-Fi phishing, is a procedure of phishing that uses a wireless network where the phisher is between the client and an illegitimate wireless Access Point using a Rouge access point (AP) [27, 63]. WI phishing is one of the most dangerous and severe attacks that deceive the user into joining a rogue access point (RAP) instead of a Legitimate Access Point (LAP) [40]. At the same time, RAP is a malicious device used by an adversary as if it is a real AP. The intruder always copies the same configurations of one or more nearby LAPs to broadcast the same Service Set Identifier (SSID) and always with even stronger transmitting power [5, 66].

DE authentication: DE authentication attack is when the attacker tries to sniff or break the connection between the victim and an AP by flooding the network with DE authentication frames to force the client to re-authenticate. Then, the attacker can save traffic during the authentication process, and this step is the base of the attack's phase one [63]. The attacker decrypts the pre-shared secret to have the secret key and bypass security encryption [5, 28]. The second step of the DE authentication attack is to force the client to connect to a RAP to sniff the whole communication, which needs special tools to be detected [53]. At the same time, RAP based on DE authentication is perhaps the most well-known assault in Wi-Fi networks [62, 66].

Karma attack: The karma attack depends on actively scanning the WLAN to collect the probe frame requests from users' devices and then generate a corresponding probe response as if the required WLAN network is nearby. With the enormous growth of the digital era more and more, many humans keep their Wi-Fi settings on their devices [40]. As it is for the device to automatically be connected to their known network, if the network is nearby, devices send probe requests in probe frames to verify the existence of a network as the device does not know physically whether the network is in range or not. The targeted device is connected to the RAP, which is made as a trap by the adversary [5]. Karma attacks can, in any case, influence customers that are using active probing authentication. Also, to perform it, the intruder can use a Pineapple AP which makes the assault a lot less difficult to achieve.

Hosts file poisoning: Replacing hostnames in the host records can override the usual process of DNS servers trying to retrieve actual IP addresses from beyond the network [27, 45]. This technique can poison the records and allow valid URLs to lead to secure sites to lead to malicious pages instead due to compromised IP associations in the server [6, 10, 24]. [9] proposes a new attack that can bypass security toolbars and phishing filters by using DNS poisoning. They use spoofed DNS cache entries to create fake results and successfully attack four renowned security toolbars and the phishing filters of three popular browsers without being detected [5, 45, 67].

Hosts file poisoning: Replacing hostnames in the host records can override the usual process of DNS servers trying to retrieve actual IP addresses from beyond the network [27]. This technique can poison the records and allow valid URLs to lead to secure sites to lead to malicious pages instead due to compromised IP associations in the server. [67], propose a new attack that can bypass security toolbars and phishing filters by using DNS poisoning. They use spoofed DNS cache entries to create fake results and successfully attack four renowned security toolbars and the phishing filters of three popular browsers without being detected [5].

3.6 URL Phishing

Hackers can inject hidden links that redirect to malicious pages into the URL, where one may not expect to find one [25]. [9] proposes a method to detect URL phishing with URL ranking [6]. They classify the URLs by their lexical and host-based features and categorize and rank them using the online URL reputation services [34]. Content injection phishing: Data collection is achieved in this technique by concatenating malicious sections within a real website [10, 26, 66, 68, 71].

3.7 Spear Phishing

Spear phishing is sending spam emails containing malware concealed inside embedded links and attachments that seem to come from reputable sources (e.g., a trustworthy and well-known corporation) [1, 2, 46, 71]. The most prevalent form of phishing attack used by hackers is spear-phishing. Spear-phishing attacks are more successful than conventional email attacks as they use specially designed emails that mimic a source known to the victim [33, 46]. The email contents are essential to the victim and do not allow the victim to suspect anything. This means there is a high chance that the victim succumbs to the attack. In 2018, Kaspersky found almost 1000 spear phishing assaults, including 83 distinct attacks targeting American-based educational institutions and 21 other UK universities that were attacked [26, 34, 36, 46]. The intention of attackers in spear-phishing is also to perform an advanced persistent threat (APT), as it is based on a targeted assault, and the phisher can launch the assault on a single person, community, or entity [15, 25, 26].

3.8 Whaling

In the context of a targeted attack, Whaling is equivalent to spear-phishing, except the target comprises high-level employees with higher levels of knowledge or services within a company [1, 18, 31, 72]. To form a more aggressive attack, the phisher, in this type of attack, spends more time sending fraudulent content via email to convince the victim to click on a link or download attachments. [33, 71] If this activity is carried out, a backdoor or keylogger is installed on the victim's machine, which sends sensitive information to the attacker. Whaling is performed on high-level targets such as high-ranked government employees or business organizers; attacks on these entities were recorded in 2008, 2010, and 2011 [15, 17, 31, 47, 55].

In 2008, several US CEOs received a subpoena with a malicious attachment that installed malware upon viewing. Other successful whaling attack victims include the Australian Prime Minister's office, Epsilon mailing list service, the Canadian government, RSA SecurID, Oak Ridge National Laboratory, and HBGary Federal in 2010 and 2011 [15, 73].

3.9 Phishing Kits

Another way to perform phishing attacks is through phishing kits. Phishing kits allow the attackers to generate malicious websites, emails, and scripts that require no advanced programming skills [11, 15, 25]. Phishing kits do not play a vital role in phishing victims' data but assist in implementing phishing scams. Phishing kits are available for free or through proper payment to cyber-criminals [27, 68]. However, free kits are not suitable for use, as they steal the user's personal information and send them back to the phisher.

There is also a competence among kit creators that relies on their kits' trustworthiness, availability, ease of use, and security concerns. Phishing kits can be used to send malicious emails or create a phishing website to deceive the victim [15]. Several attacks were reported by [74], which use phishing kits. For example, 10% of the websites active in 2013 reported phishing kits attacks, 120–160 criminal activities were detected for SMTP-based email phishing kit operators, etc. [26, 60, 65, 71, 75].

3.10 Denial of Service Attack

Denial-of-Service (DoS) assault is an unequivocal endeavor where the assailants decline the benefits to clients deserving privilege. This attack floods a computer asset with more demands than it can handle, expanding its accessible transfer speed, which over-burdens the server. This leads to the crashing of the web server so that the client is denied the resources. This causes the site to malfunction briefly or collapse, failing the framework [47, 85].

Distributed Denial of Service (DDoS) attacks are an important threat to the Internet because the number of clients is sending requests to a single server [22, 30, 79]. As a result, the server cannot provide the proper services to the clients due to excessive resource consumption [79, 81]. The terms 'Bot' and 'Botnet' originated from Internet Relay Chat (IRC) which uses a central structure on a single settled server port [23, 83, 84, 87]. They aim to spread infection through compromised systems [23].

Botnets are responsible for several security issues like executing DDoS attacks, spreading spam, organizing snap bending traps, taking individual client data (credit card information, government powerlessness information), and abusing extraordinary computational assets [22, 23, 66, 80, 83]. A Botnet is the accumulation of PCs related to the Internet which have been wrangled and controlled remotely by an intruder, usually called a Bot-master, for destructive programming [23, 30, 60, 81, 83]. Moreover, it may reasonably uncover messages being passed from the server to singular customers, which poses further security risks [74, 82] (Table 1).

Table 1. Types of DDoS attacks

Bots	Description
Agobot/Phatbot/Forbot/XtremBot [22, 23, 63, 84]	This type of malware is dangerous because it can distinguish between debugging tools like OllyDbg, SoftICE, etc., and Virtual Machines like Virtual PC, VMWare, etc. It is capable of sniffing traffic and can also hide its presence. Reverse Engineering this malware is complicated. Moreover, the Linux version can recognize the Linux distribution concerning the compromised host
SDBot/RBot/UrBot/UrXBot/... [22, 23, 30, 85]	A remarkable malware, it can provide system remote access to the adversary. SDBot is made in astoundingly poor C and passed under the GPL. It is succeeded by other malware like RBot, RxBot, UrBot, UrXBot, JrBot, etc. Although the command set is limited, and the implementation is simple, it is popular among attackers due to its catastrophic aftereffects
DSNX Bots [22, 23, 30, 79, 86]	Due to the presence of a plugin interface, an attacker may extend the features of this bot by appending scanning and spreading features. The default modification does not incorporate spreaders, although plugins are used to overcome this limitation. The plugins may also contribute to performing DDoS attacks and port scanning
Q8 Bots [22, 30, 87]	It is 926 lines of C-code written for Linux/Unix Operating systems. Like most bots, it can carry out flooding attacks and execute arbitrary commands. Some versions of this bot support spreaders
Kaiten [22, 23, 30, 88]	Kaiten does not incorporate spreaders. Written for Unix/Linux systems, this malware takes advantage of weak user authentication. Being a single file, it is easy to fetch and compile. Moreover, it offers a remote shell thus it can use IRC to discover system vulnerabilities and gain privileged access
Perl-based bots [22, 23, 30, 79, 89]	These bots contain a few hundred lines of code. They are used for Unix-based systems and have a limited set of commands

(continued)

Table 1. (*continued*)

Bots	Description
Mirai Ceron [23, 39, 51, 79, 90]	Mirai is a self-propagating worm replicating itself by finding, attacking, and infecting vulnerable IoT devices. These bots use massive DDoS attacks to take down major websites as they can compromise hundreds of thousands of IoT devices. They infect poorly secured Internet devices by using telnet to find which one still uses the factory default usernames and passwords. The harm of Mirai is due to its ability to infect a dozen of thousands of other poorly secured devices and enable them to execute a DDoS attack against a chosen target
GameOver Zeus [22, 23, 30, 34, 46, 48, 80, 84, 85, 91]	It is a peer-to-peer Botnet that derives characteristics from the ZeuS trojan. Scammers use it to control and monitor data through the Command and Control (C&C) server. The underlying idea is that the virus establishes a connection to the server once executed and disables several processes running on the system. It may cause hindrance for launching processes and downloads and may also delete files
Pushdo [22, 23, 30, 39, 90, 92]	The Pushdo Botnet is used for spamming and is observed during Distributed Denial of Service (DDoS) attacks against Secure Socket Layer (SSL) enabled websites. It may completely compromise the target system, leading to exposure of confidential information and loss of productivity

3.11 Drive-By-Download

This is a technique of unintentionally running a virus or a malicious shellcode by visiting a malicious website or responding to an email. As described by [71], this can also be carried out by maliciously using a JavaScript code to exploit the vulnerabilities in a browser-hosted using a server, or they can be injected into a website via email. Then, a malicious web page is opened containing a malicious JavaScript code that exploits the web browser vulnerabilities. In a successful exploit, the malware is downloaded on the system, which, as a result, becomes part of the botnet [15, 80, 85, 90].

Drive-by-download has become a very effective attack for several reasons. For example, there are large-scale web client flaws, composed of 15% common vulnerabilities and exposures (CVE), found in repository reports [15]. Moreover, 45% of internet users used outdated browsers with security problems. Table 2 presents a common mitigation strategy for phishing [7, 15, 20, 50, 56, 65, 71, 76].

Table 2. Stride mapping with security measures

STRIDE Threat	Description	Security violation
Spoofing	Misleading a user or system by illegally accessing authentication information	Authentication
Tampering	Maliciously modify the original information by accessing it without permission	Integrity
Repudiation	Denying the user's privileged access by performing malicious actions	Non-repudiation
Information disclosure	Exposing the sensitive information of the user without permission	Confidentiality
Denial of service	Denying the network or services access to the valid user	Availability
Elevation of privilege	Gaining the privileged resources without the user's permission to compromise the system	Authorization

4 Methodology

Using PRISMA, this study follows the principles of extracting journals from the primary scientific database aforementioned. Figure 1 presents the data collected using the PRISMA. Unique titles search resulted in 1,089, where 1,063 were relevant following the examination of titles, of which 748 were deemed relevant for this study. Sixty-three (63) articles were published from 2018 to 2022 for this study analysis aiming to evaluate phishing mitigation, published only in English. Figure 1 presents a summary of the article database search.

In this section, under our methodology, firstly, we design an iteratively coding sheet system for articles on phishing types and mitigation (see Table 3). Secondly, the scholar inputs each article based on the coding sheet after reading through it. Included in Table 3, among others, is the coding sheet system.

4.1 Results of the Study

The analysis of the current kinds of literature (2018–2022) on the types of phishing and phishing mitigation used to foster precedence and prevention of phishing by using the fascinating results from this study. This study presents finding categorized based on the country where the study was carried out; technology phished, phishing mitigation, and targeted area for phishing.

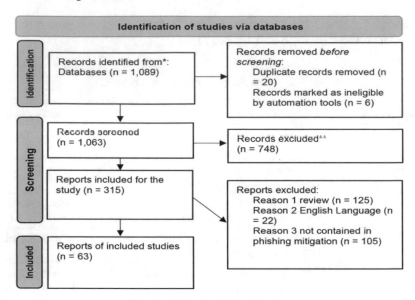

Fig. 1. Method of data collection

Table 3. Phishing mitigation/types classification and analysis coding scheme

Identification	Author(s) name (s), year of publication country and methods
Technology	Mobile devices, web technologies, social media, and others
Target area	Cloud, education, financial institution
Mitigation	Awareness, machine learning/artificial intelligence
Phishing methods	Denial of services, social engineering, and others
Type of phishing	Spear- phishing, botnet, Clone website, and others
Performance metrics	Percentage of mitigation accuracy

4.2 Phishing Mitigation Study by Year

Figure 2 presents trends per year of studies published between 2018 and 2022. This trend implies starting from the beginning of the year 2018 to 2019. And the viewer can notice an increase in the publication on phishing and phishing types rising from 11% to 28%. In 2020 there was a drop in phishing publications from 28% to 19%. It is noted [37] that during the Covid-19 lockdown, there was a spike in phishing study with 30%; however, there was a recode of high phishing attacks noted during the lockdown. This could be because of high internet traffic, where learning and working have moved online.

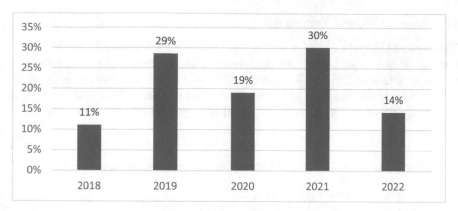

Fig. 2. Phishing types and mitigation by year

4.3 Phishing Mitigation Study by Included Database

Web of science has the most articles published during the past five years with 38%, followed by IEEE with 24%, ERIC with 11%, Scopus with 10%, and PubMed with 6%, as seen in Fig. 3.

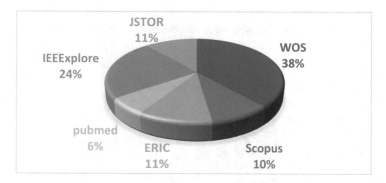

Fig. 3. Phishing types and mitigation by year

4.4 Targeted Phished Area

According to the data collected from the included study, it is found that business enterprise (10%) is the most phished. This enterprise consists of all types of businesses, such as the profit and charity enterprises across the globe for the past five years. Education (8%) is one of those highly phished, followed by banking accounts and financial institutions (6%). However, on the contrary, more than half of the targeted phished are not indicated, while 6% of the articles had a generic overview of the targeted phished area. As shown in the Fig. 4, other targeted areas are cloud with 5%, e-government, smart home, and ministry of defenses database with 3% each. Finally, the industry does have 2%.

4.5 Targeted Phished Technology

The targeted phished are as presented in Fig. 5 above are actual the database of organizations, the database are being penetrated through phishing emails (32%), clone websites or web application (22%), mobile device (10%), IOT (6%), social media (5%), wireless and LAN (3%) and finally, personal computers and the cloud (2%). Seventeen percent were not given or indicated, while 2% are other ways where penetration occurred.

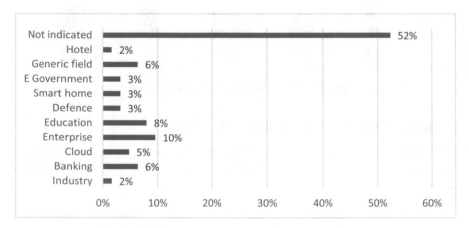

Fig. 4. Phishing targeted area based on the included study

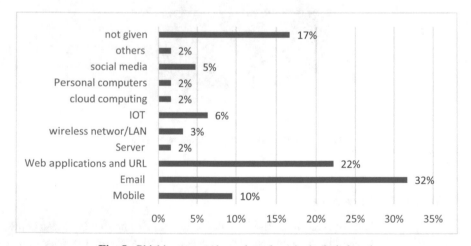

Fig. 5. Phishing targeted area based on the included study

4.6 Types of Phishing Found from This Study Included Articles

The main aim of this study is to find out the types of phishing and their mitigation. The current five-year study shows that 44% of the mitigation method were not indicated. In comparison, 37% indicated that social engineering is the most common and effective means of phishing, followed by DDoS (8%), clone websites (3%), and others (3%). It was found that spear-phishing, smishing, and convolutional neural networks each were 2%. Phishing and penetration methods were in play targeting the organizations and private entities' technology. Figure 6 presents the phishing methods.

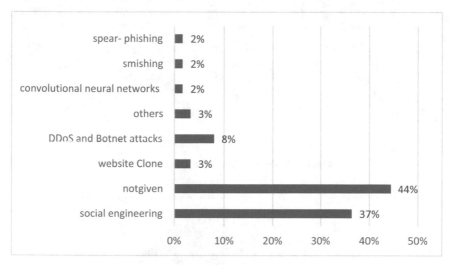

Fig. 6. Types of phishing penetration methods

4.7 Phishing Mitigation

This section provides mitigation methods from the above targeted phished area. Based on the included articles, it was found that training and awareness are effective ways for phishi0ng mitigation (38%). Machine learning/artificial intelligence and some undefined algorithms (17% each) are the second means of phishing mitigation, creating a proactive means to forecast and prevent phishing techniques [8, 9]. Five percent of the included article did not indicate the mitigation technique used. However, 10% used a proxy, also known as a virtual private network, followed by whitelists-based filtering (3%). Finally, fuzzy logic and data mining, cyber insurance, OpenPhish, software-defined network, token, and computer vision are 2% each, as seen in Fig. 7 below.

4.8 Phishing Mitigation Results

It is important to note that the current mitigation method is highly effective, especially when there is massive awareness. 73% of articles' methods of mitigation were positive, while 24% did not indicate if their result were positive or not. However, 2% are partially positive, and only 1% are negative, as indicated in Fig. 8 below.

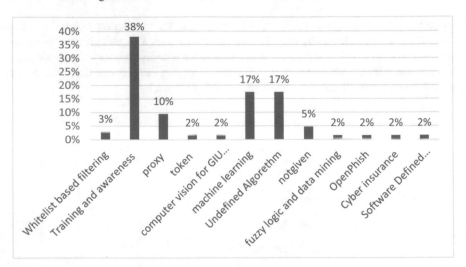

Fig. 7. Phishing mitigation methods in play

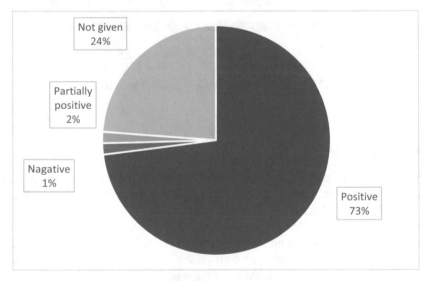

Fig. 8. Phishing Mitigation Results

4.9 Phishing Mitigation Performance Results

For more information about methods of mitigation and performance, this study tries to bring together the mitigation methods, the mitigation performance, and the results of the mitigation methods. Table 4 indicates that 12 of the mitigation methods' performance falls between 95 to 99%, and one of the methods falls at 88%.

Furthermore, it is important to note that artificial intelligence/machine learning (AI/ML) is low at 88.7%, meaning that AI/ML is not properly implemented. In contrast, other methods such as Whitelist Based Filtering, Proxy, Token, Awareness, Algorithm, Neural Network algorithm, Fuzzy logic, and OpenPhish are proven effective.

Table 4. Phishing mitigation performance results

Mitigation	Performance	Reference	Result
Whitelist Based Filtering	99.20%	[10]	Not given
Proxy	98.40%	[14]	Not given
	98.30%	[12]	Positive
Token	99.10%	[7]	Not given
Computer vision	99.20%	[8]	Positive
Awareness	96%	[21]	Positive
	96%	[31]	Positive
Machine learning/artificial intelligence	88.70%	[9]	Negative
	94%	[22]	Positive
	98%	[48]	Not given
Algorithm	95%	[6]	Positive
	98%	[44]	Positive
	94%	[57]	Not given
Neural Network algorithm	99%	[16]	Positive
Fuzzy logic	98.20%	[25]	Positive
OpenPhish	97%	[49]	Not given

5   Conclusion

This study identifies some existing phishing types that provided a good expression and understanding of the phishing methods in play. Having identified mitigation methods from the articles included, this study contribute is based on identifying the current phishing mitigation methods in play through reviewing for years (2018–2022) of research article results on phishing mitigation methods, this study, identify effective method that can be used in mitigating phishing. This study identified, phishing mitigation through; algorithms, proxy, OpenPhish, token, and awareness; it is important to note that the abovementioned mitigation methods have been implemented on a large scale, but still have some error rate where attack was eminent. Similarly, a new mitigation method through, Computer vision, Neural Network algorithms, and Fuzzy logic, provide a better mitigation, but, were not implemented on a large scale, meaning having the mitigation performance effective on a small scale or a specific area could not be effective in others or on a large scale. This study recommend a need for that mitigation method to be applied in multiple large organizations.

Similarly, from the findings in this study, we conclude that, the most effective means of mitigating phishing is through massive enlistment on phishing attach. In organizations or institutions such as universities, all users of the are to be educated or informed about the phishing methods.

Acknowledgment. This work is supported by UTM SPACE with cost center number R.K130000.7756.4J574.

References

1. Al-Hamar, Y., Kolivand, H., Tajdini, M., Saba, T., Ramachandran, V.: Enterprise Credential Spear-phishing attack detection. Comput. Electr. Eng. **94**, 107363 (2021)
2. Yeoh, W., Huang, H., Lee, W.S., Al Jafari, F., Mansson, R.: Simulated phishing attack and embedded training campaign. J. Comput. Inf. Syst. 1–20 (2021)
3. Lee, J., Lee, Y., Lee, D., Kwon, H., Shin, D.: Classification of attack types and analysis of attack methods for profiling phishing mail attack groups. IEEE Access **9**, 80866–80872 (2021)
4. Kara, I.: Don't bite the bait: phishing attack for internet banking (e-banking). J. Digit. Forensics Secur. Law JDFSL **16**, 1–12 (2021)
5. Rastenis, J., Ramanauskaitė, S., Janulevičius, J., Čenys, A., Slotkienė, A., Pakrijauskas, K.: E-mail-based phishing attack taxonomy. Appl. Sci. **10**(7), 2363 (2020)
6. Fetooh, H.T.M., El-Gayar, M.M., Aboelfetouh, A.: Detection technique and mitigation against a phishing attack. Int. J. Adv. Comput. Sci. Appl. **12**(9) (2021)
7. Azeez, N.: Identifying phishing attacks in communication networks using URL consistency features (2019)
8. Chen, S., Fan, L., Chen, C., Xue, M., Liu, Y., Xu, L.: GUI-squatting attack: automated generation of android phishing apps. IEEE Trans. Dependable Secure Comput. **18**(6), 2551–2568 (2019)
9. Hossain, S., Sarma, D., Chakma, R.J.: Machine learning-based phishing attack detection. Int. J. Adv. Comput. Sci. Appl. **11**(9) (2020)
10. Lee, H., Lee, Y., Seo, C., Yoon, H.: Efficient approach for mitigating mobile phishing attacks. IEICE Trans. Commun. **101**(9), 1982–1996 (2018)
11. Song, F., Lei, Y., Chen, S., Fan, L., Liu, Y.: Advanced evasion attacks and mitigations on practical ML-based phishing website classifiers. Int. J. Intell. Syst. **36**(9), 5210–5240 (2021)
12. Martins de Souza, C.H., Lemos, M.O., Dantas Silva, F.S., Souza Alves, R.L.: On detecting and mitigating phishing attacks through featureless machine learning techniques. Internet Technol. Lett. **3**(1), e135 (2020)
13. Wardman, B., Weideman, M., Burgis, J., Harris, N., Butler, B., Pratt, N.: A practical analysis of the rise in mobile phishing. In: Dehghantanha, A., Conti, M., Dargahi, T. (eds.) Cyber Threat Intelligence. Advances in Information Security, vol. 70, pp. 155–168. Springer, Cham (2018). https://doi.org/10.1007/978-3-319-73951-9_8
14. Chin, T., Xiong, K., Hu, C.: Phishlimiter: a phishing detection and mitigation approach using software-defined networking. IEEE Access **6**, 42516–42531 (2018)
15. Abbas, S.G., et al.: Identifying and mitigating phishing attack threats in IoT use cases using a threat modelling approach. Sensors **21**(14), 4816 (2021)
16. Mughaid, A., AlZu'bi, S., Hnaif, A., Taamneh, S., Alnajjar, A., Elsoud, E.A.: An intelligent cyber security phishing detection system using deep learning techniques. Clust. Comput. 1–10 (2022)

17. Abroshan, H., Devos, J., Poels, G., Laermans, E.: Phishing happens beyond technology: the effects of human behaviors and demographics on each step of a phishing process. IEEE Access **9**, 44928–44949 (2021)
18. Anawar, S., Kunasegaran, D.L., Mas'ud, M.Z., Zakaria, N.A.: Analysis of phishing susceptibility in a workplace: a big-five personality perspectives. J. Eng. Sci. Technol. **14**(5), 2865–2882 (2019)
19. Airehrour, D., Vasudevan Nair, N., Madanian, S.: Social engineering attacks and countermeasures in the New Zealand banking system: advancing a user-reflective mitigation model. Information **9**(5), 110 (2018)
20. McAlaney, J., Hills, P.J.: Understanding phishing email processing and perceived trustworthiness through eye tracking. Front. Psychol. **11**, 1756 (2020)
21. Ndibwile, J.D., Luhanga, E.T., Fall, D., Miyamoto, D., Blanc, G., Kadobayashi, Y.: An empirical approach to phishing countermeasures through smart glasses and validation agents. IEEE Access **7**, 130758–130771 (2019)
22. Parra, G.D.L.T., Rad, P., Choo, K.K.R., Beebe, N.: Detecting Internet of Things attacks using distributed deep learning. J. Netw. Comput. Appl. **163**, 102662 (2020)
23. Tuan, T.A., Long, H.V., Son, L.H., Kumar, R., Priyadarshini, I., Son, N.T.K.: Performance evaluation of Botnet DDoS attack detection using machine learning. Evol. Intell. **13**(2), 283–294 (2019). https://doi.org/10.1007/s12065-019-00310-w
24. Ali, G., Ally Dida, M., Elikana Sam, A.: Evaluation of key security issues associated with mobile money systems in Uganda. Information **11**(6), 309 (2020)
25. Zahra, S.R., Chishti, M.A., Baba, A.I., Wu, F.: Detecting Covid-19 chaos driven phishing/malicious URL attacks by a fuzzy logic and data mining-based intelligence system. Egypt. Inform. J. **23**(2), 197–214 (2022)
26. Althobaiti, K., Jenkins, A.D., Vaniea, K.: A Case Study of Phishing Incident Response in an Educational Organization. Proc. ACM Hum. Comput. Interact. **5**(CSCW2), 1–32 (2021)
27. Zimba, A.: A Bayesian attack-network modeling approach to mitigating malware-based banking cyberattacks. Int. J. Comput. Netw. Inf. Secur. **14**(1) (2022)
28. Meng, B., Smith, W., Durling, M.: Security threat modeling and automated analysis for system design. SAE Int. J. Transp. Cybersecur. Priv. **4**(11-04-01-0001), 3–17 (2021)
29. Shahriar, H., Zhang, C., Dunn, S., Bronte, R., Sahlan, A., Tarmissi, K.: Mobile anti-phishing: approaches and challenges. Inf. Secur. J. Glob. Perspect. **28**(6), 178–193 (2019)
30. Harikrishna, P., Amuthan, A.: Adaptive self-organizing maps inspired SDN-Based DDoS (ASOM-SDN-DDoS) mitigation framework. Int. J. Sci. Technol. Res. **8**(10) (2019)
31. Daengsi, T., Pornpongtechavanich, P., Wuttidittachotti, P.: Cybersecurity awareness enhancement: a study of the effects of age and gender of Thai employees associated with phishing attacks. Educ. Inf. Technol. **27**(4), 4729–4752 (2022)
32. Brenner, P.S.: Can phishing tank survey response rates? Evidence from a natural experiment. Field Methods **31**(4), 295–308 (2019)
33. Canfield, C.I., Fischhoff, B., Davis, A.: Better beware: comparing metacognition for phishing and legitimate emails. Metacogn. Learn. **14**(3), 343–362 (2019)
34. Perrault, E.K.: Using an interactive online quiz to recalibrate college students' attitudes and behavioral intentions about phishing. J. Educ. Comput. Res. **55**(8), 1154–1167 (2018)
35. Pollock, T., Levy, Y., Li, W., Kumar, A.: Subject matter experts' feedback on experimental procedures to measure user's judgment errors in social engineering attacks. J. Cybersecur. Educ. Res. Pract. **2021**(2), 4 (2022)
36. Weaver, B.W., Braly, A.M., Lane, D.M.: Training users to identify phishing emails. J. Educ. Comput. Res. **59**(6), 1169–1183 (2021)
37. Venkatesha, S., Reddy, K.R., Chandavarkar, B.R.: Social engineering attacks during the COVID-19 pandemic. SN Comput. Sci. **2**(2), 1–9 (2021)

38. Moustafa, A.A., Bello, A., Maurushat, A.: The role of user behaviour in improving cyber security management. Front. Psychol. 1969 (2021)
39. Chen, Y., Yang, Y.: An advanced deep attention collaborative mechanism for secure educational email services. Comput. Intell. Neurosci. 2022 (2022)
40. Dimitriadis, A., Ivezic, N., Kulvatunyou, B., Mavridis, I.: D4I-Digital forensics framework for reviewing and investigating cyber-attacks. Array **5**, 100015 (2020)
41. Mishra, S., Soni, D.: SMS phishing and mitigation approaches. In: 2019 Twelfth International Conference on Contemporary Computing (IC3), pp. 1–5. IEEE, August 2019
42. Jamil, A., Asif, K., Ghulam, Z., Nazir, M.K., Alam, S.M., Ashraf, R.: MPMPA: a mitigation and prevention model for social engineering based phishing attacks on Facebook. In: 2018 IEEE International Conference on Big Data (Big Data), pp. 5040–5048. IEEE, December 2018
43. Bikov, T.D., Iliev, T.B., Mihaylov, G.Y., Stoyanov, I.S.: Phishing in depth–modern methods of detection and risk mitigation. In: 2019 42nd International Convention on Information and Communication Technology, Electronics and Microelectronics (MIPRO), pp. 447–450. IEEE, May 2019
44. Hashim, A., Medani, R., Attia, T.A.: Defences against web application attacks and detecting phishing links using machine learning. In: 2020 International Conference on Computer, Control, Electrical, and Electronics Engineering (ICCCEEE), pp. 1–6. IEEE (2021)
45. Jin, Y., Tomoishi, M., Yamai, N.: A detour strategy for visiting phishing URLs based on dynamic DNS response policy zone. In: 2020 International Symposium on Networks, Computers and Communications (ISNCC), pp. 1–6. IEEE, October 2020
46. Khan, H., Alam, M., Al-Kuwari, S., Faheem, Y.: Offensive AI: unification of email generation through GPT-2 model with a game-theoretic approach for spear-phishing attacks. Competitive Advantage in the Digital Economy. IEEE (2021)
47. Arshey, M., Viji, K.A.: Thwarting cybercrime and phishing attacks with machine learning: a study. In: 2021 7th International Conference on Advanced Computing and Communication Systems (ICACCS), vol. 1, pp. 353–357. IEEE, March 2021
48. Shalke, C.J., Achary, R.: Social engineering attack and scam detection using advanced natural language processing algorithm. In: 2022 6th International Conference on Trends in Electronics and Informatics (ICOEI), pp. 1749–1754. IEEE, April 2022
49. Maroofi, S., Korczyński, M., Hesselman, C., Ampeau, B., Duda, A.: COMAR: classification of compromised versus maliciously registered domains. In: 2020 IEEE European Symposium on Security and Privacy (EuroS&P), pp. 607–623. IEEE, September 2020
50. Subramani, K., Jueckstock, J., Kapravelos, A., Perdisci, R.: SoK: Workerounds-Categorizing service worker attacks and mitigations. In: 2022 IEEE 7th European Symposium on Security and Privacy (EuroS&P), pp. 555–571. IEEE, June 2022
51. Niraja, K.S., Murugan, R., Prabhu, C.S.R.: Comparative analysis of security issues in the layered architecture of IoT. In: 2018 3rd IEEE International Conference on Recent Trends in Electronics, Information & Communication Technology (RTEICT), pp. 1414–1417. IEEE, May 2018
52. Mattera, M., Chowdhury, M.M.: Social engineering: the looming threat. In: 2021 IEEE International Conference on Electro Information Technology (EIT), pp. 056–061. IEEE, May 2021
53. Kikuchi, M., Okubo, T.: Power of communication behind extreme cybersecurity incidents. In: 2019 IEEE International Conference on Dependable, Autonomic and Secure Computing, International Conference on Pervasive Intelligence and Computing, International Conference on Cloud and Big Data Computing, International Conference on Cyber Science and Technology Congress (DASC/PiCom/CBDCom/CyberSciTech), pp. 315–319. IEEE, August 2019

54. Kettani, H., Wainwright, P.: On the top threats to cyber systems. In: 2019 IEEE 2nd International Conference on Information and Computer Technologies (ICICT), pp. 175–179. IEEE, March 2019
55. Vos, J., Erkin, Z., Doerr, C.: Compare before you buy: privacy-preserving selection of threat intelligence providers. In: 2021 IEEE International Workshop on Information Forensics and Security (WIFS), pp. 1–6. IEEE, December 2021
56. AlMudahi, G.F., AlSwayeh, L.K., AlAnsary, S.A., Latif, R.: Social media privacy issues, threats, and risks. In: 2022 Fifth International Conference of Women in Data Science at Prince Sultan University (WiDS PSU), pp. 155–159. IEEE, March 2022
57. Oakley, J.G.: Towards improving APT mitigation. J. Inf. Warf. **18**(1), 69–86 (2019)
58. Ogunlana, S.O.: Halting Boko Haram/Islamic State's West Africa province propaganda in cyberspace with cybersecurity technologies. J. Strateg. Secur. **12**(1), 72–106 (2019)
59. Atrews, R.: Cyberwarfare: threats, security, attacks, and impact. J. Inf. Warf. **19**(4), 17–28 (2020). https://www.jstor.org/stable/27033642
60. Merz, T., Fallon, C., Scalco, A.: A context-centred research approach to phishing and operational technology in industrial control systems. J. Inf. Warf. **18**(4), 24–36 (2019). https://www.jstor.org/stable/26894692
61. Miller, L.: Cyber insurance: an incentive alignment solution to corporate cyber-insecurity. J. Law Cyber Warf. **7**(2), 147–182 (2019). https://www.jstor.org/stable/26777974
62. Jabbour, K.: The Post-GIG era: from network security to mission assurance. Cyber Defense Rev. **4**(2), 117–128 (2019). https://www.jstor.org/stable/26843896
63. Hutton, W., McKinnon, A., Hadley, M.: Software-defined networking traffic engineering process for operational technology networks. J. Inf. Warf. **18**(4), 167–181 (2019). https://www.jstor.org/stable/26894699
64. Sapkal, V., More, D., Agme, M.: A briefed review on phishing attacks and detection approaches. Rupali, A Briefed Review on Phishing Attacks and Detection Approaches, 8 April 2022
65. Birlea, M.C.: Phishing attacks: detection and prevention (2020). arXiv preprint arXiv:2004.01556
66. Mansfield-Devine, S.: Cyber Security Breaches Survey 2022 (2022)
67. Abu-Nimeh, S., Nair, S.: Bypassing security toolbars and phishing filters via DNS poisoning. In: IEEE GLOBECOM 2008–2008 IEEE Global Telecommunications Conference, pp. 1–6. IEEE, November2008
68. Erkkila, J.: Why we fall for phishing. In: Proceedings of the SIGCHI Conference on Human Factors in Computing Systems CHI 2011, pp. 7–12. ACM, May 2011
69. Khan, A.A.: Preventing phishing attacks using one-time password and user machine identification. arXiv preprint arXiv:1305.2704 (2013)
70. Sonowal, G., Kuppusamy, K.S.: SmiDCA: an anti-smishing model with machine learning approach. Comput. J. **61**(8), 1143–1157 (2018)
71. Chiew, K.L., Yong, K.S.C., Tan, C.L.: A survey of phishing attacks: their types, vectors and technical approaches. Expert Syst. Appl. **106**, 1–20 (2018)
72. Moul, K.A.: Avoid phishing traps. In: Proceedings of the 2019 ACM SIGUCCS Annual Conference, New Orleans, LA, USA, pp. 199–208, 3–6 November 2019
73. Hong, J.: The state of phishing attacks. Commun. ACM **55**(1), 74–81 (2012)
74. Thomas, K., et al.: Data breaches, phishing, or malware? Understanding the risks of stolen credentials. In: Proceedings of the 2017 ACM SIGSAC Conference on Computer and Communications Security, pp. 1421–1434, October 2017
75. Han, X., Kheir, N., Balzarotti, D.: Phisheye: live monitoring of sandboxed phishing kits. In: Proceedings of the 2016 ACM SIGSAC Conference on Computer and Communications Security, pp. 1402–1413, October 2016

76. Cova, M., Kruegel, C., Vigna, G.: Detection and analysis of drive-by-download attacks and malicious JavaScript code. In: Proceedings of the 19th International Conference on World Wide Web, pp. 281–290, April 2010
77. Mitnick, K.D., Simon, W.L.: The Art of Deception: Controlling the Human Element of Security. John Wiley & Sons, Hoboken (2003)
78. Qin, T., Burgoon, J.K.: An investigation of heuristics of human judgment in detecting deception and potential implications in countering social engineering. In: 2007 IEEE Intelligence and Security Informatics, pp. 152–159. IEEE, May 2007
79. Bhushan, K., Gupta, B.B.: Distributed denial of service (DDoS) attack mitigation in software defined network (SDN)-based cloud computing environment. J. Ambient Intell. Humaniz. Comput. 10(5), 1985–1997 (2018). https://doi.org/10.1007/s12652-018-0800-9
80. Nadji, Y., Antonakakis, M., Perdisci, R., Dagon, D., Lee, W.: Beheading hydras: performing effective botnet takedowns. In: Proceedings of the 2013 ACM SIGSAC Conference on Computer & Communications Security, pp. 121–132, November 2013
81. Al-Jarrah, O.Y., Alhussein, O., Yoo, P.D., Muhaidat, S., Taha, K., Kim, K.: Data randomization and cluster-based partitioning for botnet intrusion detection. IEEE Trans. Cybern. 46(8), 1796–1806 (2015)
82. Pillutla, H., Arjunan, A.: Fuzzy self-organizing maps-based DDoS mitigation mechanism for software defined networking in cloud computing. J. Ambient Intell. Humaniz. Comput. 10(4), 1547–1559 (2019)
83. Karim, A., Salleh, R.B., Shiraz, M., Shah, S.A.A., Awan, I., Anuar, N.B.: Botnet detection techniques: review, future trends, and issues. J. Zhejiang Univ. Sci. C 15(11), 943–983 (2014). https://doi.org/10.1631/jzus.C1300242
84. Sarwar, S., Zahoory, A., Zahra, A., Tariq, S., Ahmed, A.: BOTNET—threats and countermeasures. Int. J. Sci. Res. Dev. 1(12), 2682–2683 (2014)
85. Gu, G., Yegneswaran, V., Porras, P., Stoll, J., Lee, W.: Active botnet probing to identify obscure command and control channels. In: 2009 Annual Computer Security Applications Conference, pp. 241–253. IEEE, December 2009
86. Erbacher, R.F., Cutler, A., Banerjee, P., Marshall, J.: A multi-layered approach to botnet detection. Secur. Manag. 2008, 301–308 (2008)
87. Meyer von Wolff, R., Hobert, S., Schumann, M.: How may i help you? –state of the art and open research questions for chatbots at the digital workplace. In: Proceedings of the 52nd Hawaii International Conference on System Sciences, January 2019
88. Lu, W., Tavallaee, M., Ghorbani, A.A.: Automatic discovery of botnet communities on large-scale communication networks. In: Proceedings of the 4th International Symposium on Information, Computer, and Communications Security, pp. 1–10, March 2009
89. Gupta, S., Borkar, D., De Mello, C., Patil, S.: An e-commerce website based Chatbot. Int. J. Comput. Sci. Inf. Technol. (IJCSIT) 6(2) (2015)
90. Ceron, J.M., Steding-Jessen, K., Hoepers, C., Granville, L.Z., Margi, C.B.: Improving IOT botnet investigation using an adaptive network layer. Sensors 19(3), 727 (2019)
91. Andriesse, D., Rossow, C., Stone-Gross, B., Plohmann, D., Bos, H.: Highly resilient peer-to-peer botnets are here: an analysis of Gameover Zeus. In: 2013 8th International Conference on Malicious and Unwanted Software: The Americas"(MALWARE), pp. 116–123. IEEE, October 2013
92. John, J.P., Moshchuk, A., Gribble, S.D., Krishnamurthy, A.: Studying spamming botnets using botlab. In: NSDI, vol. 9, no. 2009, April 2009

93. Jain, A.K., Gupta, B.B.: Feature based approach for detection of smishing messages in the mobile environment. J. Inf. Technol. Res. (JITR) **12**(2), 17–35 (2019)
94. Goel, D., Jain, A.K.: Smishing-Classifier: a novel framework for detection of smishing attack in mobile environment. In: Bhattacharyya, P., Sastry, H., Marriboyina, V., Sharma, R. (eds.) Smart and Innovative Trends in Next Generation Computing Technologies. NGCT 2017. CCIS, vol. 828, pp. 502–512. Springer, Singapore (2018). https://doi.org/10.1007/978-981-10-8660-1_38

Review of Smart Home Privacy-Protecting Strategies from a Wireless Eavesdropping Attack

Mohammad Ali Nassiri Abrishamchi[✉] [iD] and Anazida Zainal [iD]

Faculty of Computing, Universiti Teknologi Malaysia, 81310 Johor Bahru, Johor, Malaysia
anamohammad2@graduate.utm.my, anazida@utm.my

Abstract. Increasing concerns about the potential for privacy breaches cast doubt on the future of smart homes. Specifically, wireless snooping-based attacks that target home networks have demonstrated their capacity to illegitimately infer daily activities within the home. This paper reviews the fundamental strategies for safeguarding the personal data of the home residents and evaluates the efficacy of existing privacy-protecting solutions that are built upon the reviewed strategies. The study will show that, while some solutions established a reliable level of home data privacy protection, their negative effects on other system characteristics are significant, emphasizing the need for an ideal compromise between these elements. These factors are the provided privacy rate, impact on the system's response time, and energy consumption of privacy-protecting approaches. This overview of current research will aid in understanding the existing drawbacks and indicate potential avenues for future research.

Keywords: Smart home · Privacy · Fingerprinting attack · Wireless snooping attack · Side-channel attacks · In-home daily activity monitoring

1 Introduction

Smart homes are Internet of Things (IoT)-based systems that provide occupants with a variety of comfort and control-enhancing services. The services integrated into an intelligent building range from smart management applications for regulating the home's temperature, lighting, air conditioning, and energy consumption to surveillance systems for safety and health monitoring [1]. To deliver such services, smart home systems must collect detailed information about occupants using a variety of sensors embedded in smart appliances and devices in order to feed the system's algorithms. However, the transmission and storage of the collected data are processes that make a smart home vulnerable to data leaks [2, 3]. In a data breach incident, cyber attackers violate the privacy rights of victims by getting unauthorized access to their personal information, such as their lifestyle, family affairs, political views, health, and finances, setting the foundation for upcoming criminal acts, which including blackmailing, discrimination, identity theft or fraud, reputational damage, and financial loss [4, 5]. Addressing these sorts of consequences in an appropriate and timely manner, is difficult and costly.

© The Author(s), under exclusive license to Springer Nature Singapore Pte Ltd. 2023
Y. B. Wah et al. (Eds.): DaSET 2022, LNDECT 165, pp. 154–167, 2023.
https://doi.org/10.1007/978-981-99-0741-0_11

Using conventional ways, such as cryptographic solutions, to handle the security issue is insufficient for resolving privacy-related issues of smart homes, since encryption methods only protect the contents of data packets [6]. In considering the fact that adversaries have developed powerful techniques to exploit the physical features of cyber-physical systems, such as side-channel attacks (SCA) [7], preserving home data necessitates a holistic solution to protect both packet contents and their related contextual data.

The Fingerprint And Timing-based Spoofing (FATS) attack is a passive side-channel attack that infers home activities from wireless signal eavesdropping data by analyzing contextual characteristics of data packets. The authors of [8] who introduced this attack report an 80 to 95% accuracy rate for the FATS attack. In this work, the FATS attack mechanism is explained, and the main privacy protection strategies to encounter this attack are investigated. Moreover, the existing protection methods have been evaluated, emphasizing the strengths and weaknesses of the proposed approaches to discover the remaining research gaps.

2 Smart Home Activities Detection Attack

The Fingerprint And Timing-based Snooping (FATS) attack [8] is a multi-tier intrusion tool that locates in a nearby place within the coverage of the home signals (e.g., a neighbor's house). As illustrated in Fig. 1, the attack eavesdrops on the emitted signals from the home network to provide the required input for the attack's pattern recognition model. A combination of machine learning techniques such as classification, clustering, and feature-based matching, enables the attack to analyze the captured data packets based on fingerprints of signals and transmission timestamps to identify smart devices, rooms, and events. Inference of the activities of daily living (ADL) depends on the accuracy of the events' traffic pattern detection, discovering the temporal correlations of signals, and matching model performance.

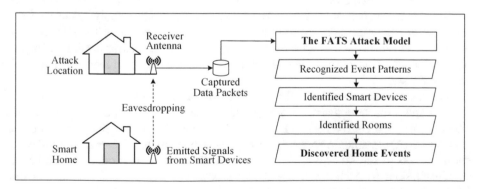

Fig. 1. Overview of the FATS attack procedures.

Once the eavesdropping is done, the attack goes through four steps to gradually finish its recognition tasks. The desired outcome consists of the occupancy state of the home, the number and type of smart devices, rooms, the deployment of devices, and the types of activities that took place. Four steps of the FATS attack are as follows:

- **Step 1** classifies smart devices based on their unique radio signatures, which are sets of RF waveform features that differentiate the signal sources, even if they have similar manufacturers and models. A few primary events, such as home occupancy or sleeping, are identifiable in this step.
- **Step 2** clusters devices based on the time correlations of signal transmissions. The produced clusters represent either the location of their member devices, e.g., a room, or their purposes, e.g., cooking events, assuming that devices that are spatially located close to each other fire their packets at a proximate time.
- **Step 3** labels the clusters using their extracted features with the priority of logical categorization rather than device locations. For example, the washing machine will be assigned to the kitchen, regardless of its actual location.
- **Step 4** performs the device's labeling based on their feature vectors' similarities with the known vectors by the attack model. For example, the attack assumes the stove to be in the kitchen; therefore, it matches the feature vectors of device clusters included in the kitchen cluster with the actual stove's features vector; in the case of a proper match, the unknown device cluster would be recognized as the stove.

The FATS attack procedure combines several statistical and machine learning techniques to discover the identities of rooms, devices, and activities. First, it clusters the records based on the radio fingerprints. Then, using the associated timestamps creates a temporal matrix representing the time proximity between the transmissions. Next, it converts the temporal matrix to a metric distance matrix using Dijkstra's shortest path algorithm [9]. Applying the classical non-parametric multi-dimensional scaling (CMDS) produces a position matrix based on the distance matrix [10]. Finally, the K-mean clustering algorithm [11] clusters device clusters that are temporally correlated. Up to this point, the attack is able to determine whether the house is occupied or vacant, referred to as the "home" and "away" events, respectively.

In the next step, the attack labels the unknown room clusters by applying the maximal min-cost bipartite mapping method [12]; A room features vector consists of the room's number of transmissions per day, the total number of transmissions during day and night, the median inter-transmission time within a room, and the median length of temporal activity clusters.

The subsequent step is to identify the device clusters. Similar to the previous phase, first, the attack makes associated feature vectors to each device cluster, then attempts to match them with feature vectors of known devices using the standard linear discriminant analysis (LDA) classifier. Afterward, creating temporal activity clusters for every device cluster results in several activity feature vectors. Start time, duration, and the number of transmissions by each device are elements of every vector. The attack applies the LDA classifier for the second time to match the unknown activity feature vectors with the actual activities feature vectors. Ultimately, the identities of all rooms, devices, and activities reveal to the adversary; since then, she can monitor all in-home activities and obtain the residents' private information [8]. Figure 2 demonstrates the FATS attack's flowchart.

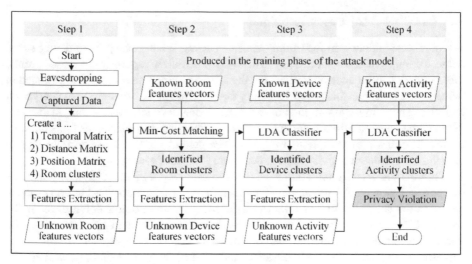

Fig. 2. Flowchart of the FATS attack.

3 Privacy Protection Strategies

The investigation of the proposed solutions established to secure smart homes against the FATS attack threat reveals that researchers attempted different defensive strategies, and in some instances, they were remarkably successful in enhancing the privacy rate. Despite the fact that preserving privacy is the major objective of every proposed solution, the effects of the used approaches on the other system parameters are crucial. For instance, solutions that impose latency on system communications degrade its quality of service; this side effect is particularly unsatisfactory for a few delay-sensitive home systems, such as healthcare monitoring systems and smart fire extinguisher systems. In addition, the energy requirement of the solutions must be justifiable. The optimal balance between three system parameters—privacy rate, communications delay, and energy consumption—is required for a viable solution.

The primary objective of defensive strategies to limit the risk of a data breach is to confuse the FATS attack's pattern recognition mechanism; the ideal approach for this reason is to obfuscate the signals traffic of the smart home's wireless network. Since the core of the attack algorithm is discovering temporal correlations between sent signals, altering the real patterns undermines the performance of the attack.

Strategies for developing a protection method fall into three categories. In the first category, solutions delay sending the data packets; the delay durations are randomly determined; therefore, the attack would have difficulty finding the actual time correlations between the transmitted signals. On the other hand, methods in the second category randomly inject some fake packets into the network traffic; the characteristics of these dummy packets are identical to the actual ones; therefore, the attack cannot differentiate them. As a result, the correctness rate of the FATS attack diminishes because it will be challenging for the attack to understand actual events occurring in the home properly. The third way is a combination of the previous approaches [13].

3.1   Late Packet Injections Technique

Temporal manipulation in forwarding the data packets means every smart device sends out its messages with a random delay instead of immediately reporting the sensed events. The system's central controller filters out the fake messages using indicators included in the encrypted packets; this is impossible for the FATS attack due to the lack of decryption facilities. The primary assumption of the attack is that devices collaborating in an event fire their messages at a proximate time; therefore, changing the order of the transmissions makes the attack's presumptions irrelevant. The pattern recognition algorithm fails in forming the unknown activity frames, decreasing the attack accuracy.

Figure 3, shows a sample for the delayed reporting strategy in which the system has postponed messages. A random number generator algorithm embedded in every device produces the delays before each packet forwarding. In the home's defenseless mode, the attack forms event frames based on the time proximity of the signals, while the transposed signal firings alter the traffic patterns and prevent the exposure of the actual temporal correlations; therefore, the attack fails in detecting two existing actual events.

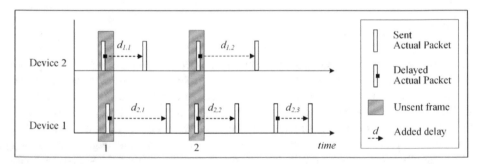

Fig. 3. An illustration of adding random delays to manipulate traffic patterns.

3.2   Fake Packet Injections Technique

Injecting a series of dummy data packets into the home's wireless traffic is an alternative approach to mislead the FATS attack. In this strategy, the system tasks smart devices to generate a random number of data packets identical to actual ones and transmit them in random intervals. The forged messages are encrypted; therefore, the attack cannot distinguish them from actual packets; therefore, it considers them in its pattern recognition processes; this mistake leads the attack algorithm to make false conclusions.

The falsehood of the results is either due to the attack's failure to detect the patterns of actual events (false negative cases) or to report some activities that never occurred (false positive cases). Both of these mistakes hit the correctness rate of the attack. Figure 4 depicts a schematic representation of network traffic in which the attack has detected six activity frames; the first and fifth frames correspond to actual events, while the sixth frame is a false positive. In addition, frames 2 and 3 exhibit true event patterns that the FATS attack was unable to detect due to interference from the fake patterns.

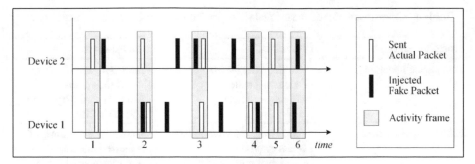

Fig 4. An example of injecting random faked patterns to alter traffic patterns.

3.3 Hybrid Technique

As shown in Fig. 5, combining the previous defensive strategies into a single strategy combines their protection benefits. Nonetheless, this also holds true for their flaws. Overall, protection solutions developed based on this strategy yield modest and more balanced results. Randomness has a substantial impact on the effectiveness and efficiency of the final results. In the demonstrated traffic scenario, frame 1 corresponds to a real event, but the attack is unable to identify it since the forwarding time of the data packets has altered. On the other hand, the attack recognized frames 2, 3, and 4 as possible actual activities; nevertheless, in each of these instances, injected dummy packets misled the attack; hence, none of these frames could be matched to any known activities by the trained model.

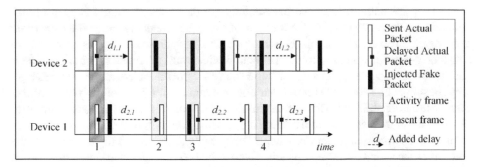

Fig. 5. An instance of traffic manipulation via hybrid strategy.

Despite the fact that all of these strategies aim to maximize the smart home's level of privacy, they all have their limitations. They have distinct effects on the system's other critical factors. Delaying actual reports, for instance, is detrimental to the system's reliability since it compromises the response time of delay-sensitive home services. The lack of control over the system's energy consumption is caused by a random number of fabricated event patterns in the alternative scenario. Table 1 summarizes the advantages and disadvantages of the presented defensive approaches and illustrates how the major system characteristics relate to each strategy.

Table 1. Comparison of the fundamental strategies for protecting privacy of smart homes

Protection strategy	Advantages	Disadvantages
Late packet injections	- Highly effective in concealing patterns - Energy-efficient	- Increasing the response time of the system - Contrary to the System's Service Quality
Fake packet injections	- Relatively effective in hiding actual events - Timely reporting of actual events	- Uncontrolled volume of counterfeit injections - Significant energy usage
Hybrid of late and fake packet injections	- Maximizes the network traffic obfuscation - Highly capable of providing powerful privacy protection	- Moderate system response time lag - Arbitrary amount of additional energy overhead

4   Existing Methods for Privacy Protection

Data privacy in the IoT-based systems has been studied extensively in recent years in which several research works focused on concerns related to the smart home, such as source anonymity [14], data eavesdropping [15], and false data injection [16] attacks. This section reviews privacy-preserving methods for smart homes encountering the FATS attack; these approaches mainly attempt to cope with the risk of passive wireless snooping by obscuring the traffic patterns of in-home daily activities.

In [17], the authors proposed that sending the data packets should happen in predefined injection slots to unify the traffic patterns. This scheme is called the ConstRate because, in this method, the injection intervals and the injection windows' duration are constant. According to the instructions, smart devices must postpone forwarding of their packet until the upcoming injection window; moreover, in the lack of any actual message, they must inject a fake packet into the traffic. An argument for the effectiveness of this scheme is that establishing a fixed transmission framework results in the signals' uniform distribution in the network traffic; therefore, finding any temporal correlations in such traffic is nearly impossible for the attack. Consequently, the incapability to recognize the patterns of the actual events completely disarms the FATS attack. The simulation results of the ConstRate scheme support this claim by showing a near-perfect privacy protection rate. However, the ConstRate scheme's drawbacks stem from its detrimental effects on the system's response time and energy efficiency. This method determines the waiting intervals randomly; the shorter intervals imply a higher number of fake packet injections that surges the system's energy consumption; conversely, the longer intervals increase the delays in transmitting the actual messages, which prolongs the system's overall latency. Neither of these consequences complies with the required optimal trade-off for the system's critical factors. Figure 6(a) depicts an example of ConstRate-manipulated traffic.

In response to the ConstRate scheme's shortcomings, in [16], the ProbRate scheme has been proposed in which every waiting interval is a random value that fits into a designated exponential distribution. With this strategy, the produced waiting intervals get shorter over time, and the overall delay in the system reduces. The provided privacy rate by this approach is near-perfect because the attack only observes a series of repetitive patterns on the network traffic, which are indistinguishable. On the negative side, although the ProbRate scheme reduces the delays, it does not eliminate the overall system latency; this issue is problematic for delay-sensitive services in a smart home. Additionally, reducing the waiting durations before sending the packets injections increases the number of injection windows; this expands the likelihood of dummy packet injections, resulting in a potential surge in energy consumption. Figure 6(b) illustrates how the ProbRate algorithm modifies network traffic to mask true patterns.

The FitProbRate (FPR) scheme [16] is another privacy-preserving approach for smart homes facing the FATS attack. This scheme is the upgraded version of the ProbRate scheme; the waiting intervals determinations are still based on an exponential distribution. Figure 6(c) shows a sample traffic pattern manipulated by the FitProbRate scheme; in this example, sending of an actual packet has been postponed to the third injection window, and the injection of a fake packet has been rescheduled to the fourth injection slot.

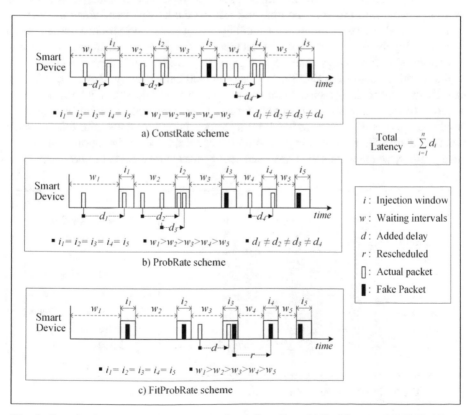

Fig. 6. Sample altered traffics by the methods: a) ConstRate, b) ProbRate, and c) FitProbRate

However, the FPR scheme employs the Anderson-Darling Test to fit the randomly produced intervals into the designated exponential distribution. In addition, it controls the deviation between the measured sample means and the actual mean of the designated distribution to avoid their significant difference. The method prioritizes forwarding of the actual packets; therefore, it sends out the data packet after the shortest waiting time that fits into the given distribution and stops the fake packet injection until the subsequent injection window. The reported results state that the FitProbRate scheme's caused latency is approximately one-tenth of the ProbRate scheme, indicating a remarkable improvement. However, the proposed scheme does not help with the energy overhead issue. For example, a long-term silence in the home enforces a massive energy consumption for injecting unnecessary fake packets. Similar to prior schemes, the number of injected fake packets is an uncontrollable random value.

In the other work [18], the authors have proposed utilizing the events' behavioral semantics to train the decision-making model. The aim is to predict the occurrence likelihood of the actual events to enable the system for purposeful injection of fake packets that increases the chance of blending the actual patterns with faked ones; altering the activity patterns diminishes the FATS attack's event detection accuracy. The provided privacy rate is lower than the prior works, although this method reduces the system's energy overhead. Moreover, this solution does not resolve the latency issue since the process uses predefined injection windows for sending data packets. Furthermore, the method's success is highly dependent on the correctness of the predictions; wrong forecasting wastes the system's energy resources. Figure 7 shows how this defense mechanism works. In this example, the FATS attack has detected three activity. However, it fails to identify them since the detected frames do not match any of the attack's known activity patterns. In the first and third injection windows, the protection method has correctly predicted the events' occurrence and, by injecting dummy packets, altered the actual events' patterns. Conversely, the prediction result for the second injection window is a false positive, which wastes the consumed energy for the fake packet injections.

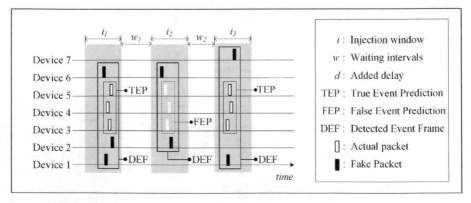

Fig. 7. Privacy protection method based on the event's behavioral semantics

Reference [19] has proposed the Sample Data and Supervised Learning (SDASL) method, a real-time adaptive approach using supervised learning techniques. In the

SDASL, the central controller periodically computes decision parameters for every smart device and applies the logistic regression model to send out the fake packet. Thus, this method requires frequent communication between the central controller and smart devices. The authors claimed that the method provides low latency, low energy consumption, strong adaptability, and adequate privacy protection for smart homes.

The SDASL consists of two phases, sample data analysis and supervised learning. In the first step, the model simulates the dissemination of fake messages using the distribution of the radio frequencies (RF); the output has referred to as FDR; the central controller updates the smart devices with the produced FDR. The similarity in the extracted frequency rate from the sample dataset and dummy messages is critical. In the second phase, a supervised learning model performs three tasks: data collection, labeling, and learning model parameters parameter updating. Every device in the home network must be upgraded with the final prediction model; the required inputs are time and the network's traffic status. The logistic regression model makes the final decision using the real-time inputs and a threshold set on 0.5; a computation result that is bigger than the threshold indicates that the fake packet must be injected. Based on the reported results, the SDASL method declines the FATS attack accuracy by 30% after 13 days of the model training which means a 70% privacy rate for the home. The achieved privacy protection is significantly lower than the aforementioned statistical-based schemes, but it resolved the need for injection delays. Energy-wise, 13 fake packets have been injected per actual data packet, meaning the energy cost of the SDASL method is thirteen times more than an unprotected home.

Table 2 provides a summary of the studied privacy protection methods for smart homes, highlighting their benefits and drawbacks.

Table 2. Smart homes privacy protection methods

Method/scheme	Benefit	Drawback
ConstRate	- Nearly perfect privacy rate	- High latency - Extreme energy overhead
ProbRate	- Nearly perfect privacy rate	- Above average latency - High energy overhead
FitProbRate	- Nearly perfect privacy rate	- Moderate latency - High energy overhead
Events Behavioral Semantics Analysis	- Average privacy rate	- Heavily Dependent on the accuracy rate of event prediction
SDASL	- Above average privacy rate - Delay free	- Limited privacy rate - High energy overhead

5 Discussion

The FATS attack efficiently enables attackers to discover private information about residents' in-home activities and pushes unwanted repercussions on users. The essential characteristics of this attack can be summarized as follows:

- Intrusion detection systems fail to detect this attack because of its passive approach.
- Encryption solutions are not effective in countering this attack since it targets the contextual aspects of the data packets.
- Attack infers in-home activities based on minimally provided inputs such as signals fingerprints and communications timings.

Since the attack is undetectable in real time, a protection solution should be proactive and conceal the true event patterns in network's traffic; it seems reasonable to assume that the attack's malfunctioning pattern recognition sabotages the performance of its subsequent phases and prevents the precise detection of actual events. To this aim, techniques such as signals temporal manipulation and fake packet injections effectively alter the traffic's patterns. Nonetheless, they might affect the system response time and energy overhead.

The home privacy rate is in a reverse relationship with the attack's accuracy; hence, declining the attack's correctness rate equals to increment of the privacy rate. Primary metrics to compute the attack's correctness rate are the Event Detection Rate (EDR) and the True Positive Rate (TPR) [8]. The EDR refers to the proportion of correctly detected in-home events from all actual activities. e.g., identifying 75 actual events in a home where 100 activities occurred produces an EDR of 75%. In addition, the TPR represents the report's correctness percentage. e.g., In 100 labelled events by the attack, if 60 items are false, the TPR would be 40%. Finally, the attack accuracy is the multiplication product of the EDR and TPR, which is 33.75% for the above example values. Secondly, concealing the actual activity patterns via the fake packet injections due to the randomness of the injections makes the number of injected dummy packets uncontrollable. The FVR is a variable that refers to the ratio of the fake packets to the actual ones, and in the reviewed protection methods always has been a positive number, meaning the number of fake injections was multiple times the actual messages. Given that the consumed energy for both types of packets is equal due to their identicality, the probability of massive energy overhead in the system is evident; this issue undermines the affordability of such solutions. Moreover, concealing the actual activities and traffic patterns demands the injection of numerous fake packets; the number of these injections is not deterministic to comply with the necessity of randomness in the protection procedures. Results have shown that to provide an adequate privacy rate, the number of fake packets is multiple times more than the actual packets. In this regard, the notation of FVR refers to the ratio of the dummy packets to the actual ones. Since transmitting both types of data packets equally consume energy resources, the system's energy overhead can be massive, which undermines the affordability of the solutions. Table 3 displays the state of attack's performance metrics for every protection method.

Table 3. FATS attack performance metrics countering protection methods

Method/scheme	EDR	TPR	Accuracy	FVR
ConstRate	Low	High	Near Zero	Very High
ProbRate	Low	High	Near Zero	Very High
FitProbRate	Low	High	Near Zero	Very High
Events Behavioral Semantics Analysis	Average	High	Moderate	High
SDASL	Average	High	Below Average	High

6 Conclusion and Future Work

Smart homes are vulnerable to wireless snooping-based side-channel attacks; therefore, protecting residents' personal information via a robust privacy-preserving mechanism is important to prevent potential data breach incidents. The FATS attack effectively exploits the home's wireless traffic to infer the in-home activities passively. The following are the results of a review of the current FATS protection methods:

- Obfuscation of the home network's wireless traffic is the most common method for confounding the attack algorithm, which reduces the Event Detection Rate (EDR) and, as a result, the attack's precision. This objective is achieved through modifying wireless traffic patterns. Delay in event reporting and injection of fake packets are employed for this purpose.
- Randomness is vital for obfuscating signal traffic; this feature makes it difficult for an attacker to determine the actual correlations between transmitted packets.
- Delay-based approaches are inadequate for delay-sensitive home services, and their imposed latency degrades the service quality of the system.
- Utilizing privacy-preserving measures increases the energy consumption of a home. To achieve a secure privacy rate, the number of false packets should be several times greater than the number of true packets; this value is variable due to randomization. Consequently, this issue may result in a significant energy expenditure.
- The established trade-off between privacy rate, system delay, and energy usage provides a decisive metric for assessing the viability of a privacy-preserving method.

To create a trustworthy, practical, efficient, and effective privacy protection solution for smart homes against the threat of the FATS attack, it is worth taking into account the following elements:

- Increasing the level of privacy should not result in a delay in the system's communications or response time.
- The energy cost associated with privacy enhancement must be viable, optimized, and reasonable.
- In addition to reducing the EDR, the TPR is another aspect of the attack's accuracy that can be targeted by security mechanisms. Decreasing the TPR effectively reduces the attack's accuracy rate.

As Fig. 8(a) depicts, in a normal condition, improving the privacy rate leads to an increment in the system's latency, energy demand, or both. On the contrary, Fig. 8(b) illustrates the desired relationship between factors of the PLE triad (privacy rate, system latency, and method energy consumption). Based on the reviews and analysis provided in this article, this trade-off should be the focus of future efforts to develop energy-efficient zero-delay privacy protection methods.

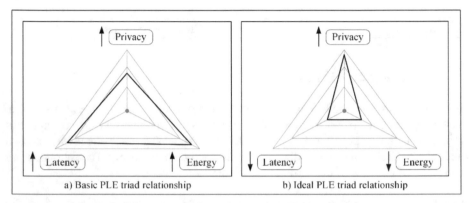

a) Basic PLE triad relationship   b) Ideal PLE triad relationship

Fig. 8. Trade-off between privacy, latency and energy

Acknowledgment. This research was financially supported by the Research Excellence Consortium in IoT Security fund from Ministry of Higher Education Malaysia. The research grant number: JPT(BKPI)1000/016/018/25(49).

References

1. Xu, W., et al.: The design, implementation, and deployment of a smart lighting system for smart buildings. IEEE Internet Things J. **6**(4), 7266–7281 (2019)
2. Haney, J.M., Furman, S.M., Acar, Y.: Smart home security and privacy mitigations: consumer perceptions, practices, and challenges. In: Moallem, A. (ed.) HCII 2020. LNCS, vol. 12210, pp. 393–411. Springer, Cham (2020). https://doi.org/10.1007/978-3-030-50309-3_26
3. Tabassum, M., Kosinski, T., Lipford, H.R.: "I don't own the data": end user perceptions of smart home device data practices and risks. In: Fifteenth Symposium on Usable Privacy and Security (SOUPS 2019) (2019)
4. Dasgupta, A., Gill, A.Q., Hussain, F.: Privacy of IoT-enabled smart home systems. In: Internet of Things (IoT) for Automated and Smart Applications, p. 9. IntechOpen, London (2019)
5. Abrishamchi, M.N., et al.: A probability based hybrid energy-efficient privacy preserving scheme to encounter with wireless traffic snooping in smart home (2017)
6. Naru, E.R., Saini, H., Sharma, M.: A recent review on lightweight cryptography in IoT. In: 2017 International Conference on I-SMAC (IoT in Social, Mobile, Analytics and Cloud) (I-SMAC). IEEE (2017)
7. Abrishamchi, M.A.N., et al.: Side channel attacks on smart home systems: a short overview. In: IECON 2017–43rd Annual Conference of the IEEE Industrial Electronics Society. IEEE (2017)

8. Srinivasan, V., Stankovic, J., Whitehouse, K.: Protecting your daily in-home activity information from a wireless snooping attack. In: Proceedings of the 10th International Conference on Ubiquitous Computing (2008)

9. Noto, M., Sato, H.: A method for the shortest path search by extended Dijkstra algorithm. In: SMC 2000 Conference Proceedings. 2000 IEEE International Conference on Systems, Man and Cybernetics. 'Cybernetics Evolving to Systems, Humans, Organizations, and Their Complex Interactions'. IEEE (2000)

10. Saeed, N., et al.: A survey on multidimensional scaling. ACM Comput. Surv. (CSUR) **51**(3), 1–25 (2018)

11. Teknomo, K.: K-means clustering tutorial. Medicine **100**(4), 3 (2006)

12. Roughgarden, T., Cs261: A second course in algorithms, lecture# 5: Minimum-cost bipartite matching (2016)

13. Nassiri Abrishamchi, M.A., et al.: Smart home privacy protection methods against a passive wireless snooping side-channel attack. Sensors **22**(21), 8564 (2022)

14. Alomair, B., et al.: Toward a statistical framework for source anonymity in sensor networks. IEEE Trans. Mob. Comput. **12**(2), 248–260 (2011)

15. Zou, Y., et al.: A survey on wireless security: technical challenges, recent advances, and future trends. Proc. IEEE **104**(9), 1727–1765 (2016)

16. Jeba, S., Paramasivan, B.: False data injection attack and its countermeasures in wireless sensor networks. Eur. J. Sci. Res. **82**(2), 248–257 (2012)

17. Yang, Y., et al.: Towards statistically strong source anonymity for sensor networks. ACM Trans. Sens. Netw. (TOSN) **9**(3), 1–23 (2013)

18. Park, H., Park, T., Son, S.H.: A comparative study of privacy protection methods for smart home environments. Int. J. Smart Home **7**, 85–94 (2013)

19. He, J., et al.: An adaptive privacy protection method for smart home environments using supervised learning. Future Internet **9**(1), 7 (2017)

Development of Graph-Based Knowledge on Ransomware Attacks Using Twitter Data

Abdulrahman Mohammed Aqel Assaggaf[1], Bander Ali Al-Rimy[1],
Noor Lees Ismail[2(✉)], and Abdulaziz Al-Nahari[3]

[1] Faculty of Computing, Universiti Teknologi Malaysia, 81310 Johor Bahru, Malaysia
`m.aqel@graduate.utm.my, bander@utm.my`
[2] Faculty of Business and Technology, UNITAR International University, 47301 Petaling Jaya,
Selangor, Malaysia
`Lees@unitar.my`
[3] UNITAR Graduate School, UNITAR International University, 47301 Petaling Jaya, Selangor,
Malaysia
`abdulaziz.yahya@unitar.my`

Abstract. Ransomware is constantly being developed on underground market-places, and spreads through Internet, causing damage to individuals' and businesses' data. The purpose of this study is to investigate the current issue related to knowledge graphs on ransomware attacks using Twitter data. To Construct a knowledge graph from informal text, three steps need to be followed. Namely, data collection and cleaning, entity extraction, and relation extraction. Although Natural Language Processing techniques are widely used for text representation and modeling, there exist some limitations related to the lack of a dedicated Named Entity recognizer for extracting Ransomware-related entities from unstructured data such as text. Therefore, this article relies on using the ontology approach to construct a ransomware knowledge graph from unstructured data. An improvement to the ontology is done to make it fit the ransomware attack representation based on data captured from the tweets. The Knowledge Graph was developed by extracting relations between entities. In the end, the accuracy of the Knowledge Graph was evaluated using the formal method.

Keywords: Ransomware ontology · Twitter · Knowledge graph · NER

1 Introduction

A malware ontology may aid in the development of models capable of detecting and tracking assaults from their inception through their conclusion. An ontology serves as the blueprint for a domain, encompassing important classes and their associated attributes. Class restrictions are used to specify the class's scope, which is subsequently inherited by the instances. As a result, it enables the aggregation, representation, and sharing of threat information that would be difficult to recreate, reuse, and evaluate on a broad scale otherwise. Humans and software agents alike may use an ontology to deduce the structure of data included in a paper, a report, a blog, a tweet, or any other structured,

© The Author(s), under exclusive license to Springer Nature Singapore Pte Ltd. 2023
Y. B. Wah et al. (Eds.): DaSET 2022, LNDECT 165, pp. 168–183, 2023.
https://doi.org/10.1007/978-981-99-0741-0_12

semi-structured, or unstructured information source [1]. An ontology establishes a shared language for scholars working in an area. It contains machine-readable definitions of the domain's fundamental concepts and their relationships [2]. In the first quarter of 2016, Symantec researchers saw an average of nearly 4,000 ransomware attacks each day, up 300% from the same period in 2015. Moreover, in a few simple steps, anyone may create malware with a customs form, distribution area, route, and time zone that is called Ransomware-as-a-Service [3]. Therefore, the existing Malware ontology cannot fit ransomware types because of the immense diversity of ransomware and the countenance of new kinds. Twitter can be a great asset in collecting data regarding vulnerabilities because individual users, usually ethical hackers, report any newly discovered malware using Twitter [4]. However, using tweets for analysis can be challenging because the social media difficulties presently being tackled in the Natural language processing community is determining and grouping tweets or posts that describe a particular event. Thus, extracting data from Twitter may encounter one of these issues: using short contexts to keep the tweet short, conversing in an informal language that may be considered noisy at times, or using non-standard abbreviations and capitalization. Additionally, using hashtags complicates the process of information extraction from social networks [5].

2 Problem Background and Related Work

Nowadays, ransomware attacks spread between people's devices; big companies also face malware attacks [6]. Therefore, it is noticed that many Twitter accounts to tweet about ransomware attacks and provide information about these attacks. So, representing that knowledge adequately might help the cyber security specialist find a way to avoid those attacks. Meanwhile, to develop a proper knowledge graph based on data taken from Twitter, the data needs to be processed to take off the noise and extract the feature from that text. This can be challenging to analyze unstructured text that humans have written. This research will adopt a ransomware ontology to construct a Knowledge Graph on ransomware attacks using collected data from Twitter after extracting and selecting the relevant features from that data and validating the knowledge graph with experts.

2.1 Knowledge Graph

Knowledge graphs, semantic networks with a directed graph structure, have been widely employed to improve Google, Baidu, and Yahoo's search engines. Each node represents an entity in a graph, and each edge indicates the connection between two entities. The diagram illustrates the information network and semantic relationships in a domain by visualizing nodes and linkages between nodes. It has also been used to evaluate and show the citation map and other information links in academic databases such as Web of Science and Wikidata. Furthermore, several researchers, such as DBPedia, OpenIE (Open information extraction), and NELL (Never-ending language learning), strive to extract structured information from the unstructured web to construct a knowledge base and then transfer the text information into a network [7]. Knowledge Graph in Malware is an interesting topic because of the wild spread; there are many types of malware such as Ransomware, spyware, APT, and Trojans. Most of the research was based on the famous

Undercoffer Knowledge Graph for Intrusion detection, which has been remodeled to suit the malware [7]. Figure 1 shows the Undercoffer framework.

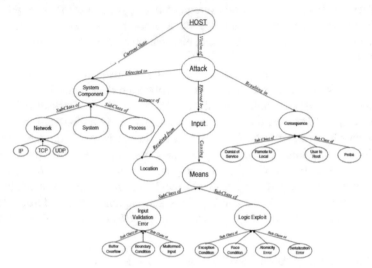

Fig. 1.   Undercoffer Ontology Retrieved From [7]

[8] developed something called TINKER, a knowledge graph constructed by hand that retrieves information from unstructured threat data. Dutta TINKER is based on RDF triples that describe entities and relations from unstructured data. TINKER was built on 83 threat reports published between 2006–2021, and it contains 3000 triples. The Ontology of TINKER was constructed on three classes that primarily describe malware behaviour Malware, Vulnerability, and Indicator. The purpose of the research was to create and train their own NER and RE models. An ensemble of machine learning and text-mining algorithms is required to automate the threat report annotation process.

In another study on Malware KG, the data source was from the After-Action Reports (AARs) because it provides an incisive analysis of cyber-incidents [9]. Piplai extracted the entities by building a customized entity recognizer called 'Malware Entity Extractor' (MEE). The MEE model has been trained to predict cybersecurity entities in an AAR based on annotated sections of cybersecurity text. Then the Relation Extractor (RelExt) indicates how the MEE extracted entities are related [10]. Figure 2 below describes the System Architecture for Piplai.

Ransomware attacks might vary depending on the identity of the attackers and the nature of the assault [11]. WannaCry and other viruses developed by North Korea. Researchers discovered a connection between the WannaCry ransomware and software produced by the Lazarus Group, a group of North Korean hackers. Neel Mehta, a Google security researcher, initially discovered the standard code. Symantec and Kaspersky found code commonality between WannaCry and malware created by the Lazarus Group [12]. A study in UTM done by [16] has successfully found a classification for ransomware entity for informal text with supervised learning. Ariffini proposed a scheme that can extract ransomware entities from casual text Fig. 3 shows that.

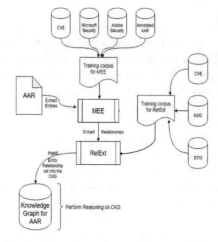

Fig. 2. Piplai F/W retrieved from [10]

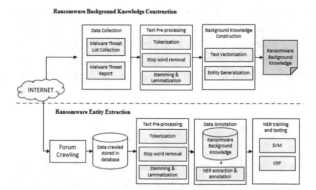

Fig. 3. Ariffini scheme retrieved from [13]

2.2 Name Entity Recognition

Name Entity Recognition (NER) can be categorized as part of NLP, NER works by finding and recognizing named entities included in unstructured text into common categories such as human names, places, organizations, time expressions, quantities, monetary amounts, percentages, and codes. A great example of NER models that are used in our case is CyNER model which is a Python module for cybersecurity named entity recognition that is open source (NER). CyNER combines transformer-based models for extracting entities pertaining to cybersecurity, heuristics for extracting various signs of compromise, and publically accessible NER models for generic entity types. User-friendly models trained on a varied corpus are made available. Events are categorized as classes in prior research - MALOnt2.0 [14] and MALOnt (Rastogi et al., 2020) - and are used to extract a broad variety of malware attack data from a threat intelligence corpus. The user may mix forecasts from numerous distinct methods to meet their own requirements.

2.3 Machine Learning

When knowledge graphs and machine learning are combined, new possibilities emerge. KG can supplement training data when there is a data shortage. Explaining predictions made by ML systems is one of the most challenging tasks in ML. By mapping explanations to appropriate nodes in the graph and describing the decision-making process, knowledge graphs can assist in overcoming this problem. Data classification using Deep Learning is the workhorse of the Machine Learning revolution. We construct subsets of data points belonging to the same class by classifying data. This relationship did not exist before the classification, but it can now build a Knowledge Graph.

The potent combination of graphs and machine learning has many applications. Working on two complex development components is required to gain access to data and identify the classes that will lead to the desired conclusion. The first is primarily an organizational, legal, and frequently an ethical concern, whereas the second necessitates domain understanding. While this was traditionally solely offered by subject-matter specialists before the ML revolution, now ML systems can support this activity, lowering entry barriers.

Creating a Knowledge Graph is a substantial undertaking requiring data access, domain and Machine Learning skills, and the necessary technical infrastructure. However, once these requirements for one Knowledge Graph have been established, additional can be constructed for different domains and use cases. Knowledge Graphs are a disruptive approach to extracting value from current unstructured data since new insights can be discovered.

3 Methodology and Design

Based on previous knowledge graph ontologies, I propose a framework for developing Graph-based Knowledge on Ransomware Attacks Using Social Media Data. The proposed methodology has four steps: social media data collection, entity extraction, relation extraction, and building a knowledge graph, which can be done in 4 phases as depicted in Fig. 4. With some changes, the ontology will be adopted from [13], but this paper will focus on ransomware attacks using social media data.

3.1 Data Collection

The data collection phase is to develop a crawler to collect tweets from Twitter related to ransomware attacks. The data has been collected by either contacting Twitter APIs or scraping the tweets because of the limitation in Twitter API based on a group of keywords related to ransomware. While collecting data, Twitter API assisted with Tweepy (python library to contact Twitter API) and twint (python library to scrape tweets without getting Twitter API) are used. The collected data were cleaned by WordNet, a huge lexical database for English [15]. The outcome of this stage was a large dataset containing tweets related to ransomware which will be used to create and train the NER model to ransomware extract entities.

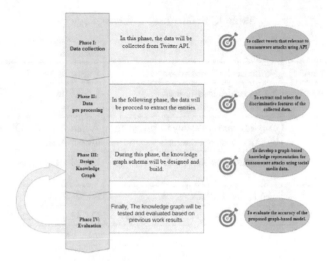

Fig. 4. F/W

3.2 Data Preprocessing and Entity Extraction

This phase aims to get the NER model to extract entities from the text. These two steps in this phase are data cleaning & pre-proccing and entity extraction, where the model is built and trained to pull the ransomware entities from the informal text.

3.2.1 Data Cleaning and Preprocessing

In this stage, we aimed to clean the text from the noise and preprocess the text. Text preprocessing is performed to prepare the text data for model development. It is the first stage in all NLP initiatives. NLP Text preprocessing is a technique for cleaning text to make it suitable for feeding to models. Textual noise includes emojis, punctuation, and other case variations. All these sounds are useless to machines; thus, they must be eliminated. This step's outcome will help us do accurate text analysis and make proper conclusions for our case difficulties with cleansed data. Therefore, text preparation is essential for machine learning and model building.

3.2.2 Entity Extraction

In the preprocessing activity, we will extract the ransomware entity based on [16] ontology from the collected data in the following phase. The training begins once the data has been fully tagged and translated into a machine-readable format. Ariffini used three classification techniques to train the NER model Conditional Random Field (CRF) [17], Support Vector Machine (SVM) [18] and Naive Bayes [19].

The adopted ontology will have some changes to suit the source of data. Table 1 below shows the entities on the ontology.

Table 1. Entities used in the ontology

Entity	Description	Example
Ransomware name	Name of the ransomware	Intermittent
Ransomware family	Name of the family that ransomware belongs to	LockFile
Target	The infected company or program by ransomware attacks	Windows, Mac OS, GitHub, XX Government
Technique	The technique that ransomware uses	Novel technique
Date	The Date of attack	2020 MAY
Damages	Short description of the damages	Data breach cause 10M lost

3.2.3 Design Knowledge Graph

During this phase, the extracted entity from the cleaned data will be sent to RRelExt, a proper function to Extract the ransomware entity relation. This function will be built to extract basic relations between entities since we have only a few ransomware entities, we can figure out those entities and link them without ML. Therefore, we move to develop the knowledge graph activity, which uses Python to build a knowledge graph using the KGEMs library and PyKEEN. These libraries are famous for Knowledge graph, and it shows effective and high accuracy while using them [20]. Figure 5 below shows the schema of the expected KG.

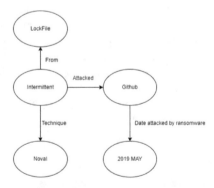

Fig. 5. Expected schema of KG

3.2.4 Evaluation

In this section, the project will discuss the approach for performance measurement of the generated Knowledge Graph in terms of accuracy. The project will follow [21] Technical Report on Knowledge Graph Accuracy Evaluation to create a KG form. Geo stated, "KG accuracy can be defined as the percentage of triples in the KG being correct". Also, he is claiming that triple is considered accurate if the matching connection is congruent with the underlying reality. Which, in my opinion, is correct? Geo proposes an evaluation model; this model starts with Entity Identification which begins by providing a small set of information regarding the subject of this task. Subsequent Relation Validation, this method requires annotators to conduct cross-source verification, which entails locating evidence of the subject-object connection in several sources (if available) and ensuring that the information about the fact is accurate and complete. Then need to conduct these steps:

- Step 1: Sample Collector selects a small batch of samples from KG using a specific sampling design D.
- Step 2: Sample Pool contains all samples drawn so far and asks for manual annotations when new models are available.
- Step 3: Given accumulated human annotations and the sampling design D, the Estimation component computes an unbiased estimation of KG accuracy and its associated MoE.
- Step 4: Quality Control checks whether the evaluation result satisfies the user-required confidence interval and MoE. If yes, stop and report; otherwise, loop back to Step 1 (Fig. 6).

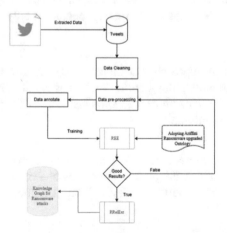

Fig. 6. System design

3.3 Data Collection

In his phase, we aimed to achieve the objective (a) of collecting tweets relevant to ransomware attacks using API. This stage could be done in two ways: using Tweepy (python library to contact Twitter API). Tweepy is a Python library that enables developers to access Twitter's publicly accessible APIs seamlessly and transparently. Without Tweepy, the user would be responsible for various low-level aspects, including HTTP requests, rate restriction, authentication, serialization, etc. Tweepy manages all of this chaos on the user's behalf, making the program error-prone. Tweepy is a simple Python library that allows developers to interact with the Twitter API. Keep in mind, however, that Twitter imposes a rate restriction on the number of API queries. Twitter permits a maximum of 900 queries every 15 min; anything over that is considered an error.

3.4 Data Pre-processing and Name Entity Recognition

This step seeks to extract entities from the text using the NER model. This phase consists of two steps: data cleaning and preprocessing and entity extraction, where the model is constructed and trained to extract ransomware entities from unstructured text.

Date Pre-Processing
Noise Removal and Remove Punctuations

Textual data may contain extraneous letters and punctuation, such as URLs, HTML elements, non-ASCII characters, and other special characters (symbols, emojis, and other graphic characters). The punctuation adds noise to the phrase, creating uncertainty while training the model.

Replace the Typos, Slang, Acronyms, or Informal Abbreviations
Since we collected data from social media, especially Twitter, because of letters limitation, we have to do extra cleaning because people usually use slang or informal abbreviations.

Tokenization
Tokenization is a typical approach that divides a phrase into tokens, which may be characters, words, phrases, symbols, or other vital items. By dividing phrases into smaller chunks, it would be easier to examine the words inside a sentence and future steps in the NLP pipeline, such as stemming.

Remove Stop Words
Stop words are common words in any language that appear often but contribute nothing to the overall meaning of the phrase. Examples of stop words are "a", "around", "above", "across", "after", "afterward", "again", etc. Traditionally, we could eliminate them during text preparation.

Stemming and Lemmatization
Stemming is the process of extracting a root word - recognizing a standard stem among distinct forms (such as single and plural noun forms) of a word; for instance, the terms "gardening", "gardener", and "gardens" all share the same stem, garden. Stemming eliminates suffixes from words to combine words with similar meanings into their primary stem.

Lemmatization identifies those two words have the same root despite their apparent distinctions. The standard lemma of the words am, are, and is be, whereas the lemma of the phrases dinner and dinners is dinner. By lemmatizing each of these forms to the same lemma, it will be possible to locate all instances of terms in Russian, such as Moscow.

Therefore, the lemmatized version of a statement such as He is reading detective tales would be He be read detective story.

Name Entity Recogntion
In this step for NER model build we going to use spaCy which is Named Entity Recognition, spaCy is a free, open-source toolkit for sophisticated Natural Language Processing (NLP) in Python. It is meant for production usage and enables the development of systems that analyse and "understand" massive amounts of text. It may be used to construct information extraction and natural language comprehension systems, as well as to prepare text for deep learning.

Text Annotation and Training
In this activity, we anointed the text to train the spaCy model to get better results while extracting the entities. The entities are mentioned in the methodology with a brief explaining. The process of this activity can be summarized in five steps.

I.   Annotate the data for model training.
II. Convert the annotated data to an object of type spaCy bin.
III. Create the configuration file using the spaCy website.
IV. The model is trained using the command line.
V.   Load and test the previously stored model.

For the first step, we used spaCy NER annotation tool, which is an online tool to annotate the text figure.

Evolution
The trained Spacy model will be assessed using the precision, recall, and F1-score for each class after training.

Precision, Recall, Accuracy, and F1-Score serve as the primary evaluative markers for NER training outcomes. These indicators are characterized by their true positives, false positives, and false negatives. Following is the definition:

- True Positive (TP): This is a collection of class members appropriately classified as belonging to a particular class.
- False Positive (FP): A collection of class members incorrectly identified as belonging to a particular class.
- False Negative (FN): This is a collection of things that are not labeled for any class by the system but belong to some class.
- True Negative (TN): events are not detected when the circumstances are absent.

So, we will use those indicators to get Precision, Recall, Accuracy, and F1-Score we can calculate them as follow:

$$\text{Accuracy} = TP + TN/(TP + TN + FP + FN)$$

$$\text{Precision} = TP/(TP + FP)$$

$$Recall = TP/(TP + FN)$$

$$F1 = 2(Precision)(Recall)/(Precision + Recall)$$

The closer the score is to 1, the more accurate the classification model is. High recall and low precision indicate that most positive cases are accurately identified (low FN), but there are many false positives. Low recall but high accuracy means that many positive cases are missed (high FN), but the ones predicted as positive are truly accurate (low FP).

Design Knowledge Graph

To construct The Knowledge Graph, we must build the Relation Ransomware Entity Extractor function. To develop the function, we have to state the entity pairs and the relations between them in one condition which is having at least 2 entities in single tweet. Table 2 display the entity pairs and the relation between them.

Table 2. Entity relations distribution

Ransomware name	Infect	Target
Target	InfectedOn	Date
Ransomware name	BelongTo	Ransomware family
Ransomware name	Uses	Technique
Target	Damages	1TB of data

So, in the begging we stored all entities in single csv file with the date of the tweet so date entity will be in all nodes but we will not use it unless we have a target entity so we can link them together. To start developing the Knowledge Graph we used Neo4j which is an online database to simplify the storage and administration of confidential information by allowing you to monitor which systems, apps, and individuals have access to it. The graph data architecture facilitates the visualization of personal data and enables data analysis and pattern recognition. Figure 7 shows the basic schema for the Knowledge graph in Neoj4.

Fig. 7. Neoj4 KG schema

Knowledge Graph Evaluation

In the Evaluation stage, we cannot evaluate the accuracy of the Knowledge Graph by our self we have to handle it with experts to evaluate it for us. However, we proposed an approach to evaluate the Knowledge graph will be as follow. Figure 8 displays the flow chart of the evaluation process.

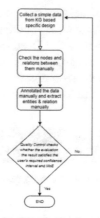

Fig. 8. Evaluation flowchart

Data Collection

To achieve the first goal, we collected Data using Twitter API we faced many problems, but we managed to collect up to 1M tweets during that process we faced many problems like the quality and the pattern of the tweets. So, this stage started with early filtration by searching queries. Moreover, we had to optimize the dataset, so we choose the most relevant 85K tweets to our topic (Fig. 9).

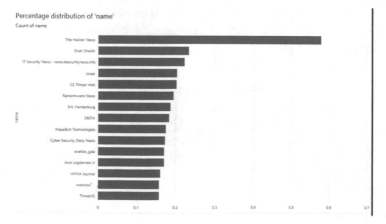

Fig. 9. The most accounts that we extracted the data from

Data Preprocessing and Entity Extraction

The purpose of this stage is to train the NER model to successfully extract entities from text. Phase two entails building and training a model to extract the ransomware entities from unstructured text, and it consists of two sub-steps: data cleaning and preprocessing, and entity extraction.

Data Cleaning and Preprocessing

In this phase, we sought to remove background noise from the text and do preliminary processing. To get the text data ready for model building, text preprocessing is carried out. All NLP projects begin with this phase. NLP The term "text preprocessing" refers to the process of preparing raw text for use as input to models. Textual noise consists of things like emojis, punctuation, and capitalization mistakes. Because of the ineffectiveness of these noises when produced by machines, they must be eradicated (Figs. 10 and 11).

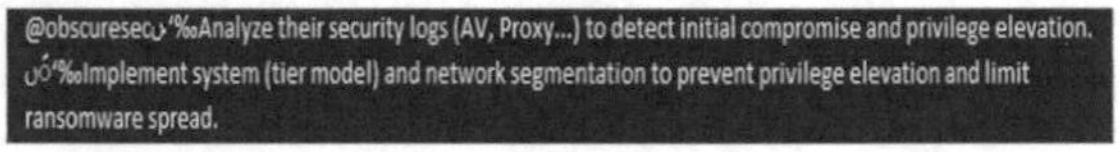
@obscuresecꞌ‰Analyze their security logs (AV, Proxy...) to detect initial compromise and privilege elevation. ‘‰Implement system (tier model) and network segmentation to prevent privilege elevation and limit ransomware spread.

Fig. 10. Text before cleaning and preprocessing

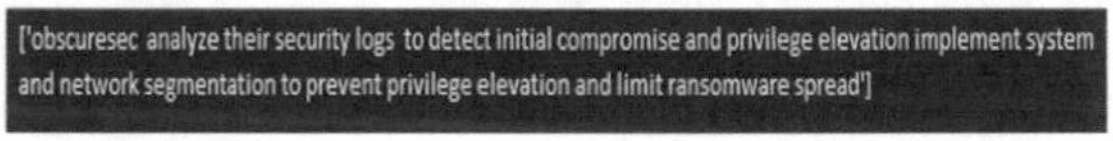
['obscuresec analyze their security logs to detect initial compromise and privilege elevation implement system and network segmentation to prevent privilege elevation and limit ransomware spread']

Fig. 11. Text after cleaning and preprocessing

We can notice the big difference between the text before and after cleaning. Text before cleaning can cause many problems regarding training the model because there is some character that the machine cannot understand. Moreover, some sentence that is written in the wrong format.

Name Entity Recognition

To evaluate the NER model, we took 100 tweets that passed to the model and evaluate it manually that 100 tweets we can extract 92 entities. Table 3 shows the confusion matrix table for the predicted entitles.

Table 3. Confusion matrix table

N = 91	Predicted 1	Predicted 0	Total
Actual 1	TP = 46	FN = 11	57
Actual 0	FP = 20	TN = 14	34
Total	66	25	91

From Table 3, it can be observed that the true positive is 51%, the false negative is 12%, the true negative is 15% and the false positive is 22%. The values are used to

Table 4. Evaluation performance

Measure	Equation	Value
Sensitivity	$TPR = TP/(TP + FN)$	0.8070
Specisicity	$SPC = TN/(FP + TN)$	0.4118
Precision	$PRC = TP/(TP + FP)$	0.6970
Negative predictive value	$NPV = TN/(TN + FN)$	0.5600
False positive rate	$TPR = FP/(FP + TN)$	0.5882
False discovery rate	$FDR = FP/(FP + TN)$	0.3030
False negative rate	$TPR = FN/(TP + FN)$	0.1930
Accuracy	$Acc = (TP + TN)/(P + N)$	
F1 Score	$TPR = 2TP/(2TP + FP + FN)$	0.6953

calculate precision, recall, and f1-score as shown in Fig. 12. The value of the f1- score is 74% which means the clusters of the predicted entities are quite acceptable (Table 4).

Knowledge Graph

We used neoj4 to construct the Knowledge graph. In the beginning, we had to get the best cases that can be displayed. So, in the begging, we had to do some manual filtering and combine all the entities in one csv file. Moreover, we combine the related entities to some events so it can be linked once. We imported 615 rows of data, which are entities for each tweet, and replaced the blank entities with unknowns after creating the nodes and their relation as seen in the table above. Duplicate nodes were automatically purged by Neoj4. Finally, we had 536 nodes and 1437 connections.

4 Conclusion and Future Work

4.1 Conclusion

The first object achieved in this research is the collection of 1 million tweets from Twitter API using some queries that can support our case. The second goal was completed when we finally got our model ready to extract the required entities from preprocessing and cleaned informal text. The model got 74% which means it is quite acceptable and useable. Moving on to the third objective, which was achieved by having the final knowledge graph describing some ransomware attacks reported on Twitter, we had to manually map the relations between entities since we already knew the connections and there was no need to build an extractor. Although, Neoj4 helped us develop our Knowledge Graph after the manual group of all the entities in one excel file, each row in that file illustrates a ransomware attack. After removing the duplicate events, we got up to 536 nodes and 1437 relations. Finally, the last objective is evolution; we proposed an evaluation schema that can be used to evaluate the Knowledge Graph; since we are not experts in that field, we cannot assess the output ourselves.

4.2 Future Work

Future researchers can enhance the quality of the obtained dataset annotation for the purpose of advancing the state of the art. Future researchers can annotate tweet-related attributes using automated methodologies as opposed to hand annotation, resulting in a more accurate and exact annotation. The next improvement is that future researchers can use a different platform or forms that are specified for ransomware news and update as in my experience Twitter alone might not be sufficient for constructing a comprehensive knowledge-base graph. Data from other platforms like Linkedin and Facebook can be used and correlated with the Twitter data to improve the representativeness of the model.

Acknowledgment. The authors would like to thank UNITAR for the support of the publication of this paper. Additionally, This project was funded by UTM Transdiciplinary Research Grant number PY/2018/03477. The authors would like to than UTM for the support provided.

References

1. Rastogi, N., Dutta, S., Zaki, M.J., Gittens, A., Aggarwal, C.: MALOnt: an ontology for malware threat intelligence. In: Wang, G., Ciptadi, A., Ahmadzadeh, A. (eds.) MLHat 2020. CCIS, vol. 1271, pp. 28–44. Springer, Cham (2020). https://doi.org/10.1007/978-3-030-596 21-7_2
2. Noy, N.F., Mcguinness, D.L.: Ontology development 101: a guide to creating your first ontology (2001). www.unspsc.org
3. Olaimat, M.N., Maarof, M.A., Al-rimy, B.A.S.: Ransomware anti-analysis and evasion techniques: A survey and research directions. In: 2021 3rd International Cyber Resilience Conference (CRC), pp. 1–6. IEEE, January 2021
4. Mittal, S., Das, P.K., Mulwad, V., Joshi, A., Finin, T.: CyberTwitter: using Twitter to generate alerts for cybersecurity threats and vulnerabilities. In: Proceedings of the 2016 IEEE/ACM International Conference on Advances in Social Networks Analysis and Mining, ASONAM 2016, Nov. 2016, pp. 860–867 (2016). https://doi.org/10.1109/ASONAM.2016.7752338
5. Virmani, C., Pillai, A., Juneja, D.: Extracting information from social network using NLP (2017). http://www.ripublication.com
6. Maseer, Z.K., Yusof, R., Mostafa, S.A., Bahaman, N., Musa, O., Al-rimy, B.A.S.: DeepIoT. IDS: hybrid deep learning for enhancing IoT network intrusion detection. CMC-Comput. Mater. Contin. **69**(3), 3945–3966 (2021)
7. Undercoffer, J., Joshi, A., Pinkston, J.: Modeling computer attacks: an ontology for intrusion detection (2003)
8. Dutta, S., Rastogi, N., Yee, D., Gu, C., Ma, Q.: Malware Knowledge Graph Generation (2021). https://brat.nlplab.org/
9. Piplai, S. Mittal, A. Joshi, T. Finin, J. Holt, and R. Zak, "Creating Cybersecurity Knowledge Graphs from Malware after Action Reports," IEEE Access, vol. 8, pp. 211691–211703, 2020, doi: https://doi.org/10.1109/ACCESS.2020.3039234
10. Pingle, A., Piplai, A., Mittal, S., Joshi, A., Holt, J., Zak, R.: RelExt: relation extraction using deep learning approaches for cybersecurity knowledge graph improvement (2019)
11. Urooj, U., Maarof, M.A.B., Al-rimy, B.A.S.: A proposed adaptive pre-encryption crypto-ransomware early detection model. In: 2021 3rd International Cyber Resilience Conference (CRC), pp. 1–6. IEEE, January 2021

12. Ahmed, Y.A., Koçer, B., Huda, S., Al-rimy, B.A.S., Hassan, M.M.: A system call refinement-based enhanced minimum redundancy maximum relevance method for ransomware early detection. J. Netw. Comput. Appl. **167**, 102753 (2020)

13. Ariffini, N., Zainal, Maarof, A., Kassim, M.N.: Cyber Resilience Conference (CRC). IEEE, 2018 (2018)

14. Christian, R., Dutta, S., Park, Y., Rastogi, N.: An Ontology-driven, Dynamic Knowledge Graph for Android Malware; An Ontology-driven, Dynamic Knowledge Graph for Android Malware (2021). https://doi.org/10.1145/3460120

15. Miller, G.A., Beckwith, R., Fellbaum, C., Gross, D., Miller, K.J.: Introduction to wordnet: an on-line lexical database. Int. J. Lexicogr. **3**(4), 235–244 (1990). https://doi.org/10.1093/ijl/3.4.235

16. Ahmed, Y.A., et al.: A weighted minimum redundancy maximum relevance technique for ransomware early detection in industrial IoT. Sustainability **14**(3), 1231 (2022)

17. Tseng, H., Chang, P., Andrew, G., Jurafsky, D., Manning, C.: A Conditional Random Field Word Segmenter for Sighan Bakeoff 2005 (2005)

18. Awad, M., Khanna, R.: Support vector machines for classification. In: Efficient Learning Machines Theories, Concepts, and Applications for Engineers and System Designers, pp. 39–66. Apress Berkeley, CA (2015). https://doi.org/10.1007/978-1-4302-5990-9_3

19. Rish, R.I.: An Empirical Study of the Naïve Bayes Classifier Predicting conversion to psychosis in clinical high risk patients using resting-state functional MRI features View project Clinical Machine Learning based on Cardiorespiratory models and simulation View project An empirical study of the naive Bayes classifier (2021). https://www.researchgate.net/publication/228845263

20. Ali, M., et al.: PyKEEN 1.0: A Python Library for Training and Evaluating Knowledge Graph Embeddings (2021). http://jmlr.org/papers/v22/20-825.html

21. Gao, J., Li, X., Xu, Y.E., Sisman, B., Dong, X.L., Yang, J.: Efficient Knowledge Graph Accuracy Evaluation (Technical Report Version) *. Efficient Knowledge Graph Accuracy Evaluation. PVLDB, vol. 12, pp. xxxx-yyyy (2019). https://doi.org/10.14778/xxxxxxx.xxx xxxx

Big Data Analytics

BigMDHealth: Supporting Multidimensional Big Data Management and Analytics over Big Healthcare Data via Effective and Efficient Multidimensional Aggregate Queries over Key-Value Stores

Alfredo Cuzzocrea$^{(\boxtimes)}$

iDEA Lab, University of Calabria, 87036 Rende, Italy
alfredo.cuzzocrea@unical.it

Abstract. This paper introduces **BigMDHealth**, *a composite framework for supporting multidimensional big data management and analytics over big healthcare data*. **BigMDHealth** adheres to the innovative paradigm that predicates engrafting *multidimensionality* into big data analytics processes. At a proper algorithmic layer, **BigMDHealth** focuses the attention on the issue of *supporting multidimensional aggregate queries over key-value stores*, as a fundamental operator of the main multidimensional big data management and analytics paradigm predicated by the framework.

Keywords: Big Data · Healthcare Big Data · Big Data Management · Big Data Analytics · Multidimensional Big Data Management · Multidimensional Big Data Analytics

1 Introduction

Nowadays, *big data management and analytics* (e.g., [23–26]) is gaining momentum in big data research, due to the evident great impact these methodologies and techniques have in the context of real-life applications and systems, ranging from *smart cities* to *social networks*, from *graph analysis tools for e-science* to *bio-informatics*, and so forth.

In this scientific context, this paper introduces **BigMDHealth**, *a composite framework for supporting multidimensional big data management and analytics over big healthcare data*. **BigMDHealth** focuses the attention of the relevant issue of defining models, techniques and algorithms for supporting *multidimensional* big data management and analytics tasks over *big healthcare data*.

Big healthcare data are of relevant interest at now, as healthcare management and analytics is one of the most relevant societal challenges of the future. In fact, it is easy to understand how much heavy is the impact of this topic in next-generation societal dynamics.

© The Author(s), under exclusive license to Springer Nature Singapore Pte Ltd. 2023
Y. B. Wah et al. (Eds.): DaSET 2022, LNDECT 165, pp. 187–194, 2023.
https://doi.org/10.1007/978-981-99-0741-0_13

Inspired by this main motivation, **BigMDHealth** aims at achieving the well-understood big data management and analytics goal over big healthcare data via innovative *multidimensional paradigms*. Born in the wider umbrella of *OnLine Analytical Processing* (OLAP) and *Business Intelligence* (BI) scientific areas, according to these paradigms, the target data domain is modeled and analyzed on the basis of fortunate multidimensional metaphors founded on the principal *dimension, measure* and *level* concepts. Thanks to this nice abstraction, decision makers are allowed to support decision-making processed via multidimensional abstractions where, for instance, the *number* of COVID-19 infections (the singleton measure) can be analyzed *simultaneously* with respect to the *zone* (one of the dimensions), *time* (one of the dimensions), *other-disease* (one of the dimensions), *patient-age* (one of the dimensions), *patient-sociality* (one of the dimensions), and so forth.

Within the so-delineated scenario, **BigMDHealth** addresses specific research problems, with a special focus on algorithmic aspects. In particular, **BigMDHealth** focuses the attention on the issue of *supporting multidimensional aggregate queries over key-value stores*, as a fundamental operator of the main multidimensional big data management and analytics paradigm predicated by the framework.

The remaining part of this paper is organized as follows. In Sect. 2, we provide an outline of relevant related work. Section 3 focuses the attention on the anatomy and algorithmic aspects of **BigMDHealth**. Finally, Sect. 4 reports on conclusions and future work of our research.

2   Related Work

Several related work can be identified in active literature. Here, we provide an outline on some of them. In particular, our related work can be organized in two main areas. The first area focuses the attention on proposals that support multidimensional big data management and analytics (as general analytical frameworks). The second area, instead, focuses the attention on proposal that deal with the issue of supporting multidimensional aggregate query evaluation over key-values stores.

As regards the first related research area, in [19], authors acknowledge that big data analytics in challenging healthcare situations is currently receiving a lot of attention. *Fetal growth curves*, a typical example of big healthcare data, are used in prenatal medicine to identify potential issues with fetal growth early, predict the fate of the *pregnancy*, and treat abnormalities as soon as they arise. However, due to their lack of precision, the currently used curves and the associated diagnostic tools have come under fire. In the literature, fresh methods built on the concept of *personalized* development curves have been suggested. In this light, this study discusses the challenge of *creating individualized or customized fetal growth curves* using big data techniques. The idea of summarizing the enormous volumes of (input) big data through *multidimensional views*, on top of which well-known Data Mining techniques like *clustering* and *classification* are implemented, is introduced by the proposed framework. In general, this describes a *multidimensional mining strategy* geared toward challenging healthcare situations. Also, authors suggested a preliminary evaluation of the framework's efficacy.

Using *multidimensional AI tools*, [22] offers *CORE-BCD-mAI, a COmposite Framework for REpresenting, Querying, and Analyzing Big Clinical Data*. In the healthcare application scenario, the suggested framework integrates *Artificial Intelligence* (AI) and big data with the goal of accessing *large clinical datasets* for managing and decision-making reasons. The CORE-BCD-mAI framework's intrinsic anatomy and methodology supporting it are presented in depth by the authors. The work also lists pertinent research problems resulting from the CORE-BCD-mAI proposal as a whole.

As regards the second related research area, *LSM-tree* is frequently employed in key-value stores for huge data storage, but [6] claims that it suffers from write amplification brought on by frequent compaction operations. *Key-value separation*, which dissociates values from the LSM-tree and saves them in a separate value log, is an efficient approach to this issue. However, as previous key-value separation strategies concentrate on reducing write amplification but ignore the SSD's access characteristics, they perform poorly for *range queries*, especially for small key-value pairs. In order to improve range query performance while retaining reasonable update performance for workloads that require a lot of updates, authors suggest *FenceKV*, a framework. In order to accomplish efficient update and range query, FenceKV uses a new partition approach to map data to the storage space depending on the key-range. To reduce garbage collection costs and preserve sequential access for range queries, it also implements a *key-range garbage collection policy*. Results demonstrate that FenceKV may greatly enhance range query speed while retaining respectable update performance when compared to the existing designs of key-value separation. Authors compare FenceKV with modern key-value stores with varied workloads.

Finally, the authors offer *a highly effective method for processing aggregate queries for massive amounts of multidimensional data* in [13]. Many other types of multidimensional data, including sensor data, have recently been produced as a result of recent advancements in network technologies. In assessing such data, aggregation queries are crucial. Increasing data size may cause problems even though relational databases (RDBs) allow effective aggregation queries with indexes that speed up query processing. The use of a *distributed key-value store* (D-KVS), on the other hand, is essential to achieving scale-out performance for data insertion throughput. However, due to its inadequate support for indexes, searching multidimensional data occasionally necessitates *a full data scan*. The suggested approach combines an RDB and a D-KVS to utilize their benefits in addition to one another. Additionally, a novel approach is described in which the aggregated values for each grid are precalculated and the data are separated into a number of subsets termed *grids*. By minimizing the amount of scanned data, this method enhances query processing speed. By reducing the amount of scanned data, this method boosts query processing performance. By comparing the proposed method's performance to that of current cutting-edge techniques, the authors determined how efficient it is and demonstrated that it outperforms them in terms of query and insertion performance.

3   BigMDHealth: Anatomy and Algorithmic Aspects

Key-value stores (e.g., [1, 2]) is a specific representation scheme for big data (on the Cloud), where so-called *rows* are univocally identified by two attributes: *key* and *value*. Thanks to this easy representation model, big data can be easily partitioned over a target Cloud, simply via *horizontal data partitioning techniques* (e.g., [3]). Therefore, computing a query against a Cloud-distributed key-value store can fully take advantage from the classical *MapReduce* computational framework, thus achieving higher performance. Actual literature has devoted a lot of attention to the issue of effectively and efficiently supporting *query optimization issues over key-value stores* (e.g., [4–6]). Our proposed framework focuses the attention along the same line of research, by specifically considering multidimensional aggregate queries. Multidimensional aggregate queries apply an aggregation operator (like SUM, COUNT, MIN, MAX, etc.) to a given multidimensional data domain. Such class of queries are traditionally recognized as computational-expensive queries when high-dimensional, large-side datasets are considered (e.g., [7–9]). This problem gets worse when they are evaluated against big data repositories, as initially recognized by some proposals in the field (e.g., [10, 11]).

Executing multidimensional aggregate queries over big data is a really-challenging task, first because, due to its proper nature, key-value stores support *one-dimensional* indexing structures only, so that data access during the standard, naive evaluation of multidimensional aggregate queries over key-value stores become prohibitive. Given this main evidence, **BigMDHealth** aims at studying and devising *intelligent pre-computation techniques* over Clouds as to reduce the cost of evaluating multidimensional aggregate queries over key-value stores, thus significantly gaining into efficiency. Pre-computation techniques for query optimization over distributed settings has a good research tradition (e.g., [12]), but it is a novel research issue in multidimensional aggregate query evaluation over big data (e.g., [13]). Pre-computation is very beneficial for multidimensional aggregate queries, due to the fact that some *partial aggregations* can be pre-computed on local Cloud nodes, and then exploited at query time to speed-up evaluation of multidimensional aggregate queries that "contain" such (partial) aggregations. While this paradigm is rather clear in traditional distributed settings, applying it over Cloud-based key-value stores is still an open research issue, which deserves the efforts of the **BigMDHealth** framework.

Basically, the main approach addressed by **BigMDHealth** consists in:

- applying intelligent data partitioning techniques on the target big dataset, and distributing it over the reference Cloud;
- compute partial aggregations over so-partitioned big dataset;
- devising innovative *optimized* multidimensional aggregate query evaluation algorithms for executing the input query over the reference Cloud, by fully exploiting ad-hoc *metadata services* available in the reference Cloud.

It should be noted that all the previous three achievements are open research challenges for state-of-the-art big data research, and the final goal of **BigMDHealth** is that of providing effective and efficient solutions to these challenges, by also specializing achieved solutions to the target big healthcare data management and analytics application scenario.

Figure 1 depicts the conceptual framework proposed by the **BigMDHealth** framework, along with the main query evaluation workflow. As shown in Fig. 1, the **Big-MDHealth** framework comprises several intelligent big data management components for achieving the goal of supporting effective and efficient multidimensional aggregate

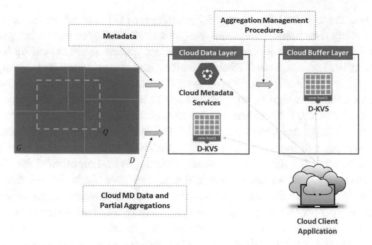

Fig. 1. BigMDHealth: conceptual framework.

query evaluation over key-value stores. First, the target big dataset is partitioned into several *big multidimensional data chunks*, according to a given *partitioning grid* defined by somewhat *partitioning criterion* (e.g., [14]). These chunks are then distributed over the reference Cloud, and still indexed and located via ad-hoc metadata Cloud services, which are implemented within specific API (*Application Programming Interface*) available at the reference Cloud platform (e.g., *Hadoop*, *Azure*, etc.). Cloud data are thus represented as key-value elements, stored within proper Cloud-based data stores, called D-KVS (*Distributed Key-Value Stores*) On top of these so-obtained chunks, in the local Cloud nodes, partial aggregations are computed and stored. It should be noted that, locally, partial aggregations are still computed on top of key-value Cloud data, hence specific approaches are necessary (e.g., [15, 16]). In turn, so-computed partial aggregations are stored in terms of key-value elements, and become officially part of the *Cloud loop* (i.e., they could be accessed at query time when partial aggregations are necessary during the evaluation of a multidimensional aggregate query that "contains" them). At query evaluation time (an example query is shown as a dashed rectangle in Fig. 1), given a multidimensional aggregate query Q, the Cloud metadata services first recognize which sub-queries q_k of Q refer to already-computed partial aggregations, so that these latter can be excluded from the Cloud-based evaluation. This allows us to gain a significant speed-up during the global query evaluation process. Then, the "*uncovered*" sub-queries of Q are evaluated based on common *MapReduce* paradigms. Still looking at Fig. 1, at the client-side, the **BigMDHealth** framework is commonly accessed via ad-hoc Cloud middleware that is in between of the reference Cloud and the client applications. A proper *buffering layer* is also envisioned, as to deal with the computational requirements dictated by processing large-size, large-scale big data repositories.

Still looking at system-oriented aspects, a critical question addressed by the **BigMDHealth** proposal is the following: *how the **BigMDHealth** framework can be successfully applied and exploited in the context of modern intelligent big data applications and systems (e.g., [27])?* Indeed, **BigMDHealth** can be fruitfully applied as pursuing the goal of migrating the classical *OLAP and BI paradigms* [17] towards most recent multidimensional big data management and analytics methodologies and techniques, being acting as baseline computational layer for more complex systems and tools. This is one of the critical questions that **BigMDHealth** aims at addressing and solving, with a particular focus on the applicative setting represented by the healthcare domain. First of all, this challenge is significantly characterized by the special requirement of being tightly related to Cloud infrastructures, since, due to high-performance requirements, big data processing frameworks are usually deployed on top of such infrastructures. Not by chance, one of the main pillars of the **BigMDHealth** framework is being based on Clouds. In turn, this means that models, techniques and algorithms for supporting multidimensional big data management and analytics over big healthcare data must be designed as to cope with distributed nodes of a Cloud, for instance, as predicated by the **BigMDHealth** proposal, on top of the emerging MapReduce computational model. Second, classical algorithms for managing and analyzing multidimensional data (e.g., *OLAP data cubes* [18]) cannot be applied *"as it is"* to scenarios addressed by **BigMDHealth**, since these are not designed for enormous and heterogenous data repositories like big data and, most importantly, not ready to run on top of big data processing platforms, where dataset distribution across Cloud nodes cannot be easily governed according to a pre-fixed data partitioning scheme. It should be noted that, not by chance, the latter is one of the fundamental axioms of the **BigMDHealth** framework.

Indeed, looking at actual solutions available in literature, there are only few research results that focus the attention on the specific aspect of supporting multidimensionality in the big data management and analytics tasks (e.g., [19–22]), despite the clear evidence of utility and support in real-life applications (for which, indeed, the healthcare setting is just one of the most relevant instances). Therefore, it is easy to envision that principles and goals of **BigMDHealth** are of significant relevance in emerging trends of big data research. Another non-trivial goal of the **BigMDHealth** proposal consists in developing proper case studies in the healthcare setting, which can be used not only to verify the applicative requirements the framework pushes for, but also to derive new insights and new challenges for still magnifying the impact of multidimensional big data management and analytics methodologies in the investigated research context.

Implementation-wise, the **BigMDHealth** framework is developed on top of a reference Cloud-based big data processing platform, such as Hadoop or Azure. Specialized algorithmic solutions are developed within the core layer of these reference platforms in a domain-specific programming language, such as *Java* or *C#*.

4 Conclusions and Future Work

Within the consolidated big data analytics research area, this paper has introduced a composite framework for supporting multidimensional big data management and analytics over big healthcare data, called **BigMDHealth**. Algorithm-wise, **BigMDHealth** supports multidimensional aggregate queries over key-value stores, as a fundamental operator of the main multidimensional big data management and analytics paradigm predicated by the framework. We provided complete anatomy of the framework as well as the main algorithm, along with discussion on how to effectively and efficiently exploit the proposed framework in real-life big healthcare data applications and systems.

Future work is oriented towards embedding innovative features in our proposed framework, such as *security and privacy-preservation ones*, which are actually of great interest in modern big data research (e.g., [28–30]).

References

1. Lakshman, A., Malik, P.: Cassandra: a decentralized structured storage system. ACM SIGOPS Oper. Syst. Rev. **44**(2), 35–40 (2010)
2. Cooper, B.F., et al.: PNUTS: Yahoo!'s hosted data serving platform. In: Proceedings of the VLDB Endow, vol. 1, no. 2, pp. 1277–1288 (2008)
3. Gencturk, M., Sinaci, A.A., Cicekli, N.K.: BOFRF: a novel boosting-based federated random forest algorithm on horizontally partitioned data. IEEE Access **10**, 89835–89851 (2022)
4. Sfakianakis, G., Patlakas, I., Ntarmos, N., Triantafillou, P.: Interval indexing and querying on key-value cloud stores. In: Proceedings of 29th IEEE International Conference on Data Engineering, pp. 805–816 (2013)
5. Borkar, D., Mayuram, R., Sangudi, G., Carey, M.: Have your data and query it too: from key-value caching to big data management. In: Proceedings of the 2016 International Conference on Management of Data, pp. 239–251(2016)
6. Tang, C., Wan, J., Xie, C.: FenceKV: enabling efficient range query for key-value separation. IEEE Trans. Parallel Distrib. Syst. **33**(12), 3375–3386 (2022)
7. Toruńczyk, S.: Aggregate queries on sparse databases. In: Proceedings of the 39th ACM SIGMOD-SIGACT-SIGAI Symposium on Principles of Database Systems, pp. 427–443 (2020)
8. Hu, X., Yi, K.: Parallel algorithms for sparse matrix multiplication and join-aggregate queries. In: Proceedings of the 39th ACM SIGMOD-SIGACT-SIGAI Symposium on Principles of Database Systems, pp. 411–425 (2020)
9. Wang, Y., Khan, A., Xu, X., Jin, J., Hong, Q., Fu, T.: Aggregate queries on knowledge graphs: fast approximation with semantic-aware sampling. In: Proceedings of 38th IEEE International Conference on Data Engineering, pp. 2914–2927 (2022)
10. Wang, Z., Luo, T., Xu, G., Wang, X.: The application of Cartesian-join of bloom filters to supporting membership query of multidimensional data. In: Proceedings of the 2014 IEEE International Congress on Big Data, pp. 288–295 (2014)
11. Qin, Y., Guzun, G.: Faster.: multidimensional data queries on infrastructure monitoring systems. Big Data Res. **27**, 100288 (2022)
12. Peng, J., Zhang, D., Wang, J., Pei, J.: AQP++ connecting approximate query processing with aggregate precomputation for interactive analytics. In: Proceedings of the 2018 ACM International Conference on Management of Data, pp. 1477–1492 (2018)
13. Watari, Y., Keyaki, A., Miyazaki, J., Nakamura, M.: Efficient aggregation query processing for large-scale multidimensional data by combining RDB and KVS. In: Hartmann, S., Ma, H., Hameurlain, A., Pernul, G., Wagner, R.R. (eds.) DEXA 2018. LNCS, vol. 11029, pp. 134–149. Springer, Cham (2018). https://doi.org/10.1007/978-3-319-98809-2_9
14. Rong, K., Lu, Y., Bailis, P., Kandula, S., Levis, P.: Approximate partition selection for big-data workloads using summary statistics. In: Proceedings of the VLDB Endow, vol. 13, no. 11, pp. 2606–2619 (2020)
15. Xu, C., Sharaf, M.A., Zhou, M., Zhou, A., Zhou, X.: Adaptive query scheduling in key-value data stores. In: Meng, W., Feng, L., Bressan, S., Winiwarter, W., Song, W. (eds.) DASFAA 2013. LNCS, vol. 7825, pp. 86–100. Springer, Heidelberg (2013). https://doi.org/10.1007/978-3-642-37487-6_9

16. Huang, C., Hu, H., Qi, X., Zhou, X., Zhou, A.: RS-Store: RDMA-enabled skiplist-based key-value store for efficient range query. Front. Comput. Sci. **15**(6), art. 156617 (2021)
17. Chaudhuri, S., Dayal, U.: An overview of data warehousing and OLAP technology. SIGMOD Rec. **26**(1), 65–74 (1997)
18. Gray, J., et al.: Data cube: a relational aggregation operator generalizing group-by, cross-tab, and sub-totals. Data Min. Knowl. Discov. **1**(1), 29–53 (1997)
19. Bochicchio, M., Cuzzocrea, A., Vaira, L.: A big data analytics framework for supporting multidimensional mining over big healthcare data. In: Proceedings of 15th IEEE International Conference on Machine Learning and Applications, pp. 508–513 (2016)
20. Orphanidou, C., Wong, D.: Machine learning models for multidimensional clinical data. In: Khan, S.U., Zomaya, A.Y., Abbas, A. (eds.) Handbook of Large-Scale Distributed Computing in Smart Healthcare. SCC, pp. 177–216. Springer, Cham (2017). https://doi.org/10.1007/978-3-319-58280-1_8
21. Cuzzocrea, A.: Innovative paradigms for supporting privacy-preserving multidimensional big healthcare data management and analytics: the case of the EU H2020 QUALITOP research project. In: Proceedings of the 4th International Workshop on Semantic Web Meets Health Data Management, Co-located with the 20th International Semantic Web Conference, pp. 1–7 (2021)
22. Cuzzocrea, A., Bringas, P.G.: CORE-BCD-mAI: a composite framework for representing, querying, and analyzing big clinical data by means of multidimensional AI tools. In: Proceedings of 17th International Conference on Hybrid Artificial Intelligence Systems, pp. 175–185 (2022)
23. Tsai, C.-W., Lai, C.-F., Chao, H.-C., Vasilakos, A.V.: Big data analytics: a survey. J. Big Data **2**, art. 21 (2015)
24. Cuzzocrea, A., Leung, C.K.-S., MacKinnon, R.K.: Mining constrained frequent itemsets from distributed uncertain data. Future Gener. Comput. Syst. **37**, 117–126 (2014)
25. Balbin, P.P.F., Barker, J.C.R., Leung, C.K., Tran, M., Wall, R.P., Cuzzocrea, A.: Predictive analytics on open big data for supporting smart transportation services. Procedia Comput. Sci. **176**, 3009–3018 (2020)
26. Leung, C.K., Braun, P., Hoi, C.S.H., Souza, J., Cuzzocrea, A.: Urban analytics of big transportation data for supporting smart cities. In: Ordonez, C., Song, I.-Y., Anderst-Kotsis, G., Tjoa, A.M., Khalil, I. (eds.) DaWaK 2019. LNCS, vol. 11708, pp. 24–33. Springer, Cham (2019). https://doi.org/10.1007/978-3-030-27520-4_3
27. Coronato, A., Cuzzocrea, A.: An innovative risk assessment methodology for medical information systems. IEEE Trans. Knowl. Data Eng. **34**(7), 3095–3110 (2022)
28. Cuzzocrea, A., Martinelli, F., Mercaldo, F., Vercelli, G.V.: Tor traffic analysis and detection via machine learning techniques. In: Proceedings of 2017 IEEE International Conference on Big Data, pp. 4474–4480 (2017)
29. Campan, A., Cuzzocrea, A., Truta, T.M.: Fighting fake news spread in online social networks: actual trends and future research directions. In: Proceedings of 2017 IEEE International Conference on Big Data, pp. 4453–4457 (2017)
30. Wang, N., et al.: Collecting and analyzing key-value data under shuffled differential privacy. Front. Comput. Sci. **17**(2), art. 172606 (2022)

Design and Implementation of Data Warehouse Solution at Kumpulan Wang Persaraan (KWAP)

Mohamad Fairul Hussein[1], Paridah Daud[1(✉)], Omar Musa[2], Normaiza Mohamad[1], and Noor Lees Ismail[1]

[1] UNITAR International University, Petaling Jaya, Selangor, Malaysia
paridah69@unitar.my
[2] Jaycorp Berhad, Kuala Lumpur, Malaysia

Abstract. The Retirement Fund (Incorporated), also known as KWAP, is a statutory body that manages the pension scheme for Malaysia's public employees. KWAP has over 550K members. Before the implementation of online data transfer (ODS) in KWAP, finance and management reporting were extracted from the production table. Because everything (dashboard and reporting) was extracted from the main system, the main system or application was impacted and became slow or crushed, so they decided to implement ODS. The data was gathered by the ODS from the main database. The management requires a live dashboard to access the target files or cases to be processed and then display how many cases the operation team has processed. This dashboard will be used to track all progress. However, the live dashboard is not available. However, the live dashboard is inconvenient for reporting because of many inaccurate data in the ODS module. The data warehouse provides a solid platform for integration of the historical data to facilitate information processing systems to produce the information that can be used to support the analytical purpose. Currently, KWAP implementing the ODS concept to handle pensioners' personal and data transactions. For that, this research focuses on how to implement a data warehouse by proposing a data warehouse design to handle pensioner data at KWAP by moving all reporting part that uses the ODS to the data mart.

Keywords: Data warehouse · Data mart · KWAP

1 Introduction

Kumpulan Wang Persaraan (KWAP) is the Retirement Fund (Incorporated) a statutory body which manages the pension scheme for Malaysia's public employees. Before online data transfer (ODS) was implemented in KWAP, finance and management reporting were extracted from the production table. Because everything (dashboard and reporting) was extracted from the main system, the main system or application was impacted and became slow or crushed, so they decided to implement ODS. The ODS is used for both dashboard and reporting, which is inconvenient and unreliable. ODS design is unsuitable for large data transactions, such as payment information. ODS in KWAP initially was

© The Author(s), under exclusive license to Springer Nature Singapore Pte Ltd. 2023
Y. B. Wah et al. (Eds.): DaSET 2022, LNDECT 165, pp. 195–208, 2023.
https://doi.org/10.1007/978-981-99-0741-0_14

implemented for operation dashboard purposes. On the business analysis side, they still need to generate reports requested by another department. Business analysis report mostly involves transaction data such as payment transaction and here they face another problem where the ODS extraction process to produce a report take a very long time to finish more than 5 h for the certain report. The report is also not accurate and very difficult to troubleshoot when facing an issue.

This ODS concept use trigger and view, where a trigger was created to capture any changes made to the main system table and store to change data capture (CDC) table. A structure query language (SQL) view is created to query changes captured in the CDC table and transfer to the ODS database using Extract, Transform and Load (ETL). ETL will push the CDC data every hour to the ODS table since this concept will process data row by row so it cannot process too much data (<1k rows). If more than 1k rows of data need to be processed, then this CDC will need to be stopped. Pensioner data reporting and dashboarding (around 70 million payment data in a single table) are handled by the KWAP business analysis department. Currently, KWAP doesn't have a data warehouse, thus, this study is to find either any performance improvement if using a data warehouse for business analysis reporting purposes, to propose What is the best data warehouse design to be implemented to handle pensioner data and to identify the Extract, Transform and Load (ETL) tool as a useful for data warehouse implementation.

What is the impact of not having such a data warehousing solution for KWAP?

a) No flexibility in designing reports and dashboards because data resources come from various sources.
b) Authentication or access issue when ETL needs to touch the production database.

The rest of the paper is organized as follows: Sect. 2: presents the Literature Review to get some overview of data warehouse implementation. Section 3: The proposed methodology is discussed. Section 4: the design and implementation. The results and discussion. Section 5. Finally, the conclusions and future work directions are presented.

2   Related Works

A system based on a data warehouse has been used to improve the pharmaceutical distribution sector decisions by forecasting drug sales. The predicted sales using a neural network (NN) and an autoregressive model time series are applied to historical data [22]. Many other machine learning algorithms (such as naive Bayes, decision trees (DT), and clustering) have shown model accuracy in the unimplemented prediction sector. This study enhanced the data mart abilities and adapted its characteristics to satisfy the management requirements for enterprises to make convenient and correct decisions. Data warehousing can be defined as a repository, where a huge amount of data are stored [30]. The data cleansing is required from the source system and this process comes after data warehousing and should be carried out to be ready for the data for final load into DW dimensions [23]. The data mart structure can simplify the reporting structure into a series of base tables and then create several reporting schemas each around a specific concept based on the cases, which contains the data required for reporting on various metrics. This structure allows centralized definitions with simplified reporting by many individuals who access only the Reporting Schemas [31]. Data Mart was suggested to DW and data mining for administrative prediction purposes [24]. The star schema is chosen as the storage schema for the DW architecture [21]. Various assessment approaches, including static and dynamic methods, are utilized during data mart evaluation and assessment to evaluate the knowledge, and understanding in the data mart [27]. Experts have reviewed data mart to identify faults in the knowledge base. Experts use the static technique to assess the knowledge and understanding and analyze the data mart performance. This paper employs a model-driven data mart that makes use of OLAP and KPI to enable managers to look for trends in online reviews and display them at different levels and viewpoints of analysis [28, 29]. ETL details techniques to deal with unstructured data and then benefits managers by providing a distinct dimensional model way. As a result, the approach may be used to deliver data from many sources for further evaluation on a dashboard.

Data warehousing research has concentrated on design issues, data maintenance strategies related to relational view materialization, implementation issues and query performance issues [1, 10]. Several researchers have investigated data warehouse ETL tools and discovered alternatives. There has also been a substantial of research work addressing the issues of data inconsistency, data quality, and technical factors that influence data warehouse implementation effectiveness [3, 4]. A data warehouse is intended to provide reporting and business intelligence solutions to companies seeking to remain competitive [5, 17]. The data warehouse also assists business organisations in gathering strategic and timely business information to support corporate strategies to stay competitive and increase revenue [7]. For the last decade, business organisations have increasingly relied on data warehouses to serve as the central repository for their enterprise data. The study attempted to investigate how understanding data warehouse users' requirements affect the successful implementation of a data warehouse project. The data quality issues are primarily caused by source data, and to address the issue, they developed a framework for defining relevant metrics that comprise a profile's quality and from the perspectives of a lack of system development capability and a process for identifying and resolving data integration issues [18]. System quality, information quality, and user satisfaction are all interrelated and interdependent. Previous research has identified the following key factors for data warehouse success such as improved data availability to meet business needs, improved capability of satisfying analysis and reporting needs, improved data quality for better decision-making, and improved information timeliness for business users' use [6, 8].

3 Research Methodology

3.1 Research Approach and Design

The research design for this study was based on Fábio Rilston Silva Paim's BUILDING the Data Warehouse (1998). The methodology is presented as shown below (see Fig. 1). The process is surrounded by Requirements Management Control, a backbone activity that performs a continuous quality assessment of requirements changes in the background.

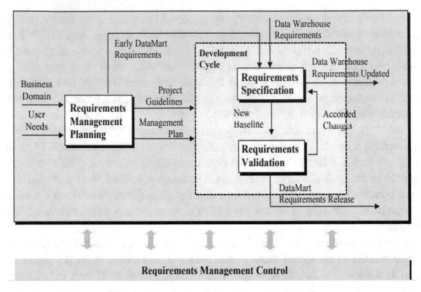

Fig. 1. Requirements management control

Requirements Management Planning

An effective requirements management process must define the guidelines, in terms of business rules, apply to the acquisition, documentation and control of specific requirements, procedures and processes that have been agreed upon to clarify the goals of the project, types of DM, the integration structure to clear rules for data sharing across systems must be set and cantered, responsible team for development.

Requirements Specification

The modelling of data warehouse systems needs a particular consideration for the reuse of previously agreed-upon requirements, because establishing and integrating separate Data Marts is a time-consuming operation, and the only way to achieve such integration is to describe common factual and dimensional criteria such that they imply the same thing across all Data Marts. The established needs must comply to the overall data warehouse requirements set to eliminate redundancy and assure conformity to the broader.

Requirements Elicitation

This phase is to adopt a multidimensional needs discovery process. The data warehouse elicitation step, like conventional systems, requires application domain and organizational experience from both users and systems analysts. The strategies were proven to be the most successful for eliciting data warehouse requirements are interviews, prototyping to replicate system behavior and providing a crucial change to consolidate the views regarding system needs, and scenarios (Creating a collection of interaction scenarios helps developers explain and specify system requirements in the form of use cases).

Analysis and Negotiation of Requirements
Analysis and Negotiation are subjected to a comprehensive examination to look for omissions, conflicts, overlaps, and contradictions. The produced papers must be reviewed to verify that the specification complies with quality standards and main multidimensional constraints and that a proper balance between architectural and conceptual concerns is struck.

Requirements Documentation
This stage is critical to our approach. The purpose here is to generate a thorough, detailed description of the elicited system requirements that all stakeholders can comprehend. Requirement's documentation is developed throughout the data warehouse construction process, not only during the documentation phase (see Fig. 2). In this study a collection of templates is created to satisfy all data warehouse functional and non- functional requirements.

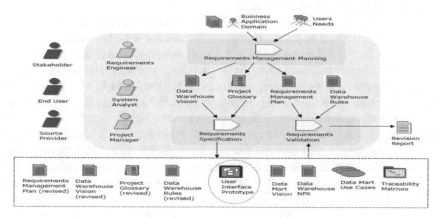

Fig. 2. Documentation phase

Requirements for Complying
DW initiatives can only be effective if the defining components in two interrelated dimensions, the subject-driven DM vision, and the global corporate data warehouse architecture, are good. In our method, the idea is extended to the greatest degree of abstraction, where all common system criteria are confirmed. If a need is the same in each Data Mart view of the enterprise data warehouse, it is called compliant. More than merely multidimensional aspects, conformed requirements react to every system development function, characteristic, or constraint that has the same logic across the project and so must be stated separately. Conformed requirements improve the data warehouse

specification in the ways of avoiding ambiguity and redundancy in requirements applicable to the whole data warehouse, enabling the use of common dimensional features, including such dimension tables, on various facts in the same database space, to increase quality, encourage reusability of agreed-upon information in the project, in combination with a scenario-based approach, improve the reliability of interfaces and data while utilizing the conformed model, allow drill-across actions between Data Marts, ensure the essential integration of Data Marts for the business multidimensional architecture to work, make data warehouse evolution considerably easy, makes adhering to the style and organizational standards easier.

Requirements for Validation

When evaluating a data warehouse program certain misunderstandings about the analytical features to be supplied will persist even after all the previously described processes have been completed. Suggested the external subject matter experts who were not engaged in the requirements definition process involved during this phase. These external evaluators provide a fresh viewpoint to the construction environment because they are not restricted by prior preconceptions about the solution.

Requirements Management Oversight

This analysis will be extended to the database structures to establish the influence of the underlying modification on the multiple schemas. Some tools enable both types of investigations. Because the creation of data warehouse apps necessitates dealing with a huge number of needs and database attributes, the usage of such tools becomes vital. For visualizing requirement dependencies, matrices are the most widely utilized component in special- purpose tools. CASE tools may be quite useful in accomplishing this process because they offer automatic search capabilities to locate the relevant dimension needed in the database and establish how many (and to what degree) database elements are affected by changes. These capabilities simplify the work required to evaluate and conduct essential tables in the database field maintenance, ensuring the solution's security. Suggested to use of ETL tools such as Talend. Open Studio by Talend, is an open-source data integration solution that is available in both free and premium editions. Open Studio is appropriate both for professional and non- technical users because it links on-premises and cloud-based sources of data.

4   Design, and Implementation

4.1   ETL and Data Warehouse System Design

To demonstrate how the system works through local design. The design will be divided into two parts: the first will be an ETL workflow design, and the second will be a Data warehouse design.

ETL Workflow Design

ETL tool's functions are to capture, modify, filter, and load into a data warehouse. The ETL method used for data warehouses may not only provide batch data load but would also support data append for daily operations. This approach allowed them to organize the mapping diagram into four levels as shown in Fig. 3.

1) Database level 0: This level represents all the types of schemas of DW (source conceptual schema; DW conceptual schema) along with their mapping through a single mapping package. A Single mapping package encapsulates all the lower-level mappings among different schemata.

2) Data flow level 1: In this level, the relationship between different source tables and the target DW is represented. To achieve this target the mapping diagram is expanded to provide more detailing and thus capture the flow of data from source to target.

3) Table level 2: All the intermediary transformation details of the flow of data from source to target are detailed at this level. The process of checking the contents also takes place at this level. A set of packages is required to segment complex data mappings in sequential steps.

4) Attribute level 3: Inter attribute mapping is maintained to set this level. This implies that the mapping diagram must be explored till the provider and consumer attribute mapping is not depicted. It could involve transformation and cleaning details.

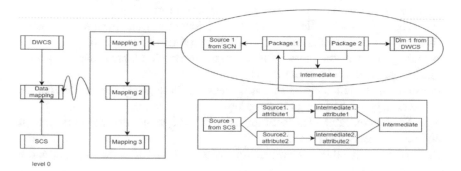

Fig. 3. Mapping diagram

Data Warehouse Design
The proposed data warehouse in this study was designed using Kimball and Ross's nine-step methodology [30], which is as follows:

1) Choose the Process - Summary Demographic Report, presented to KWAP once a month.

2) Choose the Grain - Grains are pension information, and pensioner information (age, address, etc..).

3) Identify and Conform the Dimension -Dimensions are Age range Dimension, Pension range Dimension, State Dimension and Holiday Dimension

4) Choose the Facts - Facts is Pensioner payment Fact.

5) Store Pre-calculations in the Fact Table - There is no pre-calculation, so the design process is continued to the next stage.
6) Round Out the Dimension Tables - Next step is to add text that is intuitive and easy to understand by users in the dimension table.
7) Choose the Durations of the Database - Database duration is set on month, which is the 1st day to the last day of the previous month.
8) Determine the Need to Track Slowly Changing Dimension - To anticipate data changes that might occur, the field of refresh data was added to each table. This field is used to reset the whole table.
9) Decide the Physical Design - The last stage is the physical design of the database which contains tables of facts and dimensions. The database is designed using Star Schema.

Fact and dimension tables contained in the schema are as follows:

1) F demographic Aggregate Fact Table - The monitoring fact table contains 112 attributes that summarize all pensioner payment information.
2) F demographic Fact Table - The monitoring fact table contains attributes assigned from all Dimension tables to each Pensioner payment information.
3) Age range Dimension Table - The age range dimension table group the pensioner age by age group and assign group key
4) Pension Dimension Table - Pension Dimension Table group pension amount range by pension group and assign a group key.
5) Race Dimension Table - The race dimension table contains attributes race description and race category and race id.
6) State Dimension Table - The usage dimension table contains the state key and state name attributes.

Source data information:

1) Master Payment table - Master Payment contains all pensioner payment information gathered from multiple tables in the operational database and stored in a single table. The primary data stored in this table were the payment number, pension account number, payment amount, and deduction amount.
2) Master Person table - Mater Person table stores all pensioner personal, account, and service information, which comes from multiple tables in the operation database and is also stored in this table. The primary data stored in this table were the pensioner ID number, address, age, race, account, and last service information.

Figure 4 shows the final KWAP Demographic data mart structure.

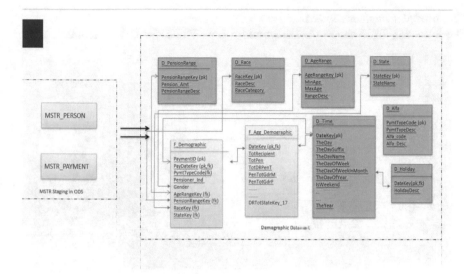

Fig. 4. KWAP demographic data mart structure

4.2 Data Warehouse Implementation

This report will summarize recipients of pension payments based on state, gender, race, and age range. Currently, they generate this report using the existing MSTR PERSON and MSTR PAYMENT tables, which is inconvenient and takes a long time to generate almost one hour per report. List of tables created to implement the proposed system based on DW design.

Star Scheme Diagram
The relationship between the tables is created as shown in Fig. 5.

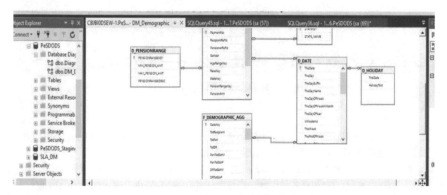

Fig. 5. M_DEMO_AGG Star Schema

ETL Workflow Implementation
The following Fig. 6 shows final diagram for KWAP ETL for DW structure.

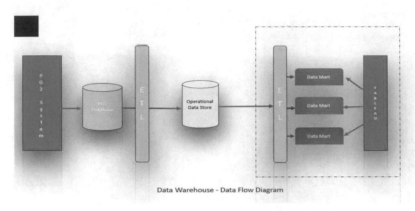

Data Warehouse - Data Flow Diagram

Fig. 6. Data workflow diagram

The three ETL workflows were created to load data into the two fact tables that were created in phase one, and by using Talend Open Studio software the flow is shown below.

Create a One-Time Flow to Load the F_DEMOGRPHIC Table
This ETL job will full-load all joining data from MSTR_PERSON, MSTR_PAYMENT and all Dimension tables. ETL job name: F_DEMOGRAPHIC Data to pick: All (Fig. 7)

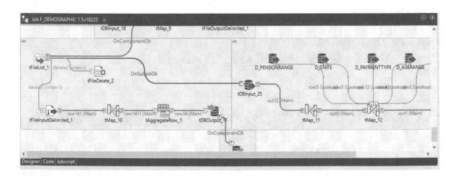

Fig. 7. F_DEMOGRAPHIC ETL

Create Incremental Workflow to Load the F_DEMOGRPHIC Table Monthly Basis
This ETL job will load incremental joining data from MSTR_PERSON and MSTR_PAYMENT tables on monthly basis. The logic for this workflow is just the same as full load job but the major difference is this will pick the previous month's data to load. ETL job name: F_DEMOGRAPHIC_INC Data to pick: last month's date. Ex: 2021-11 (Fig. 8).

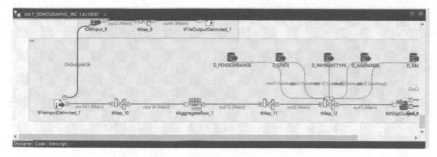

Fig. 8. F_DEMOGRAPHIC_INC ETL

Create a Workflow to Load the F_DEMOGRPHIC_AGG Table Monthly Basis
This ETL job will load incremental aggregate data from F_DEMOGRAPHIC to the
F_DEMOGRPHIC_AGG table on monthly basis.

ETL job name: M_DEMO_AGG. Data to pick: last month's date. Ex: 2021-11
(Fig. 9).

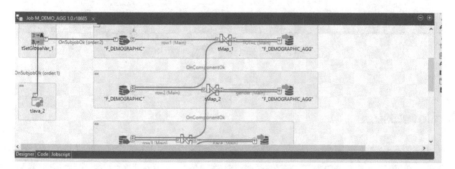

Fig. 9. M_DEMO_AGG ETL

b) Result and Evaluation
Final Table Result

Figure 10 shows the snapshot results of loading F DEMOGRAPHIC AGG for October
and November data.

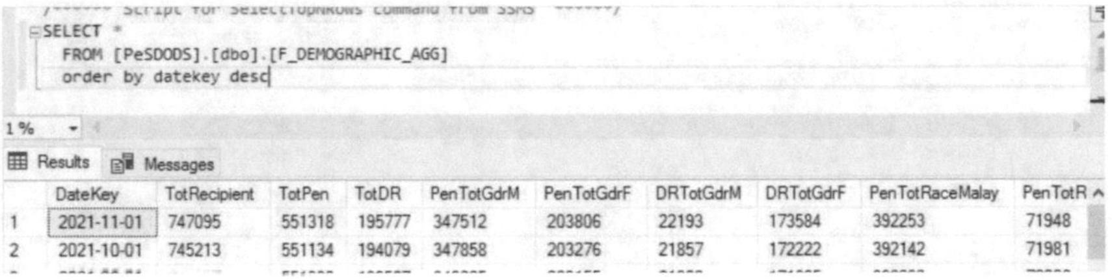

Fig. 10. Snapshot of loading demographic AGG

Query Time Taken

Figure 11 shows that the total query time for table F DEMOGRAPHIC AGG when retrieve from many sources in the data mart table takes less than one second for 28 rows. This is very fast because all of the large and complex data has already been compressed and aggregated, resulting in a significant improvement in table performance over the previous version using ODS, which took nearly an hour to display the information.

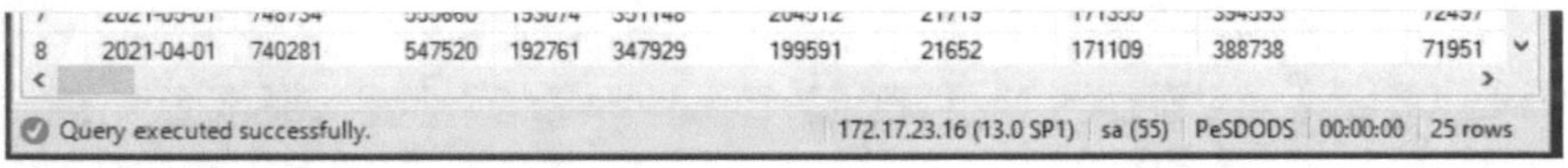

Fig. 11. Result of total query time

5 Conclusion

KWAP only manages contributions from permanent government staff with pensionable status and who are in service with Regulatory Bodies and Local Authorities. KWAP has different constraints and requirements for the system. This study is to suggest an improvement to the current operation in KWAP. Why does KWAP need a data warehouse? Customer demand, competitive pressure, and better decisions need to be solved. Today's decision-makers require data that is updated several times per day. The volume of data is increasing. Data volumes in operational data stores grow very fast when the business expands. Therefore, the more resources and time the ETL processes required. The data warehouse provides a solid platform for integration of the historical data to facilitate information processing systems to produce the information that can be used to support the analytical purpose. Business intelligence (BI) is at the heart of the world's best organizations, allowing them to understand business trends, make better decisions, and support day-to-day operations. ETL (extract, transform, and load) is the process by which businesses create the consolidated data stores (e.g., data warehouses and data marts) needed to support the effective BI. The current business conditions need a new approach to data integration and running in real time. The pensioner data continually grow over time. Then, the KWAP need the architecture to manage the technical design of the DW and ETL. Both are fundamental pillars to sustain the DW. The architecture was established by merging proprietary ETL tools. An organization of data structure and data mart will improve the performance. This study suggested an acceptable framework can be used in any organizational. This method was successfully increasing the performance of proses the data from an existing structure. The approach is being used in various industries, including manufacturing, health care, insurance, and finance based on the organization's requirements.

Acknowledgement. The publication of this paper was supported by UNITAR International University, Malaysia and Jaycorp Berhad industry grant.

References

1. Akhter, S., Rahman, N.: Building a customer inquiry database system. Int. J. Technol. Diffus. (IJTD) **6**, 59–76 (2015)
2. Akhter, S., Rahman, N., Rahman, M.N.: Competitive strategies in the computer industry. Int. J. Technol. Diffus. (IJTD) **5**, 73–88 (2014)
3. AlMabhouh, A., Ahmad, A.: Identifying quality factors within data warehouse. In: Proceedings of the Second International Conference on Computer Research and Development, pp. 65–72. IEEE (2010). https://doi.org/10.1109/ICCRD.2010.18
4. Ballou, D.P., Tayi, G.K.: Enhancing data quality in data warehouse environments. Commun. ACM **42**, 1 (1999)
5. Brohman, M.K., Parent, M., Pearce, M.R., Wade, M.: The business intelligence value chain: data-driven decision support in a data warehouse environment: an exploratory study. In: Proceedings of the 33rd Hawaii International Conference on System Sciences (HICSS), pp. 1–10. IEEE (2000)
6. Cataldo, M., Herbsleb, J.D.: Coordination breakdowns and their impact on development productivity and software failures. IEEE Trans. Software Eng. **39**, 343–360 (2013)
7. Cooper, B.L., Watson, H.J., Wixom, B.H., Goodhue, D.L.: Data warehousing supports corporate strategy at first American corporation. MIS Q. **24**, 547–567 (2000)
8. DeLone, W.J., McLean, E.R.: Information systems success revisited. In: Proceedings of the 35th Hawaii International Conference on System Sciences (HICSS). IEEE Computer Society Press (2002)
9. Gefen, D., Wyss, S., Lichtenstein, Y.: Business familiarity as risk mitigation in software development outsourcing contracts. MIS Q. **32**, 531–551 (2008)
10. Golfarelli, M., Rizzi, S.: Data Warehouse Design: Modern Principles and Methodologies. McGraw-Hill Osborne Media, New York (2009)
11. Idris, N., Ahmad, K.: Managing data source quality for data warehouse in manufacturing services. In: Proceedings of the International Conference on Electrical Engineering and Informatics, 17–19 July 2011, Bandung, Indonesia (2011)
12. Isik, O., Jones, M.C., Sidorova, A.: Business intelligence success: the roles of BI capabilities and decision environments. Inf. Manage. **50**, 13–23 (2013)
13. Keil, M., Carmel, E.: Customer-developer links in software development. Commun. ACM **38**, 33–44 (1995)
14. Leitheiser, R.L.: Data quality in health care data warehouse environments. In: Proceedings of the 34th Hawaii International Conference on System Sciences (HICSS), pp. 1–10. IEEE (2001)
15. Lu, Y., Ramamurthy, K.: Understanding the link between information technology capability and organizational agility: an empirical examination. MIS Q. **35**, 931–954 (2011)

16. Mithas, S., Ramasubbu, N., Sambamurthy, V.: How information management capability influences firm performance. MIS Q. **35**, 237–256 (2011)
17. Miyamoto, M.: Application of competitive forces in the business intelligence of Japanese SMEs. Int. J. Manage. Sci. Eng. Manage. (IJMSEM) **10**, 273–287 (2015)
18. Rahman, N.: Measuring performance for data warehouses - a balanced scorecard approach. Int. J. Comput. Inf. Technol. (IJCIT) **4**, 1–7 (2013)
19. Paim, F.R.S., Carvalho, A.E., Castro, J.D.: Towards a methodology for requirements analysis of data warehouse systems. In: Proceedings of the XVI Simpósio Brasileiro de Engenharia de Software (SBES2002), Gramado, Rio Grande do Sul, Brazil (2002)
20. Wijaya, G.: Perancangan data warehouse nilai mahasiswa dengan kimball nine-step methodology. J. Inform. **4**(1) (2017)
21. Kimball, R., Ross, M.: The Kimball Group Reader: Rentlessly Practical Tools for Data Warehousing and Business Intelligence. Wiley Publishing Inc., Indianapolis (2010)
22. Elazeem, N.M.A., Labib, N.M., Abdella, A.K.: A proposed data warehouse framework to enhance decisions of distribution system in pharmaceutical sector. Egypt. Comput. Sci. J. **43**(2), 43–60 (2019)
23. Girsang, A.S., Arisandi, G., Elysisa, C., Michelle, Saragih, M.H.: Decision support system using data warehouse for retail system. J. Phys. Conf. Ser. **1367**(1), 1–6 (2019)
24. Peng, X.: Analysis of administrative management and decision-making based on data warehouse. In: Proceedings of the 11th International Conference on Measuring Technology and Mechatronics Automation, Qiqihar, China, pp. 527–530 (2019)
25. Katkar, V., Gangopadhyay, S.P., Rathod, S., Shetty, A.: Sales forecasting using data warehouse and Naïve Bayesian classifier. In: Proceedings of the International Conference on Pervasive Computing, Pune, India, pp. 1–6 (2015)
26. AbdAlrazig, H.: Designing a data warehousing model to support forecasting and decision making for sales. Doctoral dissertation. Sudan University of Science and Technology (2018)
27. Torres-Sanchez, R., Navarro-Hellin, H., Guillamon-Frutos, A., San-Segundo, R., Ruiz-Abellón, M.C., Domingo-Miguel, R.: A decision support system for irrigation management: analysis and implementation of different learning techniques. Water **12**(2) (2020)
28. Baumeister, J., Striffler, A.: Knowledge-driven systems for episodic decision support. Knowl.-Based Syst. **88**, 45–56 (2015)
29. Hosio, S., Karppinen, J., Berkel, N., Oppenlaender, J., Goncalves, J.: Mobile decision support and data provisioning for low back pain. Computer **51**(8), 34–43 (2018)
30. Nayak, L.S.A., Das, K., Hota, S., Sahu, B.J.R., Mishra, D.A.: Implementation of data warehouse: an improved data-driven decision-making approach. In: Mishra, D., Buyya, R., Mohapatra, P., Patnaik, S. (eds.) Intelligent and Cloud Computing. Smart Innovation, Systems and Technologies, vol. 286, pp. 419-427. Springer, Singapore (2022). https://doi.org/10.1007/978-981-16-9873-6_38
31. Hofer, I.S., Gabel, E., Pfeffer, M., Mahbouba, M., Mahajan, A.: A systematic approach to creation of a perioperative data warehouse. Anesth. Analg. **122**(6), 1880–1884 (2016)

Consumer Behavior Prediction During Covid-19 Pandemic Conditions Using Sentiment Analytics

Saravanan Murugan[1], Sulaf Assi[2], Abbas Alatrany[1], Manoj Jayabalan[1], Panagiotis Liatsis[3], Jamila Mustafina[4], Abdullah Al-Hamid[5], Maitham G. Yousif[6], Ahmed Kaky[7(✉)], Danny Ngo Lung Yao[8], and Dhiya Al-Jumeily OBE[1]

[1] Faculty of Engineering and Technology,
Liverpool John Moores University, Liverpool L3 3AF, UK
[2] School of Pharmaceutical and Biomolecular Science, Liverpool John Moores University,
Liverpool L3 3AF, UK
[3] Department of Electrical Engineering and Computer Science, Khalifa University, Abu Dhabi,
UAE
[4] Kazan Federal University, Kazan, Russia
[5] Saudi Ministry of Health, Najran, Saudi Arabia
[6] College of Science, University of Al-Qadisiyah, Al Diwaniyah, Iraq
[7] University of Anbar, Ramadi, Iraq
A.J.Kaky@ljmu.ac.uk
[8] Faculty of Business and Technology, UNITAR International University, Petaling Jaya,
Selangor, Malaysia

Abstract. Covid-19 pandemic created a global shift in the way how consumers purchase. Restrictions to movements of individuals and commodities created a big challenge on day today life. Due to isolation, social media usage has increased substantially, and these platforms created significant impact carrying news and sentiments instantaneously. These sentiments impacted the purchase behavior of consumers and online retailers witnessed variations in their sales. Retailers used various customer behavior prediction models such as Recommendation systems to influence consumers and increasing their sales. Due to Covid-19 pandemic, these models may not perform the same way due to changes in consumer behavior. By integrating consumer sentiments from online social media platform as another feature in the prediction machine learning models such as recommendation systems, retailers can understand consumer behavior better and create Recommendations appropriately. This provides the consumers with appropriate choice of products in essential and non-essential categories based on pandemic condition restrictions. This also helps retailers to plan their operations and inventory appropriately.

Keywords: Covid-19 pandemic · Social media · Consumer behavior · Machine learning · Sentiment analysis · Recommendation systems

© The Author(s), under exclusive license to Springer Nature Singapore Pte Ltd. 2023
Y. B. Wah et al. (Eds.): DaSET 2022, LNDECT 165, pp. 209–221, 2023.
https://doi.org/10.1007/978-981-99-0741-0_15

1   Introduction

1.1   Background

Digitalization is touching every individual's life daily. Marketers use these Digital Platforms especially Social Media Platforms to influence buyers [1]. Almost all famous retail giants, new digital born sellers have their own ecommerce websites. As per Statistical [2], global retail e-commerce sale in 2020 was 4.28tr USD and it continues to increase. Retail organizations have used several Machine Learning (ML) Models to understand customer buying patterns and built recommendation engines to improve customer purchases [3]. The Covid-19 pandemic severely affected global economy. The World Bank forecasted that 5.2% contraction of global economy due to movement restrictions and lockdowns [4]. Hence, the pandemic caused several disruptions in supply chain and shortage of essentials. Customers who used online e-commerce apps, websites to buy luxury and non-essential items have changed to shopping for essentials as a result of the pandemic [4].

Social media sites have become primary news and views exchange media. Every information related to pandemic has been amplified much faster due to more user followers and contributions [5]. The surge in data posted over social media platforms alongside the diversity of data have made classical ML approaches ineffective [6]. This urged the need to find new ways of approaching social media data in order to understand customers' behavior and purchase patterns.

All the aforementioned factors changed the way how customer purchase patterns changed. The current Machine Learning models (such as support vector machine and Naïve Bayes) in the system are no longer effective in this new unseen data scenario [6]. All the classical prediction models and Recommendation models built using Machine Learning algorithms will not be able to give accurate results.

The main aim of this research is to explore the impact of Covid –19 and such pan-demic scenarios in customers' purchase patterns and identify important influencing factors from Twitter keywords and Trends. Subsequently, these keywords will be used to improve the prediction models for retail organizations.

1.2   Literature Review

During Covid-19 pandemic period, the global economy has been affected severely due to restricted movement and to availability of essentials across the supply chain [7]. Regular Shops were allowed to open and sell during specific time only. Most of the days they remained shut. Since people at home had access to internet much more than usual hours, they spent lot of time in Social Media Platform than ever before. Consumers who used online platforms such as Amazon to buy items on regular basis (even luxury and non-essential items) now switched to buy only food, medical and essential items only [7].

During Covid-19 pandemic period, consumer purchases were not planned [8]. Hence, purchasing behavior was ruled by panic buying due to restricted movement of people and commodities. In addition, consumers experienced feeling of anxiety and fear during the Covid-19 pandemic added with restrictions posed by government creates 'lack of control'.

Twitter, being a micro blogging social platform provides access to every individual to share their thoughts in the form of short blogs called as Tweets, which are maximum 140 characters long. A tweet can have a header of its own which can be given a unique name also. This is called "Hash Tag". There are about 500M tweets created every day with active users of around 320M [9].

Apart from classification algorithms mentioned in the above research, there are several machine learning methods have been used to perform Sentiment Analysis. Several algorithms in Clustering approaches such as Density-Based Spatial Clustering of Applications with Noise (DBSCAN) and Gaussian Mixture Models (GMM) models have been adopted [10]. Nonetheless, Bidirectional Encoder Representations from Transformers (BERT) model gave best result in text classification [11]. While this works very well at individual sentence-level sentiments, the applicability at the aspect level is still rare. Recommendation systems have shown effective tools in filtering information effectively and predict/recommend items to consumers. They have been used by online retailers to influence the consumers for up-selling and cross-selling [12].

This research reports how to create various types of BERT models to extract accurate sentiment. The paper explains Collaborative filtering methods with user-based collaborative filtering and item-based collaborative filtering methodologies. The paper also explains building a recommendation engine using hybrid approach of integrating Association Rule Mining with Collaborative filtering.

2 Methodology

The research explored sentiments of online consumers during Covid-19 pandemic period by adopting two major tracks (Fig. 1).

The first major track comprised twitter data analysis and pattern analysis. Twitter data analysis included: data collection; and text processing (NLP based Lexical Processing). Pattern analysis encompassed pattern analysis encompassed: topic modelling; sentiment analysis of consumers buying behavior; and extract keywords for panic buying and essential buying.

The second major track comprised mainly sentiment analysis as a feature in a recommendation engine. Hence, making recommendations per sentiments was more accurate. Prior to building the recommendation engine, sales data pattern was matched with twitter sentiments. This was performed to identify the effect of the pandemic and how online social media influences the consumers.

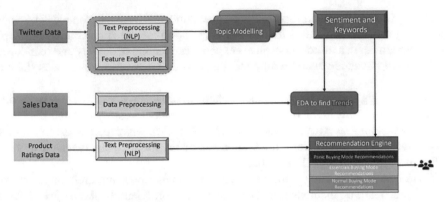

Fig. 1. Research methodology showing the two tracks. EDA refers to exploratory data analysis.

2.1 Datasets

A total of five datasets were used as follows:

Twitter Data: Kaggle Dataset with Tweets from March 29, 2020 through April 30, 2020. The dataset contains fields including Tweet ID, User ID, Creation Data and Time, Screen Name, Tweet Text, Reply ID, Is_quote etc. There are 22 features. Files are created for individual days. So, there are 33 files in total.

Amazon Sales Data: This dataset provides data about sales in Amazon platform during Covid pandemic period. There are 12 columns and 5460 rows of data. Most important features are "product name", "brand name", "mrp", "sale price" and "product description". The dataset provides details about discount price details of medical supplies and kits such as facial masks and gloves during Covid 19 period.

Canada Retail Sales Data: This dataset contains Retail Sales, Trade data for Canada and provinces as per North American Industry Classification System (NAICS). The data is aggregated as Monthly Total for every NAICS code item. The data is present from January 1991 through September 2021. The file contains 16 features and important ones are REF_DATE, NAICS, Adjustments and Value.

Amazon Product Reviews Data: This dataset is huge repository of several files. This dataset contains product reviews and metadata from Amazon, including 142.8 million reviews spanning May 1996–July 2014.

2.2 Twitter Data Processing

Very few features from the Twitter dataset were needed in this study and were related to "Created_at" (date and time of the account that tweeted) and "text" (the text of the Tweet). For pre-processing the Twitter, text a number of steps conducted include tokenization, Specifically I used 'regex tokenizer'. This gave me flexibility to define my own

"Punctuations". Second step, stop words and Stemming: After cleaning the text with only relevant information, as a next step, used nltk library's stop words function to remove the stop words from the tweets. Lastly, Feature Engineering: used CountVectorizer function sklearn library. This is one of feature engineering method which converts text corpus into an array along with the frequency value of the text based on the appearance count in the corpus.

Tokenizer
This is used to tokenize every sentence. Specifically, I used 'regex tokenizer'. This gave me flexibility to define my own "Punctuations" ('!"$%&\'()* +,−./:;<=>?[\\]^_'{|} ~ •@') which I found in iterative manner after passing the text repeatedly.
    A function is written to perform following pre-processing tasks

- Converted all the tweets to lower case alphabets. This helps in eliminating case sensitivity issues
- Using 're.sub' function, replaced all unwanted punctuations and links with spaces.
- Similarly removed double spaces
- Removed all the numbers as they will not add value for my research corpus
- Removed all the emojis (J L etc.) from the tweets

Stop Words and Stemming
After cleaning the text with only relevant information, as a next step, I used nltk library's stop words function to remove the stop words from the tweets. Once the stop words are removed, then I used Porter Stemmer from nltk library to trim the words to their root forms.

Feature Engineering
Count Vectorizer function sklearn library was used. This is one of feature engineering method which converts text corpus into an array along with the frequency value of the text based on the appearance count in the corpus. I got the words and their frequencies in the array format using this function.

2.3 Topic Modelling

Used the 'Latent Dirichlet Allocation (LDA)' from sklearn library. The numpy array which I got from my previous step, is passed to '(LDA)' algorithm. I set the total number of topics to 3. The objective is to find topics primarily in three different scenarios for online consumers. One scenario is normal buying behavior. Second scenario is about 'Panic buying' and third scenario is about 'Essentials buying'.

2.4 Experiments

Once the LDA algorithm is run on the corpus, the algorithm identifies important keywords and scores them across all the three topics. Each keyword is scored based on the occurrence. This helps in identifying important keywords for each category. Once we run the LDA algorithm for a large corpus comprising of several days' tweets, we will get a comprehensive list of keywords for each sentiment. A 'word dictionary' is built based on these keywords. Thus, we will have comprehensive set of keywords for each sentiment namely 'Panic buying', 'Essential buying' and 'Normal buying'. This is the process of building the dictionary. The LDA algorithm when run on daily tweets, the output will give top 5 keywords based on their occurrence score. These keywords are then compared against the 'word dictionary' and then based on the topic where majority of these keywords fall, sentiment for that day is chosen. This sentiment is fed into the Recommendation Engine.

During comparison of top keywords with 'word dictionary', if for any corpus, majority of the keywords are not found in any of the sentiments, then currently algorithm will default to 'Normal Sentiment'. But this also means model has to be revisited for these new keywords. This can be considered similar to 'feedback' to rebuild the 'word dictionary' with additional new keywords. Suppose a new pandemic situation arises, a new set of keywords will emerge. For example, in covid pandemic conditions, 'Covid', 'Virus', 'lockdown' etc. are keywords identified. During new pan-demic situation such as 'war', 'flood' etc. a set of new keywords will appear in social media. Those keywords must be identified and added to the 'word dictionary' for appropriate recommendations.

Recommendation Engine

Figure 2 gives the flow of Keywords and sentiments from LDA and how Recommendation engines are integrated with these keywords. Recommendation engines are built using filtering algorithms. The Amazon product reviews dataset used for this step.

Twitter text for a day is passed through pre-processing steps. After processing sent to LDA algorithm and top 5 keywords based on their occurrence context are identified. These top keywords are searched under respective sentiments. Whichever sentiment dictionary has most of the keywords (at least 3 keywords), that is voted as the sentiment for the corresponding day's tweets.

All the recommendations are sequenced, and final outcomes will be as per following steps:

- The 'Recommended' products are validated against 'user' purchases and the product which was not purchased by the 'user' before is kept as the top recommendation.
- Then product belonging to the same category is prioritized followed by items from other categories
- For my study, I have taken following categories 'Grocery', Medical' and 'General'
- Then finally, 'Sentiment' from LDA is checked
- If the Sentiment is 'Normal', then the top product is recommended irrespective of the 'category' it belongs. So, in this mode, recommendations can be from any of the three categories.
- If the Sentiment is 'Panic Buying' mode, then product recommendation is restricted to 'Grocery' and 'Medical' categories only.
- If the Sentiment is 'Essential Buying' mode, then product recommendation is restricted to only the same category which is either 'Grocery' or 'Medical' based on the product purchased.

The difference between 'Panic buying' and 'Essential buying' mode is, when consumers buy a product from 'Grocery' during 'Panic buying' mode, then can get recommendations from both 'Grocery' and 'Medical' categories. But products from 'General' categories are not recommended. But in 'Essential buying' mode, if the consumer buys a product in 'Grocery' category, then recommendation is strictly restricted to 'Grocery' category only. If the consumer buys from 'Medical' category, then recommendation also is suggested from 'Medical' category. In this mode, 'General' category is not allowed for recommendation.

So, during 'Normal buying' mode and 'Panic buying' mode, the 'Item Similarity Matrix' is checked for recommendation followed by 'User Similarity Matrix'. This allows to get product recommendations across categories. But during 'Essential buying' mode, only 'Item Similarity Matrix' is used for Recommendation and 'User Similarity Matrix' is ignored.

Post Covid Pandemic, when restrictions are lifted, normalcy will return. Twitter will be able to reflect them in the tweets indicating consumer behavior changes. So, our word dictionary will be able to capture this with top 3 words reflecting to normal buying pattern. When a new Pandemic appears, then the word dictionary will not return the sentiment for top 3 words. This 'Data Drift' gives indication that the LDA model has to be retrained and a new set of keywords to be identified. These new keywords which reflect the new Pandemic scenario will be added to the Word Dictionary based on the LDA model output. Then the model is ready to give recommendations based on the new Pandemic situation as well.

3 Results and Discussion

The research problem has a series of steps to be followed and in each step a best approach was taken. This section explains the results obtained and how these results explain the research problem.

3.1 EDA

Canada Sales Data
Canada Sales Data is public dataset about Sales volume across various categories (Fig. 2).

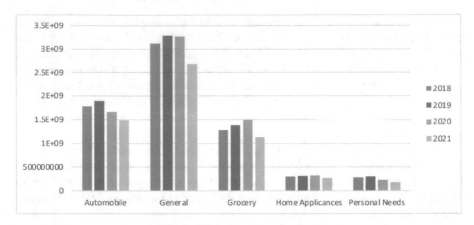

Fig. 2. Category-wise sales data.

The dataset analysis in Fig. 2 gives the following inferences:

- In the year 2020 when the Covid Pandemic has affected the consumer buying, there is an increase in 'Grocery' category. The nine-month sales for this in 2021 is also high and full year projection surpasses 2020 sales value.
- Similarly, there is a marginal increase in 'Home Appliances' category in the year 2020 and the trend continues in 2021 with a healthy rate.
- Similarly, 'Automobile' category saw a dip in the year 2020. But this is recovered in 2021.
- 'General' category also had a marginal dip in 2020. Same is the case for 'Personal Needs' category also.

This EDA proves that there is a change in consumers buying pattern during Covid Pandemic year where most of the purchase shifted towards 'Grocery' and 'Essentials' rather than 'Non-Essentials' or 'Comfort/Luxury' items. This holds good across the globe where there are significant online shoppers.

Amazon Sales Data

Table 1 displays the top sellers who provided highest discounts along with the category where they provided the discounts. We can see that mask are given a maximum of 97% discount offered by several sellers. This answers the research question about how online sellers used the pandemic time to attract consumers by providing huge discounts on necessary items. Since Mask and other medical items are essential items during pandemic period and there is a huge demand for such items during this period, online retailers used the opportunity to provide huge discounts to attract consumers. This highlights how the consumer behavior is influenced by the online retailers during Pandemic period.

Table 1. Top discounted items on Amazon

Seller name	Category	Discount percentage
Any DesignerÂ®	N95 mask	97
	Surgical mask	98
BLUE WISHÂ®	Respiratory mask	95
Bhagirath fab	Surgical mask	89
		99
Om Fab	Surgical mask	96
Siddharth fab	Surgical mask	90
VPINDUSTRIES	Mask	92
ultrasav	Surgical mask	88

3.2 Twitter Analysis

The pre-processing done with regex functions worked very well and cleaned the corpus for analysis. The count vectorizer was able to split the words appropriately and get the count of the words as well. This includes the important symbols as well. To split the words based on context, the LDA algorithm gave very good result with 3 topics. After several runs and training with the corpus, finally the 3 topics are the ones which gives good results. To measure the performance of the topic modelling and the top key words, impurity measure was used (Table 2). The latter measure showed wise words that overlapped across topics. The first topic contained 9 words out of top 20 words which are shared with other topics. The weighted Gini Index for the first four topics was 0.28070707.

The latter three topics in Table 2 showed 5 words out of top 20 words for the first topic; which were overlapping with other topics. The Weighted Gini Index for the latter three topics was 0.16692314. The average Gini impurity index for three topics was the lowest. Also, the top 20 words in the first four topic modeller has words such as 't, ', and space. So, amongst all these, Topics model algorithm 3 topics was the most accurate for this data.

Table 2. Impurity measures for topics

Topics/parameter	Topic 1	Topic 2	Topic 3	Topic 4
First four topics/Overlapping words	9	6	8	3
First four topics/Gini impurity	0.99	0.88	0.97	0.6
Three topics/Overlapping words	5	5	6	NA
Three topics/Gini impurity	0.81	0.81	0.88	NA

Therefore, LDA will give three topics and relevant words for each topic. The LDA algorithm gave three topics which were sentiments that were passed to the Recommender engine. Topic 1 corresponded to 'Normal Buying', Topic 2 corresponded to 'Panic Buying', and Topic 3 corresponded to 'Essential Buying'.

There are opportunities here to improve quality (precision) of the classification by reducing false positives and false alarms (negatives). The future research could build on 'Focus Attention Module (FAM)' baseline design and incorporate a multi-headed attention neural network. These attention-heads have individual tasks where one attention-head is focused on recognizing the noise and clutter to avoid, while the second attention-head is tasked with spatially identifying the disaster specific regions and dynamically developing disaster image mask filters. The third attention-head uses these masks to filter the portion of the original image and direct the neural network classification based on the filtered image.

This is akin to a design proposed by Guan et al. 2020 for Thorax Disease Classification [13]. The results in the research paper by Guan et al. 2020 are interesting and reemphasizes the original premise for this research that a vision driven attention models can aid better in disaster image classification [13]. Understandably, the future solution is not just limited to this recommendation but can build on this research for end-to-end crisis management pipeline.

3.3 Recommender Engine

Two models were constructed in the recommender engine. The first was Market-basket analysis. In this model, since there were no transactions with multiple products that can be associated with a customer, in the customer – product matrix, every product count was increased by one, whenever the customer reviewed that product. If the customer had not reviewed the product, then the value remained zero. This matrix provided the list of products bought together by a customer, even though the purchase might not have happened at the same time. After attempting several combinations for 'lift' and 'confidence' values, the lowest threshold values for which the output came was lift ≥ 1.0 and confidence ≥ 0.8. The model was very weak since the data did not reflect the transactions. Also, the recommendations given by this model could be related to any product across any category. This was purely dependent on the customer purchases across the categories. The underlying assumption of this algorithm was to analyze Transactions containing 'products bought together' and then find 'closest products' among them. But the data frame created as input for this algorithm did not fall under this assumption. By using Collaborative filtering method, we achieved the best results for recommendation.

Twitter text for a day was passed through pre-processing steps. After processing sent to LDA algorithm and top five keywords based on their occurrence context were identified. These top keywords were searched under respective sentiments. Whichever sentiment dictionary had most of the keywords, that was voted as the sentiment for the corresponding day's tweets. When the Recommendation engine algorithm uses the sentiments from Twitter, it chose the recommendations accordingly. Figure 3 shows recommendation given by algorithm during 'Normal conditions'. The products were from all three categories. Top recommendations are sorted from all categories and final top five recommendations are given.

Figure 4 provides recommendations during 'Panic buying' sentiment for a customer. Here the recommendations were from both 'Grocery and Gourmet Food' category as

```
Below are the recommended items for user(user_id = A1249PZ3T4C5Y4)

Item code          Item Name
B00GUGYA9U         Set of 6 Ottoman Peshtemal Turkish Towel...
B000Q5I70Y         M&M Milk Chocolate Candy Singles Size 1.69-Ounce...
B0015MGC1I         Lavazza Qualita Rossa Ground Coffee Blend, Mediu...
B005CPVTHG         Mango Pineapple Fruit Tea | MAUI MANGO 6 Ounce Ti...
B00C1ZNH5M         5 Mixed Plastic Black Banana Ponytail Holders Hai...
```

Fig. 3. Recommendations during 'Normal' Sentiment.

well as 'Home and Kitchen' items were selected. These items come across as some that we may need to refill in advance anticipating upcoming lockdown period.

```
Below are the recommended items for user(user_id = A10ZBR6O8S8OCY)

Item code          Item Name
B00ECOSY08         Strawberry Pineapple Green Tea | FRUITY PEBBLE...
B00014JNI0         YS Organic Bee Farms CERTIFIED ORGANIC RAW HONEY
B00DT4757A         Thermal Polar Ear Muffs Warmer for men
B00IZL255O         Sherpa Pink Himalayan Salt, 2lbs Extra-Fine Grain
B00B91RZ84         Partanna Premium Select Whole Olives, Green Cerig...
```

Fig. 4. Recommendations during 'Panic buying' Sentiment.

Figure 5 gives recommendations during 'Essential buying' sentiment for a customer. Here the recommendations are only from 'Grocery and Gourmet Food' category only. During 'Essential buying' sentiment period, lockdown is in progress and Sales are strictly enforced. In places only essential items are sold. During this period, when customer purchases through online channel, the options are provided for essentials such as 'Food items' or 'Household Medical items'. So, the recommendations are given only in the respective category to the customer. If the customer buys a product in Food item, then recommendations are given only from Food item.

```
Below are the recommended items for user(user_id = A100WO06OQR8BQ)

Item code          Item Name
B00016XLEA         Frontier Turmeric Root Ground, 1.92-Ounce Bottle
B00014JNI0         YS Organic Bee Farms CERTIFIED ORGANIC RAW HONEY
B00028MI1O         La Tourangelle Roasted Almond Oil 16.9 Fl. Oz,...
B000028Q450        Darjeeling Tea Boutique
B0001M11RS         Spice Jungle Whole Nutmeg - 4 oz.
```

Fig. 5. Recommendations during 'Essential' sentiment.

To measure the performance of the model, I want to measure the performance same way as a classification model. Since the model doesn't have any labelled column, I used another column in the dataset which gives 'also_buy' which gives the other products the user already bought. If the Recommendation given by the model is present then it is positive. Also, the model should not recommend the product for which the user gave review. With these criteria, the model's accuracy is 90.76% and recall value of 0.75.

The Recommendation Engine without providing any sentiment always returns the recommendations which is similar to 'Normal buying' sentiment scenario. The items recommended by this engine during Pandemic period will not be accurate since online consumers needs shifted to essential items. My Recommendation Engine, during 'Normal' Sentiment, provides recommendations from all the three product segments which I considered. The product segments are 'Food and Gourmet Items', 'Household and Kitchen Items' and 'Luxury Items'. The engine recommends from all the three categories.

By incorporating additional feature 'Sentiment', the Recommendation Engine can recommend items that will be useful to consumers based on the current sentiment. The research considered two other sentiments. One is 'Panic Buying' period, where there is a anxiety with people about upcoming lockdown period. During this period, they start their shopping to buy essential items such as 'Food Items' but slowly their anxiety drives them to buy other items such as 'Household items', 'Medical Items' etc. to keep them on stock. But they do not spend time in buying unnecessary items such as 'Luxury Items', 'Automotive' etc. Our Recommendation Engine gives output in only two categories 'Food and Gourmet Items' and 'Household Medical Items'.

The Third Sentiment taken is 'Essential Buying'. This happens during lockdown period where movements are restricted, and supply chain is strictly operational for essential items only. So, the consumers will be shopping only for essential item which is 'Food and Gourmet Items' or 'Medical Items'. So, whenever they buy an item in one of these two categories, the recommendation will be given only from that category only. This is clearly seen from my Recommendation engine's output. When a customer bought item from 'Food and Gourmet Item', the recommendations are given only from that category. When the sentiment keywords are not identified in the list, then it means the recommendation engine gives default status which is 'Normal buying' behavior. But if there are any repeated keywords identified, then the model must be retrained to understand the sentiments given by these words and appropriately the keyword list library must be configured so that recommendation engine works according to the latest sentiment.

4   Conclusions

The approach performed in this research with steps on Twitter extract for current hot topics and applying sentiments using LDA model is one a very powerful method to understand people behavior during any given period. Since social media platform is currently reflecting majority of the people's thoughts in near real-time, collective understanding of this is useful for any analysis relevant to business. The research here taken is for the benefit of online retailers where this can help in understanding consumers mindset and stock up only those items which consumers will buy, which is the outcome from Social Media Sentiment. This can be applied for any other business also. An insurance organization can provide property insurance based on hurricane and bad weather prediction topics discussed in social media.

Acknowledgment. The authors thank UNITAR for the support of the publication of this paper.

References

1. Mason, A.N., Narcum, J., Mason, K.: Social media marketing gains importance after Covid-19. Cogent Bus. Manage. **8**(1), 1870797 (2021)
2. Chevalier, S.: Global retail e-commerce sales 2014–2024. Statista (2021). https://www.statista.com/statistics/379046/worldwide-retail-e-commerce-sales/
3. Policarpo, L.M., et al.: Machine learning through the lens of e-commerce initiatives: an up-to-date systematic literature review. Comput. Sci. Rev. **41**, 100414 (2021)
4. Kwon, S., Kim, E.: Sustainable health financing for COVID-19 preparedness and response in Asia and the Pacific. Asian Econ. Policy Rev. **17**(1), 140–156 (2022)
5. Modgil, S., Singh, R.K., Gupta, S., Dennehy, D.: A confirmation bias view on social media induced polarisation during Covid-19. Inf. Syst. Front. 1–25 (2021)
6. Amutha, J., Sharma, S., Sharma, S.K.: Strategies based on various aspects of clustering in wireless sensor networks using classical, optimization and machine learning techniques: review, taxonomy, research findings, challenges and future directions. Comput. Sci. Rev. **40**, 100376 (2021)
7. Rasul, G., et al.: Socio-economic implications of COVID-19 pandemic in South Asia: emerging risks and growing challenges. Front. Sociol. **6**, 629693 (2021)
8. Barnes, S.J., Diaz, M., Arnaboldi, M.: Understanding panic buying during COVID-19: a text analytics approach. Expert Syst. Appl. **169**, 114360 (2021)
9. Adwan, O., Al-Tawil, M., Huneiti, A., Shahin, R., Zayed, A.A., Al-Dibsi, R.: Twitter sentiment analysis approaches: a survey. Int. J. Emerg. Technol. Learn. (iJET) **15**(15), 79–93 (2020)
10. Kersten, J., Klan, F.: What happens where during disasters? A workflow for the multifaceted characterization of crisis events based on Twitter data. J. Contingencies Crisis Manage. **28**(3), 262–280 (2020)
11. Gao, Z., Feng, A., Song, X., Wu, X.: Target-dependent sentiment classification with BERT. IEEE Access **7**, 154290–154299 (2019)
12. Biswas, A., Vineeth, K.S., Jain, A.: Development of product recommendation engine by collaborative filtering and association rule mining using machine learning algorithms. In: 2020 Fourth International Conference on Inventive Systems and Control (ICISC), pp. 272–277. IEEE (2020)
13. Guan, Q., Huang, Y., Zhong, Z., Zheng, Z., Zheng, L., Yang, Y.: Thorax disease classification with attention guided convolutional neural network. Pattern Recognit. Lett. **131**, 38–45 (2020)

Big Data Application on Prediction of HDD Manufacturing Process Performance

N. G. Meng Seng[1], Abdulaziz Al-Nahari[1(✉)], Noor Azma Ismail[2], and Azlin Ahmad[3]

[1] UNITAR Graduate School, UNITAR International University, Petaling Jaya, Selangor, Malaysia
Abdulaziz.yahy@unitar.my
[2] Faculty of Business and Technology, UNITAR International University, Petaling Jaya, Selangor, Malaysia
[3] Faculty of Computer and Mathematical Sciences, Universiti Teknologi MARA, Shah Alam, Malaysia

Abstract. Every industry including manufacturing is going through digital transformation in this digital era we are in today. The digitization has transforming manual manufacturing to a highly connected end-to-end manufacturing leveraging on IoT and sensors. One of the common focuses in manufacturing is product yield and quality. However, lack of supplier's data due to the manufacturing data generated is not shared and integrated into a common data platform becomes challenge among manufacturer to improve product's yield and quality. The objective of this paper is to demonstrate the application of big data analytics on Hard Disk Drive (HDD) manufacturing and supplier data to predict HDD product yield/performance. In this research, the supplier data is focused on Head (read/write data) and Media (store data) component because it is critical to HDD's performance in achieving the targeted storage capacity (i.e., Terabytes). The scope of this study includes data integration requirements and methods between supplier and HDD manufacturer, data platform to store and retrieve the integrated data, and the feasibility of predicting HDD product yield/performance using analytics with supplier's data using Machine Learning. At the end of this research, target user groups will gain the ability to store and retrieve the integrated data between supplier and manufacturer from a big data platform, and to optimize manufacturing process for HDD performance using the insights and predictions generated by machine learning models.

Keywords: Product yield · Data analytics · Big data

1 Introduction

Digital era has transformed almost every industry from 1.0 to 4.0. This digital transformation concerned with interconnectivity, automation, machine learning and real-time data that could led to creation of better-connected ecosystem for organizations. Similarly, manufacturing industry is also going through such transformation [1]. As every organization operating is unique, they possibly face with different challenge in terms of

© The Author(s), under exclusive license to Springer Nature Singapore Pte Ltd. 2023
Y. B. Wah et al. (Eds.): DaSET 2022, LNDECT 165, pp. 222–236, 2023.
https://doi.org/10.1007/978-981-99-0741-0_16

implementation, but the objectives to achieve is to adopt an industry 4.0 model for their businesses to improve manufacturing efficiency.

One of the challenge faces by manufacturers to achieve greater products improvements is lack of supplier's data due to the manufacturing data generated is not shared and integrated into a common data platform. The data is owned by the respective organization. Hence, the purpose of this study is to demonstrate the application of big data analytics on Hard Disk Drive (HDD) manufacturing and supplier data to predict HDD product yield and performance. The supplier data is focused on this study is on Head and Media component because it is critical to HDD's performance in achieving the targeted storage capacity (i.e., Terabytes) [2].

2 Related Works

Many researchers made attempts to understand the benefits of Big Data Analytics in manufacturing [3]. Applying analytics produces new insights and intelligence to improve manufacturing efficiency and productivity for growth. [4] These Analytics intelligence includes Descriptive, Diagnostics, Predictive and Prescriptive Analytics (Fig. 1) [5].

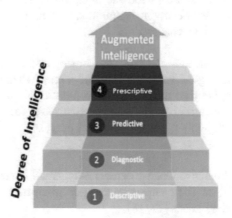

Fig. 1. Analytics degree of intelligence

The key enabler to this generating these analytics insights is data. Data has always been the soul of manufacturing. Without data, there is no gauge or visibility to the health of the manufacturing process and products being manufactured [6]. In the past, manufacturers have limited ability to collect data because of the high cost of sensors and storage platform to keep data. Moreover, there is also lack of computation power to generate analytics insights. [7] Today, with the progression in 4IR technologies, all these constraints and limitation are no longer concerns. The cost of sensors continues to reduce, and it is cheaper to implement everywhere in the manufacturing. Whereas Big Data platform (for example, Hadoop) allows massive storage capability with redundancy and computing that is more cost effective compared to conventional Datawarehouse [8].

As a result of this, many manufacturers are rapidly adopting 4IR technologies to transform their business to improve productivity by producing more with lower cost to stay competitive [9]. This is especially crucial in response to various global challenges like supply chain breakdowns and economic instabilities. Not only that those manufacturers who adopted 4IR managed to overcome these challenges, they also generated new growth for their businesses. [10] It took more than 50 years for the transition of Industry 3.0 to 4.0 and for its technology to mature [11, 12]. So, the time is now to reap the benefits.

Some of the common technical challenges to establishing such IIoT stack (aka platform) are:

1. Most manufacturers have legacy and outdated systems. These systems are already deeply rooted into the production process that cannot be unplugged and replaced easily without causing outages and impacting business [13].
2. Right sizing the platform for current and future at optimal ROI (Return of Investment) [14]. An easy approach is to deploy large platform with all technologies, but this requires high investment which is not business friendly.
3. Selecting the right technology from the many options that will address Big Data and Fast Data [14]. What works for one does not mean it will work for another. Determining the right ever-evolving technology and partner to work with to build an ideal solution stack can be challenging.

Being able to store, process and compute big or fast data through the enabling technologies is just the beginning step. The value comes when new analytics insights can be generated from the data to seize improvement opportunities like productivity and efficiency in manufacturing. [15] There are many big data analytics use cases in manufacturing. The commonly deployed use cases are: [16].

1. Asset Utilization
2. Product Quality: Predictive Quality detects even small deviation from normal and predict potential failures during manufacturing from the many process variables thus avoiding product yield issues [17, 18].
3. Supply Chain Optimization

3   System Design

The model development approach used for this research consist of 6 stages which are Business Understanding, Data Collection, Data Preparation, Modelling, Evaluation and Deployment as shown in Fig. 2.

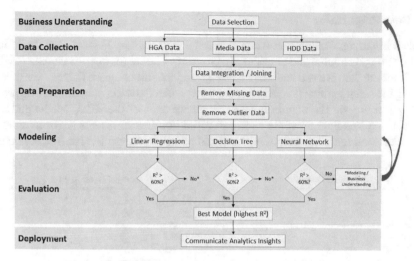

Fig. 2. Model development approach

3.1 Business Understanding

HGA and Media data from the supplier was selected as they are presented the magnetic performance of the read/write operations in the HDD which correlates to ACC and yield. [2] For the reason of data confidentiality, the HGA data is renamed and denoted by HGA1, HGA2, HGA3 etc. while Media data is denoted by Media1, Media2, Media3 etc.

3.2 Data Collection

The 4 weeks of manufacturing data were collected and extracted from Big Data platform (Hadoop) which contains about 460,000 rows of data as shown in Fig. 3. The dataset contains a variety of manufacturing process and test data from Media, HGA and HDD.

week	N
1913	41671
1914	161824
1915	189935
1916	74958
N	468388

Fig. 3. Total weeks and data rows

3.3   Data Preparation

The data is extracted using Presto query from HDFS (Hadoop Distributed File System). Presto is an open source and distributed SQL query engine for queries on data in HDFS [19]. After the data is extracted, the first step is to join and integrate the datasets together by using the serial number (i.e., adds) of each part collected during manufacturing process as shown in Fig. 4. The dataset is flattened where each of the HGA, Media and HDD data represents as column while each part serial number represents a row as shown in Fig. 5. For confidentiality purposes, the actual names in the dataset have been renamed.

HGA data query

```
select
hddsn slidersn
,HGA1
,HGA2
,HGA3
from vqaa.fact_hdd_hga
where product = 'product'
and enddt>='20211102'
and enddt<'20211202';
```

hddsn = HDD serial number
slidersn = HGA serial number

Media data query

```
select
hddsn mediasn
,media1
,media2
,media3
from vqaa.fact_hdd_media
where product = 'product'
and enddt>='20211102'
and enddt<'20211202';
```

hddsn = HDD serial number
mediasn = Media serial number

Fig. 4.   Example of Presto data query on HDFS

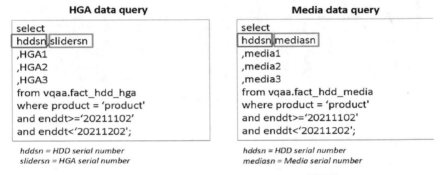

Fig. 5.   Example of integrated data from HGA, Media and HDD

The next step after data integration is data cleansing. The dataset is examined for missing data and outlier data. If missing data or outliers are found, the row of data is removed. It is important to reduce the influence of outliers in the response variable. For detection of outliers in the dataset, Robust Fit using Huber M-estimation method is applied. Given a robust estimate of the center and spread, outliers are those values that are 4 times (i.e., K) the robust spread from the robust center. Huber M-estimation finds parameter estimates that minimize the Huber loss function.

3.4 Modelling

The algorithm used for this research are Linear Regression, Decision Tree and Neural Network representing regression, classification, and deep learning types of algorithms respectively. The variety of algorithms are considered to evaluate types of algorithms that would produce highest prediction accuracy.

Linear Regression
Linear Regression is the most common type of least squares fitting for continuous response variables. Configuration of the regression model is shown in Fig. 6:

a) Model: Standard Least Squares\
b) Input Variables: HGA and Media data
c) Target Response: HDD's ACC
d) Training/Validation data split: 70/30

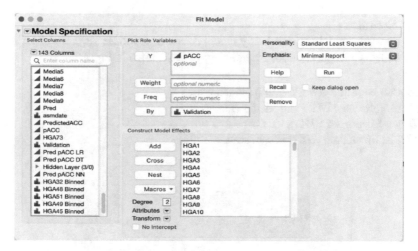

Fig. 6. Regression model configuration setup

Decision Tree
It recursively splits data according to a relationship between the input variables and target response until the desired fit is reached. Configuration for decision tree is shown in Fig. 7:

a) Model: Decision Tree with Depth = 6
b) Input Variables: HGA and Media data
c) Target Response: HDD's ACC
d) Training/Validation data split: 70/30

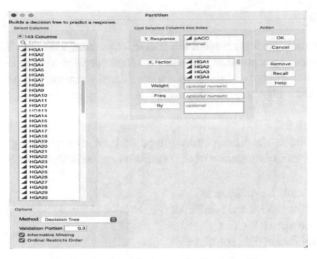

Fig. 7. Configuration setup for Decision Tree

Neural Network

Neural network model can efficiently model different response surfaces given enough hidden nodes. Configuration for Neural Network as shown in Fig. 8:

e) Model: Neural Network with 3 hidden nodes
f) Input Variables: HGA and Media data
g) Target Response: HDD's ACC
h) Validation: KFold = 8 (random seed = 12345)

Fig. 8. Configuration setup for Neural Network

3.5 Evaluation

Since the prediction of the target response is a continuous variable (i.e., HDD's ACC), R^2 is used to evaluate the accuracy of the prediction. R^2 is a statistical measure that represents the proportion of the variance for a dependent variable that's explained by an independent variable. For example, a R^2 value of 0.5 (or 50%) means that half of the variance in the outcome variable is explained by the model. Typically, R^2 between 0.3 and 0.5 is considered weak prediction, between 0.5 and 0.7 is considered moderate, and above 0.7 (maximum of 1) is considered good [24].

In this paper, the targeted R^2 for Predicted ACC versus Actual ACC is 0.6 based on validation dataset. Although the HDD's ACC performance is primarily driven by HGA and Media performance, there are other influencing factors from the mechanical components in the HDD. So, a high R^2 is not expected. If any of the three models are unable to achieve at least R^2 of 0.6, then the model configurations (modelling stage) or data selection (business understanding stage) need to be revisited. However, if there are more than one model achieved higher than R^2 of 0.6, then the model with the highest R2 is selected for deployment.

3.6 Deployment

Once the best prediction model is selected, the following analytics insights is generated for engineers optimize the manufacturing process to improve HDD's performance.

1. Variable Importance ranks from the prediction model
2. Top ranked variable importance sensitivity against HDD's ACC
3. Top ranked variable importance sensitivity against HDD's Yield

4 Result

4.1 Data Preparation Result

In this paper, the integrated data is profiled for missing data and outliers. Any missing or outlier data rows are removed. In the initial datasets, a total of 468,388 rows of data with 85 columns were extracted. After missing data is removed, the remaining total rows of data became 463,930 in the dataset. When the outlier data is also removed, the remaining total rows of data became 418,936 in the dataset. The final dataset contains 418,936 rows and 85 columns as shown in Fig. 9. The 85 columns consist of 82 variable features from HGA, Media and HDD data while the remaining 3 columns are the part serial numbers.

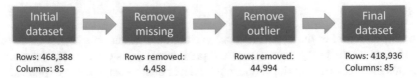

Rows: 468,388   Rows removed: Rows removed: Rows: 418,936
Columns: 85   4,458   44,994   Columns: 85

Fig. 9. Data preparation steps and outcomes

4.2 Prediction Model Results

In this section, we will discuss the results of applying the machine learning models on the prepared dataset and discuss the findings.

Linear Regression Model

The R2 for HDD's ACC is 0.728 and 0.727 for training and validation dataset respectively. This indicates the training is consistent with validation with no signs of potential overfitting. A total of 40 out of total 82 variable features were selected by the Linear Regression model as shown in Fig. 10.

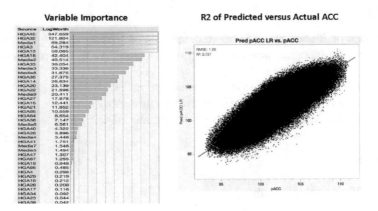

Fig. 10. Prediction results from Linear Regression

Decision Tree Model

The R2 for HDD's ACC is 0.608 and 0.609 for training and validation dataset respectively. This indicates the training is consistent with validation with no signs of potential overfitting. A total of 16 out of total 82 variable features were selected by the Decision Tree model as shown in Fig. 11.

Neural Network Model

The R2 for HDD's ACC is 0.739 and 0.740 for training and validation dataset respec-

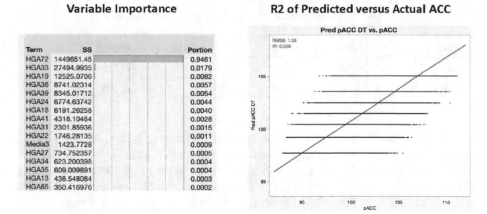

Fig. 11. Prediction results from Decision Tree

tively. This indicates the training is consistent with validation with no signs of potential overfitting. A total of 19 out of total 82 variable features were selected by the NeuralNetwork model as shown in Fig. 12.

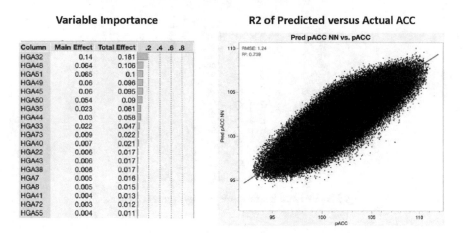

Fig. 12. Prediction results from Neural Network

4.3 Model Accuracy Evaluation and Selection

The R^2 produced from the training and validation dataset is relatively consistent for all the three types of algorithms as shown in Table 1. This indicates no potential sign of overfitting. Overfitting occurs when the model trained too well but could not achieve the same prediction accuracy in the validation as the training. An overfitting model will result in high prediction error if it is deployed.

Table 1. Summary of R^2 results from the prediction models

Dataset	Linear regression	Decision tree	Neural network
Training R^2	0.728	0.608	0.739
Validating R^2	0.72	0.609	0.740

Based on the R^2 from all the three types of algorithms, Neural Network model produced the best R^2, and prediction accuracy indicated in Table 1. Neural Network model achieved the highest R^2 of 0.74 (i.e., Champion model) with 19 variable features identified by the model. R^2 above 0.7 is considered good with strong effect size. The prediction result from the Neural Network model is used to generate analytics insights for engineers to optimize the manufacturing process to improve HDD's performance.

19 of 82 total variable features were selected by Neural Network model as shown in Fig. 13. The top 5 variable features are HGA32, HGA48, HGA51, HGA49, HGA45.

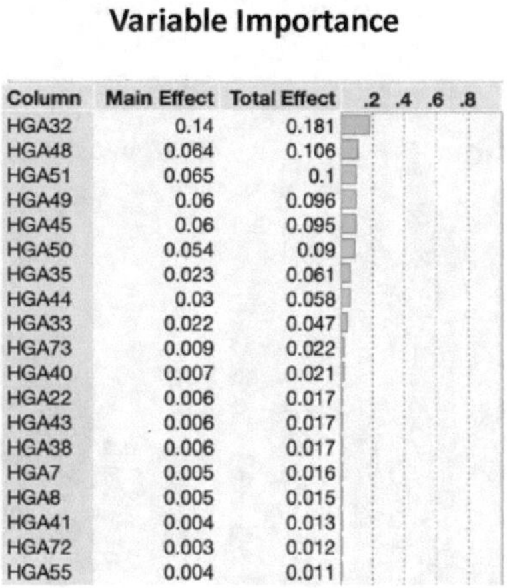

Variable Importance

Column	Main Effect	Total Effect	.2 .4 .6 .8
HGA32	0.14	0.181	
HGA48	0.064	0.106	
HGA51	0.065	0.1	
HGA49	0.06	0.096	
HGA45	0.06	0.095	
HGA50	0.054	0.09	
HGA35	0.023	0.061	
HGA44	0.03	0.058	
HGA33	0.022	0.047	
HGA73	0.009	0.022	
HGA40	0.007	0.021	
HGA22	0.006	0.017	
HGA43	0.006	0.017	
HGA38	0.006	0.017	
HGA7	0.005	0.016	
HGA8	0.005	0.015	
HGA41	0.004	0.013	
HGA72	0.003	0.012	
HGA55	0.004	0.011	

Fig. 13. Variable features selected by Neural Network

The top ranked variable features selected by Neural Network model clearly shows the sensitivity towards HDD's ACC as shown in Fig. 14. For example, the lower the HGA32 value, the higher the HDD's ACC. However, the higher the HGA48 value, the higher the HDD's ACC. This analytics insight allows engineers to understand the optimal supplier's process variables to achieve higher ACC.

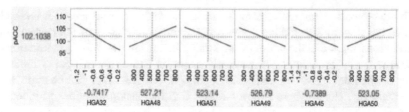

Fig. 14. HDD's ACC sensitivity against top ranked variable features

The top ranked variable features selected by Neural Network model clearly shows the sensitivity towards HDD's yield as depicted in Fig. 15. For example, the lower the HGA32 value, the higher the yield. However, the higher the HGA48 value, the higher the yield. This analytics insight allows engineers to understand the optimal supplier's process variables to achieve higher yield.

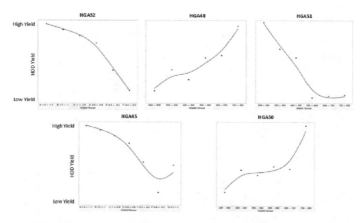

Fig. 15. Yield sensitivity against top ranked variable features

4.4 Neural Network Implementation and Evaluation for Higher Prediction Accuracy

Neural Network is a computational model that mimics human's biological brain in processing information. A basic neural network usually has only one node where it receives input from some other nodes and then computes an output as. Each input has an associated weight which is assigned based on the relative importance to the inputs.

A feedforward neural network contains multiple nodes arranged in layers. Nodes from adjacent layers have connections or edges between them where weights are applied. The hidden nodes have no direct connection with the outside world because computation is done and transferred from input nodes to output nodes. A collection of hidden nodes forms a hidden layer. The higher the number of hidden nodes increases the expressive power of the network which may further improve the prediction accuracy but significantly increases the computation resources [25].

The two layers neural network model implemented in this project uses three hidden nodes in each layer. The hidden nodes are nonlinear functions of the original inputs. The functions applied at the nodes of the hidden layers are the activation functions. The activation function is a transformation of a linear combination of the input variables. With an optimal number of hidden nodes and layers, response surface can be approximated to any accuracy. Hence, there is a possibility that increasing number of hidden nodes may further increase the accuracy of the model. Additional study is performed on the neural network by varying the number of hidden nodes to see if the R2 continues to improve further. As seen in Fig. 16, the R2 (prediction accuracy) continues to improve as the number of hidden nodes increases. However, the prediction accuracy increments are small. The highest achieved R2 is 0.748 with 7 hidden nodes. Comparing to R2 achieved with 3 hidden nodes, the difference is only 0.008. With the consideration of increasing computation resources as number of hidden nodes increases, a small increase in R2 accuracy of only 0.008 is not justifiable for this project. Hence, it is concluded that 3 hidden nodes are sufficient in this case.

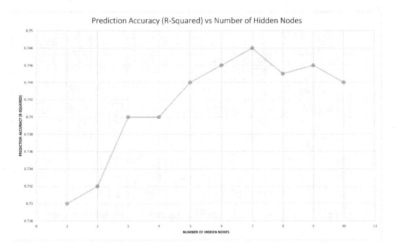

Fig. 16. Prediction Accuracy (R-Squared) versus Number of Hidden Nodes

In summary, the additional study conducted on varying the neural network's number of hidden nodes concluded that the R2 prediction accuracy does improve but at a very negligible and small magnitude. Hence, it is concluded that the original setup configuration of the neural network with 3 hidden nodes is sufficient.

5 Conclusion

This paper has successfully demonstrated the possibility of improving manufacturer's product yield and performance with supplier's manufacturing process data. The approach demonstrated in this paper is a valuable add to the research topic related to manufacturing analytics. While there are limitations observed related to yield improvement realization time because of the long cycle time between supplier and manufacturers, this capability is already valuable to reduce number of trial-and-error iterations at the suppliers refining their manufacturing process. There is also an opportunity for possible improvement of the prediction model accuracy by identifying and collecting even more supplier manufacturing process data.

Acknowledgement. The authors thank UNITAR for the support of the publication of this paper.

References

1. Mahiri, F., Najoua, A., Souda, S.: Data-driven sustainable smart manufacturing: a conceptual framework. IEEE (2020)
2. Frank, P., Wood, R.: A perspective on the future of hard disk drive (HDD) technology (2006)
3. Villareal, G., Na, J., Lee, J.: Advantages of using big data in semiconductor manufacturing. IEEE (2018)
4. Berges, C., Bird, J., Shroff, M., Rongen, R., Smith, C.: Data analytics and machine learning: root-cause problem-solving approach to prevent yield loss and quality issues in semiconductor industry for automotive applications (2021)
5. Vater, J., Harscheidt, L., Knoll, A.: Smart manufacturing with prescriptive analytics. IEEE (2019)
6. Moyne, J., Samantaray, J., Armacost, M.: Big data capabilities applied to semiconductor manufacturing advanced process control. IEEE (2016)
7. Shah, S., Soriano, C.B., Coutroubis, A.D.: Is big data for everyone? The challenges of big data adoption in SMEs. IEEE (2017)
8. Gokalp, M.O., Kocyigit, A., Kayabay, K., Eren, P.R., Zaki, M., Neely, A.: OpenSource big data analytics architecture for businesses. IEEE (2019)
9. Chiang, L., Lu, B., Castillo, I.: Advanced in big data analytics at the Dow chemical company (2017)
10. Zhou, J., Yao, X., Zhang, J.: Big data in wisdom manufacturing for industry 4.0. IEEE (2017)
11. Khan, M., Wu, X., Xu, X., Dou, W.: Big data challenges and opportunities in the hype of industry 4.0 (2017)
12. Hurta, M., Noskievicova, D.: Literature review, research issues and future perspective of relation between industry 4.0 and lean manufacturing. IEEE (2021)
13. Thumati, B., et al.: Large-scale data integration for facilities analytics: challenges and opportunities (2020)
14. Illa, P., Padhi, N.: Practical guide to smart factory transition using IoT, big data and edge analytics (2018)
15. Ramdasi, P., Ramdasi, P.: Industry 4.0: opportunities for analytics. IEEE (2018)
16. Sheth, A.: Manufacturing analytics and industrial internet of things (2017)
17. Fan, X., Zhu, X., Kuo, K.C., Lu, C., Wu, J.: Big data analytics to improve photomask manufacturing productivity. IEEE (2017)
18. Li, X., Tu, Z., Jia, Q., Man, X., Wang, H., Zhang, X.: Deep-level quality management based on big data analytics with case study. IEEE (2017)
19. Bansal, S., Mammo, M.: Distributed SPARQL over big RDF data. IEEE (2015)
20. Efroymson, M.A.: Multiple regression analysis. In: Ralston, A., Wilf, H.S. (eds.) Mathematical Methods for Digital Computers. Wiley, New York (1960)
21. Breiman, L., Friedman, J.H., Olshen, R.A., Stone, C.T.: Classification and Regression Trees. Chapman and Hall/CRC, UK (1984)

22. Schmidhuber, J.: Deep learning in neural networks: an overview. Neural Netw. **61** (2015)
23. Phung, N.M., Mimura, M.: Data augmentation of JavaScript dataset using DCGAN and random seed. IEEE (2021)
24. Moore, D.S., Notz, W.I., Flinger, M.A.: The Basic Practice of Statistics, 6th edn. W. H. Freeman and Company, New York (2013)
25. Fletcher, L., Katkovnik, V., Steffens, F.E., Engelbrecht, A.P.: Optimizing the number of hidden nodes of a feedforward artificial. IEEE (2002)

Visualising Economic Situation Through Malaysia Economic Recovery Dashboard (MERD)

Wan Ahmad Ridhuan Wan Jaafar[✉], Mazliana Mustapa, Fatin Ezzati Mohd Aris, Noradilah Adnan, and Ahmad Najmi Ariffin

Core Team Big Data Analytics, Department of Statistics Malaysia, Block C6, Complex C, Putrajaya, Malaysia
ridhuan@dosm.gov.my

Abstract. This paper emphasis on an interactive dashboard to visualise the economic situation in Malaysia by providing input and insights for better decision-making using the Malaysia Economic Recovery Dashboard (MERD). Most of the economic indicators are on different platforms, and it is difficult to make references and monitor the economic situation. To address this issue, MERD (www.dosm.gov.my/economydb) which is a web-based dashboard, has been developed as a user-friendly tool that collects economic indicators from many sources. This dashboard provides a quick overview of the country's economic position in the short term. MERD was developed using Tableau software to visualise data into graphs and charts. MERD consists of nine key economic indicators and 39 sub indicators, namely Gross Domestic Product (GDP), Labour Force, Manufacturing, Prices, Services, Commodities, Balance of Payments, International Trade and Other indicators. The latest version of MERD known as 'version 2' has been optimised through web-based technology. The version 2 are able to reduce the loading time of the dashboard. MERD serves as a catalyst for economic monitoring and provides input for decision-making processes.

Keywords: Economy · MERD · Visualisation · Dashboard · Malaysia economic recovery

1 Introduction

The COVID-19 pandemic has impacted the global economy and hampered global supply and demand in 2020 [1]. Malaysia is no exception in experiencing the impact of this pandemic that has significantly affected the country's economic situation throughout the year due to the Movement Control Order (MCO) and business closure. The economic uncertainty can be visualised and monitored through a dashboard. It is feasible to develop the analytical skills required for the integrated visualisation of measurements and variables in a dashboard [2, 3]. Visualisation helps users to understand the data and provide meaningful insights [4, 5].

© The Author(s), under exclusive license to Springer Nature Singapore Pte Ltd. 2023
Y. B. Wah et al. (Eds.): DaSET 2022, LNDECT 165, pp. 237–249, 2023.
https://doi.org/10.1007/978-981-99-0741-0_17

Prior to now, we've only used conventional methods, including obtaining printed materials from the appropriate website. Data sources are also available for download from the web-site. There are multiple sources of data, and they are not organised. As a result, users find it difficult to access data. As more people are relying on online services during the COVID-19 pandemic, notably during the movement control orders (MCO), hence there is a need for data access in a digital format. Thus, this project has utilised visualisation techniques in developing a dashboard for monitoring current economic situations in one platform. By using tools from visual culture, ad imagery, or visual art that goes beyond the bounds of normal communication [4, 6]. The MERD project aims to build a solution to support decision-making regarding economic situations based on the visualisation by providing meaningful insights. The objective of the MERD is: (1) to provide an overview of the Malaysian economy based on selected economic indicators. (2) to display selected economic indicators interactively for monitoring the current economic situation (3) to provide input for decision making related to the Malaysian economy.

The Malaysia Economic Recovery Dashboard (MERD) was introduced on 2 December 2021 which consists of 39 sub-indicators in the form of monthly, quarterly, and annual time series for the nine core economic indicators namely Gross Domestic Product (GDP), Labour Force, Commodities, Prices, Manufacturing, Services, International Trade, Balance of Payments and Other Indicators.

MERD is the initiative by DOSM to increase public awareness and understanding of the key economic indicators and the relationship between sub-indicators that provide the current economic and recovery scenario. Besides that, it also assists the government's efforts to revive and revitalise the economy and the rakyat's well-being due to the COVID-19 pandemic. By offering current, precise, and comprehensive information, this dashboard may deliver information to stakeholders in the shortest amount of time, cutting expenses and enhancing the decision-making process [6, 7].

During the COVID-19 pandemic, MERD emerged as a new DOSM initiative which enables the government to make the right intervention at the right time and at the same time create public awareness on the economic indicators. The economic situation and economic recovery in the country is reflected in MERD through the analysis of the relationships between the indicators. In July 2022, DOSM improved the MERD into version 2 by upgrading the visualisation using the combination of web development tools such as HTML and CSS with Tableau as compared to the previous version which only used Tableau. Tableau has the ability by recommending a right visualisation depending on the types of fields that choose by the user [8, 9].

2 Dashboard Design

Data visualisation aims to maximise understanding and deliver important economic information to users effectively. All the information is gathered and visualised in a dashboard. The following diagram shown in Fig. 1, represents the architecture of the dashboard.

There are four (4) phases in designing the dashboard. Starting from Phase 1, which is gathering the data source. Then continue with Phase 2 (Data Preparation), Phase 3 (Data Visualisation) and Phase 4 (Data Output).

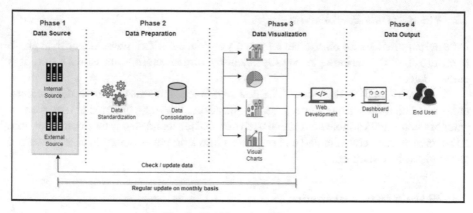

Fig. 1. Dashboard architecture.

2.1 Phase 1: Data Source

There are seventeen (17) economic indicators included in the dashboard which represent 17 data sources. The economic-related data were gathered from eleven (11) internal sources and six (6) external sources. The internal source includes DOSM data while external data includes data from other agencies. All 17 economic indicators data were collected from published data on a monthly, quarterly and yearly basis from the year of 2019 until the current month. The list of economic indicators is shown as below:

Internal Data

1. Gross Domestic Product
2. Labour Force
3. Commodities
4. Prices
5. Manufacturing
6. Services
7. Trade
8. Balance of Payments
9. Leading Index
10. Income of e-Commerce
11. Business Tendency Statistics

External Data

1. Business Registration
2. Credit Card
3. Utilities (Electricity)
4. Vehicles
5. Ports
6. Tourist Arrivals

2.2 Phase 2: Data Preparation

Data preparation is often referred to informally as data prep. This phase is a critical phase to ensure all of the data are placed properly in the dimensions and measures before it is consolidated.

There are 2 steps in this phase. The first step is to standardise the multidimensional data collected from various sources. Standardisation of the data in terms of date format, category name and variable name have been done. The second step is data consolidation. All data have been consolidated and organised into a proper structure in a database and are ready to be visualised.

2.3 Phase 3: Data Visualisation

The process of designing the dashboard starts with mental imagery whereby all the ideas were sketched first and continued with visualising each chart using a visualisation tool.

Business intelligence tools are utilised in the creation of the dashboard. Tableau was used as the visualisation tool in designing the dashboard. The dashboard has to be designed in an effective and usable way to make sure that the visualisation created conveys the message of the data clearly to the audience [10]. Multiple dashboards were created for each selected indicator. Web interface was then developed as the domain hosting for all the created dashboards.

2.4 Phase 4: Data Output

In this last phase, the dashboard for each economic indicator was checked to make sure that all the values and dashboard functions (navigation button) work well to avoid any error for user experience. The dashboard visualisations are ready to be published for the public reference once approved. Details for the Phase 3 and Phase 4 were discussed in the following section.

3 Dashboard for Data Visualisations

The national economy circuit shown in Fig. 2 depicts how income is distributed both within a nation's economy and between that economy and the rest of the globe. The economic circuit defines economic flows (different sorts of transactions) carried out by the resident economic players as producers or consumers, based on symmetrical transactions with a financial counterpart [11]. Each flow is represented by an indicator to monitor the economic situation in a country. In order to make sure each significant economic indicator can be monitored thoroughly, MERD has been proposed to be developed. This visualisation aims to provide quick and accurate information on the short-term indicator of the overall economic position in line with the economic cycles in one view.

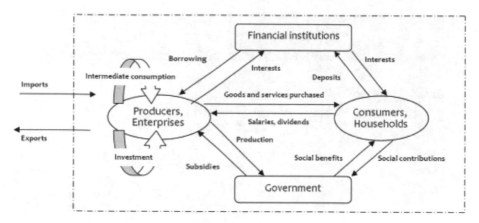

Fig. 2. National economy circuit [11].

Developing this dashboard is a crucial process as it is important that each requirement of the visualisations was aligned with its purpose. This section discusses all the dashboard design and what is available in each dashboard. MERD consists of nine (9) main indicators with one (1) overall indicators page.

All dashboards developed contain information and time series charts. Users can filter in terms of time series periods according to their respective needs. In addition, users can also see the growth and value of each indicator either monthly, quarterly or annually.

3.1 Overall Dashboard

Overall section presents the summary of each indicator consisting of current month/quarter/annual growth and time series data for each indicator at the national level as shown in Fig. 3. The information in the dashboard has been segregated between monthly, quarterly and annual series of data. The growth is calculated by using the following formula:

$$Growth = \left(\frac{Current\ Value}{Previous\ Value} - 1 \right) \times 100 \tag{1}$$

3.2 Dashboard by Indicators

Details for each indicator are displayed in each dashboard according to the respective indicator. There are 8 main indicators in this dashboard and 9 sub-indicators under the other indicator section.

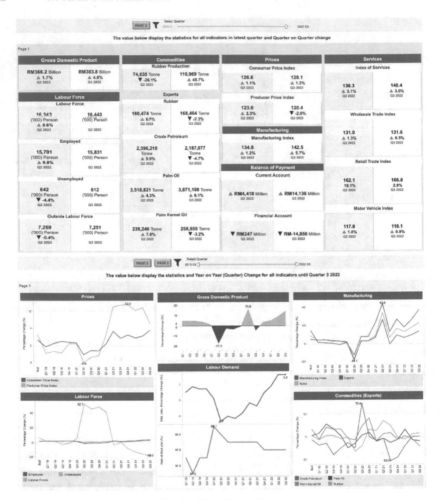

Fig. 3. Overall dashboard.

Gross Domestic Product: The Gross Domestic Product (GDP) dashboard shown in Fig. 4 displays GDP on a quarterly basis at constant 2015 prices. Time series for GDP growth on a monthly, quarterly and annual basis is also provided to observe the country's economic situation at a certain period. In addition, the details of each GDP by the main sector for both production and expenditure approaches are also displayed and detailed in a separate tab.

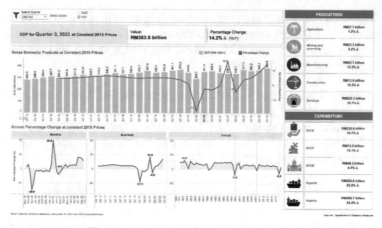

Fig. 4. Gross domestic product dashboard.

Labour Force: Labour force dashboard presents 4 main labour indicators:

- Labour force and labour force participation rate
- Outside labour force
- Employed labour force
- Unemployed and unemployment rate

These four indicators display monthly, quarterly and annual growth with annual time series starting from 1982. On top of that, the visualisation of labour demand is also presented quarterly.

Commodities: Commodities indicators produce a dashboard on production and exports of rubber as well as exports of crude petroleum and palm oil. These three (3) commodities were among the top commodities in Malaysia that consistently contribute to the national economy.

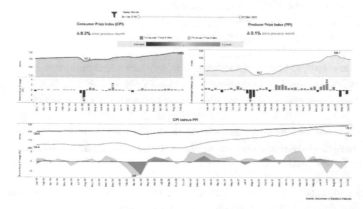

Fig. 5. Prices dashboard.

Prices: The Price indicator is important in reflecting what happens to the economy and purchasing power. Figure 5 shows the dashboard contains two (2) main indicators on prices, the Consumer Price Index (CPI) and Producer Price Index (PPI) to compare both trends, especially on a monthly basis.

Manufacturing: Manufacturing dashboard as presented in Fig. 6 focused on manufacturing sectors consist of Manufacturing Index, Sales of Manufacturing Products and Exports of Manufacturing Products. All three (3) sub-indicators displayed in this section were published monthly through Index of Industrial Production, Monthly Manufacturing Statistics and Malaysia External Trade Statistics.

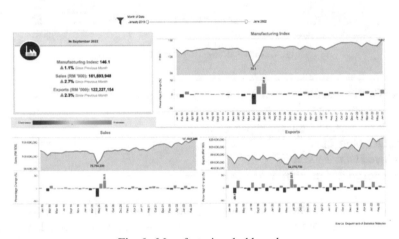

Fig. 6. Manufacturing dashboard.

Services: The services sector is the largest sector in Malaysia with a contribution of 58.2% to the Malaysian economy [12], majorly supported by the Wholesale and Retail subsector. This section mainly displays the visualisation of Wholesale and Retail Trade Index (including Motor Vehicle) on a monthly basis. Apart from that, the quarterly the section also shows time series visualisation for Index of Services which has been published on a quarterly basis by DOSM.

Trade: Trade statistics is one of the important elements in a country's economic activities, connecting the cycle of economy of a country with the rest of the world. Trade statistics dashboard was built to monitor the exports and imports sectors in Malaysia. Time series of export and import data were displayed to continue providing information on how well trade sectors in Malaysia perform. Through this dashboard, major food imports were also visualised as shown in Fig. 7 to monitor the dependencies of Malaysia on imported food which can be related to the food security and readiness in the country.

Balance of Payments: Balance of Payments dashboard displays the situation of current account and financial account both on a quarterly and annual basis.

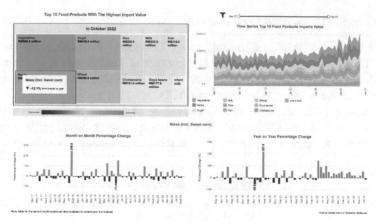

Fig. 7. Trade dashboard.

Other Indicators: Other indicators include a dashboard related to eight (8) sub-indicators (Fig. 8), which are: -

- Leading Index
- Business Tendency Statistics
- Utilities
- Credit Card Transactions
- Vehicles
- Business Registrations
- Income of e-Commerce
- Tourist Arrivals
- Port Statistics

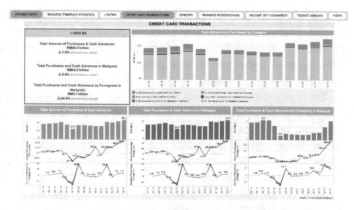

Fig. 8. Other indicators dashboard.

3.3 Optimising Dashboard Performance

MERD was initially developed entirely using Tableau and published on the DOSM website. The increase in data and the large number of Dashboards result in relatively long loading times and this can interfere with user satisfaction in obtaining preliminary information on the economic situation from MERD.

Therefore, MERD development is improved by combining elements of web development using HTML and CSS with Tableau to ensure more optimal loading times and provide benefits to users. The effectiveness of this improvement shown in Table 1.

Table 1. Dashboard loading time

Section	Loading time	
	Version 1	Version 2
Home page	2 min	5 s
Overall dashboard	2 min	25 s
GDP	30 s	5 s
Labour force	15 s	5 s
Commodities	15 s	5 s
Prices	10 s	5 s
Manufacturing	10 s	6 s
Services	10 s	7 s
Trade	10 s	7 s
Balance of payments	10 s	6 s
Other indicators	2 min	10 s

Migrating to a digital real-time dashboard does not have to be difficult. According to recent research [13, 14], a simple but comprehensive dashboard focused on significant indicators and created with limited resources (including long hours of usability testing) increased satisfaction among government executives and stakeholders in communication and coordination. A legend for interpreting the graphics should be provided to ensure understanding. The government may want to think about going beyond fact data and incorporating regional and national benchmarks into the dashboard and systematically translating those data into ways to compare local situations in order to drive performance excellence. Furthermore, it is possible that an indicator already used in business operations will be useful in assessing community needs or progress in real-time. The evolution of data dissemination toward the initiative of digitalizing insight in real-time dashboard from Reporting/Publication approach is simplified in Fig. 9.

As the big data, analytic and communication era evolves, innovative initiatives involving information creation in previously untapped areas can prove to be a bold new market. As new types of data variables emerge in the big data era, analysts must more intensely exhaust data resources that may provide descriptive elements of a given process. Analysts must also consider whether a process needs to be monitored in real time (velocity and volume) to uncover strategic insights in the information creation and decision-making process.

Reporting/
Publication

Near Real-time
Dashboard

Real-time
Dashboard

Fig. 9. Evolving data dissemination to real-time dashboard.

Management and government should take a step-by-step approach to creating a dashboard. Having a digital real-time dashboard in mind from the start will make the process easier in the long run. Information should be finely tailored to the needs and tasks associated with official statistics goals and the overall organisational vision. As economic indicators, care-fully selected economic datasets to fit the goals and monitor the business or provide services. A variety of community dashboard possibilities exist, just as there are a variety of industries and functions where near real-time dashboards can be found.

Table 2. Comparison between dashboard and reporting/publication

Description	Dashboard	Reporting/Publication
Purpose	Performance at a glance	Managing performance/report
Timeliness of data	Current(often real time)	Periodic(often monthly)
Data freshness	Depends on data point accuracy	Depends on frequency of data collection
Users	Lower level managers	Upper level managers
Linked to system	Almost always - continuously	Sometimes

The useful dashboard's emphasis on the right data, summaries, graphics and exception visuals help provide the clarity and conciseness to bridge the gap between objectives - margin and allow the government to demonstrate that their work within the community makes a difference based on significant indicators. The comparison between Dashboard and Reporting/Publication as showed in Table 2.

4 Conclusion

The MERD successfully developed an interactive web-based dashboard. The new version of MERD is able to speed up the loading time as compared to the previous version. MERD has outperformed as tools in order to support decisions especially on economic situation by providing a quick overview of Malaysian economy by selecting 39 sub indicators. For future study, MERD is recommended to upgrade by improving the selected economic indicators and timeline of the indicators.

4.1 Future Work

Future work will focus on discussion about the user dash-board's relevance to needs. This discussion includes additional research that is required, owing to the rapid adoption of mobile devices as a physical platform for decision support. The study differentiates between various types of research studies, as well as various aspects of interactive dashboards and their maturity in terms of evaluation. Because the research field is still in its early stages, many of the studies are exploratory or proof-of-concept in nature. This study identifies the need for longitudinal research in authentic settings, as well as studies that systematically compare different dashboard design options, as among the major open issues and future lines of work in the field of interactive dashboards.

When it comes to evaluation, any user usability evaluation must start with validating the best approach guidelines. The following stage of our research will involve users in various usability evaluations of dashboard interaction designs. We discussed dashboard design elements, such as the various types of visuals and interactions available, and offered advice on their situational utility. Dashboard designers have many options when it comes to data visualisation methods, as described in previous sections. As with any effective design, they must allow the users' goals, the context of use, and the constraints—or lack thereof—of the underlying data to guide their design. For instance, if it demonstrates that different indicators are reported in various ways. If all of these indicators were in the same column, a user might get the wrong impression if they interpreted rate or incident as a percentage rather than a count.

Most dashboards cater to a variety of users, including executives who want the big picture, functional data users who require detailed information to perform their duties, and analysts. It may be advantageous to present the same data in different ways so that all perspectives are understood. Drill-down features are another way the dashboard supports different types of users and decisions. Providing information on the definition of the indicator and how it is calculated is always a good idea, especially for indicators that are not part of the organisation's daily nomenclature.

The interface design has evolved into the process of defining the product itself [15]. The next step will be to identify dashboard items by reviewing features in existing dashboard users' engagement behaviour. The dashboard items will be chosen based on their ability to support the needs of the user on the dashboard. A proposed model of the user's dashboard in NSO can be established using these dashboard items. The availability and reliance on current, business-critical data necessitates a broader set of designer skills. Instead of simply designing a visual display and specifying interactive behaviour, the task of designing itself is morphing and broadening.

Acknowledgment. The authors gratefully acknowledge Department of Statistics Malaysia (DOSM) for the support and visualisation tools provided. Special thanks to Ms. Jamaliah Jaafar and Dr. Mohamad Shukor Talib for the guidance and recommendations during this paper writing.

References

1. Nicola, M., et al.: The socio-economic implications of the coronavirus pandemic (COVID-19): a review. Int. J. Surg. **78**, 185–193 (2020). https://doi.org/10.1016/j.ijsu.2020.04.018
2. de Castro Neto, M., Nascimento, M., Sarmento, P., Ribeiro, S., Rodrigues, T., Painho, M.: A dashboard for security forces data visualization and storytelling. In: Ramos, I., Quaresma, R., Silva, P., Oliveira, T. (eds.) Information Systems for Industry 4.0. LNISO, vol. 31, pp. 47–62. Springer, Cham (2019). https://doi.org/10.1007/978-3-030-14850-8_4
3. de Castro Neto, M., Nascimento, M., Sarmento, P., Ribeiro, S., Rodrigues, T., Painho, M.: Implementation of a dashboard for security forces data visualization. Atas Da Conferencia Da Associacao Portuguesa de Sistemas de Informacao, 2018-October (2018)
4. Catalin, S.: The culture of visuals and the visuals of culture. Eur. Sci. e-J. **2022** (2022). https://doi.org/10.47451/art2022-05-01
5. Gandomi, A., Haider, M.: Beyond the hype: big data concepts, methods, and analytics. Int. J. Inf. Manage. **35**(2), 137–144 (2015). https://doi.org/10.1016/j.ijinfomgt.2014.10.007
6. Juhair, S.N.K., Daud, N. A., Noordin, N., Redzuan, F., Wan Adnan, A., Ashaari, N.S.: An interactive dashboard for information visualization on the air pollution index in Malaysia (Apimas) (2021)
7. Vahedi, A., Moghaddasi, H., Asadi, F., Hosseini, A.S., Nazemi, E.: Applications, features and key indicators for the development of Covid-19 dashboards: a systematic review study. Inform. Med. Unlock. **30**, 100910 (2022). https://doi.org/10.1016/j.imu.2022.100910
8. Hoelscher, J., Mortimer, A.: Using Tableau to visualize data and drive decision-making. J. Account. Educ. **44**, 49–59 (2018). https://doi.org/10.1016/j.jaccedu.2018.05.002
9. Akhtar, N., Tabassum, N., Perwej, A., Perwej, Y.: Data analytics and visualization using Tableau utilitarian for COVID-19 (Coronavirus). Glob. J. Eng. Technol. Adv. **3**(2), 028–050 (2020). https://doi.org/10.30574/gjeta.2020.3.2.0029
10. Pappas, L., Whitman, L.: Riding the technology wave: effective dashboard data visualization. In: Smith, M.J., Salvendy, G. (eds.) Human Interface 2011. LNCS, vol. 6771, pp. 249–258. Springer, Heidelberg (2011). https://doi.org/10.1007/978-3-642-21793-7_29
11. Eurostat. Essential SNA: Building the basics 2013 (2013)
12. Gross Domestic Product Q1 2022 (First Quarter) (2022). dosm.gov.my, https://www.dosm.gov.my/v1/index.php?r=column/cthemeByCatcat=100bulid=Zm8xRXoyVitzKzVUbG9Cc0pPQ0s3Zz09menuid=TE5CRUCblh4ZTZMODZIbmk2aWRRQT09. Accessed 28 July 2022
13. Dixit, R.A., et al.: Rapid development of visualization dashboards to enhance situation awareness of COVID-19 telehealth initiatives at a multihospital healthcare system. J. Am. Med. Inform. Assoc. **27**(9), 1456–1461 (2020). https://doi.org/10.1093/jamia/ocaa161
14. Khajeh Goodari, S., Rahdar, M.A.: Designing a management dashboard for healthcare professionals and managers in the COVID-19 epidemic. J. Milit. Med. **22**(10), 1013–1024 (2020). https://doi.org/10.30491/JMM.22.10.1013
15. Cooper, A., Reimann, R., Cronin, D.: About Face 3, p. 357. Wiley, Hoboken (2013)

Machine/Deep Learning

Lung Nodules Classification Using Convolutional Neural Network with Transfer Learning

Abdulrazak Yahya Saleh[(⊠)] and Ros Ameera Rosdi

FSKPM Faculty, University Malaysia Sarawak (UNIMAS), 94300 Kota Samarahan, Sarawak, Malaysia
ysahabdulrazak@unimas.my

Abstract. Healthcare industry plays a vital role in improving daily life. Machine learning and deep neural networks have contributed a lot to benefit various industries nowadays. Agriculture, healthcare, machinery, aviation, management, and even education have all benefited from the development and implementation of machine learning. Deep neural networks provide insight and assistance in improving daily activities. Convolutional neural network (CNN), one of the deep neural network methods, has had a significant impact in the field of computer vision. CNN has long been known for its ability to improve detection and classification in images. With the implementation of deep learning, more deep knowledge can be gathered and help healthcare workers to know more about a patient's disease. Deep neural networks and machine learning are increasingly being used in healthcare. The benefit they provide in terms of improved detection and classification has a positive impact on healthcare. CNNs are widely used in the detection and classification of imaging tasks like CT and MRI scans. Although CNN has advantages in this industry, the algorithm must be trained with a large number of data sets in order to achieve high accuracy and performance. Large medical datasets are always unavailable due to a variety of factors such as ethical concerns, a scarcity of expert explanatory notes and labelled data, and a general scarcity of disease images. In this paper, lung nodules classification using CNN with transfer learning is proposed to help in classifying benign and malignant lung nodules from CT scan images. The objectives of this study are to pre-process lung nodules data, develop a CNN with transfer learning algorithm, and analyse the effectiveness of CNN with transfer learning compared to standard of other methods. According to the findings of this study, CNN with transfer learning outperformed standard CNN without transfer learning.

Keywords: Deep learning · Convolutional Neural Network · Lung nodules · CT scan

1 Introduction

Lung cancer is a well-known disease with a significant mortality rate in the modern era. It starts in the lungs, where cancer cells can move to lymph nodes or other organs such as the brain [1]. In 2018, lung cancer had the highest percentage of new cases and deaths

© The Author(s), under exclusive license to Springer Nature Singapore Pte Ltd. 2023
Y. B. Wah et al. (Eds.): DaSET 2022, LNDECT 165, pp. 253–265, 2023.
https://doi.org/10.1007/978-981-99-0741-0_18

worldwide by 11.6% [2]. According to the statistics by World Health Organization, lung cancer has the second highest rate of cases in Malaysia in 2018 after breast cancer [2]. About 16.6% of the patients are male while 5.4% are female. Based on The Malaysia National Cancer Registry Report 2012–2016, lung cancer stands as the third highest cancer detected among Malaysians which affects male more than females [3]. Deep learning is no longer a strange concept in the field of medical image analysis [4, 5, 6, 7]. It is a growing trend, and there is a growing demand for the use of deep learning to achieve accurate and precise outcomes [8]. Deep learning involves imitating how the human brain works in dealing with data and recognizing patterns for the decision-making process. With the emergence of technology and better algorithms, more and more machines give high rates of accuracy and reliability for medical analysis. Detection of cancerous or malignant cells are crucial for treatment of lung cancer. Applications of image analysis deep learning on computed tomography (CT) scan images help to detect malignant cells early before they develop and become lethal [9, 10]. Deep learning has been proven to have significant performance in image processing especially in object detection and localization [11]. Deep learning, specifically CNN, is known to have a high success rate if a large amount of data is implied [12]. CNN requires a large amount of well-labeled training data, such as ImageNet, to perform well, which medical images lack. Large datasets are not always available due to several factors such as the costly expert explanatory notes, ethical issues, and shortage of disease images [12]. Models with large number of parameters fail to learn the patterns if supplied with small datasets and can easily cause overfitting [13]. Most traditional CNN architectures' performance is heavily dependent on the size of the data because they initially have a large number of parameters, and state-of-the-art CNN models trained with large datasets such as ImageNet are unsuitable for datasets with hundreds or thousands of instances [14]. This is a challenge that researchers must address in order to improve model performance while working with a large amount of labelled data. However, standard CNN is unsuitable for medical imaging processes with small datasets (hundreds to thousands number of data). A comprehensive review of various types of CNN for pulmonary nodules detection, false positive reduction and classification has been done in [15, 16, 17, 18]. Table 1, on the other hand, focuses on CNN with advanced implementation approaches rather than conventional methods or traditional CNN.

Table 1. A comprehensive review of various types of CNN for pulmonary nodules detection

	Ref.	Model	Data sets	Key Points
Hybrid CNN	[12]	2D LeNet + 2D AlexNet	LIDC/IDRI	Layers of LeNet settings are combined with AlexNet parameter settings to create a model for malignancy prediction

(continued)

Table 1. (*continued*)

	Ref.	Model	Data sets	Key Points
Transfer Learning Based System	[19]	2D CNN, SVM, MLP, KNN, RF, Naïve Bayes	LIDC/IDRI	Eleven 2D CNN models are used for features extraction. The models are Xception, VGG16, Inception-ResNet-V2, VGG19, DenseNEt201, MobileNet, InceptionV3, DenseNet169, ResNet50, NASNetLarge and NASNetMobile. SVM, MLP, KNN, RF, and Naïve Bayes are trained separately with collected features
	[20]	VGG16	LIDC/IDRI Private Data set	Comparison of DCNN models for features extraction, features engineering based model and transfer learning model
Multiscale Feature with Transfer Learning	[21]	3D U-Net CNN, Transfer learning	TIANCHI17 LUNA16	Using 3D U-Net structure for feature extraction. Transfer learning is introduced and fine-tuning the structure helps to extract features from image input. Experimental results showed that layer by layer transfer training method improved the accuracy of image detection under condition of small samples
Advanced Off-The-Shelf CNNs	[22]	3D CNN	LUNA16	Pioneer application of reinforcement learning to medical image analysis for detecting pulmonary nodules in CT scan images
	[23]	3D U-Net and 3D DenseNet	LUNA 16	2D U-Net, 3D DenseNet and Region Proposal Network (RPN) is employed for automatic detection of pulmonary nodules by utilizing multitask residual learning and online hard negative example mining approaches. 3D U-Net is for nodule candidates' generation and 3D DenseNet is mainly for false positive reduction
	[24]	3D Faster R-CNN and 3D DCNN	Tianchi AI competition	Two-phased framework for nodules identification and false positive reduction. 3D Faster RCNN is employed to create nodule samples and 3D DCNN model identified nodules candidates

Despite numerous studies attempting to improve the accuracy of early detection of lung cancer using deep learning, there remains a gap in detection and implementation of these algorithms in the real medical field. CNN deep learning system with transfer learning approach was proposed in this study to classify malignant and non-cancerous nodules. The proposed method is compared to a conventional CNN model that does not employ transfer learning.

2 Materials and Methods

This study focuses on classification of benign cells at its early-stage manifestation in the form of pulmonary nodules using CNN with transfer learning on a subset of LIDC-IDRI data set, the LUNA16 data set. Data can be found in [25]. According to [26], screening of CT scans has been proven to be effective in diagnosing lung cancer by analyzing the pulmonary nodules present which decreasing mortality rate.

In implementing CNN, a huge number of parameters is required to be estimated and some hardware and software are required to use. Data collection and analysis are critical for achieving higher accuracy and better results for CNN algorithms. This study focuses on increasing the accuracy and lower the loss model for CNN with transfer learning and prove that CNN model with transfer learning is better to tackle data set with small amount and gives out better results than standard CNN without transfer learning. This study makes use of a commonly used lung cancer CT image data set. The Lung Nodules Analysis 2016 (LUNA16) dataset is a subset of the LIDC-IDRI Data Set. The LUNA16 dataset is a subset of the LIDC-IDRI dataset, with the heterogeneous scans filtered using different factors. Since pulmonary nodules can be very small, a thin slice should be chosen. Therefore, scans with a slice thickness greater than 2.5 mm were discarded. Furthermore, scans with inconsistent slice spacing or missing slices were also excluded. This led to 888 CT scans, with a total of 36,378 annotations by radiologists. In this dataset, only the annotations categorized as nodules $\geq$ 3 mm are considered relevant, as the other annotations (nodules $\leq$ 3 mm and non-nodules) are not considered relevant for lung cancer screening protocols. Nodules found by different readers that were closer than the sum of their radii were merged. In this case, positions and diameters of these merged annotations were averaged.

The files containing all the CT scan images have been divided into two subsets files. Subset1 files are for training the proposed model and subset2 is for testing the model. CT images are stored in MetaImage (mhd/raw) format in each subset and each mhd file is stored with a separate raw binary file for the pixeldata. Next, annotations csv file contains the annotations that are used as reference standard for nodule detection track. Each line holds SeriesInstanceUID of the scan, the location and position of x, y, and z in world coordinates and the diameter in mm. The file contains 1186 nodels. Candidates_V2.csv file contains set of candidate locations in the CT scans for false positive reduction and lung segmentation is a directory containing lung segmentation of CT scan images computed using automatic algorithms [29]. The lung segmentation images are not intended to be used as the reference standard for any segmentation study. CT scan data has an alternative format which is stored in DICOM (. dcm) format. The original data set from LIDC-IDRI data set is in DICOM.

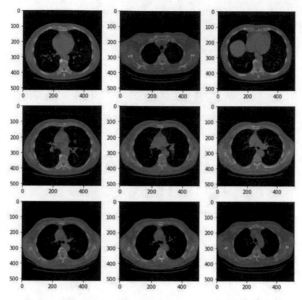

Fig. 1. Original CT scan DICOM slices from LUNA16 dataset

These results in a set of 2290, 1602, 1186 and 777 nodules annotated by at least 1, 2, 3 or 4 radiologists, respectively. In Fig. 1, different slices from a LUNA16 CT scan.

There are four main phases conducted in this study: data set preparation, research design, application and implementation and performance analysis. LUNA16 data set is the input for the algorithm. CNN with transfer learning, the proposed algorithm, was trained and tested using Python programming language then evaluated using performance evaluation. Figure 2 shows the flow of research methodology of the study. Following the preparation of the data set, the data set was utilised to train and test the proposed algorithm. The results of the models are then analyzed and compared in terms of model performance. The proposed method has been processed using the same computer system with processor Intel® Core™ i7-10750H CPU with 16 GB of installed RAM. The proposed method is implemented using Python programming software as Python is most suitable for computational processes. The data sets used in models are divided into two groups: training set and testing set. Most of data set will be used for training and the rest of data set will be used for testing the models.

The training data is used to train the model and includes 711 patients out of 1595 total. The rest of the patients' CT scans are used for testing the proposed model.

Transfer learning is a popular approach where pre-trained models developed for a task is reused as starting point for another model on second task [27]. There are two common approaches for transfer learning which are develop model approach and pre-trained model approach [28, 29]. For this study, pre-trained VGG16 model that been trained with ImageNet which trained on 1.2 million natural images. Transfer learning is also proven to increase discriminatory power of a generic dataset which increase generalization ability to perform other tasks [19].

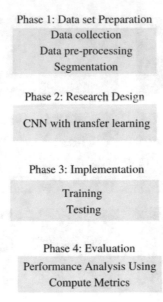

Fig. 2. The flow of research methodology of the study

An overall framework for the proposed methods has been created. CNN algorithm has been used to leverage the use of data set. To propose intended method of classifying of lung nodules, the LUNA16 data set has been utilized in this study (Fig. 3).

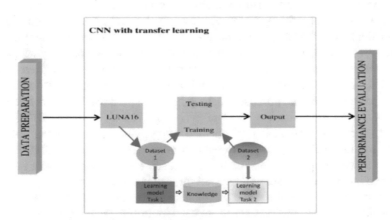

Fig. 3. The overall framework for the proposed methods

3   Findings and Discussion

Visualization of data set is using pydicom library to display the image array and metadata information stored in CT scan images. Images are stored in DICOM file and cannot use standard browser to access DICOM images. Figure 4 shows an example on metadata that is stored in the file.

```
(0008, 0005) Specific Character Set          CS: 'ISO_IR 100'
(0008, 0016) SOP Class UID                   UI: CT Image Storage
(0008, 0018) SOP Instance UID                UI: 1.2.840.113654.2.55.247817952625791837963403492891187883824
(0008, 0060) Modality                        CS: 'CT'
(0008, 103e) Series Description              LO: 'Axial'
(0010, 0010) Patient's Name                  PN: '00cba091fa4ad62cc3200a657aeb957e'
(0010, 0020) Patient ID                      LO: '00cba091fa4ad62cc3200a657aeb957e'
(0010, 0030) Patient's Birth Date            DA: '19000101'
(0018, 0060) KVP                             DS: None
(0020, 000d) Study Instance UID              UI: 2.25.86208730140539712382771890501772734277950692397709007305473
(0020, 000e) Series Instance UID             UI: 2.25.115758773296352289258085968002699747408935194517846260466614
(0020, 0011) Series Number                   IS: '3'
(0020, 0012) Acquisition Number              IS: '1'
(0020, 0013) Instance Number                 IS: '134'
(0020, 0020) Patient Orientation             CS: ''
(0020, 0032) Image Position (Patient)        DS: [-145.500000, -158.199997, -356.200012]
(0020, 0037) Image Orientation (Patient)     DS: [1.000000, 0.000000, 0.000000, 0.000000, 1.000000, 0.000000]
(0020, 0052) Frame of Reference UID          UI: 2.25.83033509634441686385652073462983801840121916678417719669650
(0020, 1040) Position Reference Indicator    LO: 'SN'
(0020, 1041) Slice Location                  DS: '-356.200012'
(0028, 0002) Samples per Pixel               US: 1
(0028, 0004) Photometric Interpretation      CS: 'MONOCHROME2'
(0028, 0010) Rows                            US: 512
(0028, 0011) Columns                         US: 512
(0028, 0030) Pixel Spacing                   DS: [0.597656, 0.597656]
(0028, 0100) Bits Allocated                  US: 16
(0028, 0101) Bits Stored                     US: 16
(0028, 0102) High Bit                        US: 15
(0028, 0103) Pixel Representation            US: 1
(0028, 0120) Pixel Padding Value             US: 63536
(0028, 1050) Window Center                   DS: '40.0'
(0028, 1051) Window Width                    DS: '400.0'
(0028, 1052) Rescale Intercept               DS: '-1024.0'
(0028, 1053) Rescale Slope                   DS: '1.0'
(7fe0, 0010) Pixel Data                      OW: Array of 524288 elements
```

Fig. 4. Metadata stored in a single DICOM file.

For segmentation of data set, watershed algorithm is used. The watershed is a classical algorithm used for segmentation and for separating different objects in an image. Watershed is a transformation defined on a grayscale image. The name refers metaphorically to a geological watershed, or drainage divide, which separates adjacent drainage basins. The watershed transformation treats the image as if it were a topographic map, with the brightness of each point representing its height, and finds the lines that run along the tops of ridges. "The topological watershed" was introduced by M. Couprie and G. Bertrand in 1997. Starting from user-defined markers, the watershed algorithm treats pixels' values as a local topography (elevation) [30].

The algorithm floods basins from the markers until basins attributed to different markers meet on watershed lines. In many cases, markers are chosen as local minima of the image, from which basins are flooded. Watershed algorithm highlights the lung part and makes binary masks for lungs using semantic segmentation approach.

Firstly, internal and external markers from CT scan images are extracted with the help of binary dilations and add them with a complete dark image using watershed methods. It also removes image noise and provides a watershed marker of lungs and cancer cells.

As illustrated in Fig. 5, External noise is removed by using a watershed marker and applies a binary mask on the image, black pixels in lungs represent cancer cells. For better segmentation, integrated sobel filter was applied to the with watershed algorithms to remove external layers of lungs.

After removing the outer layer, the internal marker is used, and outline created to generate lungfilter using bitwise_or operations of numpy. It also removes the heart from CT scan images. Next step is to close off the lung filter with morphological operations and morphological gradients. It provides better segmented lungs than the previous process.

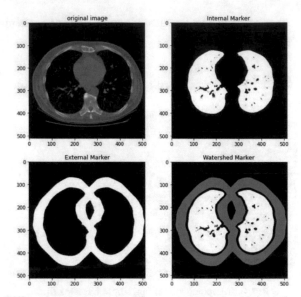

Fig. 5. Different markers extracted from watershed algorithm on the CT scan.

Figure 6 shows the segmented lungs after the application of sobel filter. In total generated 1002 images with related labels which includes almost 12 patients CT scan data in which there are almost the same number of cancer and non-cancer patients.

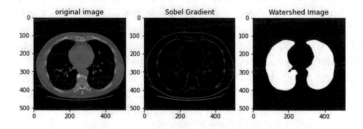

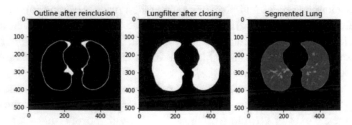

Fig. 6. Lungs segmentation visualization.

The proposed model is a CNN with transfer learning approach based on lung segmentation on CT scan images. Two different models of CNN are built for comparative study; the first model is a standard CNN model, and the proposed model is CNN with transfer learning algorithm. First model is the standard CNN without transfer learning is the basic simple approach of using the convolution layers, flatten fully connected layers, max pooling and dropout in the middle layers, which performs significantly well on the number classification problem. Table 2 shows the summary of the first model without transfer learning.

Furthermore, the proposed model approach is using transfer learning on pretrained model with changes in fully connected layers. Transfer learning on VGG-16 with some changes in the last three layers which are fully connected. This model gives appreciable results in object classification. Using Adam optimizer, learning rate applied on the model is 0.0001. Table 3 shows the summary of the proposed model.

After building the two models, without transfer learning and with transfer learning, the proposed model and comparison model are implemented and train models on segmented lungs with a batch size of 32 for image data generator and using 100 images in each epoch for 30 epochs with exception of 500 images in each epoch for CNN with transfer learning. Training images with the shape of (512, 512, 1) for the first model and the shape of (512, 512, 3) for proposed model.

For a better result data, augmentation is used to train models on different augmentations like shear range, zoom range, horizontal flip, rotation range, and center shift etc.

A single node for binary classification is used in the end layer to differentiate between cancer and non-cancer lungs. Moreover, callbacks from tensorflow keras are used to save the best accuracy model to run a complete 50 epoch training session to plot the comparison graphs.

Table 2. Model summary of standard CNN without transfer learning.

Model: "sequential_1"		
Layer(type)	Output shape	Param #
conv2d_1 (Conv2d)	(None,510,510,32)	320
maxpooling2d_1(Maxpooling2	(None,255,255,32)	0
conv2d_2 (Conv2d)	(None,253,253,32)	9248
maxpooling2d_2(Maxpooling2	(None,126,126,32)	0
conv2d_3 (Conv2d)	(None,124,124,32)	9248
droput_1 (Dropout)	(None,124,124,32)	0
conv2d_4 (Conv2d)	(None,122,122,32)	9248
flatten_1 (Flatten)	(None,476288)	0
dense_1 (Dense)	(None,128)	60964992

(*continued*)

Table 2. (*continued*)

Model: "sequential_1"		
Layer(type)	Output shape	Param #
droput_2 (Dropout)	(None,128)	0
dense_2 (Dense)	(None,128)	16512
dense_3 (Dense)	(None,1)	129

Total Params: 61,009,697
Trainable params: 61,009,697
Non-trainable params: 0

Table 3. Model summary of CNN with transfer learning.

Layer(type)	Output shape	Param #
input_3(InputLayer)	(None,224,224,3)	0
block1_conv1 (Conv2d)	(None,224,224,64)	1792
block1_conv2 (Conv2d)	(None,224,224,64)	36928
block1_pool (Maxpooling2)	(None,112,112,64)	0
block2_conv1 (Conv2d)	(None,112,112,128)	73856
block2_conv2 (Conv2d)	(None,112,112,128)	147584
block2_pool (Maxpooling2)	(None,56,56,128)	0
block3_conv1 (Conv2d)	(None,56,56,256)	295168
block3_conv2 (Conv2d)	(None,56,56,256)	590080
block3_conv3 (Conv2d)	(None,56,56,256)	590080
block3_pool (Maxpooling2)	(None,28,28,256)	0
block4_conv1 (Conv2d)	(None,28,28,512)	1180160
block4_conv2 (Conv2d)	(None,28,28,512)	2359808
block4_conv3 (Conv2d)	(None,28,28,512)	2359808
block4_pool (Maxpooling2)	(None,14,14,512)	0
block5_conv1 (Conv2d)	(None,14,14,512)	2359808

Table 4 shows the comparison of model training accuracy and model testing accuracy.

Table 4. Summary of accuracy of the two models.

Models	Training accuracy	Testing accuracy
Standard model	90.77%	81.19%
Proposed model (with Transfer Learning)	99.84%	88.00%

The Proposed model with transfer learning achieved better results than the standard CNN without transfer learning. Due to overfitting, the standard model performed worse than the proposed model without achieving the desired testing accuracy. However, it can be used to solve a classification problem. The proposed model outperformed the original model, however it still need more improvements. There are numerous elements influencing the performance of the proposed model. To improve testing accuracy, for example, the number of epochs offered for training can be increased. Although it can be beneficial, it is also important to avoid overfitting the model, since this can reduce the model's performance and hence the accuracy attained. Next, data pre-processing is essential for the model's training phase. Although the watershed algorithm can aid in segmentation, alternative methods can be used in future research.

Proposed model with transfer learning achieved better result than the standard CNN without transfer learning. The proposed model's testing accuracy was lower than its training accuracy. This is owing to the restricted number of datasets available, as well as the overall scarcity of medical data sets. The study's key findings focus on the implementation of a pretrained model utilising transfer learning on CNN. It has been demonstrated that by incorporating transfer learning into a CNN model, better accuracy in classification problems can be attained when data sets are minimal.

Several challenges were encountered during the study's development, which had an impact on the study's outcomes and outputs. The data set used was in dmc format, which is different from standard image processing formats like jpg; this format stores all of the patient's information as well as a 3D CT-Scan image of the patient, which when converted into a 2D numpy image results in the shape of (n, m, m, 1), which is not supported by the models for transfer learning. Furthermore, it is in float64 data format, which is difficult to handle and process when using standard libraries like as CV2, skimage, and matplotlib.

Furthermore, when the data is converted to a 2D grayscale image, it is difficult to train it with the Tensorflow and Keras frameworks because they do not allow such shapes (n, m, m) during model creation. When the images are transformed back into RGB or BGR format using different libraries, some or many attributes that are useful in the classification process are lost.

4 Conclusion

The medical industry is critical to the improvement of daily life. Deep learning can be used to gather more deep knowledge and assist healthcare workers in learning more about a patient's disease. Deep neural networks and machine learning are becoming more popular in the healthcare industry. The benefit they provide in terms of improved detection and classification has a positive impact on healthcare. CNN is widely used in imaging detection and classification tasks such as CT and MRI scans. Although CNN has advantages in this industry, the algorithm must be trained with a large number of data sets in order to achieve high accuracy and performance. CNN with transfer learning was created to improve the classification of lung nodules. There is still space for improvement, but CNN combined with transfer learning can assist in more properly classifying data. To improve model accuracy, more efficient pre-processing and segmentation can be performed. In addition, new data processing, training, and classification methods and architecture based on other transfer learning methods and design might be considered to aid the models in more accurate data classification.

Acknowledgment. The authors acknowledged the financial support from the Ministry of Higher Education Malaysia through Fundamental Research Grant Scheme (FRGS) F04/FRGS/2151/2021. Moreover, we want to thank the Faculty of Cognitive Sciences and Human Development, Universiti Malaysia Sarawak (UNIMAS), for their support and funding of the publication.

References

1. Prevention, C.f.D.C.a., What is Lung Cancer? (2020)
2. WHO, I.A.f.R.o.C., New Global Cancer Data: GLOBOCAN 2018. World Health Organization (2018)
3. Malaysia., M.o.H., Malaysia National Cancer Registry Report (MNCR) 2012–2016, Putrajaya (2019)
4. Alom, M.Z., et al.: A state-of-the-art survey on deep learning theory and architectures. Electronics **8**(3), 292 (2019)
5. Salahuddin, Z., et al.: Transparency of deep neural networks for medical image analysis: a review of interpretability methods. Comput. Biol. Med. **140**, 105111 (2022)
6. Arabahmadi, M., Farahbakhsh, R., Rezazadeh, J.: Deep learning for smart healthcare—a survey on brain tumor detection from medical imaging. Sensors **22**(5), 1960 (2022)
7. Zakaria, R., Abdelmajid, H., Zitouni, D.: Deep learning in medical imaging: a review. Appl. Mach. Intell. Eng. 131–144 (2022)
8. Marr, B.: The top 10 artificial intelligence trends everyone should be watching in 2020 (2020)
9. Society, A.C.: About lung cancer (2019)
10. Primakov, S.P., et al.: Automated detection and segmentation of non-small cell lung cancer computed tomography images. Nat. Commun. **13**(1), 1–12 (2022)
11. Singh, G.A.P., Gupta, P.K.: Performance analysis of various machine learning-based approaches for detection and classification of lung cancer in humans. Neural Comput. Appl. **31**(10), 6863–6877 (2018). https://doi.org/10.1007/s00521-018-3518-x
12. Zhao, X., Liu, L., Qi, S., Teng, Y., Li, J., Qian, W.: Agile convolutional neural network for pulmonary nodule classification using CT images. Int. J. Comput. Assist. Radiol. Surg. **13**(4), 585–595 (2018). https://doi.org/10.1007/s11548-017-1696-0
13. Pawan, S.J.: Learning from small data (2019)
14. Keshari, R., et al.: Learning structure and strength of CNN filters for small sample size training. In: Proceedings of the IEEE Conference on Computer Vision and Pattern Recognition (2018)
15. Sharif, M.I., et al.: A comprehensive review on multi-organs tumor detection based on machine learning. Pattern Recogn. Lett. **131**, 30–37 (2020)

16. Halder, A., Dey, D., Sadhu, A.K.: Lung nodule detection from feature engineering to deep learning in thoracic CT images: a comprehensive review. J. Digit. Imaging **33**(3), 655–677 (2020)
17. Nakrani, M.G., Sable, G.S., Shinde, U.B.: A Comprehensive review on deep learning based lung nodule detection in computed tomography images. In: Satapathy, S.C., Bhateja, V., Janakiramaiah, B., Chen, Y.W. (eds.) Intelligent System Design. Advances in Intelligent Systems and Computing, vol. 1171, pp. 107–116. Springer, Singapore (2021). https://doi.org/10.1007/978-981-15-5400-1_12
18. Hosseini, H., Monsefi, R. Shadroo, S.: Deep learning applications for lung cancer diagnosis: a systematic review. arXiv preprint arXiv:2201.00227. (2022)
19. Da Nóbrega, R.V.M., et al.: Lung nodule classification via deep transfer learning in CT lung images. In: 2018 IEEE 31st International Symposium on Computer-Based Medical Systems (CBMS). IEEE (2018)
20. Yamashita, R., Nishio, M., Do, R.K.G., Togashi, K.: Convolutional neural networks: an overview and application in radiology. Insights Imaging **9**(4), 611–629 (2018). https://doi.org/10.1007/s13244-018-0639-9
21. Tang, S., Yang, M., Bai, J.: Detection of pulmonary nodules based on a multiscale feature 3D U-Net convolutional neural network of transfer learning. PLoS One **15**(8), e0235672 (2020)
22. Ali, I., et al.: Lung nodule detection via deep reinforcement learning. Front. Oncol. **8**, 108 (2018)
23. Qin, Y., et al.: Simultaneous accurate detection of pulmonary nodules and false positive reduction using 3D CNNs. In: 2018 IEEE International Conference on Acoustics, Speech and Signal Processing (ICASSP). IEEE (2018)
24. Tang, H., Kim, D.R., Xie, X.: Automated pulmonary nodule detection using 3D deep convolutional neural networks. In: 2018 IEEE 15th International Symposium On Biomedical Imaging (ISBI 2018). IEEE (2018)
25. Camp, B.V.B.: Data from the lung image database consortium (LIDC) and image database resource initiative (IDRI): a completed reference database of lung nodules on CT scans (LIDC-IDRI). Cancer Imaging Archive (TCIA) Public Access (2022)
26. Monkam, P., et al.: Detection and classification of pulmonary nodules using convolutional neural networks: a survey. IEEE Access **7**, 78075–78091 (2019)
27. Goodfellow, I., Bengio, Y., Courville, A.: Deep Learning. MIT Press, Cambridge (2016)
28. Zhuang, F., et al.: A comprehensive survey on transfer learning. Proc. IEEE **109**(1), 43–76 (2020)
29. Peirelinck, T., et al.: Transfer learning in demand response: a review of algorithms for data-efficient modelling and control. Energy AI **7**, 100126 (2022)
30. Bertrand, G.: On topological watersheds. J. Math. Imaging Vis. **22**(2), 217–230 (2005)

Plant Growth Phase Classification Using Deep Neural Network (Case Study of ASF in Poso District, Central Sulawesi Province)

Kevin Agung Fernanda Rifki and Kartika Fithriasari[✉]

Department of Statistics, Sepuluh Nopember Institute of Technology, Arief Rahman Hakim, Surabaya 60111, Indonesia
kevinrifki.18062@mhs.its.ac.id, kartika_f@statistika.its.ac.id

Abstract. An innovation developed through a combination of satellite data with official data to provide a solution to the limitations of the Area Sample Framework (ASF) survey where surveyors have to go directly to places that are sometimes difficult to reach and require a relatively long time, the Central Statistics Agency (BPS) suggested using Landsat-8 satellite imagery with Deep Neural Network Method (DNN) to classify rice plant growth phases. Data from Landsat-8 which has the characteristics to see land cover, especially plants. Apart from the band, the variables in this study were added to the vegetation index calculated from satellite data and combined with official data. One of the classification methods used is Deep Neural Network. This study aims to compare the methods between Artificial Neural Network (ANN) and DNN in classifying rice growth phases and predicting rice growth phases using DNN. With split data stratified 5-fold cross validation and data normalization using a robust scaler, the classification results show the average performance in terms of accuracy, precision, sensitivity, f1-score, Cohen Kappa index and Average Precision (AP) values. Based on several performance evaluations of the two methods from both ANN and DNN there is no significant difference.

Keywords: Area sampling framework · Classification · Deep neural network · Landsat-8

1 Introduction

Indonesia is an agricultural country where most of the population works in agriculture and places the agricultural sector as the second largest contributor to Gross Domestic Product (GDP) value after the manufacturing sector [1]. Agriculture is the most important sector in Indonesia, especially rice plant. However, the inaccuracy of rice production data has been suspected by many parties since 1997. To correct the inaccuracy of the data, BPS made an innovation in order to produce accurate food data, namely in collaboration with Badan Pengkajian dan Penerapan Teknologi (BPPT) and related agencies to develop the ASF Survey in 2018 to estimate rice harvested area by observing the rice growth phase [2].

© The Author(s), under exclusive license to Springer Nature Singapore Pte Ltd. 2023
Y. B. Wah et al. (Eds.): DaSET 2022, LNDECT 165, pp. 266–281, 2023.
https://doi.org/10.1007/978-981-99-0741-0_19

The ASF Survey itself has a weakness, namely that it requires a relatively high cost because increasing the level of estimation requires additional samples, which means adding human resources and other resources and going directly to the place. Then there is the lack of information on non-sample individual variations, considering that the ASF estimation is only based on the selected sample [3]. So it is proposed to use satellite imagery, one of which is Landsat-8. The Landsat-8 imagery has 2 sensors, namely the Onboard Operation Land Imager (OLI) which consists of 9 bands including a high resolution panchromatic band and the Thermal Infrared Sensor (TIRS) with 2 thermal bands. The use of band data in agriculture can also be added with a vegetation index. Vegetation index is the value of the greenness of vegetation obtained from digital signal processing of brightness value data (brightness) of several satellite sensor data channels in order to examine vegetation cover above the earth's surface. By utilizing band 1 to band 7 and several types of vegetation indices, study on the growth phase of plants, especially rice plants can be carried out [3].

The use of methods in machine learning to analyze big data from Landsat-8 satellite imagery has been widely carried out, where machine learning methods have good capabilities for classification or prediction. In the previous study, we estimated the rice growth model based on the day of its growth using the Artificial Neural Network (ANN), Gene-Expression Programming (GEP), and Simple Regression Model (REG) methods and the best was obtained using ANN [4]. The type of land cover classification using Landsat-8 imagery, a neural network method based on Multilayer Perceptron (MLP) was chosen and produced a fairly good accuracy [5]. Classification using neural networks can also produce good accuracy for image data such as the use of UNet-VGG16 for MRI-based brain tumor segmentation [6] and the use of MLP and Convolutional Neural Network for face identification [7].

This study uses a method with remote sensing data that is inexpensive and real-time compared to conventional field survey methods that require substantial human resources and capital. Where surveyors will go directly to survey locations and destinations that have road terrain that is quite difficult to pass without an adequate vehicle. So with the help of the Landsat-8 satellite along with DNN and ANN methods are proposed to create a classification model using band features and vegetation indices. Deep Neural Network and Artificial Neural Network are supervised methods that are suitable for multiclass data and will be compared later. By using this method, it is expected to have a high level of accuracy to predict the growth phase of rice by engineering spatio-temporal features. In this study, site selection for the classification of rice growth phases was carried out in Poso District, Central Sulawesi Province, Indonesia as an example of a case study. Then the rice growth phase label data used comes from the official ASF Survey of BPS.

2   Literature Review

2.1   Stratified K-Fold Cross Validation

This method is used to reduce bias related to sampling from the data. This learning algorithm divides the data into two parts, namely training data used for training and testing data used to validate the model [8]. The basic form of cross-validation is k-fold cross validation. K-fold cross validation divides the data into a number of randomly determined parts, which are called folds [9]. K itself is the data partition number used for the distribution of training data and test data. Stratification is the process of rearranging data to ensure each fold is a good representation of the entire data. In this study using 5-fold for cross validation in order to get a good model performance. The choice of 5 fold cross validation is used because the data used indicates the presence of unbalanced data, the proportion of data on each label can be the same when using stratified 5 fold. If you use more than 5 folds, then the proportion of unbalanced variables will not be the same for each fold. Unbalanced and disproportionate data in each fold can cause misclassification.

2.2 Robust Scaler

Normalization is the process of scaling the attribute values of the data so that they can lie within a certain range. The normalization used is the Robust Scaler algorithm. Robust Scaler has the advantage of being robust against outlier data so it is suitable for this study. This scaler removes the median and scales the data according to the quantile range (defaults to IQR: Interquartile Range) [10]. The reason is that the data used is survey data where all the places listed must be analyzed and some data have many outliers. So to overcome this, we use the Robust Scaler as data normalization in this study.

2.3 Artificial Neural Network

Artificial Neural Network (ANN) is an information processing system with a design similar to or imitating the workings of the human brain. The creation of ANN was inspired by the awareness of the complex learning system in the brain which consists of closely related sets of neurons [11]. The Neural Network architecture is depicted in Fig. 1 which consists of 3 parts, namely the Input layer, Hidden layer, and Output layer.

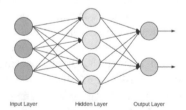

Fig. 1. Neural network architecture.

The input layer is the part that receives data or variable input, the hidden layer is the hidden part that acts as a processing unit, and the output layer is the part that produces the final result of the Neural Network process. This architecture is also called MultiLayer Perceptron (MLP) or Fully Connected Layer [12]. In one perceptron has a flow in making decisions from input to output which can be illustrated in Fig. 2.

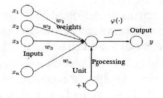

Fig. 2. Perceptron.

Based on Fig. 2, (1) is obtained which is as follows.

$$y = \varphi\left(b + \sum_{i=1}^{n} w_i x_i\right) = \varphi\left(b + w^T x\right) \tag{1}$$

where n is the amount of data, y is the output, x_i is the input, w_i is the weighting value, b is external threshold or bias, and $\varphi(\cdot)$ is the activation function.

2.4 Deep Neural Network

DNN originated from the development of ANN and works similarly to ANN, when as a supervised algorithm the observations of a number of input features are trained and combined with several operations where in the final layer for the desired prediction [13]. DNN usually uses conventional MLP with several (usually more than two) hidden layers [14]. In other words, this DNN is an MLP with more layers. In determining the number of N hidden layers, trial and error can be done to get the optimum classification results. As an example for the illustration of a DNN with N hidden layers where N is the number of hidden layers is as follows (Fig. 3).

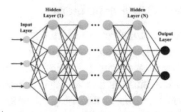

Fig. 3. Deep neural network architecture.

This study uses a backpropagation algorithm that requires three stages, the first is the feedforward input data for training in stages which calculates the forward stage of the input layer to the output layer stage with an activation function. The second stage is the backward stage to determine the error value or the difference between the network output and the desired target. The error is propagated backwards, starting from the line that is directly connected to each unit/neuron in the output layer. The last is the stage that will modify the weights to reduce the error rate that occurs [15].

The first stage is feedforward which begins with calculations on the hidden layer neurons before entering the hidden layer neurons. Then the calculation on the hidden layer neurons with the activation function. Neural Network has a model that is included with weight estimation which contains an activation function that is useful for calculating the sum of the input weights and biases, as well as being able to decide whether a neuron can be activated or not [16]. The activation function will change the linear equation of the input and weight combination into a nonlinear equation. In this study, we use the SELU activation function or the Scaled Exponential Linear Unit which aims to normalize the value between negative infinity to infinity.

The steps from one hidden layer to another are also the same, namely using the activation function always and the last one in the output layer using the softmax activation function because the classification output is more than 2. The results of the softmax calculation for classification do not produce round values (1, 2,..., G) where G is number of classes but the target obtained is a predictive value decision in the form of an opportunity value by determining the class through the threshold of the study [17]. Loss is closely related to the activation function in the output layer. The softmax activation function will produce a probability value, where the error calculation between two probability distributions uses a loss function called cross-entropy [18]. Loss or loss function is a function that is used as a criterion that must be minimized in the backpropagation algorithm which can be used as a solution to compare models [19]. Technically, cross entropy estimates the difference between the probability results from the output layer and the target results in the values of 0 and 1.

Parameter optimization is used to minimize the loss value so that it is the key in optimizing the bias and weight parameters. The optimization used in this study is adam parameter optimization. Adam (adaptive moment estimation) is an adaptive training optimization algorithm specially designed for training on deep learning methods. Adam uses a gradient, then estimates the first and second moments, and corrects with bias correction [20].

2.5   Model Evaluation and Cohen Kappa Index

Model evaluation is used to see whether the model is good or not. In this study, the confusion matrix and the Cohen Kappa Index will be used. The measure of the goodness of a predictive model consists of accuracy, precision, sensitivity, and f1-score. For the case of multiclass classification, such as this study, G classes are defined is number of classes [21]. In this study, in addition to accuracy, the model is also measured using the Cohen Kappa Index value. The Cohen Kappa Index is a comparison of the overall accuracy value with the expected accuracy. The kappa value is between $0.1 - 1.0$, where the value is getting closer to 1, meaning that the classification results class with the test data class is getting identical or the classification is more accurate.

2.6   Precision Recall Curve and Average Precision

Precision Recall (PR) is a measure of predictive success that is useful when the class label is unbalanced because it uses precision and recall or sensitivity which makes it possible to assess the classification performance of the minority class [22]. The reason for choosing the Precision Recall Curve is more recommended for high skewed domains than the Receiver operating characteristic (ROC) Curve [23].

In the classification problem, each threshold will be measured using AP. AP summarizes the PR curve with the weighted mean of the precision of each threshold with the recall difference from the previous threshold which is used as a weight. The function of this AP is to measure the area of the PR curve and the performance of each threshold.

2.7 Area Sample Framework

ASF survey is a method developed by BPS with assistance from BPPT to calculate the monthly rice harvested area and can produce an estimate of harvested area. Final vegetative phase, generative phase, land preparation phase, puso (crop failure), and estimation of the area of rice fields and fields that are not being planted with rice. The growth phase of the rice plant in Indonesian can be shown in Fig. 4.

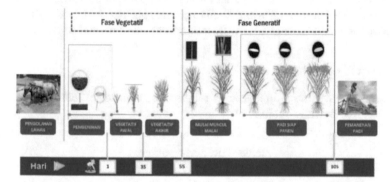

Fig. 4. Illustration of ASF survey rice plant observation phase from BPS.

2.8 Vegetation Index

Vegetation index is the greenness value of vegetation obtained from digital signal processing of brightness value data for several channels of satellite sensor data. In development, the vegetation index is used by scientists to evaluate the vegetative cover qualitatively and quantitatively with spectral measurements. The spectral response of vegetated areas has a complex mixture of vegetation, soil brightness, environmental influences, shadows, soil color and humidity [24]. Vegetation indices derived from satellite imagery to detect plant growth phases, especially rice, include Normalized Difference Vegetation Index (NDVI), Enhanced Vegetation Index (EVI), Rice Growth Vegetation Index (RGVI), Green-Red Vegetation Index (GRVI), Wide Dynamic Range Vegetation Index (WDRVI), Normalized Difference Built-up Index (NDBI), and Modified Normalized Difference Water Index (MNDWI). One of the RGVI vegetation indices is an index developed to monitor and map rice plants by utilizing the infrared reflection band from Landsat (band 4 and band 5) [25]. All of these vegetation indices were chosen because they are thought to be used as variables in this study.

3 Methodology

3.1 Data Source

The data used in this study are seven band data and seven vegetation index data for predictor variables as well as label response data for ASF observations in Poso District, Central Sulawesi Province from January 2018 to February 2021 obtained from BPS and accessed on 17 February 2022. The number of ASF survey samples calculated in the Landsat-8 data period is recorded every 16 days and adjusts to the ASF conducted in the last 1 week of each month. Landsat-8 satellite image data obtained and processed through GEE based on the coordinates of the ASF sample points to produce raw data in text file format in the form of Comma Separated Value (*.csv). Text data processed by GEE contains information on the date of image shooting, coordinates, and band data. Choosing less than 50% cloud cover and less than 30% cloud shading with Landsat-8 imagery because of the default settings and to get images that can be read without clouds. Cloud cover and cloud shadows will affect band data on the Landsat-8 satellite.

3.2 Research Variable

The research variables used in this study ranging from variables, data scales and some information are shown in Table 1.

Table 1. Research variables.

Variable		Data scale	Description
Y_p	ASF Labels	Ordinal	Code of rice growth phase ASF survey 1: Early Vegetative Phase, 2: Final Vegetative Phase, 3: Generative Phase, 4: Harvest Phase, 5: Land Preparation Phase
X_1	*Band* 1	Ratio	Wavelength *band* 1
X_2	*Band* 2	Ratio	Wavelength *band* 2
X_3	*Band* 3	Ratio	Wavelength *band* 3
X_4	*Band* 4	Ratio	Wavelength *band* 4
X_5	*Band* 5	Ratio	Wavelength *band* 5
X_6	*Band* 6	Ratio	Wavelength *band* 6
X_7	*Band* 7	Ratio	Wavelength *band* 7

(*continued*)

Table 1. (*continued*)

Variable		Data scale	Description
X_8	NDVI	Ratio	Vegetation Index NDVI
X_9	EVI	Ratio	Vegetation Index EVI
X_{10}	NDBI	Ratio	Vegetation Index NDBI
X_{11}	MNDWI	Ratio	Vegetation Index MNDWI
X_{12}	RGVI	Ratio	Vegetation Index RGVI
X_{13}	GRVI	Ratio	Vegetation Index GRVI
X_{14}	WDRVI	Ratio	Vegetation Index WDRVI

3.3 Research Step

The research steps used are as follows.

- **Step 1:** Data preparation which includes downloading the data from the ASF survey on the website http://ksa.bps.go.id to obtain the response variable, namely the ASF label for the observation phase for the period January 2018 to February 2021 and the coordinates of the ASF sample in Poso District, Central Sulawesi, Response variables and coordinates from ASF are used as a reference for downloading band data from Landsat-8. Band data download with less than 50% cloud cover and less than 30% cloud shading with Landsat-8 image sets via GEE based on predetermined periodization.
- **Step 2:** Data preprocessing which includes data cleaning processes such as detecting the presence of a column downloaded from Landsat-8 satellite imagery there is missing data caused by cloud cover that exceeds the criteria in the research area with band data and filtering the ASF label, so that the label used is input as a label growth phase which includes labels 1 to 5 as in Table 1 and calculating the value of the vegetation index based on the band data that has been obtained.
- **Step 3:** Make data exploration with several plots like pie chart and line chart.
- **Step 4:** Split ASF Survey data from January 2018 to December 2020 for model training data and ASF Survey data from January 2021 to February 2021 for prediction testing data.
- **Step 5:** Split the data for the model into training data and testing or validation data using the stratified 5-fold cross validation method.
- **Step 6:** Perform feature scaling on band data and vegetation index using Robust Scaler.
- **Step 7:** Classification with data bands and vegetation index of the Landsat-8 satellite using the Artificial Neural Network method.
- **Step 8:** Classification with data bands and vegetation index of the Landsat-8 satellite using the Deep Neural Network method.
- **Step 9:** Calculate and compare the performance evaluation of the training model using accuracy, precision, sensitivity, f1-score, cohen kappa index, and AP.
- **Step 10:** Draw conclusions from study results

4   Result and Discussion

4.1   Descriptive Preprocessing Result Data

Poso District is an area that is often cloudy, this causes an area that has cloud cover and its shadow is removed quite a lot of Landsat-8 images every period. The dataset shows that from 1,062 subsegment points, both active and inactive points from all ASF labels and these points are used in downloading band data from January 2018 to February 2021, there are 574 observations after adjusting to the criteria for the data to be analyzed (Fig. 5).

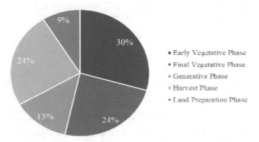

Fig. 5. Proportion of ASF observation.

There are 574 observations that are ready for further analysis in this study. The number of available observations provides information that the land preparation phase label data has the least proportion of 9% and the largest proportion is the early vegetative phase with 30%. Furthermore, exploration was carried out using graphs to see the pattern of the band data in each ASF label group including the rice growth phase. At each phase of rice growth has a certain pattern on the band value and vegetation index. Each band and vegetation index has its own characteristics and can be seen in the following results.

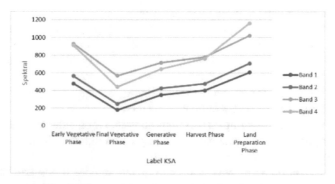

Fig. 6. Composition of the average data band 1 – 4.

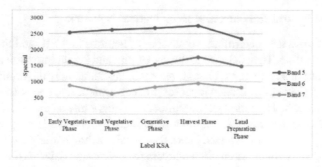

Fig. 7. Composition of the average data band 5 – 7.

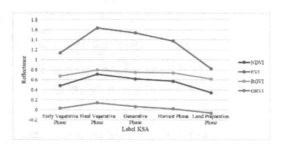

Fig. 8. Composition of the average vegetation index NDVI, EVI, RGVI, and RGVI.

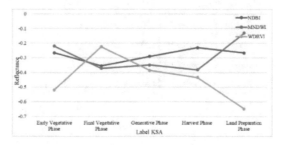

Fig. 9. Composition of the average vegetation index NDBI, MNDWI, and WDRVI.

The existence of band patterns and vegetation indices that have different characteristics in each phase. Based on Fig. 6, highest visible wavelength (Band 1 – Band 4) is in the late vegetative phase and the highest is the land preparation phase. From the late vegetative phase to the tillage phase the average wavelength continues to rise. Based on Fig. 7, NIR and SWIR wavelengths (Band 5 – Band 7) for the highest rice growth phase were in the harvesting phase and the land preparation phase had the lowest average wavelength for the ASF label. Band 6 and Band 7 from the early vegetative phase to the late vegetative phase decreased and returned to the generative phase. Based on Fig. 8 and Fig. 9, land preparation phase has different vegetation indices, especially NDVI, EVI, GRVI, WDRVI, and RGVI because they have not yet been planted with plants.

4.2 Artificial Neural Network and Deep Neural Network

The first thing to do is to determine the number of neurons in ANN, and the selection of an activation function. Then determine the use of bias and kernel initiation and optimizer configuration. In this ANN model uses three main layers, namely the input layer for the entry of input variables, one hidden layer for processing data, and one output layer for the classification results. Then the DNN architecture is the same as the ANN architecture, the difference is to determine the number of hidden layers and add dropouts for layers that have many neurons. In this DNN uses three main layers, namely the input layer for the entry of input variables, three hidden layers for processing data, and one output layer for the classification results. Here is the architecture for both models.

Table 2. ANN architecture and parameter count.

Layer (type)	Output shape	Parameter
dense (Dense)	(None, 30)	450
dense_1 (Dense)	(None, 5)	155

Table 3. DNN architecture and parameter count.

Layer (type)	Output shape	Parameter
dense (Dense)	(None, 20)	300
dense_1 (Dense)	(None, 30)	630
dropout (Dropout)	(None, 30)	0
dense_2 (Dense)	(None, 15)	465
dense_3 (Dense)	(None, 5)	80

The number of parameters in each layer on Table 2 depends on the number of kernels, biases, and previous layers. For the first layer, it has a number of parameters of 450 which is obtained from the number of input neurons, which is 14 which is multiplied by 30 kernels or neurons and added with a bias of 30. Finally, the output layer has a number of parameters of 155 which is obtained from the number of layers of the first neuron, namely 30 which is multiplied by 5 kernels or neurons and added a bias of 5. The resulting total parameters are 605 parameters.

The number of parameters for each layer on Table 3 depends on the number of kernels, biases, and previous layers. The parameter for dropout is zero because the dropout for neurons is 0.2 at 31 layers for being inactive during the feedforward process. For the first layer it has a number of parameters of 300 which is obtained from the number of input neurons which is 14 which is multiplied by 20 kernels or neurons and added a bias of 20. For the second layer it has a number of parameters 630 which is obtained from the number of layers of the first neuron which is 20 which is multiplied by 30 kernels or neurons and added a bias of 30. The third layer has a number of parameters of 465 which is obtained from the number of layers of second neurons which is 30 which is multiplied by 15 kernels or neurons and added a bias of 15. Finally the output layer has a number of parameters of 80 which is obtained from the number of the fifth layer of neurons namely 15 which is multiplied by 5 kernels or neurons and added with a bias of 5. The resulting total parameters are 1,475 parameters.

4.3 Classification Result with the Artificial Neural Network

The first thing to do is tuning the hyperparameters. Hyperparameter tuning is performed on each data fold consisting of training data for the model and testing data for validation. The next step is to select the epoch that has the optimum result with the choice of epoch in the range 0 to 2000.

The results of setting hyperparameters on each fold have their respective optimum epochs. If continued for the number of epochs then there will be potential for overfitting and the selection of the epoch is enough to be the best scheme. The following is the ANN classification performance for each fold.

Table 4. ANN classification performance on the five fold.

Fold	Accuracy	Precision	Sensitivity	F1-score	Cohen kappa index
1	0.72	0.68	0.67	0.67	0.63
2	0.70	0.68	0.66	0.67	0.61
3	0.66	0.65	0.64	0.64	0.56
4	0.68	0.68	0.63	0.65	0.57
5	0.71	0.71	0.69	0.70	0.63

Table 5. ANN classification AP performance on the five fold.

Fold	AP value				
	Label 1	Label 2	Label 3	Label 4	Label 5
1	0.81	0.72	0.80	0.79	0.49
2	0.72	0.78	0.75	0.73	0.36
3	0.63	0.68	0.68	0.76	0.32
4	0.81	0.80	0.65	0.69	0.48
5	0.79	0.82	0.84	0.78	0.55

Table 4 shows the performance of the ANN classification for each fold and the best performance is obtained in the fifth fold. In Table 5 shows the AP performance for each label in each fold and based on the results of Table 4 and Table 5, fifth fold model was chosen as the best result.

4.4   Classification Result with the Deep Neural Network

The initial stages are the same as ANN, namely tuning hyperparameters. The next step is to choose the epoch that has the optimum result with the choice of epoch in the range 0 to 1000. Each fold will have its own number of epochs where the number of epochs will produce the optimum model. The maximum selection of epochs was chosen because DNN has more hidden layers than ANN so that fewer maximums are selected.

The results of setting hyperparameters on each fold have their respective optimum epochs. If continued for the number of epochs then there will be potential for overfitting and the selection of the epoch is enough to be the best scheme. The visible difference from the ANN model is that there are several hidden layers and dropouts in a layer that has many nodes. The following is the performance of the DNN classification for each fold.

Table 6. DNN classification performance on the five fold.

Fold	Accuracy	Precision	Sensitivity	F1-score	Cohen kappa index
1	0.73	0.68	0.67	0.67	0.64
2	0.69	0.65	0.63	0.64	0.59
3	0.62	0.60	0.59	0.59	0.51
4	0.72	0.74	0.67	0.69	0.64
5	**0.72**	**0.75**	**0.71**	**0.73**	**0.64**

Table 7. DNN classification AP performance on the five fold.

Fold	AP value				
	Label 1	*Label 2*	*Label 3*	*Label 4*	*Label 5*
1	0.80	0.74	0.70	0.79	0.33
2	0.72	0.70	0.67	0.70	0.31
3	0.63	0.60	0.69	0.78	0.31
4	0.81	0.74	0.69	0.77	0.70
5	**0.83**	**0.84**	**0.85**	**0.76**	**0.67**

Table 6 shows the performance of the ANN classification for each fold and the best performance is obtained in the fifth fold. In Table 7 shows the AP performance for each label in each fold and based on the results of Table 4 and Table 5, fifth fold model was chosen as the best result.

4.5 Performance Comparison Between ANN and DNN

In this study, predictions were made of ASF observation points throughout January 2021 to February 2021. There are 46 points that are tested into the best model and the prediction results will be matched with ASF points from observations from January 2021 to February 2021 and see the performance of the model. To choose the best model, the average model performance and AP values for each model are compared and the results are as follows (Table 8).

Table 8. Comparison of the average performance measures of model.

Model	Accuracy	Precision	Sensitivity	F1-score	Cohen kappa index
ANN	0.694	0.680	0.658	0.666	0.600
DNN	0.696	0.684	0.654	0.664	0.604

Based on the results above, it can be concluded that there is no significant difference for the ANN and DNN models. Therefore, the best model will be selected from each fold and the fifth fold model from the best DNN and ANN models. The results can be displayed as follows (Table 9).

Table 9. Comparison of the average performance of AP model values.

Model	AP Value				
	Label 1	Label 2	Label 3	Label 4	Label 5
ANN	0.752	0.760	0.744	0.750	0.440
DNN	0.758	0.724	0.720	0.760	0.464

Based on Table 10 and Table 11 for the prediction of rice growth phase data, based on the comparison of the values of accuracy, precision, sensitivity, f1-score, Cohen Kappa index and AP on each label there is no significant difference as well and it is the same as the average comparison.

Table 10. Comparison of performance values of the best model of goodness.

Model	Accuracy	Precision	Sensitivity	F1-score	Cohen kappa index
ANN	0.71	0.71	0.69	0.70	0.63
DNN	0.72	0.75	0.71	0.73	0.64

Table 11. AP values performance comparison.

Model	AP value				
	Label 1	*Label 2*	*Label 3*	*Label 4*	*Label 5*
ANN	0.79	0.82	0.84	0.78	0.55
DNN	0.83	0.84	0.85	0./6	0.6/

5   Conclusion

Based on the results of the analysis and discussion, it can be concluded that there are band patterns and vegetation indexes that have different characteristics in each phase, but some values between bands and vegetation competitiveness index overlap. We need a classification method that can overcome this complexity. Using the ANN and DNN methods resulted in a classification that was not significantly different in terms of accuracy, sensitivity precision, f1-score, cohen kappa index, and AP value.

References

1. Rahman, A., Octaviani, E.: Analisis produktivitas tenaga kerja sektor pertanian dan kemiskinan di Indonesia. Seminar Nasional Variansi (Venue Artikulasi-Riset, Inovasi, Resonansi-Teori, Dan Aplikasi Statistika), 39–48 (2020)
2. Marsuhandi, A.H., Soleh, A.M., Wijayanto, H., Domiri, D.D.: Pemanfaatan ensemble learning dan penginderaan jauh untuk pengklasifikasian jenis lahan padi. Semin. Nasional Official Stat. **1**, 188–195 (2019)
3. Triscowati, D.W., Sartono, B., Kurnia, A., Dirgahayu, D., Wijayanto, A.W.: Classification of rice-plant growth phase using supervised random forest method based on landsat-8 multitemporal data. Int. J. Remote Sens. Earth Sci. (IJReSES) **16**(2), 187–196 (2020)
4. Liu, L.-W., Lu, C.-T., Wang, Y.-M., Lin, K.-H., Ma, X., Lin, W.-S.: Rice (Oryza sativa L.) growth modeling based on growth degree day (GDD) and artificial intelligence algorithms. Agriculture **12**(1), 59 (2022)
5. Rini, M.S.: Kajian kemampuan metode neural network untuk klasifikasi penutup lahan dengan menggunakan Citra Landsat-8 OLI (kasus di Kota Yogyakarta dan sekitarnya). Geomedia: Majalah Ilmiah Dan Informasi Kegeografian, 16(1) (2018)
6. Pravitasari, A.A., et al.: UNet-VGG16 with transfer learning for MRI-based brain tumor segmentation. TELKOMNIKA (Telecommun. Comput. Electron. Control) **18**(3), 1310–1318 (2020)
7. Fithriasari, K., Nuraini, U.S.: Face identification using multi-layer perceptron and convolutional neural network. Breaking News Innovative Comput. Inf. Control, **15**. ICIC International Society (2021). https://doi.org/10.24507/icicel.15.02.157
8. Pangastuti, S.S., Fithriasari, K., Iriawan, N., Suryaningtyas, W.: Classification boosting in imbalanced data. MJS **38**(Sp2), 36–45 (2019)
9. Gokgoz, E., Subasi, A.: Comparison of decision tree algorithms for EMG signal classification using DWT. Biomed. Signal Process. Control **18**, 138–144 (2015)
10. Ayu, K.S., Utaminingrum, F.: Rancang Bangun Sistem Tingkat Kemanisan Buah Sky Rocket Melon menggunakan Metode Gray Level Co-Occurrence Matrix dan Backpropagation Neural Network. Jurnal Pengembangan Teknologi Informasi dan Ilmu Komputer e-ISSN **2548**, 964X (2021)

11. Meinanda, M.H., Annisa, M., Muhandri, N., Suryadi, K.: Prediksi masa studi sarjana dengan artificial neural network. Internetworking Indonesia J. 1(2), 31–35 (2009)
12. Ananto, M.I., Winahju, W.S., Fithriasari, K.: Klasifikasi Kategori Pengaduan Masyarakat Melalui Kanal LAPOR! Menggunakan Artificial Neural Network. Inferensi 2(2), 71–79 (2019)
13. Osco, L.P., et al.: A review on deep learning in UAV remote sensing. Int. J. Appl. Earth Obs. Geoinf. 102, 102456 (2021)
14. Deng, L., Yu, D.: Deep learning: methods and applications. Found. Trends Signal Process. 7(3–4), 197–387 (2014)
15. Jumarwanto, A., Hartanto, R., Prastiyanto, D.: Aplikasi jaringan saraf tiruan backpropagation untuk memprediksi penyakit THT di Rumah Sakit Mardi Rahayu Kudus. Jurnal Teknik Elektro 1(1), 11 (2009)
16. Nwankpa, C., Ijomah, W., Gachagan, A., Marshall, S.: Activation functions: comparison of trends in practice and research for deep learning. ArXiv Preprint ArXiv: 1811.03378 (2018)
17. Sarle, W.S.: Neural networks and statistical models (1994)
18. Goodfellow, I., Bengio, Y., Courville, A.: Deep feedforward networks. Deep Learn. 1 (2016)
19. Reed, R., MarksII, R.J.: Neural Smithing: Supervised Learning in Feedforward Artificial Neural Networks. Mit Press, Cambridge (1999)
20. Kingma, D.P., Ba, J.: Adam: a method for stochastic optimization. ArXiv Preprint ArXiv: 1412.6980 (2014)
21. Sokolova, M., Lapalme, G.: A systematic analysis of performance measures for classification tasks. Inf. Process. Manage. 45(4), 427–437 (2009)
22. He, H., Ma, Y. (Eds.). Imbalanced learning: foundations, algorithms, and applications (2013)
23. Branco, P., Torgo, L., Ribeiro, R.P.: A survey of predictive modeling on imbalanced domains. ACM Comput. Surveys (CSUR) 49(2), 1–50 (2016)
24. Bannari, A., Morin, D., Bonn, F., Huete, A.: A review of vegetation indices. Remote Sens. Rev. 13(1–2), 95–120 (1995)
25. Nuarsa, I.W., Nishio, F., Hongo, C.: Spectral characteristics and mapping of riceplants using multi-temporal Landsat data. J. Agric. Sci. (2011)

The Implementation of Genetic Algorithm-Ensemble Learning on QSAR Study of Diacylglycerol Acyltransferase-1(DGAT1) Inhibitors as Anti-diabetes

Irfanul Arifa[✉], Annisa Aditsania, and Isman Kurniawan

School of Computing, Telkom University, Bandung, Indonesia
irfanularifa@student.telkomuniversity.ac.id

Abstract. Diabetes Mellitus (DM) is one of the most common chronic diseases suffered by the population in the world. DM is the fourth leading cause of death in developing countries. Treatment of diabetes is done by using drugs and blood sugar-lowering therapy. However, the use of inappropriate drugs has side effects in long-term use, such as hypoglycemia and gastrointestinal disorders. One solution that is being tried is to use an inhibitor of the Diacylglycerol Acyltransferase-1 (DGAT-1) enzymes. In the theory of computational drug discovery, Quantitative Structure-Activity Relationship (QSAR) has been successfully created to accelerate the drug discovery process of biopharmaceutical properties of compounds that have not been tested. Build a QSAR model for predicting the activity of DGAT inhibitors as an anti-diabetic target by using the ensemble method. The ensemble method used in this study is Random Forest, AdaBoost, and Gradient Boosting. The best results are obtained by using a gradient boosting model with accuracy and the f1-score are 0.80 and 0.82 respectively.

Keywords: AdaBoost · Diabetes Mellitus · Genetic Algorithm · Gradient Boosting · Quantitative Structure-Activity Relationship (QSAR) · Random Forest

1 Introduction

Diabetes mellitus (DM) is a multifactorial disease, which is characterized by chronic hyperglycemia syndrome and disorders of carbohydrate, fat, and protein metabolism caused by insufficiency of insulin secretion or endogenous insulin activity, or both [1]. Chronic hyperglycemia is said to interfere with pancreatic beta cell function by triggering beta cell apoptosis, increasing intracellular calcium to cytotoxic concentrations, and resulting in increased synthesis of protein granules in beta cells including pro-insulin and pro-islet amyloid associated peptide – (ProlAPP) triggers endoplasmic reticulum stress [2]. Uncontrolled hyperglycemia can also cause many complications such as neuropathy, stroke, and peripheral vascular disease [1].

Diabetes mellitus (DM) ranks as the fourth leading cause of death in developing countries [3]. In 2021 World Health Organization (WHO) reported an increase in the

© The Author(s), under exclusive license to Springer Nature Singapore Pte Ltd. 2023
Y. B. Wah et al. (Eds.): DaSET 2022, LNDECT 165, pp. 282–292, 2023.
https://doi.org/10.1007/978-981-99-0741-0_20

number of people with diabetes from 108 million in 1980 to 422 million in 2014. Between 2000 and 2016, there was a 5% increase in premature deaths from diabetes, and in 2019 it was estimated that 1,5 million deaths are caused directly by diabetes [4]. Indonesia is in seventh place among countries that have diabetes patients with as many as 10.7 million people [3].

Handling diabetes is done in two ways, i.e., the use of drugs and blood sugar-lowering therapy through the application of a diet. A person with diabetes will be given oral glycemic drugs (Oral Hypoglycemic Agents/OHA) to trigger insulin production. However, it has side effects with long-term use such as hypoglycemia and gastrointestinal disturbances [5]. Until now, no effective drug has been found to cure DM. Moreover, there are not many clues from preclinical and/or clinical trials because side effects are unexpected and only revealed on testing in intact biologic settings [6]. One solution that is being tried is to use an inhibitor of the Diacylglycerol Acyltransferase-1(DGAT-1) enzymes.

DGAT-1 is a microsomal enzyme that plays a central role in the metabolism of cellular glycerolipids. DGAT-1 catalyzes the final step in triacylglycerol (TAG) biosynthesis by converting diacylglycerol and fatty acyl-coenzyme-A into triacylglycerol. DGAT-1 plays a fundamental role in the metabolism of cellular diacylglycerol and is important in higher eukaryotes for physiologic processes involving triacylglycerol metabolisms such as intestinal absorption, and lipoprotein assembly, adipose tissue formation, and lactation [7].

Many drugs are usually developed using several trials which are costly, time-consuming, and fail to produce the proper results. In trait drug discovery theory, Quantitative Structure-Activity Relationship (QSAR) has been successfully developed to predict various important biopharmaceuticals, such as genotoxicity, toxicity, oral bioavailability, carcinogenicity, and mutagenicity [8]. The main objective of QSAR is to establish empirical rules to correlate the description of a chemical compound with its bioactivity. Multivariate modeling techniques have been widely used in QSAR studies such as Multiple Linear Regression, Partial Least Squares Regression, and Different Types of Artificial Neural Networks, Genetic Algorithms, and Support Vector Machines [8].

QSAR studies have been widely used to solve problems in DM. In 2019, Kumar et al. Conducted research on the design and development of novel focal adhesion kinase (FAK) inhibitors using Monte Carlo to validate QSAR. The value of r2 is 0.8398 [9]. Then in the year 2020, Eduardo et al. Conducted research on drug reuse using QSAR, Docking, and Molecular Dynamics for possible SARS-CoV-2 Inhibitors. From 20 drug candidates, some of the best potential inhibitors with compound interaction rates were more than 50% [10]. Then in the year 2021, Vinicius et al. Conducting QSAR modeling of the Mpro SARS-CoV inhibitor identified other drugs as candidates for reuse against SARS-CoV-2. The results of the developed QSAR study obtained an accuracy of 98% [11].

In 2020, Kleandrova et al. conducted cell-based multi-target QSAR model research for the design of virtual versatile inhibitors of liver cancer cell lines, where the use of QSAR opens a new horizon in the design of anti-cancer drugs compared to treating liver cancer using chemical therapy [12]. Then in the year 2021, Hammoudi et al. Conducted in-silico research on the discovery of Acetylcholinesterase and Butyrylcholinesterase enzyme inhibitor drugs based on QSAR and drug similarity. The resulting model has a high capacity, where 0.96 of the predicted compounds are outside the negative predictions for molecular weight and pH [13]. One of the challenges of QSAR is the determination of optimal features, which can be done using metaheuristic methods such as genetic algorithms and particle swarm optimization. Based on the literature study, there are not many studies that implement the metaheuristic method in the QSAR Study in the case of DGAT-1.

In this study, we aim to develop a predictive model of the Diacylglycerol Acyltransferase-1 (DGAT1) inhibitor as an anti-diabetes using a combination of Genetic Algorithms and Ensemble methods. A genetic algorithm is used to perform optimization in getting better features from a random initial sample [14]. The advantage of GA compared to optimization with the method differential is that GA can be used to determine the optimum conditions without Differentiating the data first. Therefore, for very complex data, optimization can be done easily [15]. The ensemble method is done by combining methods of machine learning for performance better than using one method. The purpose of the ensemble is to overcome the weaknesses contained in one classifier by using advantages over other classifiers. The ensemble method has proven to be able to improve classification performance [16].

2 Methodology

2.1 Genetic Algorithm (GA)

A genetic algorithm is an optimization technique based on natural selection that allows individuals to survive and adapt to the environment, as well as reproduce. Within the population, the beneficial individuals then interbreed and produce offspring with similar characteristics. By retaining favorable characteristics individuals, those unfavorable individuals are selected from the population. GA aims to improve the solution of a problem by maintaining the best combination of input variables [17].

In solving the problems faced, the GA has several stages. In the first population, random chromosomes are created. The created chromosomes represent the selected features. The main purpose of this algorithm is to determine the most optimal chromosome, which is high accuracy using a low number of features. After creating a population of chromosomes, GA looks for several pairs of parent chromosomes to perform a crossover operation. Parent selection methods can use Roulette Wheel Selection, Tournament Selection, or Rank Selection [18, 19].

If the chromosome exceeds the parent's accuracy, then it is placed in the population (replacing the parent), otherwise, the parent remains in the population. To avoid the occurrence of a local optimum is to maintain the chromosomal diversity of the population. This can be done by implementing Crossover and Mutation. The crossover will produce a set of offspring whose diversity will be maintained by mutating. Mutations are carried out by exchanging one or more genes in a chromosome with their opposite values. The result of the mutation operation is a new derivative which will then be re-tested on the fitness function to see the feasibility of the new population from the GA process as a solution to the given problem [20].

The descriptor selection process is carried out in two stages, namely statistical analysis and GA [21]. In statistical analysis, we begin by eliminating the molecular descriptor which has a value of 0 and a standard deviation of less than 0.50. The feature selection process is then carried out using a GA. The parameters used are population size which is the total size of the population, iteration as a stop operation after the given number of second(s), mutation rate as the rate of mutation in the population's gene, elitism as the number of top individuals to be considered as elites, selective pressure which is a measure of reproductive opportunities for each organism in the population. Parameters and values of GA were determined according to ref [21], as shown in Table 1.

Table 1. Parameter of genetic algorithm

Parameter	Value
Population size	20
Iteration	50
Mutation rate	0.05
Elitism	2
Selective pressure	2

2.2 Ensemble Method

In this study, we utilized three Ensemble methods, i.e., Random Forest, AdaBoost, and Gradient boosting. Random Forest was selected to represent the bagging type of ensemble, while AdaBoost and Gradient boosting were selected to represent boosting type of ensemble. Random forest is pattern recognition based on statistical learning theory. The main steps are, using bootstraps to extract some samples from the original data set to form a new calibration set, and other samples staying to form OOB samples. The sample is modelled by a decision tree, and one decision tree outputs one result. Combine multiple output results and vote to obtain the final prediction result. Many studies have shown that RF has the advantages of not being easy to overfit, faster learning process, better tolerance to noise, and the importance of evaluable variables [22].

AdaBoost is a Boosting family of algorithms. This type of learner is characterized by more attention to the wrongly categorized samples during training, adjusting the sample distribution, and repeating this operation until the number of training of the weak classifier reaches a pre-specified value, completing the learning. And exit the loop [23].

Gradient boosting is a powerful machine-learning technique that achieves state-of-the-art results in a variety of practical tasks. Gradient boosting uses decision trees as base predictors. The advantage of the algorithm is that it uses a new schema for calculating leaf values when selecting the tree structure, which helps to reduce overfitting [24].

3   Result and Discussion

The data set of 228 DGAT-1 data anti-diabetic inhibitors was taken from the chemical database ChEMBL [25]. Furthermore, the data is divided into 2 parts, the training set and the test set with a ratio of 7:3. In this study, the data will be grouped into two class labels, namely 0 and 1. A target labeled with class 1 indicates an active molecule that has a value more than the average of pIC50, while a class 0 label is an active molecule that has a value less than the average of pIC50. The data used on the class label refers to the potential value of the drug pIC50. Furthermore, the selection of molecular descriptors that represent the structure, topology, and potential properties of each inhibitor is carried out.

3.1   Model Development

To improve the performance of the model, hyperparameter tuning was performed to increase the accuracy of the prediction model. Hyperparameters consist of such max depth, max feature, min samples split, learning rate, n estimators, and max-leaf node. The best parameter search was performed using cross-validation. Random forest uses max depth, max feature, and min samples split as parameters, while AdaBoost and Gradient Boosting use learning rates and n estimators as parameters whose values for each parameter can be seen in Table 2.

Table 2. Parameter observed during hyperparameter tuning

Method	Hyper parameter	Value
Random forest	max_depth	[80, 90, 100, 110]
	max_features	[1, 2, 3]
	min_samples_split	[2, 4, 8, 1, 12]
AdaBoost	learning_rate	[0.01, 0.1, 1.0]
	n_estimators	[50, 100, 150, 200, 500]
Gradient boosting	Learning_rate	[0.01, 0.1, 1.0]
	n_estimator	[1, 2, 5, 10, 20, 50, 100, 200, 500],

3.2   Model Validation

Evaluation is something that needs to be done to assess and test the level of usability or functionality and identify specific problems in the information system [26]. In this study, the Confusion Matrix was used as a validation method for the model that was built. The Confusion Matrix is a table that contains the number of rows of test data that are predicted to be true and false by the classification model. The parameters used are accuracy (Q), recall (R), precision (PR), and F1-Score. The equation for calculating each parameter is written in Eq. (1)–(4) (Table 3).

Table 3. Confussion matrix

Category	Predicted class		
Actual class		*Yes (+)*	*No (−)*
	Yes (+)	TP	FN
	No (−)	FP	TN

$$Q = TP + TN \big/ TP + FP + TN + FN \tag{1}$$

$$R = TP \big/ TP + FN \tag{2}$$

$$PR = TP \big/ TP + FP \tag{3}$$

$$F1 - Score = 2 \times (Precission \times Recall) \big/ (Precission + Recall) \tag{4}$$

4 Evaluation

4.1 Feature Selection

The training data set, which consists of 1444 features, is further reduced by performing statistical analysis using the variance threshold. The observed variance threshold values are 0.1, 0.5, and 0.9. After the experiment, the variance threshold value resulted in accuracy at the variance threshold of 0.1 with an accuracy value of 0.837 and the number of features 610, and the variance threshold of 0.5 with the same accuracy value of 0.837 and the number of features 508, while the variance threshold of 0.9 resulted in lower accuracy with a value of 0.756 and the number of features 469. From these data, the variance threshold value is chosen with a value of 0.5. Then obtained data totaling 508 features from the selection which were used for feature selection using a genetic algorithm. The reason for using GA is because GA will always provide a solution at the end of every optimization and can continue to get a solution that is more optimal until the results are in the application and can approach the global optimum [27].

Figure 1 shows the performance of GA produces a value, where the feature selection is carried out not only considering the accuracy produced but also the most optimal number of features with a ratio of 7:3. GA using the RF method produces the highest accuracy with a value of 0.818 and is stable in the 5th iteration. Using the AB method has the highest accuracy with a value of 0.793 and starts to stabilize at the 20th iteration. When using the GB method, the same accuracy as the AB method is obtained but a stable value can be achieved in the 8th iteration. Features generated by the GA using the RF, AB, and GB method were reduced by 54%, 48%, 49%, and with an accuracy of 0.818, 0.793, and 0.793. The results of the feature selection can be seen in Table 4.

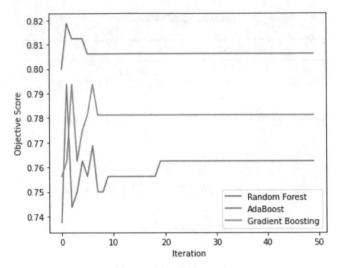

Fig. 1. Genetic algorithm performance

Table 4. Result of feature selection

Method	Number of features	Objective score
Random forest	230	0,818
AdaBoost	261	0,793
Gradient boosting	255	0,793

Table 5. Parameters generated using gridsearchcv

Method	Parameter	Value
Random forest	max_depth max_features min_samples_split	80 ["None"] 1 ["sqrt"] 4 [2]
AdaBoost	learning_rate n_estimators	0.1 [1.0] 150 [50]
Gradient boosting	learning_rate max_leaf_nodes n_estimators	1.0 [0.1] 10 ["None"] 20 [100]

4.2 Hyperparameter Tuning

To improve the accuracy of the model that has been developed, through the process of hyperparameter tuning. The parameters observed in each model can be seen in Table 2. Based on the observed parameters, accuracy using the RF method has increased which is almost the same as using the AB method, the use of hyperparameters in the GB method does not give significant results, but GB produces the highest accuracy compared to the other two methods (Table 5).

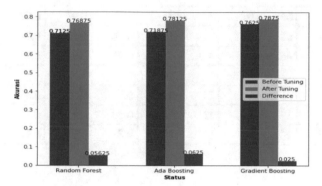

Fig. 2. Improved accuracy in the form of barplot

In Fig. 2, from the three models produced, the accuracy values are not too different. When performing the hyperparameter tuning process, the AB method experienced the highest increase in accuracy compared to the other 2 methods with an increased value of 0.062. So it can be concluded, that the hyperparameter process carried out on the three methods has a fairly good performance, especially for the AB method.

4.3 Model Validation

The validation results obtained can be seen in Table 6. In the training set, RF got the best value in 4 of 8 validation parameters including FN and accuracy. AB got the best value in 6 of 8 including recall, precision, and f1-score. GB got the best value in 4 of 8 including FN and accuracy. RF and AB produce the highest accuracy but are accompanied by a high FN value, so the resulting f-score is not better than AB.

Meanwhile, in the test set, RF got the highest value in 2 of 8 validation parameters including recall. AB got the same result, i.e., 2 of 8 parameters including FP and TN. GB excels in 4 of the 8 parameters used including accuracy and f1-score. RF managed to get the highest recall but because it got a higher FN than the other 2 methods, the resulting f1-score was not better than the others. Meanwhile, AB does not have the best value in other validation parameters, only the confusion matrix value. GB gets the best TP, accuracy, recall, precision, and f1-score values, so it can be concluded that the best model has been built compared to the other two methods.

Table 6. Result of model validation

Model	TP	FP	FN	TN	Q	R	PR	F1-_score_
Training set								
Random forest	**66**	0	**3**	**91**	**0.98**	0.97	0.67	0.79
Ada boost	**66**	2	**3**	89	0.96	**0.98**	**0.97**	**0.97**
Gradient boost	**66**	0	**3**	**91**	**0.98**	0.91	0.74	0.82
Test set								
Random forest	21	1	15	**31**	0.76	**0.97**	0.67	0.79
Ada boost	19	**6**	**17**	26	0.66	0.81	0.6	0.69
Gradient boost	**26**	3	10	29	**0.8**	0.91	**0.74**	**0.82**

5 Conclusion

We have implemented the GA-Ensemble Method to develop a QSAR model for bioactivity prediction of the Diacylglycerol Acyltransferase (DGAT-1) inhibitor. The ensemble method used in this study is Random Forest, AdaBoost, and Gradient Boosting. The use of genetic algorithms can reduce features. The hyperparameter process carried out can improve the accuracy well in the three methods. The best final result uses the gradient boosting method using 255 features with accuracy and f1-score values are 0.80, and 0.82, respectively.

References

1. Nazli, R., Kuantan, T.: Penerapan metode mamdani untuk sistem pendukung keputusan penentuan golongan obat sesuai dengan penyakit diabetes
2. Purnamasari, E., Poerwantoro, B.: Diabetes mellitus dengan penyulit kronis (2011)
3. Soewondo, P., Ferrario, A., Tahapary, D.L.: Challenges in diabetes management in Indonesia: a literature review (2013). https://globalizationandhealth.biomedcentral.com/articles/10.1186/1744-8603-9-63. Accessed 27 Nov 2021
4. Loke, A.: Diabetes (2021). https://www.who.int/news-room/fact-sheets/detail/diabetes. Accessed 27 Nov 2021
5. Syarat, S., Pendidikan, M., Studi, P.: Karya tulis ilmiah persentase penggunaan obat-obat diabetes melitus di rsu Dr. Ferdinand lumbantobing sibolga
6. Turner, N., Zeng, X.Y., Osborne, B., Rogers, S., Ye, J.M.: Repurposing drugs to target the diabetes epidemic. Trends Pharmacol. Sci. **37**(5), 379–389 (2016). https://doi.org/10.1016/j.tips.2016.01.007
7. Yu, T., et al.: Discovery of dimethyl pent-4-ynoic acid derivatives, as potent and orally bioavailable DGAT1 inhibitors that suppress body weight in diet-induced mouse obesity model. Bioorg. Med. Chem. Lett. **28**(10), 1686–1692 (2018). https://doi.org/10.1016/j.bmcl.2018.04.051
8. Pourbasheer, E., Vahdani, S., Malekzadeh, D., Aalizadeh, R., Ebadi, A.: QSAR study of 17β-HSD3 inhibitors by genetic algorithm-support vector machine as a target receptor for the treatment of prostate cancer (2017)

9. Kumar, P., Kumar, A., Sindhu, J.: Design and development of novel focal adhesion kinase (FAK) inhibitors using Monte Carlo method with index of ideality of correlation to validate QSAR. SAR QSAR Environ. Res. **30**(2), 63–80 (2019). https://doi.org/10.1080/1062936X.2018.1564067

10. Tejera, E., Munteanu, C.R., López-Cortés, A., Cabrera-Andrade, A., Pérez-Castillo, Y.: Drugs repurposing using QSAR, docking and molecular dynamics for possible inhibitors of the SARS-CoV-2 Mpro protease. Molecules **25**(21), 5172 (2020). https://doi.org/10.3390/molecules25215172

11. Alves, V.M., et al.: QSAR modeling of SARS-CoV Mpro inhibitors identifies sufugolix, cenicriviroc, proglumetacin, and other drugs as candidates for repurposing against SARS-CoV-2. Mol. Inform. **40**(1), 2000113 (2021). https://doi.org/10.1002/minf.202000113

12. Kleandrova, V.V., Scotti, M.T., Scotti, L., Nayarisseri, A., Speck-Planche, A.: Cell-based multi-target QSAR model for design of virtual versatile inhibitors of liver cancer cell lines. AR QSAR Environ. Res. **31**(11), 815–836 (2020). https://doi.org/10.1080/1062936X.2020.1818617

13. Hammoudi, N.E.H., Sobhi, W., Attoui, A., Lemaoui, T., Erto, A., Benguerba, Y.: In silico drug discovery of acetylcholinesterase and butyrylcholinesterase enzymes inhibitors based on quantitative structure-activity relationship (QSAR) and drug-likeness evaluation. J. Mol. Struct. **1229**, 129845 (2021). https://doi.org/10.1016/j.molstruc.2020.129845

14. Sudarmanto, B.S.A., Oetari, R.A.: Aplikasi deskriptor kimia kuantum dalam analisis QSAR derivat kurkumin sebagai penghambat o-dealkilasi ethoxyresorufin Application of quantum chemical descriptors in QSAR analysis of curcumin derivatives as ethoxyresorufin o-dealkylation inhibitor (2007)

15. Pemodelan dan Optimasi Hidrolisa Pati Menjadi Glukosa dengan Metode Artificial Neural Network-Genetic Algorithm (ANN-GA) (2010)

16. Onan, A., Korukoğlu, S., Bulut, H.: A hybrid ensemble pruning approach based on consensus clustering and multi-objective evolutionary algorithm for sentiment classification. Inf. Process. Manag. **53**(4), 814–833 (2017). https://doi.org/10.1016/j.ipm.2017.02.008

17. Correa, A.B., Gonzalez, A.M.: Evolutionary algorithms for selecting the architecture of a MLP neural network: a credit scoring case. In: Proceedings - IEEE International Conference on Data Mining, ICDM, pp. 725–732 (2011). https://doi.org/10.1109/ICDMW.2011.80

18. Guha, R., et al.: Deluge based Genetic Algorithm for feature selection. Evol. Intel. **14**(2), 357–367 (2019). https://doi.org/10.1007/s12065-019-00218-5

19. Rabani, F., Jondri, Rizal, A.: Klasifikasi Suara Paru Normal dan Abnormal Menggunakan Ekstraksi Fitur Discrete Wavelet Transform dengan Klasifikasi Menggunakan Jaringan Saraf Tiruan yang Dioptimasi dengan Algoritma Genetika. https://jurnal.pcr.ac.id/index.php/ele menter, vol. 7, no. 1 (2021)

20. Rajeev, B.S., Krishnamoorthy, C.S.: Discrete optimization of structures using genetic algorithms

21. Kurniawan, I., Tarwidi, D., Jondri: QSAR modeling of PTP1B inhibitor by using Genetic algorithm-Neural network methods. In: Journal of Physics: Conference Series, vol. 1192, no. 1 (2019). https://doi.org/10.1088/1742-6596/1192/1/012059

22. Safari, M.J.S.: Hybridization of multivariate adaptive regression splines and random forest models with an empirical equation for sediment deposition prediction in open channel flow. J. Hydrol (Amst), **590** (2020). https://doi.org/10.1016/j.jhydrol.2020.125392

23. He, T., et al.: The detonation heat prediction of nitrogen-containing compounds based on quantitative structure-activity relationship (QSAR) combined with random forest (RF). Chemometr. Intell. Lab. Syst. **213**, 104249 (2021). https://doi.org/10.1016/j.chemolab.2021.104249

24. Xu, B.: Institute of electrical and electronics engineers. Beijing section, and institute of electrical and electronics engineers. In: Proceedings of 2019 IEEE 8th Joint International Information Technology and Artificial Intelligence Conference (ITAIC 2019), 24–26 May 2019, Chongqing, China
25. Dorogush, A.V., Ershov, V., Gulin, A.: CatBoost: gradient boosting with categorical features support (2018). http://arxiv.org/abs/1810.11363
26. Kumar, P., Kumar, A., Sindhu, J.: In silico design of diacylglycerol acyltransferase-1 (DGAT1) inhibitors based on SMILES descriptors using Monte-Carlo method. SAR QSAR Environ. Res. **30**(8), 525–541 (2019). https://doi.org/10.1080/1062936X.2019.1629998
27. Biyanto, T.R.: Algoritma genetika untuk mengoptimasi penjadwalan pembersihan jaringan penukar panas. Jurnal Teknik Industri **17**(1), 53–60 (2015). https://doi.org/10.9744/jti.17.1.63-60

Classification of Exercise Game Data for Rehabilitation Using Machine Learning Algorithms

Zul Hilmi Abdullah[1,2]([✉]), Waidah Ismail[2], Lailatul Qadri Zakaria[3], Shaharudin Ismail[2], and Azizi Abdullah[3]

[1] Faculty of Data Science and Information Technology, INTI International University, Nilai, Malaysia
zulhilmi.abdullah@newinti.edu.my
[2] Faculty of Science and Technology, Universiti Sains Islam Malaysia, Nilai, Malaysia
[3] Center for Artificial Intelligence Technology, Faculty of Information Science and Technology, Universiti Kebangsaan Malaysia, Bangi, Malaysia

Abstract. The rehabilitation process aims to allow disabled patients to maintain optimal functioning to achieve return-to-work (RTW) progress. Exercise games are one of the virtual reality treatments (VRT) to help patients with a disability by improving the movement of their body parts. VRT is a modern interactive application that integrates computer software with hardware devices such as a screen, computer, sensor, or Kinect to create an interactive virtual environment for playing exergames (i.e., games with exercises). Artificial intelligence (AI) driven exercise games will become the future of the rehabilitation program. Artificial Intelligence (AI) is the simulation of human intelligence processes by machines. In this project, we are using machine learning (ML), a subset of AI to make useful predictions of data. The methods used are Decision Tree and Neural Network. The study aims to perform a comparison between two machine learning classification algorithms, a Decision Tree, and Neural Network. The study was carried out in the rehabilitation centre of Melaka. The Medical Interactive Recovery Assistance (MIRA) platform generate 26 attributes which are patient data files in this study containing patients' personal information such as first and last names, patient ID and birthdate. It also includes information related to the games played such as the session ID, name of the game, difficulties level, movement ID, and movement name. Before the patient plays the game in MIRA, the physiotherapist is required to set the input for specific games. The settings input required sides used (left or right), duration, difficulty, tolerance, and minimum and maximum range. The experiment was divided into 70% training data and 30% as actual data. The results show Decision Tree performs classification better than Neural Networks. The Decision Tree is good to use for clean data and supervised learning. Neural Network will perform bias in the node when required for the distribution of the data.

Keywords: Exercise game · Rehabilitation · Machine learning · Decision Tree · Neural Network

© The Author(s), under exclusive license to Springer Nature Singapore Pte Ltd. 2023
Y. B. Wah et al. (Eds.): DaSET 2022, LNDECT 165, pp. 293–304, 2023.
https://doi.org/10.1007/978-981-99-0741-0_21

1 Introduction

The rehabilitation process is very important to disable patients to maintain optimal functioning to achieve the return-to-work (RTW) intended level. Decision-making to determine the progress and next steps of the rehabilitation program for the patient is quite challenging for the practitioner. The difficulties involved in predicting the degree of recovery, and appropriate treatment for current and future based on the patient's progress and condition [1]. There are various techniques used in rehabilitation programs nowadays. Exercise games are one of the virtual reality treatments (VRT) to help patients with a disability by improving the movement of their limbs.

VRT is a modern interactive application that integrates computer software with hardware devices such as a screen, computer, and sensor (Kinect) to create an interactive virtual environment for playing exergames (i.e., games with exercises). Virtual reality therapy (VRT) has recently gained popularity for upper limb rehabilitation due to its interactivity and good performance impact on a patient's recovery process [2].

The rehabilitation dataset generated by Medical Interactive Recovery Assistance (MIRA) platform as a VRT tool will be analyzed in this paper. The analysis will be carried out with the assistance of a machine learning algorithm to classify specific data that is significantly meaningful to the patient's progress in the return-to-work (RTW) program.

Artificial Intelligence (AI) or called machine intelligence is the simulation of human intelligence processes by machines. Also, the AI term is used to relate to the machine's ability to learn and solve a problem by mimicking human cognitive functions [3]. Artificial Intelligence (AI) is widely used in various industries including the medical and healthcare industry. Rehabilitation for disabled patients is one of the areas that adapted the AI in improving the monitoring and assess the patient recovery process. Rehabilitation therapy aims to assist disabled patients to achieve and maintain their optimal functioning to obtain return-to-work (RTW) approval.

Currently, practitioners or physiotherapists measure records, and analyze assessment results manually, resulting in difficulty to predict patient progression. Also, rehabilitation data came from various sources including the rehabilitation domain knowledge, historical clinical data, medical records, and assessment as well as exercise games records. Thus, physiotherapist works more complex in analyzing the current state of the patient's progress and deciding on future rehabilitation activities effectively.

In this paper, we are using machine learning (ML) as a subset of AI that involves training to make a useful prediction using data. The algorithm or classifiers used are Decision Tree s and neural networks.

2 Introduction

2.1 Exercise Game for Rehabilitation

Exercise games such as MIRA give an alternative treatment platform for the patient's rehabilitation program. Exercise games can transform clinical exercises into interactive video games. The interactive video games may assist the patient in controlling and quantifying the intensity of the motor tasks. Also, the interactive video games and measurable scores make the rehabilitation process enjoyable, pleasing, and fun.

The exercise game platform is also integrated with patient information management that records the score of each game and movement completed by the patients. The recorded information can assist the therapist to assess the patient's condition and diagnosis. Then, it can create personalized therapy sessions and visualize statistical data on patient improvement. The exercise game for rehabilitation is also known as exergames or serious games for exercise rehabilitation (SGER) [4].

Virtual reality therapy (VRT) has been adopted by various types of rehabilitation patients due to its benefits and interactivity. Also, the capabilities of the VRT-based platforms that are equipped with various features such as patient data recording, database, integration with other systems, and others, assist the patients and practitioner tasks.

In line with the emerging technologies in the information technology domain, gamification has been applied in various fields including entertainment, education, engineering, the healthcare industry, as well as business and marketing. Exercise games or serious games for rehabilitation were already on the market in early 2000 with different types of games, platforms and hardware used. Also, there are exercise games that were designed and developed for the specific type of disease and rehabilitation of specific body parts.

2.2 Sensor-Based Exercise Games

Utilizing virtual reality (VR) technology offers more options and features for the exercise game for rehabilitation with the aid of specific hardware and software. In general, a motion tracking system that utilized sensors from hardware to capture patients' activities is widely used in exercise games for rehabilitation. The hardware sensors consist of contact sensors and non-contact sensors that are used to capture and collect the movement of the patients during therapy. The contact sensors have limitations in terms of mobility since the sensors have to be wired to the patient's body. Meanwhile, the non-contact sensors offer mobility and flexibility with wireless capability in collecting patient data [5, 6].

The flexibility offered by non-contact sensors such as Leap Motion and Kinect sensors makes it popular hardware to be used in the exercise game for rehabilitation. Also, the non-contact sensors offer broad opportunities for game developers in producing various types of games for various types of target users. The main objectives of the exercise games are to provide good engagement of patients with the games. Good engagement can be achieved with high enjoyment, meaningful goals for the specific games and movements, rewards, appropriate exercise intensity and difficulties setting [7]. The difficulty of the game developer is to provide automatic intensity and difficulty setting to the patients based on their current progress and future requirements. Therefore, the Kinect sensor become one of the popular devices used in exercise games for rehabilitation due to its flexibility and functionality [8]. The Kinect sensor is used to capture the gesture of a patient's movement and data stored for assessment of the patient's progress [9].

Several works were focusing on specific exercise games for specific patient illnesses such as stroke [2, 10, 11], spinal cord injury (SCI), traumatic brain injury (TBI) [12] and other types of illness. Dataset used in this paper consists of 12 types of diagnoses of illness such as stroke, brain injury, brain tumour, TBI and SCI.

2.3 Machine Learning in Rehabilitation

Machine learning (ML) is an important branch of artificial intelligence (AI). Machine learning refers to the algorithms that can learn and re-learn from specific data. The most popular machine learning classification algorithm used in medically related research is support vector machine, neural network, Decision Tree, and random forest [13]. Machine learning has been widely used in rehabilitation programs especially in determining patients' progress, future requirements, etc.

ML in predicting the progression of a patient's rehabilitation became a good contribution to the clinical decision support system. Requirements to automate the assessment and prediction of patient progress, the result of several proposed works by researchers. For example, Cannière and the teams propose the combination of wearable device monitoring with machine learning to track the patient's progress during the rehabilitation session [14]. Also, a bit similar project was proposed by Dester's team that combines wearable sensors with a machine-learning algorithm to track the patient's motor recovery [15].

ML also can be used in the prediction of the early stages of a patient's illness to avoid the worst-case scenario for the patient [16, 17]. Other researchers applied the machine learning modelling approaches by integrating it with a clinical decision support system or electronic medical record to assist the practitioner in assessing the patient's progression [18–20].

2.4 Current Issues in Exercise Games for Rehabilitation

In many cases, a physiotherapist or practitioner will assess the patient's progress and current condition manually. Then they will manually set the best setting for the patient's next exercise. The manual assessment and setup may time consuming and increase the burden on the physiotherapist especially when they need to deal with big numbers of patients. Therefore, the automated assessment and best setting will be required to assist the physiotherapist in patient assessment and management.

Artificial Intelligence (AI) is widely used in various industries including the medical and healthcare industry. Rehabilitation for disabled patients is one of the areas that utilize AI in improving the monitoring and assess the patient recovery process. Rehabilitation therapy aims to assist disabled patients to achieve and maintain their optimal functioning to obtain return-to-work (RTW) approval. Currently, practitioners or physiotherapists measure records, and analyze assessment results manually, resulting in difficulty to predict patient progression. Also, rehabilitation data came from various sources including the rehabilitation domain knowledge, historical clinical data, medical records, and assessment as well as exercise games records. Thus, physiotherapist works more complex in analyzing the current state of the patient's progress and deciding on future rehabilitation activities effectively.

This paper aims to assist in decision-making for the physiotherapist by automatically assessing the patient's condition and proposing the best setting for the next exercise. The machine learning-assisted decision support system will perform a comparison between two methods that include the Decision Tree and Neural Network in the classification of patient data.

3 Methods and Tools

The Medical Interactive Recovery Assistance (MIRA) platform generate 26 attributes which are patient data file in this study that contains patients' personal information such as first and last names, patient ID and birthdate. The information related to the games played such as the session, name of the game, movement ID, and movement name also recorded. Before the patient plays the MIRA, the physiotherapist is required to set the settings.

The settings input required sides used (left or right), duration, difficulty, tolerance, and minimum and maximum range. In total, there are 62,490 rows of data collected from the MIRA platform for a total of 84 patients. Each patient has the repetition of records since they need to play the same game and other games multiple times to measure their progress. However, for a specific purpose or objective of research the main dataset will be filtered to fulfil the requirement of the research. The experiment was divided into 70% training data and 30% as testing data. For the initial test, there are 17 relevant attributes tested using a Decision Tree classification algorithm.

3.1 Dataset

MIRA platform outputs a patient data file that contains personal information about the patient such as first and last name, patient id, and birth date, in addition to information related to playing the games, such as the session id and name of the game, movement id, movement name, and game parameters. Each selected game and movement have input variables in the item settings dialogue module, which refers to the sides used (left or right), duration, difficulty, tolerance, and Min and Max Ranges.

The values of these variables could be fixed based on the default values or could be adjusted by the physiotherapist after evaluating the patient. Table 1 describes the significant variables generated by the exercise game. MIRA dataset represents different types of games such as Airplane, Animals, Atlantis and others that can capture different types of movements. There are more than 30 types of movement that can be captured and measured by the games. Each movement represents a different level and difficulty that can be used to measure a patient's progress.

Table 1. Data measurement in MIRA platform.

No	MIRA application data	Description
1	Time (Duration)	Total time of the exergame (a game or a movement)
2	Still Time	The idle time is when the patient stops moving before the game finishes
3	Moving Time	Time duration of the patient's continuous movement (correctly or incorrectly) throughout a game
4	Moving Time in Exercise	The time duration of the patient's continuous correct movement (as required by the exercise) throughout the exergame
5	Average Acceleration	The average positive change rate of the velocity is divided by the overall duration of the exergame
6	Average deceleration	The average negative change rate of the velocity is divided by the overall duration of the exergame
7	Average Accuracy	The accuracy of the movement for each exergame
8	Average congruent correct answer reaction time	The movement of the objects in the game during positive reaction time where the reaction time is the positive response time for each event in the game
9	Average congruent incorrect answer reaction time	Congruent movement of the objects during a negative reaction time of responding to each event
10	Average percentage	Average Range of motion that the patient performs during the exercise
11	Average speed	The result of dividing the Distance by the overall Time duration when the movement is performed correctly
12	Average variation	The average interval of Range of Motion a patient performs during an exercise
13	Distance	Total distances traversed by a specific joint
14	Maximum percentage	Maximum Range of Motion carried out by a patient throughout the exergame
15	Minimum percentage	Minimum Range of Motion carried out by a patient throughout the exergame

(continued)

Table 1. (*continued*)

No	MIRA application data	Description
16	Repetition	The total number of correct movements throughout the exergame
17	Points	The total scores were achieved throughout the exergame

3.2 Decision Tree and Neural Network

The same dataset has been trained and tested using Decision Tree and Neural Network algorithms. Ultimately, the classifier was executed to test the correctly classified patient's diagnosis together with other attributes. Other attributes such as the maximum range of motion and difficulty setting of games were also tested using a Decision Tree classifier.

This work is based on the Decision Tree model proposed by Zainal et al. which focused on the prediction of a patient's performance based on the movement records [2]. In this paper, the Decision Tree model will be used to predict a patient's diagnosis of illness based on the data recorded. Also, applying the Decision Tree to predict patient performance (*MaxRange* and difficulty level) for a larger dataset compared to previous works by Zainal et al. [2].

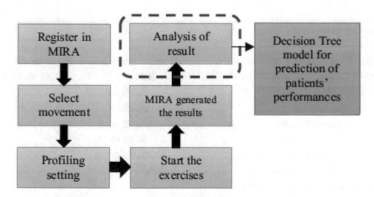

Fig. 1. Process flow of rehabilitation using MIRA [2].

Figure 1 shows the general flow of the rehabilitation process using the MIRA exercise game platform starting from the user (patient) registration until the Decision Tree modelling. We adopted a similar process flow but with a larger dataset. Meanwhile, Fig. 2 represent the sample dataset captured by the MIRA exercise game based on the patient's activities. Normally, each patient will play multiple games with multiple movements for each session.

	A	B	C	D	E	F	G	H	I	J	K
1	Gender	Value	Name	ExerciseGa	Movemen	Side	Difficulty	Tolerance	MinRange	MaxRange	Diagnosis
2	Male	74.95102	Still Time	Atlantis	Shoulder A	Left	Medium	40%	0%	100%	STROKE
3	Male	224.924	Moving Tir	Atlantis	Shoulder A	Left	Medium	40%	0%	100%	STROKE
4	Male	105.666	Distance	Atlantis	Shoulder A	Left	Medium	40%	0%	100%	STROKE
5	Male	-2.2904	Average D	Atlantis	Shoulder A	Left	Medium	40%	0%	100%	STROKE
6	Male	2.352901	Average A	Atlantis	Shoulder A	Left	Medium	40%	0%	100%	STROKE
7	Male	100	Maximum	Atlantis	Shoulder A	Left	Medium	40%	0%	100%	STROKE
8	Male	300.0211	Time	Atlantis	Shoulder A	Left	Medium	40%	0%	100%	STROKE
9	Male	60.83325	Average Vi	Atlantis	Shoulder A	Left	Medium	40%	0%	100%	STROKE
10	Male	1	Repetition	Atlantis	Shoulder A	Left	Medium	40%	0%	100%	STROKE
11	Male	28.60042	Average Pe	Atlantis	Shoulder A	Left	Medium	40%	0%	100%	STROKE
12	Male	0.34153	Average Sr	Atlantis	Shoulder A	Left	Medium	40%	0%	100%	STROKE
13	Male	43.01485	Moving Tir	Atlantis	Shoulder A	Left	Medium	40%	0%	100%	STROKE
14	Male	0	Minimum I	Atlantis	Shoulder A	Left	Medium	40%	0%	100%	STROKE

Fig. 2. Sample patient data from MIRA

4   Result and Discussion

In this work we, evaluated the classification performance of two machine learning algorithms to correctly classify the patient's diagnosis of illness. The result shows that the Decision Tree model performs better compared to the neural network model. It shows that the Decision Tree is good to use for clean data and supervised learning such as our dataset.

```
=== Summary ===

Correctly Classified Instances        18650              99.7967 %
Incorrectly Classified Instances         38               0.2033 %
Kappa statistic                        0.9959
Mean absolute error                    0.0009
Root mean squared error                0.0219
Relative absolute error                0.7141 %
Root relative squared error            8.7384 %
Total Number of Instances             18688
Ignored Class Unknown Instances                   61
```

Fig. 3. Decision Tree classification for MaxRange.

Figure 3 shows the classification result for *MaxRange* which represents the maximum range of motion (ROM) required by the patient who played the exercise game. The *MaxRange* measure the ability of the patients to move or stretch a part of their body, such as a joint or a muscle during a given exercise. The classification result uses a Decision Tree classifier score of almost 100% which considers high accuracy classification for the test dataset (30% of the total dataset). From the result, we can conclude that the patient's maximum range of motion can be accurately predicted based on their score from the previous exercise.

```
=== Summary ===

Correctly Classified Instances        18674              99.9251 %
Incorrectly Classified Instances         14               0.0749 %
Kappa statistic                       0.9986
Mean absolute error                   0.0006
Root mean squared error               0.0198
Relative absolute error               0.1493 %
Root relative squared error           4.6101 %
Total Number of Instances            18688
Ignored Class Unknown Instances                  61
```

Fig. 4. Classification result for difficulties setting of games.

To measure the accuracy of the difficulty setting for each patient and each session, we run the classification test for the three types of settings that include an easy, medium and hard difficulty level for each game. Figure 4 shows the Decision Tree classification result of the difficulties set for each session. The result shows high accuracy which only 14 out of 18688 instances incorrectly classified. This means the Decision Tree model can be used to predict the difficulty setting of specific games for a future session of rehabilitation.

```
=== Summary ===

Correctly Classified Instances        18749                 100     %
Incorrectly Classified Instances          0                   0     %
Kappa statistic                           1
Mean absolute error                       0
Root mean squared error                   0
Relative absolute error                   0        %
Root relative squared error               0        %
Total Number of Instances            18749
```

Fig. 5. Classification result for diagnosis of patient's illness using Decision Tree classifier.

Ultimately, we ran the classification for diagnosis of the patient's illness which consists of TBI, SCI, stroke, brain injury, brain tumour, muscular dystrophy, and others. Figure 5 represent the Decision Tree result of diagnosis attributes that score 100% accuracy. The result shows that the Decision Tree model can accurately classify the patient's diagnosis of illness based on the existing dataset.

Also, we tested the dataset with a Neural Network algorithm and the result shows lower accuracy in diagnosis classification. There are only 68.771% correctly classified instances for the same dataset using the Neural Network classifier as shown in Fig. 6. Neural Network classifier takes a longer time to classify the dataset compared to the

Decision Tree classifier. Also, the Neural Network classifier required pre-processing of the dataset before it can be run.

```
=== Summary ===

Correctly Classified Instances          42975              68.771   %
Incorrectly Classified Instances        19515              31.229   %
Kappa statistic                            0.3862
Mean absolute error                        0.0414
Root mean squared error                    0.162
Relative absolute error                   65.8392 %
Root relative squared error               91.3527 %
Total Number of Instances              62490
```

Fig. 6. Classification result for diagnosis using Neural Network classifier.

The results show Decision Tree s perform classification better than Neural Networks. A Decision Tree is good to use for clean data and supervised learning. Neural Network will perform bias in the node when required for the distribution of the data.

Both methods can model data that have nonlinear relationships between variables, and both can handle interactions between variables. In this project, the Decision Tree performs better than the Neural Network. In addition, the Decision Tree is easy to understand. Also, very useful as a modelling technique and provide visual representations of the data. However, the Neural Network method is more complex and required more resources to run.

In future, we are planning to embed the Decision Tree classification and optimization model into the exercise games and the clinical decision support system. Therefore, a physiotherapist can handle the patients effectively with the available tools, data, and prediction results. Also, the patient can perform their therapy session at home with the available exercise games and sensors in the market.

5   Conclusion

Exercise games like MIRA offer a good option for a therapist as a tool for a rehabilitation program. Exercise games combine physical and cognitive exercise and transform patients' exercise data into clinical evidence. The exercise game makes therapy sessions more enjoyable, convenient, interactive, and easier to follow. With the advantages of exercise games in the rehabilitation program, patient data should be managed and manipulated effectively for better patient recovery results. Effective data analysis and classification may help the therapist to effectively assess the patient's progress and future needs. It involved progress in the patient's body movement, ability to perform specific exercise/game tasks, and prediction of the type of game with specific difficulties level. This work has proven that the Decision Tree algorithm effectively classifies the exercise games data that is more accurate compared to the Neural Network algorithm. The result is due to the type of data generated by the exercise games that consider structured data. So, for the exercise game platforms that generate structured data, a Decision Tree algorithm can be utilized. Our work can be extended with a few enhancements such as focusing on a specific parameter, specific games, specific illnesses, and specific body parts from the dataset generated by the exercise games. Also, a combination of data mining algorithms and other artificial intelligent components may enhance the exercise games' capabilities. With the standard clinical terms and language, it is possible to integrate exercise games with other electronic medical systems or records.

Acknowledgement. This work is supported by Fundamental Research Grant Scheme FRGS, Ministry of Higher Education with Application ID 225085–266898, and reference code - FRGS/1/2018/ICT02/USIM/02/1, University Sains Islam Malaysia (USIM), Faculty of Science and Technology (FST), Project Code (USIM/FRGS/FST/055002/51918).

References

1. Sohn, J., Jung, I.-Y., Ku, Y., Kim, Y.: Machine-learning-based rehabilitation prognosis prediction in patients with ischemic stroke using brainstem auditory evoked potential. Diagnostics **11**(4), 673 (2021). https://doi.org/10.3390/diagnostics11040673
2. Zainal, N., Al-Hadi, I.-Q., Ghaleb, S.M., Hussain, H., Ismail, W., Aldailamy, A.Y.: Predicting MIRA patients' performance using virtual rehabilitation programme by decision tree modelling. In: Al-Emran, M., Shaalan, K., Hassanien, A.E. (eds.) Recent Advances in Intelligent Systems and Smart Applications. SSDC, vol. 295, pp. 451–462. Springer, Cham (2021). https://doi.org/10.1007/978-3-030-47411-9_24
3. Mu, P., Dai, M., Ma, X.: Application of artificial intelligence in rehabilitation assessment. IOP Conf. Ser. Earth Environ. Sci. **3**, 2021 (1802). https://doi.org/10.1088/1742-6596/1802/3/032057
4. Ning, H., Pi, Z., Wang, W., Farha, F., Yang, S.: A review on serious games for disaster relief **14**(8), 1–14 (2022). http://arxiv.org/abs/2201.06916
5. Shi, Y., Peng, Q.: A VR-based user interface for the upper limb rehabilitation. Procedia CIRP **78**, 115–120 (2018). https://doi.org/10.1016/j.procir.2018.08.311
6. Trombetta, M., et al.: Motion Rehab AVE 3D: A VR-based exergame for post-stroke rehabilitation. Comput. Methods Programs Biomed. **151**, 15–20 (2017). https://doi.org/10.1016/j.cmpb.2017.08.008
7. Goršič, M., Cikajlo, I., Goljar, N., Novak, D.: A multisession evaluation of an adaptive competitive arm rehabilitation game. J. Neuroeng. Rehabil. **14**(1), 1–15 (2017). https://doi.org/10.1186/s12984-017-0336-9
8. Tokuyama, Y., Rajapakse, R.P.C.J., Yamabe, S., Konno, K., Hung, Y.P.: A Kinect-based augmented reality game for lower limb exercise. In: Proceedings - 2019 International Conference Cyberworlds, CW 2019, pp. 399–402 (2019). https://doi.org/10.1109/CW.2019.00077
9. Miron, A., Sadawi, N., Ismail, W., Hussain, H., Grosan, C.: Intellirehabds (Irds)—a dataset of physical rehabilitation movements. Data **6**(5), 1–14 (2021). https://doi.org/10.3390/DATA6050046
10. Ahmad, N.A., et al.: Development of virtual reality game for the rehabilitation of upper limb control in the elderly patients with stroke development of virtual reality game for the rehabilitation of upper limb control in the elderly patients with stroke, vol. 4, pp. 1–10 (2020)
11. Lee, M.H., Siewiorek, D.P., Smailagic, A., Bernardino, A., Bermúdez i Badia, S.: Interactive hybrid approach to combine machine and human intelligence for personalized rehabilitation assessment. In: ACM CHIL 2020 - Proceedings 2020 ACM Conference on Health, Inference, and Learning, pp. 160–169 (2020). https://doi.org/10.1145/3368555.3384452

12. Lange, B., Chang, C.Y., Suma, E., Newman, B., Rizzo, A.S., Bolas, M.: Development and evaluation of low-cost game-based balance rehabilitation tool using the Microsoft Kinect sensor. In: Proceedings Annual International Conference of the IEEE Engineering in Medicine and Biology Society EMBS, pp. 1831–1834 (2011). https://doi.org/10.1109/IEMBS.2011.6090521

13. Ismail, W., Grosan, C., Abdullah, Z.H., Aldailamy, A.Y., Zainal, N., Hendradi, R.: Integrated-regression and whale optimisation algorithms to post-stroke rehabilitation analysis: a case study for serious games (2021). https://doi.org/10.1109/ISCI51925.2021.9633382

14. Alghatani, K., Ammar, N., Rezgui, A., Shaban-Nejad, A.: Predicting intensive care unit length of stay and mortality using patient vital signs : machine learning model development and validation corresponding author, vol. 9, no. 5, pp. 1–23. https://doi.org/10.2196/21347

15. De Cannière, H., et al.: Wearable monitoring and interpretable machine learning can objectively track progression in patients during cardiac rehabilitation. Sensors (Switzerland) 20(12), 1–15 (2020). https://doi.org/10.3390/s20123601

16. Adans-Dester, C., et al.: Enabling precision rehabilitation interventions using wearable sensors and machine learning to track motor recovery. NPJ Digit. Med. 3(1), 1 (2020). https://doi.org/10.1038/s41746-020-00328-w

17. Carriere, J., et al.: Case report: utilizing AI and NLP to assist with healthcare and rehabilitation during the COVID-19 Pandemic. In: Frontiers in Artificial Intelligence, vol. 4, pp. 1–7 (2021). https://doi.org/10.3389/frai.2021.613637

18. Choo, Y.J., Kim, J.K., Kim, J.H., Chang, M.C., Park, D.: Machine learning analysis to predict the need for ankle foot orthosis in patients with stroke. Sci. Rep. 11(1), 1–7 (2021). https://doi.org/10.1038/s41598-021-87826-3

19. Chen, W., Song, W., Chen, H., Li, Q., Zhao, P.: Motion synthesis for upper-limb rehabilitation motion with clustering-based machine learning method. ASME Int. Mech. Eng. Congr. Expo. Proc., 3 (2019). https://doi.org/10.1115/IMECE2019-10435

20. Izhar, C.A.A., Hussain, Z., Maruzuki, M.I.F., Sulaiman, M.S., Rahim, A.A.A.: Gait cycle prediction model based on gait kinematic using machine learning technique for assistive rehabilitation device. IAES Int. J. Artif. Intell. 10(3), 752–763 (2021). https://doi.org/10.11591/ijai.v10.i3.pp752-763

21. Gharaei, N., Ismail, W., Grosan, C., Hendradi, R.: Optimizing the setting of medical interactive rehabilitation assistant platform to improve the performance of the patients: a case study. Artif. Intell. Med. 120, 102151 (2021). https://doi.org/10.1016/J.ARTMED.2021.102151

SDDLA: A New Architecture for Secured Decentralized Distributed Learning

Sufyan Almajali[✉]

Princess Sumaya University for Technology, Amman, Jordan
s.almajali@psut.edu.jo
https://psut.edu.jo/users/dr-sufyan-almajali

Abstract. The proper utilization of distributed network resources solves important issues faced by machine learning algorithms and artificial intelligence in general such as the availability of high-specification processing resources and the availability of datasets. This paper proposes a new Secured Decentralized Distributed Learning Architecture (SDDLA). The new suggested architecture enables distributed learning algorithms to run on distributed datasets without compromising the privacy and security of shared datasets with unauthorized users. Also, the decentralized management approach of distributed entities simplifies the deployment, activation, and utilization of distributed learning. The proposed architecture includes a new data placement and task allocation algorithm that adds a low bandwidth overhead and low processing requirements on the distributed network.

Keywords: Distributed learning · Security · Edge computing · Machine learning

1 Introduction

In the past decade, the industry witnessed a significant increase in reliance on artificial intelligence and machine learning algorithms due to three main factors. One main factor is the improvement in hardware that made it possible to run complicated algorithms with commodity hardware. Another key factor is the improvement in machine learning techniques. Also, the number of applications and areas that can benefit from machine learning has increased as well [5–7]. Despite all these advancements, one of the main challenges machine learning suffers from is the availability of resources to perform highly intense operations required by machine learning algorithms. In addition, the availability of rich representative datasets that can feed machine-learning algorithms to generate high-accuracy models is another challenge. With distributed algorithms and distributed architectures, networks, companies, and organizations with distributed resources have a potential solution to the issues of limited processing resources and datasets availability [4].

Several entities share a common goal in terms of utilizing technology and available data to solve a specific problem related to their domain of work. For instance, police departments from different states in the US like to utilize machine learning technologies to allow car plate number recognition. Each state has its own servers and datasets of

© The Author(s), under exclusive license to Springer Nature Singapore Pte Ltd. 2023
Y. B. Wah et al. (Eds.): DaSET 2022, LNDECT 165, pp. 305–315, 2023.
https://doi.org/10.1007/978-981-99-0741-0_22

pictures taken of cars' plate numbers over time. Since cars move across different states, having datasets sufficient and representative of all the states is a challenge. Collaboration among these entities, states in this case, and allowing resource utilization among them can help each entity reach its goal by building a machine learning-based solution that is capable of working on all types of plate numbers in the US and can help build the solution quicker.

Entities could be different departments that belong to the same organization, company, or government, in this case, entities have the same ownership to some extent. Although having one owner or decision-maker body implies the simplicity of sharing resources among these entities, this is not always true. The nature of the organization in terms of size, being present in multiple countries, and laws variance such as counties belonging to some state or different states makes it difficult to arrange for cooperation among these entities. Furthermore, entities could belong to completely different organizations or companies and have something in common such as belonging to a federation, consortium, or agreement.

Resources sharing among different entities, whether same owner or not, requires careful consideration in terms of security, resource availability, and ease of management [1, 2]. This paper proposes a new Secured Decentralized Distributed Learning Architecture (SDDLA). The new suggested architecture offers three key advantages: 1) it enables distributed learning algorithms to run on distributed datasets, 2) it offers a secured way of sharing datasets and processing resources through two key security services: access permissions and encryptions, 3) if offers a decentralized way of managing the relations among the different participating entities which simplifies the deployment, activation, and utilization of distributed learning.

2 Architecture Components

Figure 1 shows the components of the new proposed architecture. The components include the following:

- Areas: The first major component of the new architecture is the concept of area. An area represents a set of resources that are owned by one entity. Different areas represent different entities that want to collaborate in resources utilization and data sharing without compromising the security of their data or the availability of their resources. Different entities might be different departments in one company, different counties in one state or country, different organizations, or different companies.
- Nodes: A node has processing capabilities and storage capabilities as well. Nodes vary in terms of processing and storage specifications. Some nodes might have datasets stored in them, or collected by them such as nodes 1,3,5,8,10,11,12, and 13. Other nodes might have data storage but are without datasets initially, such as nodes 2,6,7, and 9. According to the node's processing and storage capabilities, nodes are assigned learning tasks along with partial datasets for these tasks.
- Control Node (CN): Each area has one dedicated node to act as a control node. The control node is responsible for coordinating the process of learning along with peer control nodes in other areas. Also, the control node participates in data placement decisions and the security permission setup for each area.

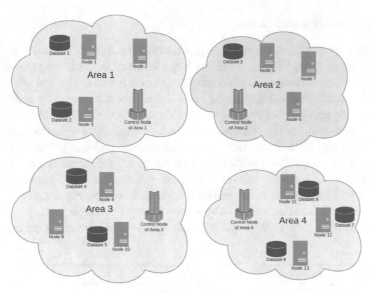

Fig. 1. The architecture components.

- Datesets (DSs): The datasets can be located in several places in different areas. The architecture is flexible as it allows learning from scattered datasets across the entire network, or splitting one giant dataset across several nodes for improved performance.
- Network Protocol: The network protocol of the new architecture is an essential component that allows all parties to communicate at every stage of the architecture. The protocol supports different types of messages that are explained in every stage in the coming sections.
- Distributed Network Services: To deliver the desired functions, the architecture requires four types of network services: Control Node Service (CNS), Node Service (NS), Aggregator Node Service (ANS), and Distributed Learning Task Service (DLTS). The CNS service allows controlling the tasks within each area. CNS service communicates with the different node services running within its area for pushing tasks to be executed or pulling information. CNS also communicates with other CNS services for other areas for coordinating tasks that will run over multiple areas. The Node Service (NS) service runs on nodes for communicating with the area CNS or with other NS services for running a task. The Aggregator Node Service (ANS) is activated on a certain node once it is selected to aggregate the results of a certain distributed learning task. The Distributed Learning Task Service (DLTS) is activated on each node selected to participate in the distributed learning task. The AN node communicates with nodes using ANS and DLTS services. In general, each service has its unique public IP address, port, and public/private key information for secured communication.

2.1 Distributed Network Setup

The first step before proceeding to run a distributed learning task is to have the network setup. Algorithm 1 explains in detail the steps involved in setting up the distributed network. As shown in lines 6–8, the control node of each area is updated with the nodes available for processing and data storage in its area. This information is communicated from each node services NS to the control node service CNS. Lines 9–12 explain the need for setting up the connections between the different control nodes in the distributed network. The setup includes the security permission from each control node perspective. For each area, Ai, the control node Ci sets up the security permissions for other areas in the network. The security permissions include which other areas are allowed to use area Ai's processing resources, and which areas are allowed to transfer area Ai's dataset records if required. Network setup can be adjusted at any later stages based on user desire, and network status. New areas might join, and other areas might leave the network.

Algorithm 1 Distributed Network Setup

1: A: The set of areas in the distributed network
2: N: The set of nodes that belong to all areas
3: C: The set of control nodes in the distributed network
4: C_i: The control node of area A_i
5: **for** each $A_i \in A$ **do**
6:      **for** each $N_i \in A_i$ **do**
7:          $C_i \leftarrow N_i's\ information$
8:      **end for**
9:      **for** each $C_j \in C\ where\ C_j' = C_i$ **do**
10:          $C_i \leftarrow (C_i, C_j)\ connectivity\ info$
11:          $C_i \leftarrow setup\ permissions\ for\ C_j$
12:      **end for**
13: **end for**
14:

2.2 Distributed Learning Task Stages

Once the distributed network is configured, the architecture can start running distributed learning tasks.

A distributed learning task includes the following steps:

1. A user initiates a distributed learning task request Ri from a certain area Ai. The request is sent to the control node Ci
2. The control node Ci assigns a certain node in the area Ni to activate the *ANS* service and act as the aggregator node and initiates a distributed learning task request Ri from a certain area Ai.
3. The data placement and task allocation algorithm is executed. This is shown in Algorithm 2 which is explained in the next section.

4. Once the datasets are in place, and the tasks are allocated for the participating nodes across the distributed network, the distributed learning task is activated by the *AN*. This is explained in the next sections.

2.3   Data Placement and Task Allocation Algorithm

Algorithm 2 explains the steps of the newly proposed data placement and task allocation algorithm.

In lines 8–9, the control node *Ci* receives a request to run a distributed learning task *Ri*, and a certain node *Ni* is assigned to do the aggregation task within that same area. The aggregation node activates the *ANS*.

The control node *Ci* updates other control nodes with the request. The update is sent to only the nodes that gave permission to area *i* to use processing resources, *CAPi*, as shown in lines 10–12.

The algorithm gives higher priorities to datasets allocations that will generate no traffic or the least amount of network traffic.

Lines 15–37 represent the overall data placement and task allocation for areas that agreed to be part of the distributed learning task initiated in *Ri*. For each node, the available resources are reserved, and the DLTS is activated per node, line 17, the *CalculateTaskDSSize(Nm)* is invoked to calculate dataset size to be allocated for each node *Nm* based on its specifications and current state, line 18. The *DSDatak* array is used to track the available dataset records on each node if any. The *TotalSize* of each area represents the maximum dataset size it can handle as an area considering the available resources. If the area is capable of processing more than what it has, a secured dataset request is performed to transfer dataset records from another area as explained in lines 23 to 32. Only areas that granted permission for data transfer will be considered, *CADk* as in line 25.

Once each area has the minimum size of dataset records, dataset records are assigned to each node in that area using the *AssignDS(DSdatak)*, see line 34. The *AssignDS* function takes into consideration the following rules:

1. First, the control node attempts to assign each node a dataset records that reside locally on the node itself if available to avoid generating local area network traffic.

    2. If a node has no enough dataset records, the control node *Ci* assigns dataset records from other nodes in area *i*, to minimize the traffic generated between areas.
    3. External dataset records from other areas are used in case local dataset records were not sufficient for the local nodes. The external data set records are added to the local area in lines 23–32.

Algorithm 2 Data Placement and Task Allocation Algorithm

1: A: The set of areas in the distributed network
2: N: The set of nodes that belong to the areas
3: C: The set of control nodes in the distributed network
4: CAD_i: The set of control nodes that granted dataset transfer permission to area i
5: CAP_i: The set of control nodes that granted processing permission to area i
6: R_i is a Distributed learning task request from area i
7: AN_i: A node in area i dedicated as an aggregator node
8: $C_i \leftarrow R_i$
9: $N_i \leftarrow AssignedANtask$
10: **for** each $C_j \in CAP_i$ **do**
11:      $C_j \leftarrow R_i$
12: **end for**
13: $TotalSize \leftarrow 0$
14: $DSSize_k \leftarrow 0$
15: **for** each $C_k \in CAP_i \cup C_i$ **do**
16:      **for** each $N_m \in A_k$ **do**
17:          $Reserve(N_m)$
18:          $S_k[m] \leftarrow CalculateTaskDSSize(N_m)$
19:          $DSData_k[m] \leftarrow GetDSSize(N_m)$
20:          $DSSize_k \leftarrow DSSize_k + DSData_k[m]$
21:          $TotalSize \leftarrow TotalSize + S_k[m]$
22:      **end for**
23:      **if** $TotalSize > DSSize_k$ **then**
24:          $TS \leftarrow TotalSize - DSSize_k$
25:          **for** each $C_j \in CAD_k$ **do**
26:              $(DSdata_k, RS) \leftarrow Request(R_i, C_j, TS)$
27:              $TS \leftarrow TS - RS$
28:              **if** $TS \leq 0$ **then**
29:                  $Exit\ for\ Loop$
30:              **end if**
31:          **end for**
32:      **end if**
33:      **for** each $N_m \in A_k$ **do**
34:          $N_m \leftarrow AssignDS(DSdata_k)$
35:          $N_m \leftarrow R_i$
36:      **end for**
37: **end for**

Finally, each node Nm involved in the distributed learning task receives the request Ri which contains detailed information about the request such as the AN address information, algorithm initialization information, etc. The AN in- formation allows each node to access the ANS service for distributed learning communication such as generated models' information.

3 Distributed Learning Algorithm

This architecture is designed to support several distributed learning algorithms. The main idea is to have one node dedicated as the aggregator node (AN), and the other nodes reserved to be part of the distributed learning tasks, where each node builds its own model, and models are combined at the aggregator node using one of the ensemble learning techniques such as in [8, 9], and [10]. Figure 2 shows an example of the structure. In this example, the request came from area 1. The control node of area 1 communicates the request across the network where the data placement and task allocation have been executed. Notice that not all nodes are selected to be part of running the distributed tasks as not all are available. Also, area 4 did not grant permission to area 1 so no are nodes selected from area 4. Each selected node, green nodes in Fig. 2 activates the DLTS service. Also, the AN node, the purple node, has the ANS service active. All selected nodes register with the AN node and communicate directly to the AN node as shown in the figure, using a secured encrypted connection.

Each node assigned for running learning tasks has its dataset assigned to it and can run and generate a model. Several models are generated by the different nodes. Using DTLS service, each node will synchronize back model results to the aggregator node.

Depending on the distributed learning algorithm, one pass over the distributed network might be sufficient. An example of this is the distributed bagged decision tree algorithm as shown in Fig. 3. In this case, each node will generate a model using its own assigned dataset. The resulting model is communicated back to the *AN* node. Several ensemble techniques are available in this such as averaging for value prediction or majority vote classification.

Other distributed learning algorithms run several iterations of model generation at each node, adjusting the weights and parameters of each node between the iterations

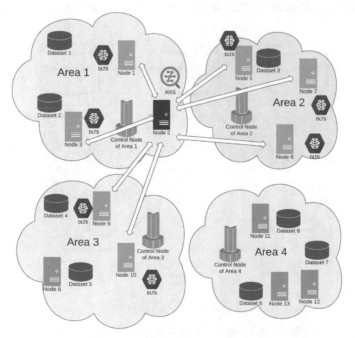

Fig. 2. A distributed learning task.

trying to reach the most optimal solution in a given time [3]. In this case, *AN* node and selected nodes communicate several versions of the models.

The new suggested architecture allows for enabling several distributed learning algorithms on distributed datasets.

4 Performance Evaluation

This section presents an analysis of the architecture's performance in terms of traffic generated and operations performed to allow a distributed algorithm to run.

4.1 Bandwidth Requirements

The architecture includes three components that generate traffic: 1) Network Setup Stage, 2) Data Placement/Task Allocation Stage, and 3) Distributed Learning Algorithm Running Stage. The network setup stage is performed once and includes a constant number of messages generated during the initial setup of peers' relations between areas, accordingly it is negligible in terms of traffic generated.

The second stage of data placement and task allocations includes the main overhead generated by the architecture. Algorithm 2 has five main types of messages:

1. M1: M(*Ri*) which includes transmitting the request information. This message is exchanged in lines 8,9, and 35.
2. M2: M(*ReserveRequest*) which includes requesting Resource reservation (line 17).

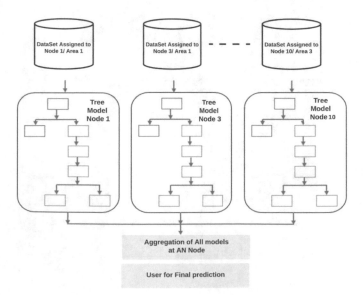

Fig. 3. A distributed bagged decision tree algorithm.

3 M3: M(*TaskDSSize*) which includes the response of task dataset size capability from nodes to control nodes (line 18).

4 M4: M(*DSSize*) which includes the response of dataset size available at each node back to control nodes (line 19).

5 M5: M(*AssignDS*) which includes the dataset assignment information to use (line 34). The dataset assignment does not reflect actual data, but rather which records to use during training as the actual dataset might already reside on the same node.

According to Algorithm 2, the total number of messages transmitted during data placement and task allocation for these 5 types of messages is

$$(2 + a - 1) \times M_1 + \sum_{i=1}^{a} \sum_{j=1}^{m_i} M_1 + M_2 + M_3 + M_4 + M_5 \qquad (1)$$

where a is the number of areas, and mi is the number of nodes in area i. The analysis can be simplified if we assume uniformly distributed areas, which means each area has the same number of nodes. In this case, $m = m1 = m2 = m3 = ... = ma$, the expression can be further simplified to

$$(2 + a - 1) \times M1 + a \times m \times (M1 + M2 + M3 + M4 + M5)$$
$$\approx a \times m \times (M1 + M2 + M3 + M4 + M5)$$

For instance, assume a large deployment scenario with 10 areas of 100 nodes each, with the default TCP message size of 1500 bytes each, this will require a total of

$$10 \times 100 \times 5 \times 1500 = 7.5 \text{ MB}$$

which is very small considering today's networks. One important part of messages that were not considered in this analysis is the traffic generated due to datasets moving from one place to another. Datasets move is considered an extra service offered by the architecture that simplifies the job of the collaborators. Also, the moved datasets can be saved for the next runs which minimize datasets move requests for future runs.

Another set of messages that require consideration is the messages transmitted during the actual distributed learning stage. Considering algorithms similar to 3 where each node is required to transmit the generated model back to the aggregator node for one time, the size of traffic generated will be $a \times m \times ModelSize$. The model size is dependent on the learning algorithm used as some algorithms generate rules such as decision tree and random forest while either generate sets of weights such as deep learning models.

Considering a small deep learning neural network model, $3 \times 6 \times 5 \times 1$ model, in this case, the architecture of 3 input neurons, 2 hidden layers of 6 nodes in the first hidden layer, and 5 neurons in the second hidden layer and 1 output, the number of bytes required assuming double representation is

$$(3 \times 6 + 6 \times 5 + 5 \times 1) \times 8 = 424 \text{ Bytes}$$

Consider the same a large deployment scenario with 10 areas of 100 nodes each, the size of traffic generated will be $a \times m \times ModelSize$ which is

$$10 \times 100 \times 424 = 424 \text{ KB}$$

which is negligible size. For the same large deployment scenario, considering a larger model generated at each node, neural network model, $50 \times 30 \times 40 \times 1$, using the same rule, the size of generated traffic during the learning model will be:

$$10 \times 100 \times 21920 = 2192000 \text{ KB} = 2.19 \text{ MB}$$

This is again a negligible size considering today's networks of high bandwidth.

In the case of distributed learning algorithms that require multiple passes of models exchange between *AN* and other nodes, the size will increase by the number of passes and what is communicated in each pass.

4.2 Processing Requirements

Similar to the bandwidth requirements analysis, the Data Placement/Task Allocation Stage is considered in this section. At this stage, the main computation is performed by the control nodes.

For main operations analysis, lines 18–21, 24, 26, and 27 have the assignment operations, while line 23 has the main comparison operations. For assignments operations, the computational complexity is

$$O(a \times m)$$

while for comparison operations it is

$$O(a)$$

For functions *CalculateTaskDSSize(Nm)*, line 18, and *GetDSSize(Nm)*, line 19, these are executed at the nodes selected for the distributed learning task, not the control nodes, and each node has to do each of these functions once per a distributed learning task.

5 Conclusion

This paper presented a new Secured Decentralized Distributed Learning Architecture (SDDLA). The new suggested architecture enables different types of distributed learning algorithms to run on distributed datasets without compromising the privacy and security of shared datasets with unauthorized users. Also, the new proposed architecture showed a decentralized peer-to-peer management approach among the distributed entities which simplifies the deployment, activation, and utilization of distributed learning. Each entity/peer is in control of the security permissions related to its resources whether computing resources or datasets resources.

References

1. Almajali, S., Abou-Tair, D.E.D.I.: Cloud based intelligent extensible shared context services. In: 2017 Second International Conference on Fog and Mobile Edge Computing (FMEC). IEEE (2017)

2. Almajali, S., Abou-Tair, D., Salameh, H.B., Ayyash, M., Elgala, H.: A distributed multi-layer MEC-cloud architecture for processing large scale IoT-based multimedia applications. Multimed. Tools Appl. **78**(17), 24617–24638 (2018). https://doi.org/10.1007/s11042-018-7049-3

3. Gao, Y., et al.: End-to-end evaluation of federated learning and split learning for internet of things. In: 2020 International Symposium on Reliable Distributed Systems (SRDS). IEEE (2020)

4. Hamdan, S., Almajali, S., Ayyash, M.: Comparison study between conventional machine learning and distributed multi-task learning models. In: 2020 21st International Arab Conference on Information Technology (ACIT). IEEE (2020)

5. Hamdan, S., Almajali, S., Ayyash, M., Salameh, H.B., Jararweh, Y.: An intelligent edge-enabled distributed multi-task learning architecture for large-scale IoT-based cyber–physical systems. Simul. Model. Pract. Theory **122**, 102685 (2023)

6. Hamdan, S., Ayyash, M., Almajali, S.: Edge-computing architectures for internet of things applications: a survey. Sensors **20**(22), 6441 (2020)

7. Kubat, M.: An Introduction to Machine Learning. Springer, Cham (2017)

8. Lu, T., Ai, Q., Lee, W.J., Wang, Z., He, H.: An aggregated decision tree-based learner for renewable integration prediction. In: 2018 IEEE Industry Applications Society Annual Meeting (IAS). IEEE (2018)

9. Zenko, B., Todorovski, L., Dzeroski, S.: A comparison of stacking with meta decision trees to bagging, boosting, and stacking with other methods. In: Proceedings 2001 IEEE International Conference on Data Mining. IEEE (2001)

10. Zhang, Z., Yin, L., Peng, Y., Li, D.: A quick survey on large scale distributed deep learning systems. In: 2018 IEEE 24th International Conference on Parallel and Distributed Systems (ICPADS). IEEE (2018)

Gated Memory Unit: A Novel Recurrent Neural Network Architecture for Sequential Analysis

Arav Kumar[1]($\boxtimes$) and Gabriel Nasrallah[2]

[1] Kingswood Oxford, 170 Kingswood Road, West Hartford, CT, USA
kumar.a.23@kingswoodoxford.org
[2] Oakland Christian School, 3075 Shimmons Road, Auburn Hills, MI, USA

Abstract. The predominant models used to analyze sequential data today are recurrent neural networks, specifically Long-Short-Term Memory (LSTM) and Gated Recurrent Unit (GRU) models, which utilize a temporal value known as the hidden state. These recurrent neural networks process sequential data by storing and modifying a hidden state through the use of mathematical functions known as gates. However, these networks hold many flaws such as limited temporal vision, insufficient memory capacity, and ineffective training times. In response, we propose a simple architecture, the Gated Memory Unit, which utilizes a new element, the hidden stack, a data stack implementation of the hidden state, as well as novel gates. This, along with a parameterized bounded activation function (PBA), allows the Gated Memory Unit (GMU) to outperform existing recurrent models effectively and efficiently. Trials on three datasets were used to display the new architecture's superior performance and reduced training time as well as the utility of the novel hidden stack compared to existing recurrent networks. On data which measures the daily death rate of SARS-Cov-2, the GMU was able to reduce losses to half that of comparable models and did so in nearly half the training time. Additionally, through the use of a generated spiking dataset, the GMU depicted its ability to use its hidden stack to store information past directly observable time steps. We prove that the Gated Memory Unit performs well on a variety of tasks and can outperform existing recurrent architectures.

Keywords: Recurrent neural network · Gated Recurrent Unit · Gated Memory Unit · Long-Short Term Memory · Sequential Analysis

1 Introduction

Recurrent neural networks, long short-term memory [10], and gated recurrent [2] neural networks are the industry standard when it comes to analyzing sequential data, particularly time-series data [5]. Existing recurrent architectures, such as the GRU, have made great strides in creating successful models for this type of data which can accurately forecast future results. These recurrent neural networks (RNN) accomplish this via the storage, mutation, and application of a value known as the hidden state. They generate a sequence of hidden states known as h_t, by utilizing the previous hidden state h_{t-1} with t being the current position in time. Because the previous information is encoded into

© The Author(s), under exclusive license to Springer Nature Singapore Pte Ltd. 2023
Y. B. Wah et al. (Eds.): DaSET 2022, LNDECT 165, pp. 316–328, 2023.
https://doi.org/10.1007/978-981-99-0741-0_23

the hidden state's value, there is an implication of short term memory. The length of this memory determines the amount of previous data used to forecast future states [7]. Regardless, the hidden state has a limit to the information it can store. During training, these networks are fed past data to imbue historical logic into the current hidden state. By comparison, Alex Graves et al. introduced Neural Turing Machines (NTM) [8], a network architecture which utilizes an external memory through attention mechanisms. NTMs use a memory bank and read and write heads to store and select portions of memory. However, the model must select which section of its stored memory to use, which can become increasingly complex and computationally difficult. This is due to the use of numerous attention mechanisms, such as cosine similarity, which ignores a much simpler solution. In this work we propose the Gated Memory Unit, or GMU, a recurrent model architecture that combines an increased memory capacity with the gates of a GRU without the computational stress of attention mechanism processing. The GMU allows for significantly more data to be stored in the hidden state, less computational pressure and therefore training time, while still reaching higher accuracies and lower losses as compared to existing RNN architectures.

2 Background

2.1 RNN

A recurrent neural network is a type of artificial neural network [6] often used to analyze sequential data such as time series. Recurrent neural networks use patterns to predict future values in a sequence by using what is known as a hidden state. As can be seen in Fig. 1, the RNN passes its previous hidden state and its current input through a TanH function to return the new output and hidden state.

2.2 LSTM

Long Short Term Memory networks (LSTM) are a type of RNN which has both a long and short term memory, created by the existence of two different hidden states and a gated system. An LSTM contains a cell state, used to store long term memories, as well as the traditional hidden state to store short term memory [7]. As can be observed in Fig. 1 there are three gates employed in an LSTM, an input gate, an output gate, and a forget gate [7]. These gates modify the data using activation functions to return the new hidden states.

2.3 GRU

The Gated Recurrent Unit (GRU) [3] is a newly popular RNN architecture that utilizes a single hidden state with a gate system designed to improve upon the LSTM's short-comings. Presented in Fig. 1, the GRU contains a reset gate and an update gate along with additional internal calculations which produce a final output. To improve upon the GRU's shortcomings is one of the purposes of the GMU.

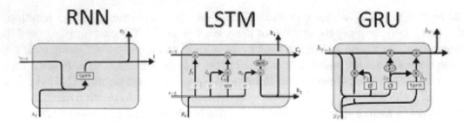

Fig. 1. RNN, LSTM & GRU models

3 Model Architecture

Most standard recurrent network architectures store their previous output in the hidden state and use recursion to make future predictions by modeling the underlying sequence. In addition, there exist other architectures which attempt to represent memory such as the Neural Turing Machine which has external memory resources in place of a hidden state [8], which uses attention mechanisms to use stored data. In the architecture shown in Fig. 2, we propose a new recurrent neural network, the Gated Memory Unit.

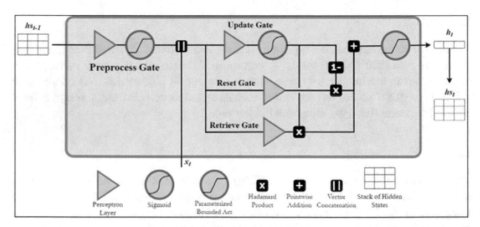

Fig. 2. Gated Memory Unit Architecture

3.1 Hidden Stack

A hidden stack is a stored array of previous hidden states from the Gated Memory Unit. The hidden stack was created as a more robust way to encode sequential data in place of a single hidden state. In the GMU, its logical understanding of the data is strengthened through the volume of information stored in its hidden stack. The hidden stack is a matrix with dimensions determined at initialization. It's size is $[\alpha * \beta]$, where α is the size of each stored hidden state, and β is the number of past hidden states stored. β is similar to that of the step size in a traditional recurrent network where β controls the previous timesteps directly inputted into the model. α is referred to as the hidden stack size while β is the hidden size, or size of the output. Each value in the hidden stack is initialized with zeros, then, with each subsequent pass through the model, the stack is updated, thus removing the oldest hidden state and pushing the newest hidden state to the top of the stack. With each iteration, the entire hidden stack is fed through the model, allowing it to see multiple past timesteps with each prediction.

As the hidden stack travels throughout the network, it is first preprocessed, then combined with the input and processed further to determine which information should be retrieved or updated. This information is then summed and processed through a final layer to create an output hidden state, h_t, which is pushed to the top of the hidden stack. This internal processing is implemented using gates.

3.2 Remodeled Gates

The GMU contains four different gates: the preprocess, reset, update, and retrieve gates. These gates transform the input of the network, x_t, and previous hidden stack, hs_{t-1}, by using perceptron layers, PBA functions, and additional computation to produce a new hidden state to be pushed onto the hidden stack. In comparison, a Gated Recurrent Unit has two gates, a reset and an update gate. Both gates contain weights which are not utilized to simplify data, and are given fixed activations, with similar additional computation to the GMU.

The Gated Memory Unit gate's outputs are referred to as states, corresponding to the title of the gate that produced them. The internal calculations for these gates are displayed in (1)

$$
\begin{aligned}
p_t &= PBA(W_{ir}h_{t-1} + b_{ir})||x_t \\
u_t &= (W_{ir}p_t + b_{ir}) \\
r_t &= W_{ir}p_t \\
n_t &= W_{ir}p_t \\
h_t &= PBA((((1 - u_t) \odot r_t) + (u_t \odot n_t)) + b_{ir})
\end{aligned}
\tag{1}
$$

with p_t, u_t, r_t, and n_t being the preprocess, update, reset, and retrieve states. Additionally, the new hidden state h_t, the output of the network, is calculated using the previously calculated states.

The first gate of the GMU is the preprocess gate which compresses the hidden stack to a size of β through the use of a perceptron layer. This output is concatenated with the input x_t and then fed through a PBA function, resulting in the preprocess state, p_t. The GMU requires a preprocess gate in order to simplify its hidden stack into a more usable compact form. In comparison, a GRU's hidden state does not utilize preprocessing as it is not necessary to maintain lower training times or computational efficiency [4, 13].

The update gate in the GMU is used to determine the ratio of information to keep from the outputs of the retrieve gate and remove from the reset state, thus allowing the model to learn to utilize its memory. The update gate passes p_t, the output of the preprocess gate, and current input through a perceptron layer and a sigmoid function to determine its output, the update state, u_t.

The reset and retrieve gates are both comprised of perceptron layers. The retrieve gate determines what information to take from memory, while the reset gate determines what should be removed from memory. Unlike the GRU, both gates do not contain an activation function and pass the preprocess state, p_t, through their perceptron layers to get their relative states, r_t and n_t [9]. Because the hadamard product is taken between the states, the impact the bias has on the final hidden state is impacted. Thus the biases, b_{ir}, are added after the intermediate states are calculated, yet before the PBA activation function. This process can be observed in the equations above, and is crucial to the GMU's performance, due to the existence of these intermediate states.

The intermediate states are combined using vector addition and passed through the PBA function to produce the final new hidden state h_t, the output of the GMU. Lastly, we push h_t to the top of the hidden stack and pop the oldest value to create the new hidden stack h_{st}.

3.3 Parameterized Bounded Activation

After experimenting with many existing set activation functions, a parameterized bounded activation function similar to HardTanH [12] was created to better suit the needs of the GMU. To allow the model to adjust its upper and lower bounds, the bounds were parameterized [1]. The bounds allow the model to optimize the ratio of positive to negative values passed through the function. This changes the impact of the activation function on the training process and final output. In this way, the model trains its bounds just as it would train any other parameter.

$$PBA = f(x, u, l) = \begin{cases} u, & if\ x \geq u \\ x, & if\ l < x < u \\ l, & if\ x \leq l \end{cases} \tag{2}$$

In Fig. 3, multiple variations of the PBA function are shown. Displayed in 3A the function has an upper and lower bounds set to a value of 1, the default similar to HardTanH. In 3B, the lower bound is set to 0 producing a function similar to HardSigmoid. In 3C, we observe that the lower bound is set to 2 and the upper bound is set to 1.5, creating an elongated function that allows for a completely unique ratio between the negative and positive values' impact. Lastly, in 3D, by setting the lower bound above the upper bound, the network can train the PBA to be utilized as a step function at any height or horizontal position. This allows the values to be truncated between an upper and lower bound so that the model's memory does not end up growing infinitely which resolves the common exploding gradient problem. Additionally, this allows for more freedom when processing data of different proportions.

The PBA is used after the preprocess gate, as well as after concatenating the outputs of the retrieve and update gates. Its purpose in these scenarios is to prevent the outputs from becoming too large or small and to assist in creating non-linear computation, as is the nature of most activation functions. Additionally, a parameterized activation function which can mimic the properties of other successful activation functions helps the novel GMU architecture reach peak performance.

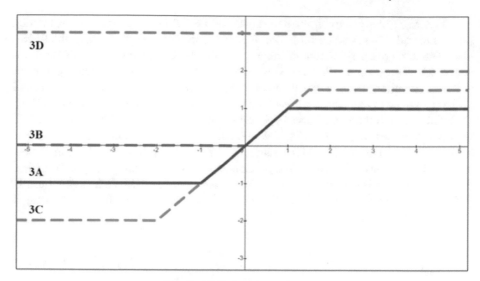

Fig. 3. PBA function variations

4 Results

The GMU's performance was compared to its predecessors, the GRU and LSTM. Each architecture was evaluated on its ability to lower the Root Mean Squared Error Loss, also known as RMSE. To ensure fair results, all models were trained using the Adam optimizer with a learning rate of 0.001, and all training data was split with a ratio of 80% training data and 20% validation data. The following model architecture was used to run all of the following experiments unless stated otherwise. For the GMU, the model contains a perceptron layer, a GMU, and an output perceptron layer with no additional activations. By comparison, the GRU's model contains a GRU followed by an output perceptron layer. The same architecture was used for the LSTM with the corresponding recurrent layer replaced. For consistency, a hidden size of 10 was used across all models and an additional hidden stack size of 5 was used for the GMU. The models were trained on a variety of datasets consisting of different features, sizes, and complexities. Additionally, multiple optimizers were tested to find which are most successful for the novel architecture. Lastly, a generated dataset was used to demonstrate the architecture's unique strengths. The models were trained via Google Cloud Computing services using an NVIDIA Tesla K80 GPU. All models were created and trained with the use of the PyTorch library.

4.1 Covid-19 Death Prediction

In the past three years, the World Health Organization [11] has collected, stored, and published data on the total daily deaths due to Covid-19 every day since the virus was declared a pandemic. This dataset contains 881 data points for 881 recorded days, and the corresponding new deaths due to the pandemic. The models, therefore, have only a singular input and output value.

The GMU greatly outperformed the GRU and LSTM in every aspect of the training process in addition to end performance. Figure 4 contains a graph showing the RMSE loss of each model per iteration of training. As can be seen, the GMU's RMSE drops significantly below the existing models after the 1st iteration, then hits a relative minimum at the 9th iteration much earlier than the other models. By comparison, the GRU and LSTM converge to much higher losses and do so very slowly as compared to the GMU.

Table 1 shows the RMSE loss values and training time for each model. The GMU obtains a minimum loss which is 54.15% better than the GRU and outperforms the LTSM by 55.78%. The speed of the GMU's training was remarkably faster than the comparable models as well. The GMU displayed superior performance by training 78.38% faster than the GRU and 81.65% faster than the LSTM. In addition, the GMU's loss significantly drops within the first or second iteration of every dataset it ran on, thereby allowing it to converge much faster than the existing state-of-the-art RNNs.

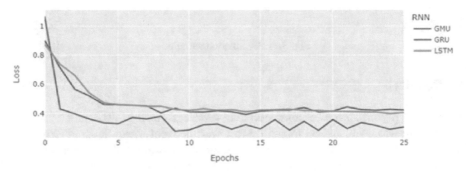

Fig. 4. RMSE Loss comparison across GMU, GRU & LSTM

Table 1. RMSE, epochs/s, and Standard deviation of a Gated Memory Unit, Gated Recurrent Unit, and Long-Short Term Memory network on Covid-19 daily death dataset

RMSE	GMU	GRU	LSTM
Starting Loss	1.066	0.900	0.874
Min Loss	**0.256**	0.394	0.398
Final Loss	0.280	0.403	0.478
epochs/s	**1.980**	1.110	1.090
Std. Deviation	0.146	0.108	0.112

A notable difference in the measured losses' volatility is observed irrespective of the Adam optimizer's equivalent learning rate. This is seen throughout the remaining experiments where the model continues to display this pattern of behavior. Across this dataset, the standard deviation of the loss across all 3 test models is calculated, ignoring the noise from the first two epochs. In this dataset, the GMU's loss had a standard deviation that was 35.28% and 31.15% greater than the GRU and LSTM, respectively.

4.2 Performance on Different Optimizers

Due to the fact that the Gated Memory Unit is a novel model architecture, we felt it was necessary to explore which optimizers best minimized training time and loss. We tested some of the most commonly used optimizers including Adam, Adagrad, Adamax, NAdam, RAdam, Stochastic Gradient Descent, and RMSprop. After testing it was revealed that SGD and RMSprop caused the loss to diverge, increase, or simply not improve from the initial loss, and thus are deemed unfit to optimize a GMU. The same model was trained on the previously used Covid-19 death data seen in Sect. 4.1. To test each optimizer's abilities to their maximum extent, an early stopping callback was employed. This allows us to also see how quickly each optimizer was able to reduce the loss.

Table 2. RMSE and Training time performance of varying Optimization algorithms on a Gated Memory Unit when trained on Covid-19 Data

Optimizers	Adam	Adagrad	Adamax	NAdam	RAdam
Starting loss	1.066	1.066	1.066	1.066	1.066
Min loss	0.279	0.256	0.262	0.304	0.261
Final loss	0.358	0.266	0.285	0.390	0.289
epochs/s	0.475	0.854	0.694	0.629	0.917
Training time (s)	40.0	41.0	36.0	35.0	24.0
Iterations	19.0	35.0	25.0	22.0	22.0

In Table 2, we find that two optimizers truly stood out: RAdam and Adagrad. Not only did these optimizers train almost twice as fast as Adam and roughly 50% faster than Adamax and NAdam, but they also had the two lowest measures of RMSE consistently. Both optimizers are highly effective, however, RAdam reached a comparable value 70.83% faster than Adagrad. In Fig. 5, the loss is plotted along each iteration of the model, which is scaled from 5 to 25 iterations. The chart was scaled to display more information directly related to each optimizer's individual performance. On a broader scale, the optimizers performed similarly with NAdam and Adam showing large levels of volatility and coming out with final losses much higher than the minimums they reached. As can be seen in the graph, Adagrad and RAdam consistently hold the lowest loss from epochs 5 to 25. RAdam tends to see more volatility in its optimization process from epoch to epoch while Adagrad's reducing learning rate allows it to follow a more stable path. Because of this, RAdam tends to reach lower losses due to its more sporadic nature while Adagrad takes much longer to produce an equivalent result.

4.3 China Data and Training Time

The next dataset used was collected from The World Bank [13], containing the annual growth rate of China and the world's total population as a percentage. Unlike the World

Health Organization's pandemic data, this dataset results in two variables for the model to be fed and predicted. Additionally, it is much smaller in size. To test training time and to deal with smaller amounts of data, this dataset containing 122 data points across 61 timesteps was used to evaluate the GMU and its predecessors. In Fig. 7 below, we see that even after 50 iterations the GMU clearly outperforms all other RNNs.

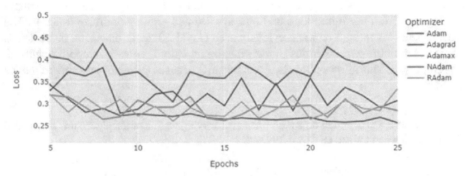

Fig. 5. Optimizer scaled performance

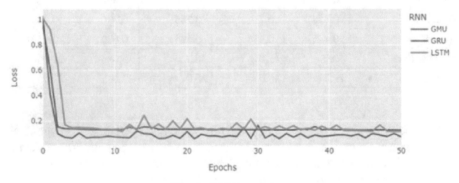

Fig. 6. RNN loss values

The results displaying training time however speak volumes on the GMU's theoretical efficiency. As seen in Table 3, the GMU reached less than half the loss of the other models in half the time. This greater computational efficiency allows for much larger datasets to be run in a comparable amount of time. Additionally, it is clear that the GMU is sufficiently better at fitting datasets of varying sizes than a GRU or LSTM (Fig. 6).

4.4 Periodic Pulse Dataset

Our last and most important dataset is one we generated. For each timestep t, the following equation generates the corresponding y values: $y = t \pmod{10}$ creating a pulse every 10 timesteps. This type of data is difficult for RNNs if the timestep of the last pulse is not fed into the model. To demonstrate the GMU's ability to store information well past the timesteps directly fed into it, all recurrent models tested were given a hidden size of

5. This means they were only capable of seeing the past 5 timesteps while the pulses in this generated dataset were 10 spaces apart. As can be seen in the graphs below, Fig. 8 displays the GMU's predictions, while Fig. 9 displays the GRU's predictions. The GMU outperforms the GRU with an RMSE of 0.0616, compared to the GRU's astonishing RMSE of 0.9989.

Table 3. RMSE, epochs/s, and Standard deviation of a Gated Memory Unit, Gated Recurrent Unit, and Long-Short Term Memory on the World Bank's Chinese Population dataset

RMSE	GMU	GRU	LSTM
Starting loss	1.031	0.961	1.012
Min loss	**0.062**	0.128	0.117
Final loss	0.074	0.127	0.119
epochs/s	**31.870**	16.320	16.050

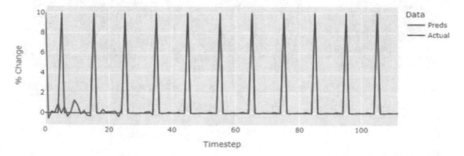

Fig. 7. GMU Predictions on generated periodic pulse dataset

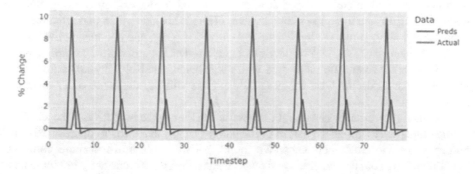

Fig. 8. GRU Predictions on generated periodic pulse dataset

4.5 Memory Heatmap

To understand the impact of the hidden stack, a heatmap is used to display the values stored in the GMU's memory stack after training on the previously mentioned pulse dataset. As shown in Fig. 9.1, each row of values is one of the GMU's hidden states, which represents the array of values it stores in memory. The bottom row represents the most recent hidden state, and the top row represents the oldest hidden state in the stack. In Fig. 9.1, the leftmost heatmap has a hidden stack size of 5, while the periodic pulse data contains a pulse every 10 iterations, therefore, the model can not directly record a visible pattern that displays where the pulse is. Yet, the GMU is still able to predict the spike with extreme precision, reflected by an RMSE of 0.0616. In Fig. 9.2 and Fig. 9.3, we can see the GMU when it's been given a hidden stack size of 10 and 20, respectively. This allows the model to directly see the previous pulse(s) and this is clearly displayed with a pulse in the heatmap every 10 iterations. These models have RMSE values of 0.0523 and 0.0388, respectively.

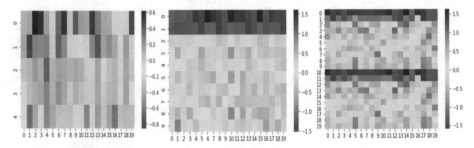

Fig. 9. Heatmap of hidden stacks on pulse data with hidden stack sizes of 5, 10 & 20

5 Discussion

Although there are many factors that allow the GMU to outperform existing RNNs, we find that there are three key elements that allowed this increase in performance and decrease in training time: the hidden stack, remodeled gates, and bounded activations.

The hidden stack is a stack of previous hidden states. It outperforms existing RNNs due to its ability to encode more data while still being relatively efficient as the network only has to decide what the next hidden state will be, similar to other memory architectures.

Another cause for its newfound success is the new system of gates. The GMU's architecture tries to utilize the nature of the gated systems in a GRU to process existing hidden states and create new ones. We found that we could utilize a more compact structure through our update, reset, and retrieve gates than what is currently implemented in existing models. Due to the varying sizes of information being fed into and used by the model, we felt it was necessary to use perceptron layers to reshape information into usable sizes. Inside of these gates, we also felt it was necessary to replace the activation functions.

The parameterized linear bounded activation function we utilize is more versatile than the standard TanH function because the model itself can adjust its horizontal and vertical bounds to best fit differently scaled data. This means the model does not have as many restrictions as we have found TanH to have during our testing. We also decided that the PBA would be linear instead of hyperbolic as this provided better results during our testing.

A lower training time is seen when training on unbatched data because the model does not require to be fed in past timesteps for each run and instead retains its state. The model has repeatedly been shown to achieve lower losses, especially in situations with limited data, and has also been shown to converge faster than previous models. During our testing, we found that the RAdam optimizer performed best in combination with the GMU and assisted in lowering training volatility while improving losses.

The GMU has displayed that it can effectively utilize its memory storage system to see data far past the length of its hidden stack. Although its stack size may only be 5, the model was able to recognize pulses in data and patterns outside of the scope of its hidden stack, showing that it may be storing info much further out than its stack size.

6 Conclusion

In this work, we presented a novel recurrent neural network architecture which demonstrates the ability to control and utilize an expandable hidden state, and therefore have memory capability. In addition, the architecture contains remodeled gated systems and a custom, parameterized activation function. The architecture replaces the limited memory of the hidden state by creating a hidden stack in which hidden states are stored and utilized throughout the network.

The gated memory unit shows improved training times, faster and earlier convergence, lower losses, and exemplary memory utility. The model was able to perform better in three tasks, featuring small, medium-sized data sets, and memory tasks. We plan to improve this model through more advanced training techniques, such as Reinforcement and Metropolis Monte Carlo optimization. We hope this model will be able to create more robust and effective solutions to current recurrent neural network tasks such as speech recognition, data forecasting, and natural language processing.

References

1. Bingham, G., Miikkulainen, R.: Discovering parametric activation functions. Neural Netw. **148**, 48–65 (2022)
2. Cho, K., van Merrienboer, B., Bahdanau, D., Bengio, Y.: On the properties of neural machine translation: encoder-decoder approaches (2014). https://doi.org/10.48550/ARXIV.1409.1259
3. Cho, K., et al.: Learning phrase representations using RNN encoder–decoder for statistical machine translation. In: Proceedings of the 2014 Conference on Empirical Methods in Natural Language Processing (EMNLP), pp. 1724–1734. Association for Computational Linguistics (2014). https://doi.org/10.3115/v1/D14-1179
4. Collins, J., Sohl-Dickstein, J., Sussillo, D.: Capacity and trainability in recurrent neural networks (2016). https://doi.org/10.48550/ARXIV.1611.09913
5. Elman, J.L.: Finding structure in time. Cogn. Sci. **14**, 179–211 (1990)
6. Bianchini, M., Scarselli, F.: On the complexity of neural network classifiers: a comparison between shallow and deep architectures. IEEE Trans. Neural Netw. Learn. Syst. **25**, 1553–1565 (2014)

7. Gers, F.A., Schmidhuber, J., Cummins, F.: Learning to forget: continual prediction with LSTM. Neural Comput. **12**, 2451–2471 (2000)
8. Graves, A., Wayne, G., Danihelka, I.: Neural turing machines (2014). https://doi.org/10.48550/ARXIV.1410.5401
9. Heck, J., Salem, F.M.: Simplified minimal gated unit variations for recurrent neural networks (2017). https://doi.org/10.48550/ARXIV.1701.03452
10. Hochreiter, S., Schmidhuber, J.: Long short-term memory. Neural Comput. **9**, 1735 1780 (1997)
11. Organization, W.H.: Covid-19 Global Death Toll. WHO Coronavirus (COVID-19) Dashboard (2019). https://covid19.who.int/data
12. Nwankpa, C., Ijomah, W., Gachagan, A., Marshall, S.: Activation functions: comparison of trends in practice and research for deep learning (2018). https://doi.org/10.48550/arXiv.1811.03378
13. Population growth (annual %) | World Bank Data. https://data.worldbank.org/indicator/SP.POP.GROW

Multi-class Classification for Breast Cancer with High Dimensional Microarray Data Using Machine Learning Classifier

Mohammad Nasir Abdullah[1], Bee Wah Yap[2(✉)], Nik Nur Fatin Fatihah Sapri[3], and Wan Fairos Wan Yaacob[4,5]

[1] Mathematical Sciences Studies, College of Computing, Informatics and Media, Universiti Teknologi MARA, Tapah Campus, Perak, Malaysia
nasir916@uitm.edu.my

[2] UNITAR International University, Petaling Jaya, Selangor, Malaysia
bee.wah@unitar.my

[3] Mathematical Sciences Studies, College of Computing, Informatics and Media, Universiti Teknologi MARA, Shah Alam, Selangor, Malaysia
nikfatinfatihah@uitm.edu.my

[4] Mathematical Sciences Studies, College of Computing, Informatics and Media, Universiti Teknologi MARA Cawangan Kelantan, Kampus Kota Bharu, Kota Bharu, Kelantan, Malaysia
wnfairos@uitm.edu.my

[5] Institute for Big Data Analytics and Artificial Intelligence (IBDAAI), Universiti Teknologi MARA, Shah Alam, Selangor, Malaysia

Abstract. Breast cancer is one of the leading causes of cancer related deaths among women. Early detection of breast cancer is very important for proper treatment and decreasing the death risk among women. Most cancer prediction study focused on binary classification of breast cancer. This study focused on multi-class classification of breast cancer with high dimensional microarray data. The dataset involved 38 cancer patients, 3 categories: normal (9), early tumour (12), and late tumor (17), and 39,426 microarray biomarkers. Boruta's feature selection algorithm selected 28 important microarray biomarkers. The performance of support vector machine, multinomial logistic regression, Naïve Bayes, and random forest were evaluated based on macro and micro accuracy, sensitivity, and precision. Results showed that multinomial logistic regression, Naïve Bayes and random forest exhibits overfitting issue. However, support vector machine performed well in multi-classification of breast cancer (macro_acc$_{test}$ = 86.7%, macro_sen$_{test}$ = 77.8%, and macro_prec$_{test}$ = 62.0%). In future work, bagging, and boosting with over sampling techniques can be considered to improve multi-class classification of breast cancer using high dimensional microarray data.

Keywords: Breast Cancer · Boruta's algorithm · Microarray · Multinomial logistics regression · Naïve Bayes · Support vector machine · Random forest

© The Author(s), under exclusive license to Springer Nature Singapore Pte Ltd. 2023
Y. B. Wah et al. (Eds.): DaSET 2022, LNDECT 165, pp. 329–342, 2023.
https://doi.org/10.1007/978-981-99-0741-0_24

1 Introduction

Breast cancer is one of the deadliest and most worrisome cancer among women, with an estimated 2.3 million incidence cases globally every year [1]. Cancer starts when cells begin to grow out of control. It has potential to be dangerous as this cancer is able to spread to other parts of the human body which later weakens the body immune system. There are many factors which contributed to breast cancer, and it is very challenging to determine the main factors that cause breast cancer. However, early detection of breast cancer able to reduce the death rate and able to increase the survival chance by 8%. [2]. Although, breast cancer is dominant among women, nevertheless, men can have breast cancer too (isolated case). It is important to have knowledge and awareness on breast cancer as this cancer is usually not detected in the early stage. Recent studies on breast cancer classification using machine learning techniques to classify breast cancer as benign or malignant [3].

Multiclass classification can be defined as a classification task with more than two classes, e.g., classify a set of images of fruits which may be oranges, apples, or pears. Multiclass classification assumes that each sample is assigned to one and only one label: a fruit can be either an apple or a pear but not both at the same time. In statistical modelling, a numerous study from researchers had applied a variety of machine learning (ML) techniques in detecting breast cancer either for binary or multiclass classification [4, 5]. Support vector machine (SVM) has been used to classify the breast cancer with cancer and without cancer using Wisconsin breast cancer dataset [5–7]. Furthermore, Goyal and Trivedi [8] used Convolutional Neural Network (CNN) to classify the breast cancer as benign tumour and malignant tumour using Breast Cancer Histopathological Data set (BreakHis). Meanwhile, Karabatak [9] proposed weighted Naïve Bayes (NB) classifier and regular NB classifier in classifying the breast cancer as benign and malignant. Moving into multiclass classification of breast cancer, [10] applied different deep learning classifiers to classify eight classes of breast cancer and CNN classifier was used to classify the breast cancer into four classes of benign and four classes of malignant tumour [11].

The aim of this study is to evaluate ML classifiers on multi-class classification of breast cancer using high dimensional microarray data. This paper is organized as follows. Section 2 covers review of previous studies on multiclass classification of breast cancer. The source of breast cancer dataset is given in Sect. 3 of this paper. Machine learning classifier and performance measures for multi-class classification are explained in Sect. 4 and Sect. 5 respectively. The results and findings based on the analysis are in Sect. 6 and Sect. 7 concludes the paper.

2 Related Work

There has been extensive literature on the development of methodology for classification of breast cancer detection. Most research focused on binary classification [5, 6, 12–14] using images and limited work begin to focus for multiclass classification (5-binary classification) [8, 10]. For two classification detection of breast cancer, several classifiers such as SVM, k-NN, and NB were commonly being used by the researchers. Rejani and Selvi [6] used SVM classifier in tumour detection to classify mammogram images into malignant or benign. The SVM model achieved 88.75% sensitivity with image enhancement and threshold segmentation to improve detection. Bihis and Roychowdhury [5] also found that SVM algorithm was the best classifier to classify breast cancer with 99.7% accuracy. Several other study also classified breast cancer tumour using different algorithm and found that SVM algorithm has promising results with higher accuracy compared to k-NN, C4.5, NB, K-Means and PAM [12, 13].

In another study [14], the authors compared the performance of four classifiers: AdaBoost, Random Forest (RF), Multi-Class SVM, and Multi-Layer Perceptron (MLP) in the classification of three different N-Grams (2-g, 3-g, and 4-g) from 168,420 proteins array peptides for breast cancer. The result of all classification models for breast cancer reached more than 80% sensitivity (SVM: [sensitivity = 88.6%, F-measure = 87.7%], RF: [sensitivity = 88.3%, F-measure = 87.3%], MLP: [sensitivity = 89.3%, F-measure = 89.0%], and AdaBoost: [sensitivity = 89.3%, F-measure = 88.7%]). Recently, researchers begin to explore more advanced techniques such as Deep Learning. Aljuaid et al. [10] used computer-aided diagnosis to classify breast cancer using deep neural networks and transfer learning. The proposed methods, ResNet (99.8%), InceptionV3Net (97.7%), and Shuf-fleNet (96.9%) has high accuracy for binary classification of benign or malignant cancer cases.

Goyal and Trivedi [8] reported CNN has accuracy rate of 98.3% in classification of benign and malignant tumour. Akay [7] used Wisconsin breast cancer dataset (WBCD) and reported that SVM-based method combined with features selection resulted in high accuracy of 99.51%. Karabatak [9] proposed weighted Naïve Bayes classifier and regular Naïve Bayes classifiers in classifying the breast cancer as benign or malignant. Their results showed that weighted Naïve Bayes classifier performed better that regular Naïve Bayes classifier with high sensitivity (99.11%), specificity (98.25%), and accuracy (98.54%). Meanwhile, Nguyen et al. [11] used the CNN of deep learning to classify 7,909 breast cancer images from public BreakHis dataset. Four benign subclasses and four malignant subclasses were classified using CNN classifier and the accuracy of the model is 73.68%.

3 Methodology

3.1 Breast Cancer Data

In this study, we used the Breast Cancer data sets from extensive Curated Microarray Database (CuMiDa) available at https://sbcb.inf.ufrgs.br/cumida which is a repository of 13 different types of cancer to perform multiclass classification and evaluate the performance of several machine learning classifiers [15]. CuMiDa datasets have been used in with more than 30,000 studies of the Gene Expression Omnibus (GEO) database. In this study, we used the breast cancer dataset GSE-89116 where 38 breast cancer patients were categorized as normal, early and late tumour patients. The dataset has 39,426 microarray biomarkers.

4 Multiclass Classification Models

Multi-class classification models can be applied to data set with a target variable that has more than two classes. In this study, four machine learning (ML) classifiers performance were evaluated in classification of normal, early, and late tumour patients based on microarray biomarkers. The following section provide some information on the four ML classifiers.

4.1 Multinomial Logistic Regression

In the context of multiclass classification, multinomial logistic regression (MLR) is a statistical model which can be used to classify more than two possible outcomes of the dependent categorical variable (known as multiclass). It is an extension of the binary logistic regression is used for prediction of a binary dependent categorical variable.

Based on Hosmer and Lameshow [16], the MLR logit model for k = 3 class would be defined as follows with respect to the based class $(Y = 0)$:

$$f_1(x) = ln\left[\frac{P(Y = 1|x)}{P(Y = 0|x)}\right]$$
$$= \beta_{10} + \beta_{11}X_1 + \ldots + \beta_{1p}X_p \tag{1}$$

$$f_2(x) = ln\left[\frac{P(Y = 2|x)}{P(Y = 0|x)}\right]$$
$$= \beta_{20} + \beta_{21}X_1 + \ldots + \beta_{2p}X_p \tag{2}$$

The equations should follow the conditional probabilities of each outcome category given by covariate vector such as:

$$P(Y = 0|x) = \frac{1}{1 + e^{f_1(x)} + e^{f_2(x)}} \tag{3}$$

$$P(Y = 1|x) = \frac{e^{f_1(x)}}{1 + e^{f_1(x)} + e^{f_2(x)}} \tag{4}$$

$$P(Y = 2|x) = \frac{e^{f_2(x)}}{1 + e^{f_1(x)} + e^{f_2(x)}} \tag{5}$$

where $P(Y = 0|x) + P(Y = 1|x) + P(Y = 2|x) = 1$. A patient will be classified into class $Y = j$ with the highest probability.

4.2 Naives Bayes

NB is a statistical classifier which is based on Bayes formula. The idea of this method is it predicts class membership probabilities in which the probability of a given sample belongs to a particular class. The class that having highest probability is taken as the best predicted class called as maximum a posteriori [17]. This can be defined as:

$$\max P(Y|X) = \max \frac{P(X|Y)P(Y)}{P(X)}$$

$$= \max(P(X|Y)P(Y)) \tag{6}$$

where $P(Y)$ is the prior probability. $P(X)$ is the probability of X, $P(X|Y)$ is the probability of X given information on Y and $P(Y|X)$ is the posterior probability.

4.3 Random Forest

Random forest (RF) is another useful ML classifier for binary or multiclass classification of breast cancer. RF classifies the dependent categorical variable by generating numerous decision trees [18]. The prediction error can be reduced as many decision trees are generated. The weighted is used where a tree with high error is given low weight, and tree with low error is given high weight.

A tree in RF is developed based on classification and regression trees (CART) methodology. Assuming a training set T with p variables, the sample of size n and T_k is a bootstrap training set sampled from T with replacement containing m random variables ($m \le p$) for n sample size [19]. From the training set T, several number of m random variables were selected at random out of the p variables, and the best variable from this set of m is used to split the node. These values of m are held constant during the forest growing. Variable selection measures for m is using Gini index. The Gini index is defined as:

$$gini(\mathrm{T}) = 1 - \sum_{j=1}^{n} p_j^2 \tag{7}$$

where the $p_j = \frac{|C_j T|}{|T|}$ is the probability of class j in dataset T. The Gini index is the minimum value when all cases fall into a single target category and the maximum value when the cases are equally distributed to all classes [19, 20].

4.4 Support Vector Machine (SVM)

SVM is another useful ML classifier for binary and multiclass classification. SVM attempts to find a linear separator (hyperplane) between the data points of two classes in multidimensional space. Such a hyperplane is called the optimal hyperplane. A set of instances that is closest to the optimal hyperplane is called a support vector. Finding the optimal hyperplane provides a linear classifier. SVM is suitable for classification of linear and non-linear separation for dealing with interactions among features and redundant features [17]. The common SVM kernels include linear, polynomial, radial basis function, and sigmoid [21]. We used radial basis function kernel (RBF) for this multiclass classification problem. The RBF kernel is defined as follows:

$$K(x_i, x_j) = e^{\frac{-||x - x_i||^2}{2\sigma^2}} \tag{8}$$

where $||x - x_i||$ is also known as Euclidean distance [21].

5 Methodology for Multiclass Classification of High Dimensional Imbalance Data

Development and evaluation of ML classifiers were carried out using R software of version 4.1.0 [22], an open-source software for statistical computing and graphics. Among the cancer datasets available in CuMiDa repository, this study selected GSE 89116 breast cancer data as patients were categorizes as normal, early tumour, and late tumour patients. This dataset is a good example high dimensional microarray data with multiclass target variable. After data cleaning, feature selection was applied. In this study, wrapper feature selection using Boruta's algorithm was employed.

Boruta's algorithm is a useful feature selection algorithm where it creates a ranking evaluation based on the similarity between the biomarkers [23]. The Boruta's algorithm is a multivariate feature selection based on RF approach to search for the most vital covariate that can predict response variable [24].

The Boruta's algorithm identified relevant features by comparing the importance of original attributes to the importance that is acquired at random. The process starts by creating the duplicate copies of all independent variables and rearrangement of added features to eliminate their correlations with the output variable (known as shadow features). Then, it trains the RF classifier using the extended dataset and calculate the Z score. Next, the variables with higher Z score compared to the maximum Z score among shadow attributes (MZSA) for breast cancer were identified [25].

After selecting the important features, the breast cancer data were partitioned into training (70%) and testing (30%) samples. Then, several selected machine learning techniques were performed such as SVM with radial basis function kernel (RBF), MLR, NB, and RF classifiers using caret package in R software of version 4.1.0 [22]. Later, the macro-averaged and micro-averaged of accuracy, precision and recall were computed for each of the multiclass classification techniques performed in the study to evaluate the model performance. Figure 1 summarizes the flowchart of the methodology for multiclass classification of high dimensional microarray data modelling process using breast cancer data which involve the phases as below.

Phase 1: Data Exploration

Step 1: Data acquisition from CuMiDa repository.
Step 2: Data cleaning (remove missing values and outliers).
Step 3: Data exploration (descriptive statistics, boxplots).

Phase 2: Feature evaluation

Step 1: Feature reduction using Boruta Algorithm.

Phase 3: Model Development and Evaluation

Step 1: Data partition (70% training, 30% testing).
Step 2: Apply ML classifiers (MLR, NB, RF, SVM).

Step 3: Evaluate ML classifier performance (Macro and Micro accuracy, sensitivity, and precision).

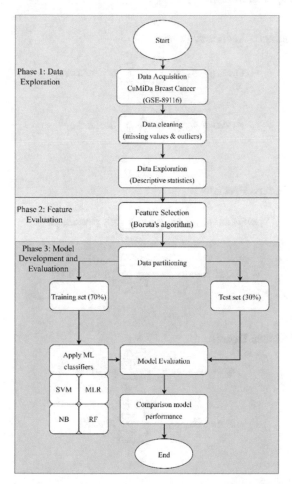

Fig. 1. Flowchart of methodology for multiclass classification of breast cancer high dimensional microarray data

6 Model Performance for Multiclass Classifier

The model performance of each multi-class classifier was compared based on micro-averaged and macro-averaged precision, recall, and accuracy. By definition, micro-averaged refers to the weighted average based on the frequency of samples from each class, while, macro-averaged refers to unweighted mean of precision, recall and accuracy [17]. Assume that N is the total number of samples and C is the total number of classes of breast cancer, then the number of samples in each class is written as $n^{(0)}, n^{(1)}, ..., n^{(C)}, \sum_{i=0}^{C} n^{(C)} = N^{(2)}$, where $N^{(2)} = 0.7N$. The calculation of precision, recall and accuracy is as follows [5].

6.1 Micro and Macro Accuracy

$$Accuracy_{micro} = \frac{1}{N^{(2)}} \sum_{i=0}^{C} n^{(i)} accuracy^{(i)} \tag{9}$$

$$Accuracy_{macro} = \frac{1}{C} \sum_{i=0}^{C} accuracy^{(i)} \tag{10}$$

6.2 Micro and Macro Precision

$$precision_{micro} = \frac{1}{N^{(2)}} \sum_{i=0}^{C} n^{(i)} precision^{(i)} \tag{11}$$

$$precision_{macro} = \frac{1}{C} \sum_{i=0}^{C} precision^{(i)} \tag{12}$$

6.3 Micro and Macro Recall

$$recall_{micro} = \frac{1}{N^{(2)}} \sum_{i=0}^{C} n^{(i)} recall^{(i)} \tag{13}$$

$$recall_{macro} = \frac{1}{C} \sum_{i=0}^{C} recall^{(i)} \tag{14}$$

7 Results and Analysis

Prior to predictive modelling, a preliminary analysis that involve data cleaning and data screening were conducted. There were no missing data or outliers. The target variable is having three classes: normal patient (class 1), early tumoral patient (class 2), and late tumoral patient (class 3). Out of 38 patients, there were 9 normal, 12 early tumoral, and 17 late tumoral patients. The dataset has 39,426 microarray biomarkers. A summary of dataset is tabulated in Table 1.

Table 1. Summary of dataset

Variables		39,426
Number of observations		38
Class	Normal patients	9
	Early tumoral patients	12
	Late tumoral patients	17

Table 2 shows the result of feature selection using Boruta algorithm. Out of 39,426 variables, Boruta's algorithm selected 28 important biomarkers. The list of the 28 important biomarkers is summarized in Table 2. Thus, the remaining 39,398 were tentatively classified as unimportant attributes from the Boruta's analysis.

Table 2. List of important biomarkers

ILMN_1712530	ILMN_1725640	ILMN_1660283	ILMN_1828034
ILMN_1821321	ILMN_1879832	ILMN_1825955	ILMN_1835958
ILMN_1845875	ILMN_1835896	ILMN_1849124	ILMN_1882185
ILMN_1912805	ILMN_1826914	ILMN_1678544	ILMN_1708709
ILMN_1770479	ILMN_2297662	ILMN_1652876	ILMN_1759708
ILMN_1654336	ILMN_1708192	ILMN_1787594	ILMN_1733346
ILMN_1761717	ILMN_1793831	ILMN_1765714	ILMN_2367530

In this study, we construct NB, SVM, RF and MLR classifiers using the 28 biomarkers for breast cancer classification. The models were evaluated based on macro and micro classification for accuracy, precision, and recall (sensitivity). The best classifier was selected based on macro-averaged accuracy, sensitivity, and precision of multiclass classification for the testing set.

Based on the results presented in Table 3 and Fig. 2 to Fig. 5, MLR, NB, and RF exhibits overfitting as the model performs well for the training set but poorly for the testing sample. However, the performance for SVM with RBF kernel was consistent for both training and testing samples. Based on testing sample results, the SVM with RBF kernel model (Macro Accuracy = 87%, Sensitivity = 78%, Precision = 62%) performed better than NB (Accuracy = 81%, Sensitivity = 66%, Precision = 56%), RF (Accuracy = 73%, Sensitivity = 64%, Precision = 53%) and MLR (Accuracy = 73%, Sensitivity = 59%, Precision = 49%). Results in Fig. 2, Fig. 3, and Fig. 5 indicate all models overfit except for SVM (Fig. 4) where the discrepancy between training and testing values is small for SVM. Although sensitivity value is bit low due to the possibility of slight imbalance class in the dataset the performance results is satisfactory. Out of

36 cases, only 9 (28.6%) for Normal patients, 12 (31.5%) for Early tumoral patients and 17 (44.7%) for late tumoral patients. The class proportion does not highly indicate imbalance class problem that require sampling techniques to balance the data.

Table 3. Comparison performance between machine learning algorithms

Classifiers	Sample	Accuracy (%)		Sensitivity (%)		Precision (%)	
		Micro	Macro	Micro	Macro	Micro	Macro
NB	Training	100	100	100	100	100	100
	Testing	80	81	70	65.56	66.25	55.83
RF	Training	100	100	100	100	100	100
	Testing	68	73.33	60	64.44	50	53.33
SVM	Training	95.28	95.24	92.86	92.13	96.43	96.3
	Testing	87	86.67	80	77.78	66.25	62
MLR	Training	100	100	100	100	100	100
	Testing	70	73.33	60	58.89	47.5	49.17

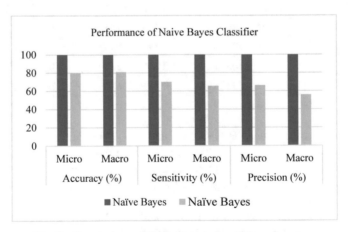

Fig. 2. Comparison of Naïve Bayes classifier performance

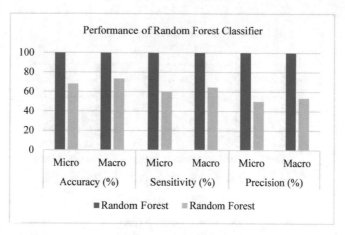

Fig. 3. Comparison of random forest classifier performance

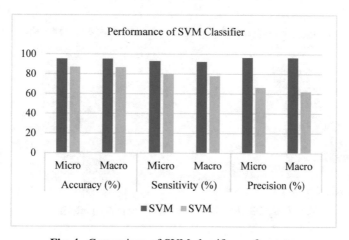

Fig. 4. Comparison of SVM classifier performance

Thus, based on Fig. 2 to Fig. 5, SVM classifier has the highest accuracy, sensitivity, and precision, compared to other classifiers.

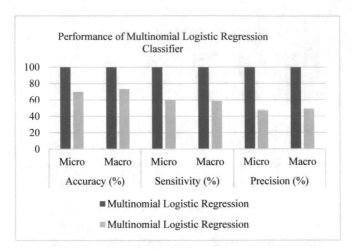

Fig. 5. Comparison of multinomial logistic classifier performance

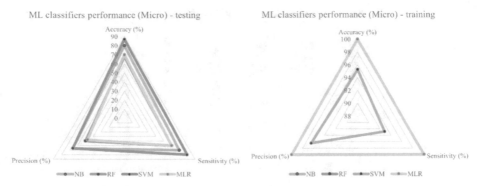

Fig. 6. Spider chart based on classifier micro performance.

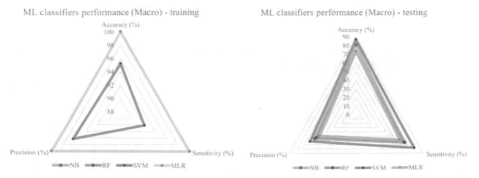

Fig. 7. Spider chart based on ML macro performance.

Based on Fig. 6 and Fig. 7, the spider chart shows the on micro and macro performance measured for the ML classifiers It clearly shows that the SVM classifier performs better compared to the other classifiers in terms of accuracy and sensitivity measures. In addition, the NB, RF, and MLR show overfitting where the accuracy, sensitivity, and precision were all 100% in the training set for both micro and macro values.

8 Conclusions

Numerous machine learning algorithms can be used for classification of medical data sets. There are several ML algorithms available for classification which depends on whether binary or multiclass data and high or low dimensionality of the dataset. With high dimensional data, the performance of ML algorithm depends on the sample size, number and type of features, complexity, and accuracy of data. This study found SVM with RBF kernel performed well in multiclass classification of breast cancer. Future work can consider ensemble model of bagging and boosting algorithms with over-sampling techniques for improving multiclass classification breast cancer.

Acknowledgements. We would like to express our appreciation and record our gratitude to UNITAR International University and Universiti Teknologi MARA for the support in conducting this research. The authors are also grateful to the reviewers for their insightful comments and suggestions.

References

1. Whittaker, A.L., George, R.P., O'Malley, L.: Prevalence of cognitive impairment following chemotherapy treatment for breast cancer: a systematic review and meta-analysis. Sci. Rep. **12**(1), 1–22 (2022). https://doi.org/10.1038/s41598-022-05682-1
2. Khan, M.H.M., et al.: Multi- class classification of breast cancer abnormalities using Deep Convolutional Neural Network (CNN). PLoS One **16**(8), 1–15 (2021). https://doi.org/10.1371/journal.pone.0256500
3. Sharma, S., Deshpande, S.: Breast cancer classification using machine learning algorithms. In: Joshi, A., Khosravy, M., Gupta, N. (eds.) Machine Learning for Predictive Analysis. LNNS, vol. 141, pp. 571–578. Springer, Singapore (2021). https://doi.org/10.1007/978-981-15-7106-0_56
4. Vohra, P.K., Bhavani, B., Gonthina, N.: Multi-class classification of breast cancer using machine learning. Int. J. Res. Signal Process. Comput. Commun. Syst. Des. **4**(2), 33–35 (2018)
5. Bihis, M., Roychowdhury, S.: A generalized flow for multi-class and binary classification tasks: an azure ML approach. In: Proceedings of the - 2015 IEEE International Conference Big Data, IEEE Big Data 2015, pp. 1728–1737 (2015). https://doi.org/10.1109/BigData.2015.7363944
6. Rejani, Y.I.A., Selvi, S.T.: Early detection of breast cancer using SVM classifier technique. Int. J. Comput. Sci. Eng. **1**(3), 127–130 (2009)
7. Akay, M.F.: Support vector machines combined with feature selection for breast cancer diagnosis. Expert Syst. Appl. **36**(2), 3240–3247 (2009). https://doi.org/10.1016/j.eswa.2008.01.009
8. Goyal, N., Chandra Trivedi, M.: Breast cancer classification and identification using machine learning approaches. Mater. Today Proc., 1–4 (2020). https://doi.org/10.1016/j.matpr.2020.10.666

9. Karabatak, M.: A new classifier for breast cancer detection based on Naïve Bayesian. Measurement **72**, 32–36 (2015). https://doi.org/10.1016/j.measurement.2015.04.028

10. Aljuaid, H., Alturki, N., Alsubaie, N., Cavallaro, L., Liotta, A.: Computer-aided diagnosis for breast cancer classification using deep neural networks and transfer learning. Comput. Methods Programs Biomed. **223**, 106951 (2022). https://doi.org/10.1016/j.cmpb.2022.106951

11. Nguyen, P.T., Nguyen, T.T., Nguyen, N.C., Le, T.T.: Multiclass breast cancer classification using convolutional neural network. In: Proceedings of the - 2019 International Symposium on Electrical and Electronics Engineering ISEE 2019, pp.130–134 (2019). https://doi.org/10.1109/ISEE2.2019.8920916

12. Asri, H., Mousannif, H., Al Moatassime, H., Noel, T.: Using machine learning algorithms for breast cancer risk prediction and diagnosis. Procedia Comput. Sci. **83**, 1064–1069 (2016). https://doi.org/10.1016/j.procs.2016.04.224

13. Rawal, R.: Breast cancer prediction using machine learning. Int. J. Emerg. Trends Eng. Res. **8**(9), 6074–6079 (2020). https://doi.org/10.30534/ijeter/2020/191892020

14. Abbas, A.R., Mahdi, B.S., Fadhil, O.Y.: Breast and lung anticancer peptides classification using N-Grams and ensemble learning techniques. Big Data Cogn. Comput. **6**(2), 40 (2022)

15. Feltes, B.I., Chandelier, B.C., Grisci, E.B., Dorn, M.: CuMiDa: an extensively curated microarray database for benchmarking and testing of machine learning approaches in cancer research. J. Comput. Biol. **26**(4), 376–386 (2019). https://doi.org/10.1089/cmb.2018.0238

16. Hosmer, D.W., Jr., Lemeshow, S., Sturdivant, R.X.: Applied Logistic Regression, 398. John Wiley, New York (2013)

17. Han, J., Kamber, M., Pei, J.: Data Mining: Concepts and Techniques. Third. Morgan Kaufmann, Burlington (2012)

18. Breiman, L.: Random forests. Mach. Learn. **45**(1), 5–32 (2001). https://doi.org/10.1023/A:1010933404324

19. Berry, M.W., Mohamed, A.H., Wah, Y.B. (eds.): SCDS 2015. CCIS, vol. 545. Springer, Singapore (2015). https://doi.org/10.1007/978-981-287-936-3

20. Rutkowski, L., Jaworski, M., Pietruczuk, L., Duda, P.: The CART decision tree for mining data streams. Inf. Sci. (Ny) **266**, 1–15 (2014). https://doi.org/10.1016/j.ins.2013.12.060

21. Karatzoglou, A., Meyer, D., Hornik, K.: Support vector machines in R. J. Stat. Softw. **15**(9), 1–28 (2006)

22. Fauvel, M., et al.: Evaluation of kernels for multiclass classification of hyperspectral remote sensing data. In: 2006 IEEE International Conference on Acoustics Speech and Signal Processing Proceedings (2006)

23. R Core Team. R: A language and environment for statistical computing. R Foundation for Statistical Computing, Vienna, Austria (2021)

24. Caraka, R.E., Nugroho, N.T., Tai, S.K., Chen, R.C., Toharudin, T., Pardamean, B.: Feature importance of the aortic anatomy on endovascular aneurysm repair (Evar) using boruta and bayesian mcmc. Commun. Math. Biol. Neurosci. (2020). https://doi.org/10.28919/cmbn/4584

25. Singla, M., et al.: Immune response to dengue virus infection in pediatric patients in New Delhi, India—Association of viremia, inflammatory mediators and monocytes with disease severity. PLoS Negl. Trop. Dis. **10**(3), 1–25 (2016). https://doi.org/10.1371/journal.pntd.0004497

26. Das, B., et al.: Comparison of bagging, boosting and stacking algorithms for surface soil moisture mapping using optical-thermal-microwave remote sensing synergies. CATENA **217**, 106485 (2022). https://doi.org/10.1016/j.catena.2022.106485

Machine Learning Techniques for Predicting Risks of Late Delivery

Ravikanth Lolla[1], Matthew Harper[1]([✉]), Jan Lunn[1], Jamila Mustafina[2], Jolnar Assi[3], Chong Kim Loy[4], and Dhiya Al-Jumeily OBE[1]

[1] Liverpool John Moores University, Liverpool L15 3AF, UK
M.L.Harper@2014.ljmu.ac.uk
[2] Kazan Federal University, Kazan, Russia
[3] UNITRA Graduate School, UNITAR International University, Petaling Jaya, Selangor, Malaysia
[4] Traders Island Ltd., London, UK

Abstract. Supply chain is a cornerstone of the eCommerce industry and is a key component in its growth. Supply chain data analytics and risk management in the eCommerce space have picked up steam in recent times. With the availability of suitable & capable resources for big data and artificial intelligence, predictive analytics has become a significant area of interest to achieve organizational excellence by exploiting data available and developing data-driven support systems. The existing literature in supply chain risk management explain various methods assisting to identify & mitigate risks using big data and machine learning (ML) techniques across industries. Although ML techniques are used in various industries, not many aspects of eCommerce had utilized predictive analytics to their benefit. In the eCommerce industry, delivery is paramount for the business. During COVID-19 pandemic, needs changed. Reliable delivery services are preferred to speedy delivery. Multiple parameters involve delivering the product to a customer as per promised due date. This research will try to predict the risks of late deliveries to online shopping customers by analyzing the historical data using machine learning techniques and comparing them by multiple performance metrics. As a part of this comparative study, a new hybrid technique which is a combination of Logistic Regression, XGBoost, Light GBM, and Random Forest is built which has outperformed all the other ensemble and individual algorithms with respect to accuracy, specificity, precision, and F1-score. This study will benefit the eCommerce companies to improve their customer satisfaction by predicting late deliveries accurately and early.

Keywords: eCommerce · Supply chain analytics · Big data · Artificial intelligence · Machine learning

1 Introduction

In recent years, there has been increasing attention to risk management in supply chain processes [1]. The increasing popularity of big data, artificial intelligence, and increased

© The Author(s), under exclusive license to Springer Nature Singapore Pte Ltd. 2023
Y. B. Wah et al. (Eds.): DaSET 2022, LNDECT 165, pp. 343–356, 2023.
https://doi.org/10.1007/978-981-99-0741-0_25

availability of computing power is helping organizations to exploit the information available and make data driven decisions [2]. Data-driven decision support systems provide hindsight analytics and also predictive analytics to estimate, assess and mitigate risks [3]. Furthermore, big data analytics has a multitude of advantages, which include reduced operational costs, improved supply chain agility, and increased customer satisfaction [4]. The eCommerce industry is a fast-growing industry and in 2020, during the COVID-19 pandemic, it grew at a rate of 33.6% [5]. Post-pandemic consumers appear to have become more comfortable buying products online than visiting stores.

Supply chains are a vital aspect of the eCommerce industry and is a key component in the growth of that industry, with to-home delivery being an advantage for eCommerce over traditional brick and mortar retailers [6]. For eCommerce, a key success factor for their business is timely delivery of goods. Efficient supply chains contribute to the timely delivery of accurate products to reach the customer, as well as guaranteeing operational efficiency, cost efficiency, and customer satisfaction [7]. Effective supply chain risk management is a key factor contributing to efficient supply chain operations and delivery. Challenges related to increasing size of eCommerce industry, various risks are imposed on eCommerce domain related to late deliveries.

One area of SCRM which is not addressed in the literature is prediction of delayed deliveries. In the field of eCommerce, existing literature has examined the use predictive analytics on certain aspects of SCRM, such as backorder prediction [8–10], lead time prediction [11], credit risk forecasting, and identifying early indicators of fraud detection [12]. However, there remains a lack of detailed study over predictive analytics for the prediction of delays in delivery and improving customer satisfaction, based on timely delivery.

The present study proposes to address the gaps related to predicting late deliveries considering aspects related to backorder prediction, credit risk forecasting and early identification of fraud detection. In this paper, machine learning (ML) models which can predict the risk of delivery delays is presented, which is trained using a public dataset of 180,519 delivery observations, including information about product information, customer information, and shipment data, among others. The models which were trained are Logistic Regression, Naïve Bayes, K nearest neighbors (kNN), Decision Tree and ensembles like Bagging, Random Forest, Gradient Boosting Machine, XGBoost, and Light GBM.

2   Background

To provide good suppleness in business decision support systems by predicting backorders with higher accuracy Islam and Amin (2020) had experimented with binary classifiers [8]. A dataset from kaggle.com with a total of 1,929,939 observations on details like lead time, inventory, sales, and forecasted sales as independent variables was used with split ratio 87.5:12.5 for training and testing models built on binary classifiers distribution random forest (DRF) and gradient boosting machine (GBM) to predict target variable indicating the product went on back order or not. The proposed model had 5 level decision trees by modifying the values under inventory, lead time, sales, and forecasts. While building the models, for better execution times and accuracy without overfitting, the parameters like the number of trees (50 to 1000), learning rate (0.1), maximum depth (10), sample rate (0.9) were optimized. Models were built and validated using both actual data and ranged data and observed that there is a significant difference in area under the curve (AUC). The tree models that showcase the probability of backorder with different range values were also provided.

In the study done by Malviya et al. (2021), a variety of machine learning algorithms were used to predict backorders and provide easily understandable backorder decision scenarios [10]. A dataset containing 1,929,939 observations with details like lead time, buffer stock, sales from kaggle.com was used to predict backorder circumstances. Classifiers like artificial neural networks, random tree, logistic regression, decision tree classifier, bayesian network, support vector machine, and discriminant analysis were built, and their performance was compared using accuracy, precision, recall, and F-score. Decision tree classifier and support vector machine-based models provided the best predictions with accuracy ranging from 0.989 to 0.997 and F-score 0.994. Gyulai et al. (2018) had dealt application of ML algorithms to predict lead times in the optics industry which has largely customized products [11]. Linear regression, tree-based models, and support vector regression are used to predict the lead times. Predictive models built were compared with conventional mathematical models (Little's Law) using performance metric normalized root mean square error (NRMSE). It was observed NRMSE was less than 10% in the case of ML models compared to >25% in the case of mathematical models. This clearly established how impactful ML models would be in promising appropriate delivery times by predicting lead times in the flow-shop environment and avoiding risks of delayed deliveries. Many players in the online retail industry have return policies that are very much customer friendly are facing various risks of fraudulent transactions.

John et al. (2020) had proposed a predictive analytics approach to mitigate this risk. Private data from an eCommerce business that has a customer base in more than 15 countries was taken for this study [9]. It had 545,037 observations with multiple attributes of refund transaction as independent variables and 'FRAUD_YES_NO' as the target variable. The authors had utilized a logistic regression model on the top 5% refund transactions with a higher risk of fraud. Kolmogorov-Smirnov (KS) statistic was used as a performance metric for the above binary classification which was observed to be more than 65%. Recall and sensitivity of this ML model were 80.2% and 73.6% respectively. 18 (Zhu et al., 2019) had developed a hybrid ensemble named RS-Multi boosting to forecast the small and medium enterprises credit risk which is a vital aspect of supply chain finance. For this study, the authors had collected data from 46 listed SME's. The performance of the predictive model built was measured using accuracy, F-measure, Type-1, and Type-II errors. It was observed that hybrid ensemble performed better than individual ML ensembles.

Pawłowski (2021) proposed a new framework to classify products using 12 classifiers in Scikit learn library for python with data sets gathered from a Polish industrial web shop [13]. Base models were compared using accuracy and 3 classifiers linear support vector classification, random forest, and ridge classifier were finalized for further experiments. User-generated data (search history data) was combined with product registry data to train the classification models which improved the accuracy and especially recall which is most crucial for better user experience while browsing the catalog.

Nguyen et al. (2018) mentioned that further research should be done on the application of BDA in each supply chain function with a special focus on predictive analytics which in turn improves the performance and accuracy of prescriptive analytics [4]. This would help to build a strong decision support system complementing SCRM. In the existing literature on the usage of predictive analytics to identify or mitigate risks in the eCommerce industry, there was not much research on how to handle imbalanced data. Also, very few aspects of eCommerce like back order prediction, credit risk forecasting, and predicting lead times in a flow-shop environment using predictive analytics were explored. There are many areas of the eCommerce industry that can leverage BD & PA to identify or mitigate risks. One such key area is to predict the risk of late deliveries to online shopping customers using ML techniques.

3   Research Problem

The eCommerce industry is one of the fastest-growing in terms of revenue. As per a survey by statista.com, the amount of sales in 2019 is around 3.5 trillion US dollars and is anticipated to breach the 6.5 trillion US dollars mark in 2022 (statista.com, n.d.). With the increasing size and volumes, there is huge scope for various kinds of risks in the eCommerce domain. Among all those risks, the risk of delivery delays is one of the key aspects which should be addressed on high priority as it directly affects the success of the eCommerce business. In the space of eCommerce, the existing studies explained the usage of predictive analytics over a few aspects like backorder prediction (Islam and Amin, 2020; John et al., 2020; Malviya et al., 2021), lead time prediction (Gyulai et al., 2018), credit risk forecasting and identifying early indicators of fraud detection (Zhu et al., 2019). There is a dearth of detailed study over predictive analytics on predicting delays in delivery and improving customer satisfaction. The pandemic situation in the year 2020 has accelerated the growth of eCommerce businesses exponentially and is expected to grow more in the coming years (Agarwal and Srivastava, 2021). The impact of the pandemic on eCommerce is very positive in terms of revenues and in the year 2020 European market had seen 10% of additional growth in revenues due to the COVID-19 situation (Jílková and Králová, 2021). A similar trend has been observed globally. These upward trends are majorly due to substantial changes in consumer behaviors embracing online shopping more than pre-pandemic times and preferring home deliveries (Unnikrishnan and Figliozzi, 2021).

This study is an attempt to predict the risk of delays in deliveries using machine learning techniques. Usage of ML techniques would fill the gaps left by mathematical models in handling big data and extracting insights. This benefits e –commerce companies to improve their customer satisfaction index. The objective of the study include:

- To analyze the public dataset containing online customer orders prepared by the company DataCo Global.
- To build and compare predictive models using (1) ML classification techniques such as logistic Regression, Naïve Bayes, K Nearest Neighbors (KNN), Decision Tree and (2) ensembles such as Bagging, Random Forest, Gradient Boosting Machine, XGBoost, and Light GBM.

- To build a hybrid model which combines ensembles to attain higher prediction accuracy.
- To evaluate the performance and compare models built using parameters like accuracy, specificity, precision, recall, and F1-score.

4 Methodology

4.1 Dataset

A public dataset containing online order information prepared by the company DataCo Global was used in this study [14]. It has 52 features relating to 180,519 deliveries, which comprise 3 years of historical data along with a label for late delivery risk (1 represents late delivery and 0 represents on-time delivery). The information included in the dataset falls into 3 general categories: customer information, sales and order information, product information, and shipment information. Some of the features were considered to be irrelevant for the methodology used in this paper, for example "customer email", "customer password", and "product image". As such, these features were removed before any further analysis.

Before progressing with further steps, it is very much necessary to check for irregularities like incorrect data types, missing values, anomalies, outliers, inconsistent data format, and invalid values under each feature. Otherwise, they may pose challenges while performing exploratory 22 data analysis and training ML models. Major approaches to treat outliers include imputation, deletion of outliers, binning of values, and capping of outliers.

4.2 Data Preparation and Exploratory Data Analysis

The datasets with independent variables having an unequal distribution of classes are pretty common in the real world. But, if the event to be predicted is a rarely occurring one like fraudulent transactions or diagnosis of rare diseases or natural disasters there will be a bias towards one class in the dataset. If the class imbalance in the datasets is not handled before training ML models, the results would be misleading and inaccurate due to overfitting or underfitting. There are multiple methods to handle class imbalance. They include data resampling techniques - random under sampling, random over sampling, cluster-based sampling, and synthetic sampling; ensemble-based techniques. Depending on the 23 dimensionality and outliers in the data, one of the aforementioned techniques should be chosen. Each of them has a set of pros and cons. It is observed that the dataset identified for this study does not have a class imbalance. Out of 180,519 observations, the dependent variable 'Late_delivery_risk' has 98,977 observations under positive class which is 54.8%. Feature scaling is another crucial step in data preparation. Multiple ML techniques are Euclidean distance-based algorithms based and require the features to be at comparable scales. Otherwise, the corresponding coefficients may become insignificant which leads to inaccurate model building. There are multiple ways to scale the features and prominent techniques include normalization and standardization. Normalization can be used when the values of the feature don't follow Gaussian distribution and standardization can be used when the feature values follow Gaussian distribution. Very often, not all independent variables in a dataset might be relevant to building an ML model. Certain features might be redundant or even functionally not related to the dependent variable. So, it is very important to choose only those which influence the dependent variable significantly. To do so, there are various methods – correlation matrix, statistical tests, recursive feature elimination, forward/backward/stepwise selection based on AIC, and elimination based on variance inflation factor. Many times, transforming a couple of given features into a new refined and meaningful feature would improve model performance to great extent.

4.3   Training and Evaluation of Models

A multitude of binary classification models were trained, including Logistic Regression, Naïve Bayes, k-Nearest Neighbors (k-NN), Decision Tree, Random Forest, Gradient Boosting Machine (GBM), Light GBM, Extreme Gradient Boosting, and Bagging.

Model Building
Each model has some key parameters when tuned can result in high performing models. These parameters are also known as hyperparameters. For individual and ensemble algorithms, basic models were built with default parameters, and then using certain optimization techniques, best performing hyperparameters are identified.

Logistic Regression
LogisticRegression() is a linear classifier model under the scikit-learn python library. It inherently utilizes the 'liblinear' library and solvers like 'lbfgs', 'saga', 'sag', and 'newton-cg' to predict binary or multi-class classification outcomes. Using the hyperopt library with accuracy as a scorer, 50 iterations of logistic regression models are analyzed to tune the hyperparameters. Below are the values for each parameter defined in the search space to identifya model with minimum loss (or best accuracy).

- 'penalty': hp.choice('penalty', ['l2', 'none'])
- 'C': hp.loguniform('C', np.log(0.01), np.log(1000))
- fit_intercept': hp.choice('fit_intercept',[True, False])
- 'solver':hp.choice('solver', ['newton-cg', 'lbfgs', 'sag', 'saga']

Naïve Bayes
In the scikit-learn library, there are a set of classifier models which apply the Bayes theorem. Depending on the kind of data under the predictor variables, Gaussian or Bernoulli classifier can be implemented. As the data in the current dataset follows Gaussian distribution, GaussianNB() classifier is implemented to make accurate predictions. The parameter 'var_smoothing' in this classifier stabilizes the calculation and its default value is $1ee-9$. Using 38 grid search hyperparameter tuning has been done with accuracy as scorer with search space defined as below.

- var_smoothing = [1e−11, 1e−10, 1e−9, 1e−8, 1e−7].

KNN

KNeighborsClassifier() in the scikit-learn library is a model which implements the k-nearest neighbors algorithm to predict the classes. For this model n_neighbors, weights and p are key parameters of this model which are defaulted to 5, 'uniform' and 2 respectively. Using the Bayesian optimization technique with accuracy as a scorer, the best hyperparameters are identified by defining search space as below and running a maximum of 25 iterations.

- 'n_neighbors': (3,20)
- 'weights': ['uniform', 'distance']
- 'p': (1,2)

Decision Tree

DecisionTreeClassifier() is a non-parametric supervised learning model in the scikit-learn library which utilizes an optimized version of CART (classification and regression trees) to predict the target variable. In this model, key parameters include criterion, splitter, max_depth, min_samples_split, and min_samples_leaf that are defaulted to 'gini', 'best', 'None', 2, and 1 respectively. Using bayesian optimization techniques with accuracy as a scorer, the best hyperparameters are identified by defining the search space as below and running a maximum of 25 iterations.

- 'criterion': ['gini', 'entropy']
- 'splitter': ['best', 'random']
- 'max_depth': (4, 15)
- 'min_samples_split': (2, 10)
- 'min_samples_leaf': (2, 10)

Bagging

BaggingCalssifier() is one of the ensemble meta-estimators provided by scikit-learn library with DecisionTreeClassifier() as the default base estimator. Key parameters for this model are n_estimators, max_samples, max_features, and bootstrap which are defaulted to 10, 1, 1, True 39 respectively. Using grid search with accuracy as a scorer, the best hyperparameters are identified by defining the search space as below.

- 'n_estimators': [10,100,1000]

Random Forest

RandomForestClassifier() is another ensemble meta-estimator provided by the scikit-learn library. As it uses DecisionTreeClassifier() as a base estimator, the key parameters are n_estimators, criterion, max_depth, min_samples_split, and min_samples_leaf which are defaulted to 100, 'gini', None, 2, and 1 respectively. Bayesian optimization techniques are used with accuracy as a scorer to identify the best hyperparameters by defining the search space as mentioned below and running a maximum of 25 iterations.

- 'n_estimators':(70, 150)
- 'criterion':(0, 1)
- 'max_depth':(4, 20)
- 'min_samples_split':(2, 10)
- 'min_samples_leaf': (2, 10)

Gradient Boosting Machine

GradientBoostingClassifier() is one of the boosting ensemble meta-estimators provided by the scikit-learn library. Key parameters include n_estimators, max_depth, max_features, subsample, and learning_rate which are defaulted to 100, 3, None, 1, and 0.1 respectively. Using Bayesian optimization techniques with accuracy as a scorer, the best hyperparameters are identified by defining search space as given below and running a maximum of 25 iterations.

- 'max_depth':(3, 10)
- 'max_features':(0.8, 1)
- 'learning_rate':(0.01, 1)
- 'n_estimators':(80, 150)
- 'subsample': (0.8, 1)

Extreme Gradient Boost (XGBoost)

XGBoost is a python package that provides a wrapper class for the scikit-learn library.

XGBClassifier() is a class with n_estimators, learning_rate, max_depth, min_child_weight, subsample, objective, gamma, colsample_bytree as key parameters. Bayesian optimization techniques are used with accuracy as a scorer, to identify the best hyperparameters by defining the search space as given below and running a maximum of 25 iterations.

- 'objective': 'binary:hinge'
- 'n_estimators':(80, 150)
- 'max_depth': (3, 15)
- 'learning_rate': (0.01, 0.5)
- 'gamma':(0, 10)
- 'min_child_weight':(3, 20)
- 'subsample':(0.5, 1)
- 'colsample_bytree':(0.1, 1)

Light Gradient Boosting Machine (Light GBM)

LGBMClassifier() is a boosting ensemble meta-estimator provide by the python package LightGBM with scikit-learn API. Key parameters of this model are num_leaves, max_depth, learning_rate, n_estimators, min_child_weight, subsample and colsample_bytree which are defaulted to 31, -1, 0.1, 100, $1ee-3$, 1 and 1. Using bayesian optimization techniques with accuracy as a scorer, the best hyperparameters are identified with search space defined as below and running a maximum of 25 iterations.

- 'min_child_weight':(1e−5, 1e−1)
- 'subsample':(0.5, 1)
- 'colsample_bytree':(0.5, 1)
- 'max_depth': (3, 15)
- 'learning_rate': (0.01, 0.5)
- 'num_leaves_percentage':(0.5, 0.9)

Hybrid Model

Multiple experiments are conducted to build a hybrid model with different combinations of individual and ensemble algorithms. In each experiment, a two-level stacking technique is used to combine multiple models. From all such experiments, the combination of Logistic Regression, XGBoost, Light GBM, and Random Forest has been finalized as a hybrid model. In this stacking mechanism, level0 estimators include Logistic Regression, XGBoost, and Light GBM models whereas Random Forest is used as the final estimator. Hyperparameters of each individual model are tuned separately and applied to tune the performance of the hybrid model.

Model Validation

The confusion matrix was used to validate the binary classification models as it provides multiple aspects of model performance. It helps to get a holistic view of model performance and error types. True positives (TP) and true negatives (TN) are the instances where both actual and prediction values match. False positives (FP) and false negatives (FN) are the instances where predictions are not matching with actuals. False positives are also known as type1 errors and false negatives are also known as type2 errors. Using the confusion matrix a number of metrics were calculated, including accuracy, sensitivity, specificity, precision, recall, F1 score, true positive rate (TPR), and false positive rate (FPR).

4.4 Hyperparameter Tuning

In machine learning, hyperparameter optimization refers to finalizing the set of hyperparameters for an algorithm to make the most of the model performance. Applying the perfect set of hyperparameters helps to obtain maximum performance. Manual hyperparameter tuning is a tedious process and cannot be practical in cases where there are many hyperparameters to check. Multiple open-source libraries available for hyperparameter optimization in python are available. Out of those, this study leverages hyperopt [15], Bayesian optimization [16] libraries for python to build the final predictive models for comparison.

5   Results and Discussion

In this section, the results of the exploratory data analysis performed on the dataset and the performance metrics of the trained machine learning models are shown and dicussed. The individual models were compared with ensemble models built to understand if predictability power increases with ensembles. Also, the performance metrics of a hybrid model built is compared with all the other models built and identify the best performing predictive model.

5.1   Exploratory Data Analysis

Before moving to further analysis of categorical and numerical variables, the class imbalance analysis is needed for any classification problem. Most of the machine learning models work best with equally balanced data. In the current dataset, 54.83% of observations fall under the late delivery risk class and 45.17% of observations fall under the no risk class. As there is not 33 much bias towards any class, there is no need for any class imbalance treatments and also accuracy can be used to analyze the performance of the ML models.

The current dataset has been analyzed by categorical features. From the analysis it is evident that LATAM, Europe, and Pacific Asia markets have a greater number of customers with the consumer segment having the highest number of customers. Most of the orders are using shipping mode 'Standard Class' whereas, 'Debit' is the most common method of transaction. Products under the departments 'Fan Shop', 'Apparel', and 'Golf' are ordered most by the customers.

Furthermore, there are around 40,000 orders with pending payments and more than 50% of orders are delivered later than the promised date. Though scheduled days to ship products is of maximum 4 days, in actual more than 50,000 orders took >=5 days to ship the products. These visualizations show that 'Days for shipment (scheduled)', 34 'Days for Shipping (real)', 'Shipping Mode' and 'Transaction Type' are the important features to predict the risk of late deliveries.

Moreover, the number of orders with risk/no-risk of late delivery has been analyzed against various categories, and from this we can make a number of observations from the data:

- In every market, the number of late deliveries is more than on-time or advance deliveries.
- Late deliveries are more frequent in the case of the consumer segment.
- The shipping modes 'First Class' and 'Second Class' have a very high percentage of late deliveries.
- Only the 'Transfer' transaction type has less than 50% of later deliveries.
- More number of late deliveries are with transaction type 'Debit'.
- Products with scheduled shipment in 1 or 2 days have higher chances of late deliveries

It was also observed that though 28.8% of orders were from the LATAM market, the sales amount is less than other markets at 6.24%. In the current dataset, there are more than 50% of orders placed with the 'Debit' transaction type, but the corresponding sales amount is less than other types at 10.94%. The majority of the sales amount is from the department 'Footwear' at 46.52%.

5.2   Machine Learning Results

Table 1 summarizes the performance metrics by each model after tuning the hyperparameters. It shows that the hybrid model built has outperformed all the other individual and ensemble algorithms with respect to accuracy, specificity, precision, and F1-score. The recall is marginally less for the hybrid model whereas the light GBM model has got highest recall value. The 5-fold cross-validation accuracy is also high for the hybrid model at 0.99 (±0.02). It is observed that Naive Bayes is the poor performing model among the lot with the lowest accuracy, specificity, precision, recall, and F1-score. The kNN and logistic regression models have poor cross-validation accuracy.

Table 1. Comparison of performance metrics

Model	Accuracy	Specificity	Precision	Recall	F1-Score	Cross validation
Hybrid (combination of LR, LGBM, XGBoost & RF)	0.9938	0.9889	0.9909	0.9978	0.9943	0.99(±0.02)
Light Gradient Boosting Machine	0.9928	0.9847	0.9874	0.9996	0.9935	0.99(±0.02
Gradient Boosting Machine	0.9910	0.9830	0.8892	0.8736	0.4052	0.98(±0.02)
Bagging	0.9887	0.9759	0.9223	0.9365	0.3565	0.97 (±0.04)
EExtreme Gradient Boosting Machine	0.9853	0.9698	0.9625	0.9501	0.2706	0.97(±0.04)
Logistic regression	0.9821	0.9627	0.9699	0.9981	0.9981	0.70(±0.03)
Decision Tree	0.9783	0.9577	0.9659	0.9659	0.9955	0.96(±0.05)
Random Forest	0.9773	0.9500	0.9601	1.0	0.9796	0.97(±0.02)
k-Nearest Neighbors	0.9723	0.9484	0.9586	0.9921	0.9751	0.62(±0.02)
Naïve Bayes	0.8445	0.8752	0.8876	0.819	0.8519	0.85(±0.06)

All the ensemble models outperformed the individual algorithms and the hybrid model is better performing than the ensemble models. Almost all the interpretable models have suggested the same set of features are important ones to predict risks of late deliveries. They include days for shipping (real), days for shipment (scheduled), transaction type, and shipping mode. This helps eCommerce companies to focus on these areas to reduce risks of late deliveries and thus improve customer satisfaction.

The best hyper-parameters for each model were also identified and are shown in Table 2.

The importance of each feature- meaning their predictive power for predicting the class of each observation (late risk/not late risk) in the classification decision of each of the interpretable machine learning models was also identified using coefficients or feature importance attributes. For the Logistic regression model, the most importnant

Table 2. Best hyperparamaters for each trained model

Model	Best hyperparameters
Light Gradient Boosting Machine	{'colsample_bytree': 0.5, 'learning_rate': 0.5, 'max_depth': 14, 'min_child_weight': 1e-05, 'subsample': 0.5, 'objective': 'binary', 'num_leaves': 8192}
Gradient Boosting Machine	{'learning_rate': 1.0, 'max_depth': 10, 'max_features': 0.8, 'n_estimators': 105, 'subsample': 1.0}
Bagging	{ 'n_estimators': 100}
Extreme Gradient Boosting Machine	{'colsample_bytree': 0.9807799500944937, 'gamma': 5.873353045089665, 'learning_rate': 0.24309974507405047, 'max_depth': 9, 'min_child_weight': 12, 'n_estimators': 92, 'subsample': 0.965399374829356, 'objective': 'binary:hinge'}
Logistic regression	{'C': 0.31049056210450227, 'fit_intercept': True, 'penalty': 'none', 'solver': 'sag
Decision Tree	{'criterion': 'entropy', 'max_depth': 14, 'min_samples_leaf': 2, 'min_samples_split': 10, 'splitter': 'best'}
Random Forest	{ 'criterion': 'gini', 'max_depth': 20, 'min_samples_leaf': 2, 'min_samples_split': 2, 'n_estimators': 79}
k-Nearest Neighbors	'n_neighbors': 20, 'p': 1, 'weights': 'distance'
Naïve Bayes	{'var_smoothing': 1ee−11}

features were "days for shipping (real)" and "order status". For the Decision tree model the most important features were "days for shipping (real)" and "shipping mode", while "type" and "days for shipment (scheduled)" were also identified as important to a lesser extent. For the random forest model, the features identified as important were "Days for shipping (real)", "Shipping mode", and "Days for shipment (scheduled)", in descending order. For the gradient boosting model, the most important attributes identified were "Days for shipping (real)", "Days for shipment (scheduled)", and "shipping mode", again in descending order. For the XGBoost model the top 3 most important features, in descending order, were "shipping mode", "days for shipping (real)", and "days for shipment (scheduled)". Finally for the light GBM model the list of features identified as having some importance is extensive, but of particular note are "customer name", "order city", "order state", and "order item profit ratio".

6 Conclusion

In this study, various algorithms were used to build predictive models for this classification problem. They include both individual algorithms and ensembles. As there is not much bias towards any class in the dataset, accuracy & F1-score have been considered as key metrics to evaluate the prediction performance. Additionally, metrics like specificity, precision, and recall are also considered to understand the performance of models built. It is observed that models based on ensembles performed well compared to those based on individual algorithms. The cross- 54 validation accuracies are poor for models based on individual algorithms. Further, the hybrid model built by stacking logistic regression, XGBoost, light GBM, and random forest algorithms achieved not just higher accuracy but also better scores in sensitivity, precision, and F1-score and 5-fold cross-validation accuracy.

This study has provided a new hybrid model that achieved higher accuracy. However, it lacks interpretability due to the stacking of the ensemble algorithms and looks more like a black box. Though the models based on ensembles like the random forest, GBM, XGBoost, and light GBM are interpretable the predictive performances are slightly lower than the hybrid model. There is always a trade-off between performance and interpretability. Along with prediction power, eCommerce companies would be benefited much if the predictive models provide the important features influencing the predictions. It would help SCRM to not only identify the risks but also help to provide directions to mitigate them. Further study should be done on building more interpretable predictive models without compromising on performance. Explainable boosting machines as utilized by Velmurugan et al., (2021) is one such area that can be researched in detail and apply it to the problems in the eCommerce space [17]. The data set used in this study has only customer, product, order, and shipment information. In future studies, the inclusion of inventory data would also benefit the businesses to understand the reasons behind the shipment delays which is the major driving factor for the late deliveries. Another area of recommended research is to study challenges and how machine learning techniques can help address them in instant delivery services like dunzo.com, blinkit.com, and zeptonow.com which is the latest segment of the eCommerce industry. Future studies can focus more on bringing the application of predictive analytics in the front-office than back-office and end-user scenarios. In that way, it will be more impactful in improving the business.

References

1. Baryannis, G., Validi, S., Dani, S., Antoniou, G.: Supply chain risk management and artificial intelligence: state of the art and future research directions. Int. J. Prod. Res. **57**(7), 2179–2202 (2019)
2. Dubey, R., et al.: Big data analytics and artificial intelligence pathway to operational performance under the effects of entrepreneurial orientation and environmental dynamism: A study of manufacturing organisations. Int. J. Prod. Econ. **226**, 107599 (2020)
3. Gunasekaran, A., et al.: Big data and predictive analytics for supply chain and organizational performance. J. Bus. Res. **70**, 308–317 (2017). http://www.springer.com/lncs. Accessed 21 Nov 2016
4. Nguyen, T., Zhou, L., Spiegler, V., Ieromonachou, P., Lin, Y.: Big data analytics in SCM: A state-of-the-art literature review. Comput. Oper. Res. **98**, 254–264 (2018)
5. Goldman, S.: Post-pandemic e-commerce: The unstoppable growth of online shopping. The future of customer engagement and experience (2021)

6. Weingarten, J., Spinler, S.: Shortening delivery times by predicting customers' online purchases: a case study in the fashion industry. Inf. Syst. Manage. **384**, 287–308 (2021). https://www.tandfonline.com/doi/full/10.1080/10580530.2020.1814459
7. Bag, S., Wood, L.C., Xu, L., Dhamija, P., Kayikci, Y.: Big data analytics as an operational excellence approach to enhance sustainable supply chain performance. Res. Conserv. Recycl. **153**, 104559 (2020)
8. Islam, S., Amin, S.H.: Prediction of probable backorder scenarios in the supply chain using distributed RF and GB ML techniques. J. Big Data **71**, 65 (2020)
9. John, S., Shah, B.J., Kartha, P.: Refund fraud analytics for an online retail purchases. J. Bus. Anal. **31**, 56–66 (2020)
10. Malviya, L., Chittora, P., Chakrabarti, P., Vyas, R.S. and Poddar, S.: Backorder prediction in the supply chain using machine learning. Mater. Today: Proc. (2021)
11. Gyulai, D., Pfeiffer, A., Nick, G., Gallina, V., Sihn, W., Monostori, L.: Lead time prediction in a flow-shop environment with analytical and machine learning approaches. IFAC-PapersOnLine **51**(11), 1029–1034 (2018)
12. Zhu, Y., Zhou, L., Xie, C., Wang, G.-J., Nguyen, T.V.: Forecasting SMEs' credit risk in supply chain finance with an enhanced hybrid ensemble machine learning approach. Int. J. Prod. Econ. **211**, 22–33 (2019)
13. Pawłowski, M.: Machine learning based product classification for eCommerce. J. Comput. Inf. Syst. **66**(4), 1–10 (2021)
14. Constante, F.: DataCo smart supply chain for big data analysis. Mendeley (2019). https://data.mendeley.com/datasets/8gx2fvg2k6/5
15. Bergstra, J., Yamins, D., Cox, D.D.: Hyperopt: A python library for optimizing the hyperparameters of machine learning algorithms. In: Proceedings of the 12th Python in Science Conference, p. 20. Citeseer (2013)
16. Nogueira, F.: Bayesian optimization: open source constrained global optimization tool for Python (2014). https://github.com/fmfn/BayesianOptimization
17. Velmurugan, M., Ouyang, C., Moreira, C., Sindhgatta, R.: Evaluating fidelity of explainable methods for predictive process analytics. In: Nurcan, S., Korthaus, A. (eds.) Intelligent Information Systems. CAiSE 2021. Lecture Notes in Business Information Processing, vol. 424, pp.64–72. Springer, Cham (2021). https://doi.org/10.1007/978-3-030-79108-7_8

Quora Insincere Questions Classification Using Attention Based Model

Snigdha Chakraborty[1], Megan Wilson[2], Sulaf Assi[2(✉)], Abdullah Al-Hamid[3], Maitham Alamran[4], Abdulaziz Al-Nahari[6], Jamila Mustafina[5], Jan Lunn[1], and Dhiya Al-Jumeily OBE[1]

[1] Faculty of Engineering and Technology, Liverpool John Moores University, Liverpool, UK
[2] School of Pharmacy and Bimolecular Science, Liverpool John Moores University, Liverpool, UK
S.Assi@ljmu.ac.uk
[3] Saudi Ministry of Health, Najran, Saudi Arabia
[4] Al-Qadisiyah University, Al-Qadisiyah, Iraq
[5] Kazan Federal University, Kazan, Russia
[6] UNITAR Graduate School, UNITAR International University, Selangor, Petaling Jaya, Malaysia

Abstract. The online platform has evolved into an unparalleled storehouse of information. People use various social question-and-answer websites such as Quora, Form-spring, Stack-Overflow, Twitter, and Beepl to ask questions, clarify doubts, and share ideas and expertise with others. An increase in inappropriate and insincere comments by users without a genuine motive is a major issue with such Q & A websites. Individuals tend to share harmful and toxic content intended to make a statement rather than look for helpful answers. In the world of natural language processing (NLP), Bidirectional Encoder Representations from Transformers (BERT) has been a game-changer. It has dominated performance benchmarks and thereby pushed the limits of researchers' ability to experiment and produce similar models. This resulted in improvements in language models by introducing lighter models while maintaining efficiency and performance. This study utilized pre-trained state-of-the-art language models for understanding whether posted questions are sincere or insincere with limited computation. To overcome the high computation problem of NLP, the BERT, XLNet, StructBERT, and DeBERTa models were trained on three samples of data. The metrics proved that even with limited resources, recent transformer-based models outscore previous studies with remarkable results. Amongst the four, DeBERTa stands out with the highest balanced accuracy, macro, and weighted f1-score of 80%, 0.83 and 0.96, respectively.

Keywords: Online platform · Question-and-answer · Language models · Natural language processing · Bidirectional encoder representations from transformers · Accuracy

© The Author(s), under exclusive license to Springer Nature Singapore Pte Ltd. 2023
Y. B. Wah et al. (Eds.): DaSET 2022, LNDECT 165, pp. 357–372, 2023.
https://doi.org/10.1007/978-981-99-0741-0_26

1 Introduction

Online question-and-answer platforms represent one of the key methods for sharing information among people. Most of the websites focus on generating content encouraging people to use the platforms regardless of the quality of content. Despite utilizing countless benefits of this medium to educate and connect with people sharing similar opinions, an emerging issue of online hatred, spam, racism, and toxicity is emerging [1]. The toxic behavior users disrupt the terms and conditions of a platform, resulting in hurting the sentiments of others. This result in causing emotional trauma to other users, impacting the health and norms of online communities [1, 2]. The study by Del Vicario et al. 2016 describes the formation of a polarized groups due to user's inclination towards misinformation [3]. As highlighted mitigating such bias from the content is equally difficult as identifying it [4, 5, 8].

Aslam et al. [8] have tackled the classification of the insincere content from questions and answers forums using different machine learning and deep learning models. The results showed that, the LSTM models gave better performance in accuracy, and decreases the loss value. Al-Ramahi [9] has tackled the issue of detecting insincere content on Quora as a case study for spotting insincere content on online social media. Based on his study, the logistic regression model is better for identify insincere questions. Rachha, A. and G.J.a.p.a. Vanmane [10] have studied the detecting of insincere questions from text using BERT and three BERT-based transformer models approach text classification.

The primary focus of this paper is to approach this problem using transformer-based language models. BERT will be used as a baseline model, and the result will be compared against other pre-trained models by fine-tuning the final linear layer as per the selected task. The reason for choosing an attention-based model is primarily that it will perform better in noisy datasets and will be able to generalize the model [6]. The models will be trained on the Quora Insincere Questions Classification dataset [5]. We anticipate that our models will perform better when it comes to classifying the insincere question data.

2 Methodology

An efficient system for classifying insincere questions on the Quora dataset was developed. The questions were classified using the CRISP-DM (Cross-industry procedure for data mining) approach, which aided in the planning, organization, and execution of data analysis initiatives [7].

2.1 Data Understanding

This study used the Quora-insincere-questions classification provided by Kaggle [5]. The dataset consists of a training and test set. The training set consists of 1.3M rows with required labels corresponding to each row. The training data includes different questions asked by users in English and are not guaranteed to be perfect. Hence, there is a certain amount of noise in the dataset. Questions are marked as 0 (Sincere) and 1 (Insincere). The columns available in the dataset are: (1) qid - unique identifier for the question; (2) question_text - question text posted in Quora 33; and (3) target - a question labelled as 1 for "insincere" and 0 for "sincere".

2.2 Exploratory Data Analysis

The training set consists of 1,306,122 unique values, whereas the test set has 375,806. Based on target variable distribution, it had been observed that the data was highly imbalanced with 1.2M belonging to sincere questions and just a small proportion of 80k belonging to target class 1 for insincere questions. Subsequently, data had been pre-processed before application of machine learning (ML) models.

2.3 Data Pre-processing

Data preprocessing approach underlined:

I.   Removing special characters, numeric characters, punctuation, and duplicate records The question text column in the dataset can contain smileys, numeric or roman literals, punctuations, or special characters as in "?". Such characters were removed with the help of lambda expressions by utilizing regex to filter the format.

II. The distinct contraction terms were enlarged with the use of python libraries such as expanding contractions using a pre-defined contractions dictionary map, such as replacing "couldn't" with "could not".

III. Removing all unwanted spaces from the text.

IV. Truncation of sequence length. Bert's maximum sequence length was 512, but for efficient memory usage, those could be truncated to a shorter length before being encoded by the tokenizer for use. After pre-processing, the cleaned question texts were tokenized and turned into vectors to suit the pre-trained transformer-based model's needed input format. To preserve the syntactical meaning of the questions, stemming or lemmatization was avoided. This was done to ensure that the precious information of the data is restored.

2.4 Tokenization, Truncation and Padding

Tokenization was one of the main steps in breaking down a text into smaller chunks of words called tokens. The tokens were then converted into vectors of numbers. The build tensor was then fed to the model along with any additional input required to make the model work seamlessly. The splitting of text was done by a tool called a tokenizer. One could either construct their tokenizer class or could directly use the existing AutoTokenizer class. It was advised to use the associated pre-trained tokenizer while using a pre-trained model. This ensured that the tokens were built the same way as done for the pre-training corpus and used the same vocab during model pre-training. The pre-training vocab could be downloaded from the AutoTokenizer class. When the batch sequences are of different lengths, padding was used to maintain the length of the sequence. Similarly, when the length was longer than needed, truncation was used to shorten the length. Both were performed by specifying the value in the max_seq_length parameter.

2.5 Modelling

In this study the performance of the four state-of-art models viz. BERT, XLNet, Struct-BERT and DEBERTa on the dataset. All the mentioned models were trained using python and its libraries in TPU (Tensor Processing Unit) or GPUs (Graphical Processing Unit). TPU was a powerful custom-built processor and provided a better performance against GPUs. To restrict memory issues, model could be evaluated on a smaller batch size. Truncating the max length of the sequence also helped to utilize the memory efficiently. Based on model performance epochs and hyper tuning were decided. The evaluation metrics of each model were recorded for the outcome.

2.6 Evaluation

Model evaluation was undertaken by considering:

I. Accuracy – This is the percentage of all the correct predictions made from the total number of predictions. This is the widely used most common metric to validate a model. However, for an imbalanced dataset, it is not a good indicator of the model's behavior as it tends to incline towards the majority classes present in the dataset. Hence, even if the model accuracy is high, it is highly possible that the model may fail to generalize the pattern and will not perform better in an unseen dataset. Balanced accuracy is the mean of specificity and sensitivity.

$$\frac{TP + TN}{TP + FP + TN + FN} \tag{1}$$

II. Precision – This is the percentage of all the correct positives out of total predicted positives. In other words, it checks how much a model is able to generalize the behaviour in terms of quality. It is preferred where false negative has less impact such as YouTube ad recommendation.

$$\frac{TP}{TP + FP} \tag{2}$$

III. Recall – This is the percentage of predicted positives out of the total number of actual positives present in a dataset. In other words, it checks how much a model failed or missed to identify the correct class in terms of quantity. It is preferred where false negative are equally important such as clinical analysis

$$\frac{TP}{TP + FN} \tag{3}$$

IV. F1-Score – This is the harmonic mean of both precision and recall. In other words, it identifies the performance by determining the actual positives predicted correctly and the model did not even misclassify positive to negative. Thus, the higher the F1 score better 39 the model. As the dataset is highly imbalanced, F1-Score will be used as a primary model evaluation metric. The Macro F1 Score is the arithmetic mean of the F1 score for each class label. A Weighted F1 Score is the sum of each class's F1 Score with respect to each class's weight.

$$\frac{2}{\frac{1}{\text{precision}} + \frac{1}{\text{recall}}} = \frac{2 * \text{precision} * \text{recall}}{\text{precision} + \text{recall}} \qquad (4)$$

2.7 Data Analytics

Data Preparation

The dataset was downloaded from Kaggle and loaded into the GPU-enabled Google Collaboratory. The implementation was done using Python libraries. The dataset contained three variables: a unique question id, actual question text, and a binary target variable indicating the sincerity of a question. The qid was used to uniquely identify a row and was not needed for our analysis, so the column is dropped. Thus, now the data contained one independent variable (the actual question posted by the user) and the target variable (sincere or insincere). There was no missing value found, and hence, missing value identification and imputation were not needed. The platform provided a respective dataset for training and testing. The training dataset was used for data analysis and data pre-processing, while the test set is used to validate the model's performance on unseen data.

Word Cloud

Words such as "work," "help," "book," "best," and so on were frequently used in standard questions. They are mostly neutral and indicate someone looking for advice or similar things. Insincere questions contain words like "atheist," "Trump," "Muslim," "black," etc. The difference is obvious. Instead of the generic advice topics (like relationships, work, education, etc.), the most frequently used words are all very political (for example, the frequent use of Donald Trump's name). This also shows why a simple sentiment analysis of words is not sufficient in this case.

N-Grams

Word clouds are a great tool to get a first impression of the word frequencies, but without looking at the actual values as well, they might be misleading. The N-grams method is used to find the most frequently used words and phrases in each question type. N-grams can be defined as continuous sequences of items in a text that help perform sentiment analysis on a word. A single word such as "bad" will be identified as discriminative. But when it gets combined with another token like "not bad," the label of sincerity changes to 44 positive.

Sentiment Analysis

Sentiment analysis helps us understand the mood and context behind the sentence. TextBlob, an NLP Python library, uses nltk to return the subjectivity and polarity of a sentence. Polarity can be defined as the orientation of the sentence. It has a range of $[-1, 1]$, with -1 denoting negative sentiment and 1 denoting positive feeling. Subjectivity refers to the semantic descriptors that aid in the fine-grained analysis of emoticons, exclamation marks, emojis, and so on. It has a range of [0, 1]. The degree of opinion and factual data in a text is measured by subjectivity. The subjectivity value towards 1 conveys that the text contains opinion rather than factual data. The package calculates it by looking into the intensity of each word in a sentence based on a lexiconbased approach. The polarity and subjectivity of 0 indicate that text is not the best.

Results higlighting polarity distribution in the insincere set shows that there is a tendency towards polarity within the insincere group. The distribution turns towards (-1). The mean polarity and subjectivity come out to be 0.03 and 0.36, respectively. When using the sentiment scores approach (from nltk.sentiment.vader), a concentration towards the negative values is found, and the mean was quite low at -0.099.

Feature Extraction

Meta-features are created as a part of feature extraction and their distribution between the classes are studied. The list of features constructed is as follows: - I. Number of words in the text II. Number of characters in the text III. Number of unique words in the text IV. Number of special characters in the text V. Number of stop words VI. Number of upper-case words VII. Mean, Max, Min length of the words. The spread of the extracted features among sincere and insincere classes is analysed with the help of box plot. It is observed that Insincere questions have a larger word count, unique word count, and special character count than sincere questions (apart from a few sincere outliers). The outliers in the boxplot shown in Fig. 1 state question_length of sincere text is approximately 800. Insincere can be spam questions, ads, and so on. Based on the number of special symbols, insincere can be latex math formulas or questions containing icons or emojis and non-English characters. It is also observed that the sincere queries are longer than disingenuous questions, with a maximum length of 134.

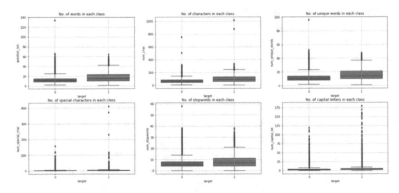

Fig. 1. Feature Boxplot of insincere and sincere text

Data Cleaning and Expansion of Contractions

Due to the self-attention mechanism, the models trained on huge corpus text data can adapt and learn features and generalize the behaviour of data on their own. Hence, the feature engineering steps are skipped and proceeded with the data cleaning process. The subsection includes different data cleaning steps performed before model training. Post data cleaning the values are stored in a new column named "cleaned_text".

Contractions are the shorthand form of text by dropping letters and replacing them with apostrophes, such as "shouldn't" or "don't". To ensure text standardization and dimensionality, the contracted words present in question_text are expanded using the Python Contractions 48 Library. The library is installed and imported, and the question text is passed as an argument to the fix function.

Punctuations, Special Characters Numbers and Spaces Removal

The question text in the dataset contains various emoticons, punctuation, emojis, symbols, and so on. These special characters are removed using regular expressions (regex). Lambda expressions with the tqdm progress_apply function are used on the question_text column to apply a regex filter to each text.

The question_text in the dataset contains numeric data. The numeric expressions are filtered out using lambda expressions and are removed from the text. After expanding the contractions and removing characters, extra spaces are found in the question text. All the leading, trailing, and double spaces resulting due to removal of punctuation are trimmed. Finally, the data is checked for presence of any duplicate records and clean data along with the target variables is retained in a new CSV file for implementation.

Data Sampling, Splitting and Hyper-Tunning

The platform provided different datasets for training and testing samples where each set was large, with millions of records. To achieve better model performance and state-of-the-art results with efficient resources in hand and limited computational power, the models were trained by considering different fractions of data samples per iteration. The random sample of items is taken from the original dataset by using the sample () function of the Pandas DataFrame class. The function takes frac as an argument to return 5%, 10%, and 50% of random samples per iteration while keeping the target variable imbalance consistent.

Data Split and Hyperparameter Tunning

The extracted fraction of the dataset was again split into train and validation sets in the ratio of 80:20 using the train test split function from the sklearn library in Python. Models are trained on a training set and tested on a validation set. The testing set was used to predict the target outcome. Due to its expensive computational nature, K-fold cross-validation was avoided.

The libraries provided a list of hyperparameters to be tuned as per the dataset and task requirement. In this study, from an entire list, only a selected set of hyperparameters was used. The value was determined after multiple iterations and thorough consideration. The reason for selecting each hyperparameter and the final value was chosen for implementation is explained below:-

- Epoch - The transformer models were pre-trained on a large set of data, and finetuning the model with 2, 3, or 4 epochs yields outstanding results. When fine-tuning with 100k+ data, researchers found that training with greater epochs did not improve the outcomes considerably. Models were initially trained for seven epochs; however, it is discovered that after the fourth epoch, the models tend to overfit. As a result, epoch = 4 was used to train our models.

- Learning Rate - For text classification tasks, a learning rate of $1e^{-5}$ or $2e^{-5}$ was chosen to avoid model overshooting from local minima.
- Dropout - After studying the literature on text classification, the dropout value was determined. The models were trained with a dropout of 0.1 at the start. The models performed well in the training data (dropout value increased from 0.1 to 0.3). When the model with 0.1 dropout and 4 epochs was implemented, it was discovered that it generalised better. As a result, 0.1 was chosen as the dropout value for all models.
- Train Batch Size - The desirable batch size was determined since a smaller value resulted in a lengthier time for training and a higher value resulted in non-convergence to global minima and memory issues. Hence, with 4 epochs, a train batch size of 32 was chosen to obtain a proper trade-off between model training duration and memory performance.

Implementation

The PyTorch Transformer, built on top of the HuggingFace (HuggingFace Transformers, 2022) and SimpleTransformer (Classification Models - Simple Transformers, 2022) libraries, provided an easy-to-use framework for pre-trained language modelling for a variety of NLP workloads. The Transformers provide APIs for easily downloading and using pre-trained models on a given text, as well as fine-tuning them on our own datasets, whereas the SimpleTransformers' ClassificationModel class was used for binary class text categorization. TensorFlow 2.0+ and PyTorch 1.1.0+ were both required for the implementation. The model's name and type were supplied as input parameters for each model trained. The implementation details of BERT, XLNet, DeBERTa, and StructBERT with tweaked hyperparameters were detailed in the subsections below.

Model Training

The pre-processed data saved in the CSV file was loaded in Google Colab. The different models were trained for three samples of data. The base models were trained with default arguments while the finetune models were trained with tuned hyperparameters. The implementation steps remain the same for all four models. The Python logging module was used to log model execution details, to help us to track any exception raised during training. The two different ways of implementation tried were HuggingFace PyTorch Transformer and Simple transformer. All the steps mentioned below were implemented for each dataset sample, i.e., 5%, 10%, and 50%. The 51 final trained model is saved, and the model was evaluated on the validation data loader while labels are predicted on an unseen test dataset.

PyTorch Transformer

The model dependencies were installed, and the PyTorch and Tensor Flow versions were validated for smooth execution. This method was used to note the performance of the pre-trained models with passed arguments. For the StructBert model the Alice Mind code repository was cloned to download weights, configuration, and pre-trained model checkpoint to be used for the binary classification tasks. All the steps outlined below remains the same. The steps followed included: I. Tokenization: At the beginning and end of each sentence, the special tokens [CLS] and [SEP] were added. One represents the class of input and the other was used to separate the two sentences in the same input. The output label of classification was decided by the last hidden state of the [CLS] token. The sentences were then tokenized with the tokenize function, and each tokenized text was converted to token ids using pre-trained tokenizers as mentioned in These tokens act as vectors for the embedding layer. II. Padding: The models expect all the input to be of the same length. To ensure the same, the maximum sequence length was initialized to 128 and the input ids were padded accordingly with the pad_sequences function. If a sentence was long, it was truncated, and if it was short, it was padded with 0. III. Attention mask: An attention mask was created with a mask of 1 s for each token, followed by 0s for padding. These masks acted as indices to determine which token the model needs to attend. IV. Data Loader: The input ids and the attention mask were split into training and validation sets, respectively. The respective training and validation inputs/masks were converted into torch tensors, the required data type for our model. The batch size was initialized to 32, and an iterator was created of our data with the torch DataLoader. This helped save memory during training because, unlike a for loop, with an iterator, the entire dataset did not need to be loaded into memory. V. Optimization and Training: The pre-trained model with a single linear classification layer was loaded and num_labels = 2 was passed as an argument. For optimization, a weight_decay_rate of 0.01 was passed along with a learning rate of 2e−5 to the AdamW optimizer with an epsilon of 1e−8. Binary Cross entropy with a dropout of 0.3 is the default selected loss function. Model arguments contained all the hyperparameter information for our training loop. The weight decay rate was applicable to all layers except all bias and layer norm layers in the AdamW optimizer. Finally, the model was trained for 2 or 4 epochs and the metrics were recorded.

Simple Transformer

The SimpleTransformers library has two task-specific classification models. In this study, the ClassificationModel class was used to train the base and fine-tuned model by modifying the default set of arguments. The class provided extensive options for configuration to update the default model argument as per the use case. The configuration was passed as a Python dictionary to the ModelArgs class. In all Simple Transformer models, CUDA was enabled by default but can be disabled by passing a Boolean to the use_cuda parameter. The steps performed are detailed below. I. Initialization: The ClassificationModel class was created by passing the model_type and model_name. The model type must be one of the supported models, and the model's name assists in obtaining the precise architecture and trained weights for use. The model code was used to specify the model type.-

- To utilize the available GPU, use_cuda = True
- To override the default arguments with the arguments passed as-
- num_train_epochs = 4, the number of epochs the model will run for
- train_batch_size = 32, the training batch size · max_seq_length = 128, the maximum sequence length for the model to support
- learning_rate = 1e−5 or 2e−5, the learning rate for training

- output_dir = output/, the directory to save model checkpoints and results
- overwrite_output_dir = True. If True, it will overwrite the output folder to store the latest model result. Due to multiple trials and to save memory space it is assigned a Boolean true in this study.
- For the rest, default arguments were used, and the parameters were not provided.

II. Split: The dataset sample is split in the ratio of 80:20 into training and evaluation data. III. Training: The initialised model is used to call the train_model () function, which instructs the model to train on the training dataset.

Evaluation of Model Performance

The classifiers were built using the training dataset and assessed using the performance evaluation matrix. The models were trained on three fractions of the dataset sample. This is a two-step procedure. In the first step, the best value in the training states across all models were picked based on performance measures. In the second step, the performance of the models was compared to find the best model. For both the class labels, the confusion matrix and classification report were utilized to extract various measures such as accuracy, true positive, false positive, true negative, false negative, sensitivity or recall, precision, f1-score, Area Under Curve (AUC), Area Under Precision recall Curve (AUPRC), and Mathews Correlation Coefficient (MCC). The class labels "insincere" (1) and "sincere" (0) are classed as "Yes" and "No," respectively.

3 Results and Discussion

3.1 Model Evaluation

Following model training, each model was evaluated against a validation set. As the data was highly imbalanced, accuracy would not be the correct metric to understand the model's performance. Hence, F1-score was used for accuracy. The macro averaged F1-score was noted as it provided the average arithmetic mean of each class. The MCC was also a reliable statistical rate to understand the obtained prediction. It returned a high score if the prediction was good in all four quadrants of the confusion matrix. The results in all matrix categories were proportional to the size of the dataset belonging to each class. Balanced Accuracy was calculated as it gave the arithmetic average of sensitivity and specificity.

3.2 Experiment 1-Transfer Learning with Original Hyper-Param Setting

The first set of experiments were performed with default settings, without modifying the values of any hyperparameter. Initially, the models were trained and evaluated on 5% of the data, and the percentage was increased in subsequent iterations. The dataset contained 52,177 training and 13,045 validation records. Keeping the imbalance ratio consistent, the test set contains 12,240 sincere and 805 insincere records. Each model was trained and validated, resulting in four executions in this experiment. Table 1 shows the standard value of the hyperparameter used in this specific execution. Table 2 depicts the performance metrics recorded from the validation set.

Table 1. Exp 1: Hyper-parameters – values

Hyper-parameter	Value
Adam-epsilon	$1e^{-8}$
Eval-batch-size	8
Learning-rate	$4e-5$
Max-seq-length	256
N-gpu	1
Num-train-epochs	1
Train-batch-size	8
Optimizer	AdamW

Table 2. Model results- default settings

SL. No	Model	Acc	Pr	Recall	F1-score	MCC	AU-ROC	AU-RPC	Train Time	Eval Loss
1	BERT	0.95	0.75	0.38	0.5	0.51	0.94	0.6	40	0.15
2	XLNET	0.95	0.62	0.52	0.56	0.61	0.95	0.68	40	0.16
3	DeBERTa	0.95	0.58	0.59	0.58	0.54	0.85	0.6	43	0.17
4	StrucctBERT	0.9	0.59	0.36	0.48	0.46	0.8	0.4	120	0.2

3.3 Experiment 2 - Transfer Learning with Modified Learning Rate (LR)

After performing the first set of experiments with the default model argument settings, the second set of experiments was done by changing the learning rate. The training batch size was kept at 32, the eval batch size at 8, the maximum sequence length was 128 and the number of training epochs was 4. The number of GPUs assigned was 1. With a 5% data sample, the distribution of data across the train and validation set remained the same as in experiment 1. In each iteration, the value of the learning rate was changed, and metrics were noted. Each model was tried with three different values, resulting in a total of 12 executions in exp-2. Table 3 shows the standard value of the hyperparameter used in this specific execution. Table 4 depicts the performance metrics recorded from the validation set for different learning rates used. Table 4 shows that BERT and DeBERTa produced better results with an LR of $2e^{-5}$. And XLNET and StructBERT produced it with an LR of $1e^{-5}$. Thus, the best-performing learning rate was chosen for the final set of evaluations with higher samples of data.

Table 3. Exp. 2: Hyper-parameters – values

Hyper-parameter	Value
Adam-epsilon	$1e^{-8}$
Eval-batch-size	8
Learning-rate	$4e^{-5}, 2e^{-5}, 1e^{-5}$
Max-seq-length	128
Num-train-epochs	4
Train-batch-size	32
Optimizer	AdamW

Table 4. LR vs model results

Model name	LR	Performance metrics								
		BAcc	Acc	Pr	Recall	F1-score	AU-PRC	AU-ROC	MCC	Eval loss
BERT	$4e^{-5}$	0.78	0.96	0.68	0.59	0.63	0.66	0.96	0.61	0.13
	$2e^{-5}$	0.80	0.96	0.70	0.62	0.66	0.67	0.96	0.67	0.12
	$1e^{-5}$	0.78	0.96	0.71	0.57	0.64	0.66	0.95	0.62	0.11
XLNET	$4e^{-5}$	0.75	0.95	0.63	0.52	0.57	0.95	0.95	0.55	0.15
	$2e^{-5}$	0.78	0.96	0.67	0.59	0.63	0.96	0.96	0.61	0.12
	$1e^{-5}$	0.79	0.96	0.69	0.61	0.65	0.95	0.95	0.63	0.11
DeBERTa	$4e^{-5}$	0.78	0.95	0.70	0.58	0.66	0.62	0.96	0.62	0.11
	$2e^{-5}$	0.80	0.96	0.71	0.62	0.74	0.65	0.97	0.65	0.12
	$1e^{-5}$	0.79	0.96	0.72	0.61	0.68	0.64	0.96	0.64	0.10
StructBERT	$4e^{-5}$	0.69	0.94	0.40	0.40	0.48	0.55	0.90	0.55	0.21
	$2e^{-5}$	0.69	0.94	0.40	0.40	0.48	0.55	0.90	0.55	0.21
	$1e^{-5}$	0.7	0.95	0.41	0.41	0.51	0.60	0.95	0.60	0.18

3.4   Experiment 3 - Transfer Learning with Modified Epochs

After performing the second sets of experiments with the different learning rates, the third set of experiments was done by changing the training epochs. The training batch size was kept at 32, the eval batch size at 8, and the maximum sequence length was 128. The number of GPUs assigned was 1. With a 5% data sample, the distribution of data across the train and validation set remained the same as in experiment 1. In each iteration, the value of the epoch was changed, and metrics were noted. Each model was tried with four different values, resulting in a total of 16 executions in experiment 3.

3.5   Experiment 4 - Transfer Learning with Modified Data Samples

After performing the third set of experiments with different epochs, the fourth and final set of experiments were done by changing the size of the dataset (Table 5). The training batch size was kept at 32, the eval batch size at 8, and the maximum sequence length was 128. The number of GPUs assigned was 1. In each iteration, the finetuned model performance on 5%, 10%, and 50% of the data with the best-chosen learning rate was noted. Each model was tried with three samples, resulting in a total of 12 executions in experiment 4. It was evident that there was only a slight to negligible increase in the model performance with the increase in the size of the data. It was observed that the model produced a macro f1 in the range of 0.74–0.82 with just 5% data. The model tends to generalize the pattern with 10% of the data within half the computational time of the entire dataset. Hence, results obtained from 10% of the data were considered for further analysis and discussion. In the following sub-section, the confusion matrix generated from each model with 5% and 10% data was explained.

Table 5. Epoch versus model results

Model name	Accuracy	Epoch			
		2	3	4	7
BERT	Training	95.41%	95.45%	96.77%	94.21%
	Validation	94.88%	94.92%	95.56%	78.13%
XLNET	Training	95.35%	95.39%	96.20%	94.02%
	Validation	94.91%	94.95%	95.49%	77.36%
DeBERTa	Training	95.73%	95.77%	97.02%	95.01%
	Validation	95.28%	95.32%	95.75%	79.38%
StructBERT	Training	94.70%	94.74%	94.82%	94.12%
	Validation	94.26%	94.30%	95.21%	73.56%

Confusion Matrix - BERT

Taking sincere as positive class and insincere as a negative class the confusion matrix obtained with a learning rate $2e^{-5}$ and data of 5% and 10%. With 5% data, the true positive of 12,065 and the true negative of 456 depicted that the model could correctly predict most of the question. 340 records were predicted as insincere while they were observed to be sincere. Whereas 184 records were observed to be insincere but predicted as sincere. With 10% data, the true positive of 24,089 and the true negative of 1,093 depicted that the model could correctly predict most of the question. 409 records were predicted as insincere while they were observed to be sincere. Whereas 498 records were observed to be insincere but predicted as sincere.

Confusion Matrix - XLNET

With 5% data, the true positive of 12,030 and the true negative of 484 depicted that the model could correctly predict most of the question. 219 records were 64 predicted as insincere while they were observed to be sincere. Whereas 312 records were observed to be insincere but predicted as sincere. With 10% data, the true positive of 24,034 and the true negative of 1,070 depicted that the model could correctly predict most of the questions. 440 records were predicted as insincere while they were observed to be sincere. Whereas 545 records were observed to be insincere but predicted as sincere.

Confusion Matrix - DeBERTa

With 5% data, the true positive of 12,058 and the true negative of 483 illustrate that the model could correctly predict most of the questions. 191 records were predicted as insincere while they were observed to be sincere. Whereas 313 records were observed to be insincere but predicted to be sincere. With 10% data, the true positive of 24,135 and the true negative of 1,098 illustrate that the model could correctly predict most of the questions. 363 records were predicted as insincere while they were observed to be sincere. Whereas 49,365 records were observed to be insincere but predicted to be sincere.

Confusion Matrix - StructBERT

With 5% data, the true positive of 12,107 and the true negative of 323 illustrated that the model could correctly predict most of the questions. 149 records were predicted as insincere while they were observed to be sincere. Whereas 466 records were observed to be insincere but predicted to be sincere. With 10% data, the true positive of 24,213 and the true negative of 645 illustrated that the model could correctly predict most of the questions. 298 records were predicted as insincere while they were observed to be sincere. Whereas 973 records were observed to be insincere but predicted to be sincere.

3.6   Comparison of MODel's Performance

Model validation resulted in the generation of models with identical values of evaluation metrics to those of trained results. It was observed that the models did not record any significant differences between the two. This demonstrated that almost all the tested models performed well when tested in the eval dataset and the predicted-on test set. The total number of data points supported for evaluation and prediction is 26,089 and 15,000, respectively. When looking at the findings of all four models, it was clear that they could achieve state-of-the-art outcomes with only 5% of the data. The high rate of false positives was due to the imbalance in the dataset. When the dataset size was increased, the model's performance improved slightly. With 50 percentage data, there was just a minor rise in the F1 score and the macro average F1 score. Some of the assessment metrics discovered in question classification were compared between the classifiers to select the best or most resilient model among the proposed models. DeBERTa produced a macro f1-score of 0.83 with a 10% data sample, followed by BERT with a core of 0.82.

The MCC value of less than 0.5 clearly showed that the base models acted as random classifiers. The MCC shifting towards 0.7 in finetuned models showed that the model could understand the question text and generalize the class labels. It was observed that most fine-tuned models record similar values across all metrics. Based on the table, the balanced accuracy level was highest for DeBERTa and BERT with 80%. The StructBERT yielded the lowest balanced accuracy of all, with 70%. Even after multiple iterations with different learning rates, an increase in the F1 score was not observed. Due to the StructBERT model sizebeing equivalent to 1.25 GB, execution needed more space. Due to the limitation of computational size, the model was not validated for higher epochs. For insincere labels, BERT and DeBERTa had the lowest false positive rate. The two models predicted the greatest precision for both class labels and macro and weighted averages. The average sensitivity for DeBERTa was given as 0.81 and for BERT as 0.80. The model DeBERTa resulted in an MCC of 0.66 and an AUPRC of 0.74 compared to BERT with 0.66 and 0.71, respectively. Thus, comparing all the performance measures, DeBERTa was chosen as the champion model over the rest.

4 Conclusion

The various types of questions posted on online social media platforms were not always intended to extend support or guidance. There are times when people use such platforms to promote hate or spread rumours to create restlessness among groups of people. This degrades the platform's quality, which in turn leads to a decline in its usage. Thus, the classification of destructive content before mass reading was equally essential and challenging. In the past, research was done using machine and deep learning models to restrain such content from posting. However, the models were highly dependent on feature engineering and the ratio of class labels in the dataset.

Starting with the most basic BERT model and progressing to the most esoteric three Bert-based transformer models, XLNET, DeBERTa, and StructBERT, the study showed how these models compete with one other. The basic model was BERTbase, which was one of the most stable models built on a big corpus of data. To classify the cleaned text and determine the outcome, all four models were applied. The entire implementation was done in Python programming language and the models were built in the Colab and Kaggle consoles using GPU and TPU for faster computation. The dataset is taken from Kaggle and processed to form cleaned text. The lexicon and tokenizer for each of these models were used to train the model with the default argument on the cleaned dataset. Both PyTorch and Simple Transformer were used to demonstrate the implementation, as well as how to alter the standard parameter settings.

Multiple iterations were run, and the hyperparameters were chosen after a comprehensive examination. The preferred learning rate was determined by the metrics, and fine-tuned models with changed hyperparameters were trained on three separate dataset samples of 5%, 10%, and 50% of the total dataset. The class label imbalance ratio was guaranteed to be consistent across dataset samples. Model development, evaluation, and prediction were done in training, validation, and test sets, respectively, to maintain quality. As label imbalance was so crucial, the major assessment metric was a mix of f1 score, MCC, and balanced accuracy. To determine the best classifier, the outcomes of pre-trained and fine-tuned models were compared. DeBERTa and BERT surpassed all other models trained in this study. The results also outscored logistic regression or other deep learning models noted in the previous studies.

The objective of generating state-of-the-art results using noisy dataset without using any class balancing approaches and with low computational capacity was met since all the models stated delivered outstanding results with only 10% of the data. When comparing DeBERTa with BERT, the former exceeded with a slightly higher F1-score, indicating that the study's research aim is justified. It also demonstrated that combining transfer learning with attention mechanisms could improve existing performance.

References

1. Hosseinmardi, H., Mattson, S. A., Ibn Rafiq, R., Han, R., Lv, Q., Mishra, S.. Analyzing labeled cyberbullying incidents on the instagram social network. In: Liu, TY., Scollon, C., Zhu, W. (eds.) Social Informatics. SocInfo 2015. Lecture Notes in Computer Science, vol. 9471, pp. 49-66. Springer, Cham (2015). https://doi.org/10.1007/978-3-319-27433-1_4
2. Maslej-Krešňáková, V., Sarnovský, M., Butka, P., Machová, K.: Comparison of deep learning models and various text pre-processing techniques for the toxic comments classification. Appl. Sci. **10**(23), 8631 (2020)
3. Del Vicario, M., et al.: The spreading of misinformation online. Proc. Natl. Acad. Sci. **113**(3), 554–559 (2016)
4. Morzhov, S.: Avoiding unintended bias in toxicity classification with neural networks. In: 2020 26th Conference of Open Innovations Association (FRUCT), pp. 314–320. IEEE (2020)
5. Quora Insincere Questions Classification | Kaggle, https://www.kaggle.com/c/quora-insincere-questions-classification/data. Accessed 02 Nov 2021
6. Kumar, A., Makhija, P., Gupta, A. Noisy Text Data: Achilles' Heel of BERT. *arXiv preprint* arXiv:2003.12932 (2020)
7. Wirth, R., Hipp, J.: CRISP-DM: Towards a standard process model for data mining. In Proceedings of the 4th international conference on the practical applications of knowledge discovery and data mining, vol. 1, pp. 29–39 (2000)
8. Aslam, I., et al.: Classification of Insincere Questions Using Deep Learning: Quora Dataset Case Study. Springer International Publishing, Cham (2021)
9. Al-Ramahi, M.A. Alsmadi, I.: Using data analytics to filter insincere posts from online social networks. a case study: Quora Insincere Questions (2020)
10. Rachha, A. Vanmane, G.: Detecting insincere questions from text: a transfer learning approach (2020)

Suicide Ideation Detection: A Comparative Study of Sequential and Transformer Hybrid Algorithms

Aniket Verma[1], Matthew Harper[1(✉)], Sulaf Assi[2], Abdullah Al-Hamid[3], Maitham G. Yousif[4], Jamila Mustafina[5], Noor Azma Ismail[6], and Dhiya Al-Jumeily OBE[1]

[1] Faculty of Engineering, Liverpool John Moores University, Liverpool L3 3AF, UK
M.L.Harper@2014.ljmu.ac.uk

[2] School of Pharmacy and Bimolecular Science, Liverpool John Moores University, Liverpool, UK

[3] Saudi Ministry of Health, Najran, Saudi Arabia

[4] College of Science, University of Al-Qadisiyah, Al Diwaniyah, Iraq

[5] Kazan Federal University, Kazan, Russia

[6] Faculty of Business and Technology, UNITAR International University, Petaling Jaya, Malaysia

Abstract. Suicide is turning out to be one of the most dangerous health hazards in today's fast paced world and is one of the leading causes of deaths among general population. Unfortunately, it also happens to be one of the most ignored factors when we compare it against other causes of fatality like road accidents, terminal illness, crimes etc. It is well and truly turning out to become a silent pandemic. Suicide ideation is commonly referred to someone having suicidal tendencies which may include, thoughts, planning, enactment, failed attempts etc. Social media platforms such as Reddit allow a relatively safe and secure space to express any sufferings without the anxiety of peer-to-peer communication or judgement and in many cases, anonymity. The study attempts to apply deep feature extraction based learning techniques on cherry-picked Kaggle dataset from r/SuicideWatch which includes reddit posts by users that contain suicide ideation which is combined with reddit posts from other domains. These modelling techniques look out for sentimental phrases, vocabulary patterns in suicidal posts, grammatical similarities and preferences of such posts like use of parts of speech and references to various entities. The end goal is to propose a model which can build upon the knowledge of existing social media content and facilitate early detection of suicide ideation in similar content in future. The study involves a comparative analysis of the most sequential and transformer-based algorithms to achieve near optimal results. The primary focus is on developing models which can correctly classify suicide ideation texts thereby minimizing false negatives to prevent loss of lives as a result of suicide.

Keywords: Suicide prevention · Suicide ideation · Reddit · Text classification · Deep feature extraction · Machine learning

© The Author(s), under exclusive license to Springer Nature Singapore Pte Ltd. 2023
Y. B. Wah et al. (Eds.): DaSET 2022, LNDECT 165, pp. 373–387, 2023.
https://doi.org/10.1007/978-981-99-0741-0_27

1 Introduction

According to the World Health Organization (WHO), 700,000 people commit suicide every year and this number is likely to see an uptick [1]. Additionally, suicide is the 4th leading cause of deaths among 15–19 year olds. According to CDC Facts About Suicide, in the United States alone, one person died of suicide every 11 min [2]. But the number of people who attempt or contemplate suicide is much higher. In 2019 alone, about 12 million adult population in the US, had thoughts about suicide, 3.5 million had planned a suicide attempt, and nearly 1.4 million went on to attempt suicide. According to NRCB, India, the second most populous country in the world saw an unfortunate demise of about 1,39,000 people out of which almost 67% were young adults between the age of 18–45 [3]. Amongst all the registered death events in 2017, suicide was the culprit behind 1.4% of the total number [4]. Some of the world's most developed and developing nations have witness higher suicide rates in worlds with countries like South Korea, United States, India, Russia, Canada, and Australia leading the charts.

Suicide ideation or suicidal thoughts include feeling of worthlessness, passing thoughts of killing oneself, role-playing, casual planning, extensive planning, unsuccessful attempts, self-harm etc. [5]. Any thought process or action that may be a causative factor behind contemplation or the actual act of suicide can be classified as suicide ideation. Suicidal thoughts usual emerges out of prolonged mental illness which may be attributed to factors like anxiety, isolation, critical illness, demise or separation from loved ones, financial issues, inferiority complex, hopelessness, trauma, peer pressure and many more. In some rare cases, it may have genetic root causes [6]. Studies have shown that suicide ideation may have a strong correlation with gender and geographies [4]. Though researchers are still trying to deep dive into the correlations, a generic opinion persists across that gender or ethnicity differences means that individuals are subjected to different challenges and environments which affects their neurotic structures within the brain thereby causing the excess or deficit of serotonin.

Although suicide ideation can be detected via many tools including pathological and psychiatric [7], keeping into consideration the challenges pertaining to cost effectiveness and ease of information mining, this study delves into social media as a repository for suicide ideation. Various researchers have established that the social media activities give a clear indication of the mental well-being of an individual [8–10]. With the lack of general companionship and expert guidance, individuals have been supremely vocal about their opinions, day to day activities, life events on public forums like social networking sites. Hence, data pertaining to suicide ideation can be easily mined from these platforms which is also aided by public APIs [11]. Using mathematical models and machine intelligence, suicide ideation detection can become more affordable by leveraging the power of cloud web applications, easily accessible on desktop or mobile based devices and may serve as a great tool for self-diagnosis [12].

Researchers have understood this opportunity have proposed various machine learning and deep learning mathematical models into order to tackle these challenges. These models have provided cutting edge performance and have highlighted various tools of interest that can be used to address suicide ideation detection [8–10]. This study builds upon the existing work while addressing some research gaps. While most of the researchers have attempted to use a either a standalone or a combination of machine learning and deep learning models, this study is a three-way comparative analysis between machine learning, sequential and deep learning algorithms. Moreover, a comparative analysis is done between standalone algorithms and its CNN hybrid to establish the efficacy of convolution layers in enhancement of model performance.

2 Background

Burnap et al. (2017) describe bulletin boards, newsgroups, chat rooms, and web forums as good sources of content extraction for suicide ideation [13]. A lot of investigation has been done with respect to social media posts as the prime repository for suicide ideation content. This is primarily attributed to the fact that social media has a wider reach and gives a "behind the screen" platform to users to express themselves. To ease things further content aggregation has been facilitated using various APIs like the PushShift API [14]. Web scraping has been another breakthrough in social media content aggregation as discussed by [15]. Data sources pertaining to suicide ideation can be broadly classified into the following categories. Online health forums (OHC) include discussion forums, peer to peer communication networks or online consultations. These forums are usually exclusively for health-related discussions and the data aggregated need lesser pre-processing time [16–18]. Websites and apps that allow users to create and share content, as well as participate in social networking. Many of these social networks have public APIs or aggregated datasets which pertain to certain topics like suicide ideation. Some of the most popular GSN are Twitter, Weibo, Facebook, Instagram, Tumblr and Reddit. A large number of publications use GSN as a single source of truth for datasets [20–24]. Suicide notes may be in the form of website blogs or handwritten notes which may be manually digitized or can be done using an OCR tool. Sentiment analysis [25, 26] and emotional detection [27] can be detected easily using this medium as suicide notes usually elaborate a person's mental state comprehensively before their unfortunate demise.

After aggregation of social media content, various data pre-processing techniques are utilized to convert textual content to numeric format which is understood by machine learning and deep learning algorithms. Word2vec by [28] is a very popular technique to implement word embeddings. Word embeddings have a capability of capturing semantic and syntactic relationships between words and also the context of words in a document. In the context of suicide ideation these techniques are used to generate vectors from social media posts. Bag of words (BOW) and TF-IDF are other popular vectorization techniques with limitations of semantic meaning capturing. LIWC proposed by [29] emphasized on capturing metadata of the document including frequencies of various figures of speech of spoken English. It provides about 90 features meant to capture textual patterns. Word cloud, a popular visualization technique captures most frequently used words in the dataset. Words like suicide, live and die are common occurrences in such content which are captured by BOW.. A more sophisticated technique includes the use of n-grams (tri-grams) to capture patterns like "want to die", "can't bear anymore" etc. Topic modelling techniques like Latent Semantic Analysis (LSA), Latent Dirichlet Allocation (LDA) proposed in [30] allow extracting important topics from a verbose text. Other techniques such as PCA [13] and NMF [24] are used less frequently.

Data aggregated from social media or health platforms need to be labelled for supervised classification. Most of the researchers use human annotators or other manual process for labelling data into target classes for classification. Human annotators are usually computer scientists or domain experts who can easily understand and quantify the risk associated to every datapoint. [13, 16, 19, 21–23, 31, 32] Unsupervised techniques like clustering may not use annotation techniques. In some cases, automation algorithms are used to annotate datapoints appropriately and serve as a great tool to reduce manual efforts.

Kiros Bitew et al. (n.d.) used an ensemble of Linear SVM along with logistic regression and emotional classifiers [33]. TF-IDF representation using n-grams, emotional features, suicide risk features were extracted using the DeepMoji model [34]. In similar works, Reddit dataset pertaining to suicide survivors was aggregated using the Python Reddit API wrapper. Since single lemma independently isn't usually enough to annotate a sentence, trigrams are used to determine suicidal patterns like "have to kill", "want to die". Standard and customized POS tagging and tokenization techniques are used to determine the distribution of 19 most frequently used words like "disease", "death" etc. which helps in generating correlations before comparing machine learning models like linear SVC, logistic regression etc. [16].

Topic modelling techniques like Latent Semantic Analysis (LSA), Latent Dirichlet Allocation (LDA), and Non-Negative Matrix Factorization (NMF) serve as a great tool to derive important keywords from a corpus of data. Researchers have attempted to use unsupervised learning techniques such as K-means clustering algorithms on a labelled dataset meant for supervised learning. [24] compared supervised learning (decision trees) with clustering 20 algorithm on a binary dataset with high-risk and at-risk categories. Other unsupervised clustering techniques include partition around medoids and hierarchal clustering. The decision on the cluster formation is usually taken on the basis of parameters like cosine similarity, Word Mover's Distance (WMD) and Latent Semantic Indexing (LSI) [21].

Kalchbrenner et al. (2014) introduces a convolutional architecture called the Dynamic Convolutional Neural Network (DCNN) that is used to model phrases semantically [35]. The network can handle a variety of phrase lengths and creates a feature graph over the sentence that can explicitly capture short and long-range relationships. The network is not based on a parse tree and may be used with any language. Yin et al. propose MVCNN, a sentence classification architecture based on convolutional neural networks (CNNs) [36]. It uses variable-size convolution filters to extract features from multi-granular phrases and mixes different versions of pretrained word embeddings. MVCNN outperforms the competition in four areas: smallscale binary, small-scale multi-class, large-scale Twitter sentiment prediction, and subjective classification [37, 38] established that when used as vectors to represent a complete sequence, single recurrent and convolutional neural networks architectures are insufficient to capture all of the critical information. Hence various researchers have tried a hybrid approach involving a combination of RNN based sequential algorithms along with CNN for more robust outcomes. He and Lin introduced a neural network architecture which is a hybrid of BiLSTM, used for context modelling of input and CNN, a 19 layer deep neural network for similarity in pattern recognition [39]. The model uses a similarity focus mechanism with selective attention to key pairwise word interactions for STS problem.

As it can be seen, many studies have tried using various techniques to achieve near optimal results. This study builds upon the existing state-of-the-art research and experimental models to explore some untapped areas. With the recent research with respect to CNN architectures for text classification such as [40] this study works on the hypothesis that hybrid sequential and transformer algorithms could provide more robust results as compared to standalone architectures. Though various researchers have attempted the use of attention-based models, not much exploration has been carried out in the usage of such models which have been pre-trained to provide flagship performance in text classification tasks. Some of these models include BERT, XLNet. In addition, in most use cases the baseline algorithm for comparison has been a vanilla machine learning model which may provide sub-optimal experience. Hence, 29 to overcome this, two baseline algorithms, one being a standalone machine learning model and the other being a sequential RNN based model are used. Therefore, in summarization, this study attempts to compare two transformer algorithms against a vanilla machine learning baseline model and a sequential RNN based baseline model along with its CNN hybrid.

3 Methodology

The methodological approach adopted in this study is shown in Fig. 1.

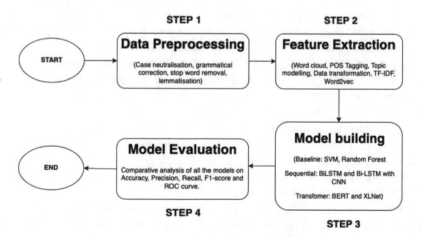

Fig. 1. Process workflow adopted in this study.

3.1 Dataset

The data is pre-aggregated from r/SuicideWatch sub-reddit1 and is loaded into the Kaggle dataset2 repository. SuicideWatch sub-reddit is a peer-to-peer online support group for individuals struggling with suicidal thoughts. As of 23rd day of February 2022, this online support group has 334K members since its inception back in December 2008. Hence, being an open public forum, the SuicideWatch sub-reddit forms an excellent data repository for suicide ideation content and this study works upon a cherrypicked subset of the available data appropriately tagged to its target class which is obtained from Kaggle. It has only two features: text and class. The dataset has 232074 datapoints with the textual column having no repetitions or missing values. The target (class) column has just 2 unique values viz.., "suicide" and "non-suicide". There are no missing values and the class distribution is fairly equal thereby rendering the use of any class balancing technique as unnecessary.

3.2 Data Preprocessing

To make the dataset ready for feature extraction, some processing techniques are applied on the dataset. For execution of these techniques, the The Natural Language Toolkit (or NLTK) library is used. The first step is case neutralization, which includes converting the original text to lowercase. This ensure proper formation of feature vectors and prevents unnecessary duplication as there are limited number of alphanumeric entities. For ex, keywords, "suicide" and "Suicide" though representing same words may map to different vectors. Moreover, without neutralization, the same two words would map to different tokens during the tokenization process thereby increasing the vocabulary size. Hence to ensure consistency, case neutralization is extremely important.

Next, A manual observation of the dataset reveals several grammatical mistakes some of which being missing whitespaces between consecutive sentences. This leads to misrecognition of tokens and the model may interpret the conjoint words as a standalone token. This may lead to loss of vital information in the topic modelling phase and the words within the conjoint tokens may individually hold paramount importance but together they may form a non-existent word which may be ignored by the topic modeler. For ex, "Recently, I've been extremely suicidalMy family has been non-supportive. In this statement, suicidal is an important keyword but its concatenation with the pronoun "my" will lead to its non-consideration.

Contractions such as "I'm", "couldn't", "wouldn't" etc. do not add much information to model building and end up complicating the vectorization process by increasing the number of unique tokens. Moreover, during punctuation removal and lemmatization, these words can prove disastrous in conveying the underlying meaning of the sentence. The contraction "can't" on punctuation removal would bifurcate into "can" and "t" which changes the sentiment of the statement from negative to positive thereby causing erroneous results.

Punctuation marks do not add any value (additional information) to the model building phase and hence can be safely eliminated from the document corpus. In addition, all the tokens which are non-alphabetical are removed for similar reasons. This is a simple step which results in efficient use of available bandwidth in time and resources.

Stop words are very commonly used words in the English vocabulary that convey minimal to no information to the classification model and are better off removed. The low-level information from text is removed by deleting these terms, facilitating more focus on the crucial information. In other words, we can say that removing such phrases has no negative impact on the model which is being trained for this task. Most common stop words include "the", "is", "in", "for", "where", "when", "to", "at" etc. These model add minimal contribution to information presented to the model and also use computational power during word embeddings.

Lemmatization techniques involve the reduction of a derivationally related form of a word to its base form. The word lemmatization is derived from the word "lemma" which means base or dictionary form of the word. The base word is a part of the vocabulary and is standalone. For example, feel, feels, and feeling have their base word as feel. This maintains a consistency 35 throughout the dataset while preserving the original meaning of the sentence. Like stop word removal, this process also ensures more optimal use of resources.

3.3 Feature Extraction

Feature extraction techniques involve the extraction of relevant and important information from plain text corpus. Word clouds (also known as text clouds or tag clouds) facilitate extraction of most important key words from any textual data. The recurrence of a word within a text corpus which maps to suicide ideation can hint to the fact that for any suicide utterance, there is a possibility of the word to occur within the text. For ex, suicide ideation texts may frequently contain words like, "kill", "die", "self" or "harm". This is primarily a visualization step and is extremely useful in manual annotations of untagged data or human classification and can assist in verifying results generated by the underlying model.

Parts of Speech (POS) tagging involves identification of each word in the document and labelling them against an appropriate part of speech. These parts of speech include noun, pronoun, adverb, adjective, preposition etc. For each post, we generate appropriate tags and get an aggregated count of each category. The hypothesis is as follows.

- Suicide ideation posts have a lot of self-referencing like "I am sad", "my life" etc.
- Suicide ideation posts usually refer a lot to their counterparts, loved ones, family, friends etc. like "my friend", "my boss", "a family member".
- Suicide ideation posts refer to certain actions multiple times like "die", "kill", "harm", "feel" etc.

Topic modelling involves extracting abstract topics in a document. They are tagging based unsupervised, dimensionality reduction technique to identify key discussion points from a large chunk of data. In the context of suicide ideation, topics like, "health", "sadness", "loneliness" can be easily extracted which gives a new perspective to the classification model. This study uses Latent Dirich-let Allocation (LDA) proposed by [41] for extracting topics from the posts. In theory, LDA is a probabilistic model which works on the assumption that each topic is a combination of underlying set of words and each document is a combination of underlying set of probabilities of topics. It happens to be one of the most used methods for topic modelling. It is a generative probabilistic model for discrete data collections like text corpora and is a three-level hierarchical Bayesian model in which each collection item is represented as a finite mixture over an underlying set of topics. Each topic is thus represented as an infinite mixture over a collection of topic probabilities. The topic probabilities give an explicit representation of a document in the context of text modelling. LDA works on two assumptions, a document is a collection of topics, and a topic is a collection of words or tokens.

Word embeddings train fixed-length dense vectors and continuous-valued vectors based on a huge text corpus using an algorithm. Each word is a point in vector space that is learned and manipulated around the target word while maintaining semantic links. Words with comparable meanings are clustered in the vector space representation of words.

TF-IDF is used to associate a quantifiable value to every word in a sentence. This value (referred to as score) measures the relevant importance for each word in the sentence or corpus. TF-IDF consists of 2 parts: Term Frequency and Inverse Document Frequency. On generation of TF-IDF vectors, each value within the vector increases in proportion to the number of times a word appears in text offset by the corpus's word frequency. Hence a higher value corresponding to a word means that the word, in reference to other words corresponding to lower values in the TF-IDF vector holds more importance.

Proposed by Pennington et al. GloVe is an unsupervised learning technique that are used to generate word embeddings by deriving relationship between words from statistics [42]. It uses vector differences to store the co-occurrence probability ratio between two words. GloVe attempts to figure out a model with its parameters to represent the syntactical meanings unlike other word embedding models which are primarily a blackbox. The underlying model used is a global log-bilinear regression model that combines the advantages of 2 modelling approaches that are Global Matrix Factorization and Local Context Window. In this study, GloVe is used to generate document vectors to be used as an input to sequential models.

3.4   Classification Models

This study attempts to use two baseline classification models, a vanilla machine learning ensemble model (SVM and Random Forest) and a sequential model (BiLSTM). Furthermore, a comparative analysis of the three models is done against the baseline models. The three models include one sequential model (BiLSTM) with a convolution layer and two transformer models (DistilBERT and XLNet). The primary object of the study is to build a model with cutting edge performance while simultaneously establishing 1) whether convolution networks which were primarily meant for image related tasks can help improving textual classification results and 2) using transformer-based algorithms known to achieve flagship performance in text classification. The evaluation metrics to be used to assess the efficacy of the models are accuracy, precision, recall, and F1 score.

4   Results and Discussion

4.1   Interpretation of Features

For dimensionality reduction and feature extraction, plaintext data is subjected to several processes. These include utilizing a POS tagger to tag all of the words in the text to the relevant part of speech, PCA to remove redundant features from the model, and LDA to extract important topics. The following are some of the most important observations.

Nouns and adjectives are the most occurrent entities within the text corpus. The occurrence of nouns outweighs out parts of speech by a long margin. This is evident from the fact that suicidal texts have a lot of referencing to a close family member, friend, colleague or any other individual. Moreover, other parts of speech such as prepositions are eliminated during the stop word removal phase and hence have minimal occurrence.

PCA is used to reduce a pre-processed dataset with 137 features down to 80 features that explain 92 percent of the variance. By expressing the dataset in a new dimension, PCA allows for the seamless separation of positive and negative classes,

Topic extraction technique (LDA) is used on the text corpus with the intention of extract important topics and their relative distribution across the dataset. Table 5–1 shows the topics and their percentage distributions in the dataset.

4.2   Model Performance

This study uses a total of 6 models divided into 3 segments, n Machine learning models: Support Vector Machines (SVM) and Random Forest (RF); Sequential models including Bidirectional LSTM (Bi-LSTM) and Bidirectional LSTM with CNN; and transformer models DistilBERT and XLNet. All these models are compared on 4 evaluation metrics: Accuracy, Precision, Recall and F1-score. The primary objective is to classify suicidal texts correctly with the possible expense of non-suicidal texts being wrongly categorized as suicidal thereby minimizing the type-II error. Some of the key observations of the implementation are as follows. Table 1 shows the relative performance of each model used in the study. Machine learning baseline models outperform expectations. SVM offers an accuracy of 89% and a sensitivity of 87.5% while Random Forest offers an accuracy of 88.7% and a sensitivity of 87.2% on the test data which is very much comparable to their neural network counterparts. This is attributed to the detailed pre-processing and feature extraction techniques used as a part of this study. Note that these results are obtained on standard hyper-parameters which when tuned can offer much better metrics thereby in most cases eliminating the use of complex architectures. The sequential models (Bi-LSTM) also offer great performance, however they do not offer a considerable improvement over the vanilla machine learning models. One thing to consider is that Bi-LSTM when used in conjunction with convolution layers offer better accuracy, precision and F1-score. Bi-LSTM offers the highest sensitivity of nearly 97% among all models. Hence, convolution layers do offer enhance performance which proves the underlying hypothesis of this study. Given the time and resources constraints, the transformer models are trained over 4 epochs, but still they outperform every other model discussed so far. DistilBERT offers the highest potential accuracy at 92.15% and the highest potential3 sensitivity at 95.24%.

Table 1. Relative performance of each model used in the study.

Metric	SVM	RF	Bi-LSTM	Bi-LSTM-CNN	DistilBERT	XLNet
Accuracy	0.89000	0.88700	0.86275	0.91450	0.92150	0.91500
Precision	0.89948	0.89678	0.79917	0.91243	0.89543	0.95907
F1-Score	0.88752	0.88439	0.87593	0.91471	0.92307	0.91071
Recall	0.87588	0.87235	0.96900	0.91700	0.95247	0.86700

4.3   Model Training Parameters

All the models used in the study are trained on a train set of 16000 datapoints and validated on 4000 which counts for 80:20 train-test split. The sequential models are trained over 20 epochs. The number of datapoints and epochs are considered with respect to the limitations in computational power available and as a risk contingency plan with respect to the available bandwidth in time. Table 2 and Table 3 show the training of Bi-LSTM and Bi-LSTM-CNN model over 20 epochs respectively.

Table 2. Training of Bi-LSTM model over 20 epochs.

Epoch	Accuracy (%)	Loss	Val accuracy (%)	Val loss
1	71.53	7.2934	61.65	6.6464
2	81.76	5.6600	82.90	4.9818
3	84.61	4.4445	84.30	3.9375
4	85.44	3.5278	88.02	3.0888
5	86.95	2.7991	87.55	2.4887
6	86.79	2.2629	88.48	1.9838
7	88.40	1.8110	89.45	1.5911
8	87.63	1.4942	88.98	1.3231
9	89.06	1.2120	83.67	1.1920
10	88.79	1.0179	87.27	0.9479
11	88.44	0.8675	90.00	0.7677
12	88.94	0.7442	82.70	0.7874
13	89.74	0.6307	90.87	0.5670
14	89.19	0.5745	89.28	0.5275
15	88.74	0.5208	86.58	0.5321
16	89.26	0.4689	90.62	0.4217
17	89.63	0.4288	89.28	0.4287

(continued)

Table 2. (*continued*)

Epoch	Accuracy (%)	Loss	Val accuracy (%)	Val loss
18	90.06	0.3902	88.82	0.4015
19	89.74	0.3800	90.25	0.4025
20	90.07	0.3602	86.27	0.4226

For transformer-based models, training is done over 4 epochs considering the limitations in computational bandwidth available. For DistilBERT, the validation accuracy does not change after 3rd epoch and for XLNet, the validation accuracy takes a dip after the 3rd epoch. These models though provide comparatively much better performance compared to other models, they tend to slightly overfit the train dataset due to training done on just 20,000 datapoints due to computational power and time constraints. With higher processing power, the 71 entire dataset of 0.2M records can be used to train these models to ensure minimal to zero overfitting.

Table 3. Training of Bi-LSTM-CNN model over 20 epochs.

Epoch	Accuracy (%)	Loss	Val accuracy (%)	Val loss
1	71.77	4.2074	81.92	3.6195
2	81.01	3.2796	76.02	2.9926
3	83.82	2.5907	86.73	2.2555
4	86.52	2.0554	88.20	1.8026
5	87.38	1.6621	83.47	1.5576
6	88.48	1.3439	89.35	1.1897
7	88.81	1.1051	87.85	1.0017
8	89.08	0.9170	88.42	0.8357
9	89.67	0.7678	88.55	0.7279
10	89.52	0.6600	88.27	0.6177
11	89.93	0.5732	90.43	0.5178
12	90.46	0.4941	90.60	0.4550
13	89.87	0.4541	90.57	0.4113
14	90.71	0.3996	85.47	0.4882
15	90.95	0.3629	90.62	0.3459

(*continued*)

Table 3. (*continued*)

Epoch	Accuracy (%)	Loss	Val accuracy (%)	Val loss
16	90.39	0.3452	90.95	0.3216
17	90.54	0.3238	88.17	0.3688
18	91.09	0.3037	91.07	0.2900
19	91.03	0.2912	91.22	0.2785
20	91.49	0.2742	91.45	0.2675

5   Conclusion

The study was conducted on a Kaggle dataset aggregated dataset from the SuicideWatch subreddit which contains posts pertaining to suicide ideation, depression and in some cases, other non-relevant topics. This study was a binary text classification study aimed at correctly classifying the suicide ideation texts from the remaining corpus. The primary consideration was to classify as many suicide ideation posts as possible correctly with a possible leverage of non-suicidal posts being mistaken as suicidal which can be rectified manually. However, missing a suicidal text could lead to disastrous consequences with respect to the author. The dataset was a two column dataset where the only independent column contained a text corpus and other column taking two distinct values indicating where a text is pertaining to suicide ideation or not. The plain text data was subjected to various data cleaning and pre-processing techniques such as stopword removal, lemmatization, rectification of grammatical mistakes etc. Post cleaning, feature extraction techniques such as POS tagging, LDA topic modelling are used to obtained critical information in numerical from a verbose text corpus. These extracted features are used as an augmentation on the vectorized texts which are generated using document embedding techniques. Two document embedding techniques are used one being TF-IDF and other being pre-trained glove embeddings. The former being used in conjunction with extracted 74 features as an input to machine learning models while the latter is used to vectorize texts which are used as an input to sequential and deep learning models. A total of 6 models are attempted 2 each from the category of machine learning, sequential and transformers. Finally, the models are compared on several metrics, two of the most prominent ones being accuracy and sensitivity. In the model training phase, the dataset was split on a 80:20 proportion for train and validation sets. The machine learning models were trained on the default hyper-parameters while the other models were trained by trying various combination of hyper-parameters manually given the computational bandwidth available. The sequential models are trained over 20 epochs while the transformer models are trained over 4 epochs. While Bi-LSTM offered the highest sensitivity at 97%, DistilBERT emerged to be the best performing model with regards to accuracy and F1- score and a close second place in terms of sensitivity at 95.2%. So, from an overall perspective, DistilBERT emerged to be the best performer amongst all the models.

This study has built upon the historical work and presented some novel approaches in suicidal text classification. There is a lot of scope of improvement and improvisation on the tools and techniques described throughout the study and also on historical works carried out in this domain. The use of more comprehensive feature extraction techniques such as LIWC can be explored to provide more information to the underlying model. For all the models used in this study hyper-parameters are chosen manually on a trial-and-error basis due to lack of appropriate computational bandwidth and time. More robust hyper-parameter tuning using tools such as Grid Search can help in choosing hyper-parameters that would enhance model performance. For similar reasons, the dataset of nearly 0.27M records in sampled to 20K with each text being capped at 128 tokens. Therefore, with the availability of higher computational power, the complete dataset can be used thereby resulting in more accurate and efficient models. Using Explainable AI tools, the neural network based black box models could be explained for human interpretation to address the causative factors and patterns in suicide ideation.

References

1. WHO Suicide (2021). https://www.who.int/news-room/fact-sheets/detail/suicide.Accessed 14 Oct 2022
2. CDC Facts about suicide, CDC.gov (2022). https://www.cdc.gov/suicide/facts/index.html. Accessed 14 Oct 2022
3. NCRB (2021) Accidental deaths statistics [Online] Avaialable at: https://ncrb.gov.in/en/accidental-deaths-suicides-in-india (Accessed 12th May 2022)
4. Ritchie, H. Roser, M. Ortiz-Ospina, E.: Suicide, Our World In Data (2015). https://ourworldindata.org/suicide. Accessed 14 Oct 2022
5. Mind Suicidal feelings (2022). https://www.mind.org.uk/information-support/types-of-mental-health-problems/suicidal-feelings/about-suicidal-feelings/. Accessed 14 Oct 2022
6. De Berardis, D., et al.: Suicide and genetic biomarkers: toward personalized tailored-treatment with lithium and clozapine. Curr. Pharm. Des. **27**(30), 3293–3304 (2021)
7. Ji, S., Pan, S., Li, X., Cambria, E., Long, G., Huang, Z.: Suicidal ideation detection: a review of machine learning methods and applications. IEEE Trans. Comput. Soc. Syst. **8**(1), 214–226 (2020)
8. Aladağ, A.E., Muderrisoglu, S., Akbas, N.B., Zahmacioglu, O., Bingol, H.O.: Detecting suicidal ideation on forums: proof-of-concept study. J. Med. Internet Res. **20**(6), e9840 (2018)
9. Sawhney, R., Manchanda, P., Singh, R. Aggarwal, S.: A computational approach to feature extraction for identification of suicidal ideation in tweets. In: Proceedings of ACL 2018, Student Research Workshop, pp. 91–98 (2018)
10. Sarsam, S.M., Al-Samarraie, H., Alzahrani, A.I., Alnumay, W., Smith, A.P.: A lexicon-based approach to detecting suicide-related messages on Twitter. Biomed. Signal Process. Control **65**, 102355 (2021)
11. Dongo, I., Cadinale, Y., Aguilera, A., Martínez, F., Quintero, Y. Barrios, S.: Web scraping versus twitter API: a comparison for a credibility analysis. In: Proceedings of the 22nd International Conference on Information Integration and Web-based Applications & Services, pp. 263–273 (2020)
12. Rajesh Kumar, E., Rama Rao, K.V.S.N., Nayak, S.R., Chandra, R.: Suicidal ideation prediction in twitter data using machine learning techniques. J. Interdisc. Math. **23**(1), 117–125 (2020)
13. Burnap, P., Colombo, G., Amery, R., Hodorog, A., Scourfield, J.: Multi-class machine classification of suicide-related communication on Twitter. Online Soc. Netw. Media **2**, 32–44 (2017)

14. Baumgartner, J., Zannettou, S., Keegan, B., Squire, M., Blackburn, J. Io, P.: The Pushshift Reddit Dataset (2020). www.aaai.org

15. Hernandez-Suarez, A., Sanchez-Perez, G., Toscano-Medina, K., Martinez-Hernandez, V., Sanchez, V. Perez-Meana, H.: A Web Scraping Methodology for Bypassing Twitter API Restrictions (2018). http://arxiv.org/abs/1803.09875

16. Ambalavan, A.K., Moulahi, B., Azė, J. Bringay, S.: Unveiling online suicide behavior: What can we learn about mental health from suicide survivors of Reddit? In: Studies in Health Technology and Informatics. IOS Press, pp.50–54 (2019)

17. Liu, X., Liu, X., Sun, J., Yu, N.X., Sun, B., Li, Q. Zhu, T.: Proactive suicide prevention online (pspo): machine identification and crisis management for chinese social media users with suicidal thoughts and behaviors. https://www.jmir.org/2019/5/e11705/. Accessed 3 Jan. 2022

18. Grant, R., Kucher, D., León, A.M., Gemmell, J., Raicu, D. and Fodeh, S.: Automatic extraction of informal topics from online suicidal ideation. www.reddit.com/r/SuicideWatch

19. Birjali, M., Beni-Hssane, A. Erritali, M.: Prediction of suicidal ideation in Twitter data using machine learning algorithms. In: International Arab Conference on Information Technology. ACIT (2016)

20. Burnap, P., Colombo, G. Scourfield, J.: Machine classification and analysis of suicide-related communication on Twitter. In: HT 2015 - Proceedings of the 26th ACM 111 Conference on Hypertext and Social Media. Association for Computing Machinery, Inc, pp.75–84 (2015)

21. Chiroma, F., Liu, H. Cocea, M.: Suiciderelated Text Classification With Prism Algorithm; Suiciderelated Text Classification With Prism Algorithm (2018)

22. Desmet, B., Hoste, V.: Online suicide prevention through optimised text classification. Inf. Sci. **439–440**, 61–78 (2018)

23. Du, J., et al.: Extracting psychiatric stressors for suicide from social media using deep learning. BMC Med. Inform. Decis. Making **18**, 77–87 (2018)

24. Fodeh, Set al.: Using machine learning algorithms to detect suicide risk factors on twitter. In: IEEE International Conference on Data Mining Workshops, ICDMW. IEEE Computer Society, pp.941–948 (2019a)

25. Pestian, J.P., et al.: Sentiment analysis of suicide notes: a shared task. Biomed. Inform. Insights, **5s1**, BII.S9042 (2012)

26. Wang, W., Chen, L., Tan, M., Wang, S. Sheth, A.P.: Discovering fine-grained sentiment in suicide notes. Biomed. Inform. Insights, **5s1**, BII.S8963 (2012)

27. Liakata, M., Kim, J.-H., Saha, S., Hastings, J. Rebholz-Schuhmann, D.: Three hybrid classifiers for the detection of emotions in suicide notes. Biomed. Inform. Insights, **5s1**, BII.S8967 (2012)

28. Mikolov, T., Chen, K., Corrado, G. Dean, J.: Efficient Estimation of Word Representations in Vector Space (2013a). http://arxiv.org/abs/1301.3781

29. Pennebaker, J.W.: Linguistic inquiry and word count (LIWC) Personality, Images, and Text View project Measurement and psychosocial factors of well being View project. https://www.researchgate.net/publication/246699633

30. Hassan, S., Al-Augby, M., Mohammed, S.H. Al-Augby, S.: LSA & LDA topic modeling classification: comparison study on e-books sentiment analysis on social media view project. Indonesian J. Electr. Eng. Comput. Sci. 191 (2020)

31. O'Dea, B., Wan, S., Batterham, P.J., Calear, A.L., Paris, C., Christensen, H.: Detecting suicidality on twitter. Internet Interv. **22**, 183–188 (2015)

32. Jung, H., Park, H.A. Song, T.M.: Ontology-based approach to social data sentiment analysis: detection of adolescent depression signals. J. Med. Internet Res. 19(7), e259 (2017). https://www.jmir.org/2017/7/e259. Accessed 2 Jan 2022

33. Kiros Bitew, S., et al.: Predicting suicide risk from online postings in Reddit the UGent-IDLab submission to the CLPysch 2019 Shared Task A. www.reddit.com

34. Felbo, B., Mislove, A., Søgaard, A., Rahwan, I. Lehmann, S.: Using millions of emoji occurrences to learn any-domain representations for detecting sentiment, emotion and sarcasm (2017). http://arxiv.org/abs/1708.00524

35. Kalchbrenner, N., Grefenstette, E. Blunsom, P.: 重要基础!!!Dcnn动态卷积 nn. In: Proceedings of the 52nd Annual Meeting of the Association for Computational Linguistics, (Volume 1: Long Papers), pp. 655–665 (2014). http://aclweb.org/anthology/P14-1062. Accessed 6 Jan 2022

36. Yin, W., Schütze, H. Schütze, S.: Multichannel Variable-Size Convolution for Sentence Classification

37. Hill, F., Bordes, A., Chopra, S. Weston, J.: The Goldilocks Principle: Reading Children's Books With Explicit Memory Representations. http://fb.ai/babi/. Accessed 6 Jan 2022

38. Moritz, K., et al.: Teaching machines to read and comprehend. http://www.github.com/dee pmind/rc-data/. Accessed 6 Jan 2022

39. He, H. Lin, J.: Pairwise word interaction modeling with deep neural networks for semantic similarity measurement, pp.937–948 (2016)

40. Kim, Y.: Convolutional Neural Networks for Sentence Classification (2014). http://arxiv.org/abs/1408.5882

41. Blei, D.M., Ng, A.Y., Edu, J.B.: Latent dirichlet allocation. J. Mach. Learn. Res. **3**, 993–1022 (2003)

42. Pennington, J., Socher, R. Manning, C.D.: Glove: global vectors for word representation. In: Proceedings of the 2014 Conference on Empirical Methods in Natural Language Processing (EMNLP), pp. 1532–1543 (2014)

Well Log Data Preparation and Effective Utilization of Drilling Parameters Using Data Science Based Approaches

Rahul Talreja[1], Thomas Coombs[2], Sulaf Assi[3(✉)], Noor Azma Ismail[7],
Manoj Jayabalan[1], Panagiotis Liatsis[4], Mohamed Mahyoub[5], Abdullah Al-Hamid[6],
and Hoshang Kolivand[1]

[1] Faculty of Engineering and Technology, Liverpool John Moores University, Liverpool, UK
[2] University Hospital Dorset, Bournemouth BH7 7DW, UK
[3] School of Pharmacy and Biomolecular Science, Liverpool John Moores University, Liverpool, UK
s.assi@ljmu.ac.uk
[4] Department of Electrical Engineering and Computer Science, Khalifa University, Abu Dhabi, UAE
[5] eSystem Engineering Society, Liverpool, UK
[6] Saudi Ministry of Health, Najran, Saudi Arabia
[7] Faculty of Business and Technology, UNITAR International University, Petaling Jaya, Malaysia

Abstract. Researchers in the early 1990s utilized neural networks for different applications in the oil and gas sector. Gradually, with the development of powerful algorithms and computing power applications of random forests, clustering, isolation forests, XGboost increased. The digital transformation in the oil and gas sector began because of two downturns. Literature review suggests that not much work has been published on outlier identification in the geophysical well logs. The proposed workflow uses a machine learning (ML) approach to clean and condition the data and handle missing values. The planned workflow will construct the cleaned composite log data in quick time which can then directly be consumed by 1D geomechanical workflows. The proposed study has two-fold objectives to utilize drilling parameters. Firstly, use an unsupervised machine learning (ML) algorithm like K-means and Hierarchical clustering to classify drilling data of 6 wells into meaningful clusters. These clusters are further classified based on their relationship with rate of penetration feature which defines the drillability. Secondly, the proposed workflow will utilize predictive machine learning algorithms like XGboost and random forest to estimate geophysical logs indicating rock type using the drilling parameters. The results and conclusions of this study are expected to save time in data preparation steps of 1D geomechanical modelling and lead to effective utilization of drilling parameters. This study will redound to the benefit of geoscientists and drilling data analysts in their day-to-day work. Geoscientists spend an enormous amount of time in data preparation before getting to the modelling stage. This workflow saves more than 50% of the time invested in the

---

The original version of this chapter was revised: the first author's name has been corrected. The correction to this chapter is available at
https://doi.org/10.1007/978-981-99-0741-0_39

© The Author(s), under exclusive license to Springer Nature Singapore Pte Ltd. 2023, corrected publication 2023
Y. B. Wah et al. (Eds.): DaSET 2022, LNDECT 165, pp. 388–402, 2023.
https://doi.org/10.1007/978-981-99-0741-0_28

identification of outliers and hence reducing the overall turnaround of time of the modelling. Furthermore, the study gives insight of hidden trends in drilling parameters to the drilling optimization engineers. The clustering algorithms helped in the identification of high-performance and low-performance clusters and knowledge of their characteristics can optimize drilling operations in the field. Lastly, the predictive algorithms will aid in the understanding of downhole rock lithofacies in near-real-time.

Keywords: Machine learning · Data science · Norwegian oil drilling · Data manipulation

1 Introduction

1.1 Background

The well logs are the record of different geological sub-surface layers encountered in a borehole. A well log is a type of numerical structured data representing properties of different layers in the sub-surface. They are the basis for most of the reservoir characterization workflows in the oil and gas industry [1, 2]. In the field of 1D geomechanical modelling workflows, as explained in [3] the data preparation step is the most time-consuming step. It forms the first step in any of the modelling processes especially when the data is known to have been affected by different reasons [4]. The geophysical well log data acquired in a borehole is usually affected by casing shoe, bad hole intervals, downhole noise, and many other factors. Geoscientists invest hours of their time to clean the bad data and handle the resulting missing values. Most of the methods used to clean these outliers' values are manual. The methods used for the prediction of the resultant missing values are based on linear or power or exponential algorithms derived in a laboratory setup. One of the most common approaches to construct a bulk density log is using the algorithm proposed by [5]. Other key well logs are compressional and shear slowness which requires cleaning as it is taken as direct input in the modelling. The work of Iwuoha [6] and many others have proposed algorithms to reconstruct the shear slowness in the missing intervals using established algorithms and neural networks.

The applications of machine learning approaches are gaining popularity in the oil and gas industry for data preparation stages of modelling. In recent times both industry and institutes have successfully applied machine learning algorithms for various applications including the prediction of synthetic logs using available data. The study has demonstrated the use of Z-score, Density-Based Spatial Clustering of Applications with Noise (DBSCAN) [7], Isolation Forrest [8], K-means and Hierarchal algorithm to identify the outliers in the geophysical logs acquired in each section. The workflow will then construct the composite well log data by merging data of each section which can then be directly consumed by 1D geomechanically workflows.

The other aspect of the study intends to effectively utilize drilling parameters acquired by a mud logging unit installed on a rig site. Drilling parameters are measured by a mud logging unit (MLU) by putting sensors on a different part of the rig. These drilling parameters are acquired in almost every well but are rarely utilized for any interpretation. These datasets are primarily affected by rig-site activities, drilling-related noise, faulty sensors, and drillers bias. In addition, the relationship between geophysical logs and drilling parameters is complex making it difficult to directly interpret the results. Recently published work by researchers [9–11] shows that the drilling parameters can be utilized for the selection of drilling bit and BHA design. These applications suggest that the ML algorithms can be applied to interpret drilling parameters.

Therefore, the purpose of the study was established to understand the feasibility of ML techniques in the identification of outliers in well logs and to improve utilization of drilling parameters. The ML approaches proposed through the study would benefit geoscientists and drilling engineers by saving 40 to 50% of their time invested in data preparation. The ideas and ML workflow presented will also assist in the documentation of high-performance clusters and their characteristics to express how drillability can be improved. Hence, the overall study would be of great importance to the geoscientists, drilling engineers and operation geologists adding value in their day to day ask.

1.2 Rationale

Machine learning applications are gaining more traction in the Oil and Gas sector in recent times. This journey is briefly discussed by [12, 13]. A great amount of focus today is to utilize the existing archived data and streamline the existing workflows by reducing the turn-around time. For example, the outlier removal technique has been manual for various reasons in the oil and gas industry. Geoscientists invest a good amount of time to clean the acquired well logs before feeding them into different geophysical and geomechanical workflows.

Statistically, outliers are the data points that are significantly away from other data points in the overall trend of the distribution. Multiple researchers have given their definitions and some references can be found in [14, 15]. These observations are easily marked by visual inspection in the crossplots and log plots, but the process will consume a good amount of time. The origin of these outliers can be attributed to different reasons linked with operations and technology being used. Automating this process and using it for operation optimization can be seen in [16, 17] where authors apply this technique in production optimization and increasing oil and gas production. Similar ML techniques are being attempted in data cleaning and preparation stages by different Exploration and Production (E&P) companies to reduce the turn-around time (TAT) associated with different geological and geophysical (G&G) workflows.

Drilling parameters are recorded by different rig-site sensors in almost every well drilled around the world. It might occur that the wild-cat and exploratory wells will have more data in comparison to appraisal and development wells. These data points are usually recorded and made available in real-time to understand drilling progress and associated challenges. An extensive literature review suggests the used cases of drilling parameters using ML techniques [10, 13, 18]. The published work suggests drilling parameters can be in the identification of suitable bit types, BHA and prediction of ECD. The work of [19] discusses the utility of drilling parameters for the prediction of geophysical logs using the Inception-based Convolutional Neural Network (CNN) algorithm which resulted in a decent model with lower correlation coefficients varying between 0.4 to 0.6 (R−0.16 to 0.36).

1.3 Aim and Objectives

The first aim of this research is to propose an ML based systematic workflow to enable log data preparation and streamline the geomechanical modelling process. This research further aims to explore ways to improve the utilization of drilling parameters. The first application works to identify clusters giving the best drilling performance and the second application work on the prediction of rock facies in a geological area. The identification of the higher drillability clusters and lithofacies will reduce non-productive time leading to lower operating costs (OPEX) and help operation geologists to take decisions on time. The research objectives are framed based on the aims of this study which are as follows:

- To suggest a suitable machine learning technique to remove outliers in geophysical logs to reduce the requirement of manual intervention
- To check for the presence of hidden high and low performing clusters using drilling parameters
- To compare the performance of predictive machine learning models in predicting the lithologies.
- To evaluate the performance and applicability of the predictive algorithms on a hidden well dataset.

2 Methodology

2.1 Data Selection

The dataset used in this study belongs to the Volve field located in the Norwegian North Sea and was operated by Equinor [20–23]. The field was finally decommissioned in 2016 after nearly 8.5 years of production. It's an offshore field with an average water depth of 80 m. The primary target formation is locally named as Hugin Formation and belongs to the Middle Jurassic age. The producing interval is a sandstone type rock. The well drilled in the field varied between 2.7 km to 3.2 km below the seabed. Figure 1 below shows the location of the Volve field. Equinor released a complete set of data to foster research and development under Equinor Open Data License (Fig. 1).

The study research plan required the use of well log data and drilling parameters. These datasets were collected for 6 wells and included a total of 26 features with actual 16 measured features and rest are derived features. All the features are continuous numerical variables except formation tops and these features can be broadly classified into three categories. The first category consists of well logs which are acquired by a process called borehole logging. In this process, different tools are lowered to measure the properties of the formations drilled. The second category is drilling parameters which are measured by a mud logging unit using sensors on the rig. The last category of features includes information regarding the geological age and derived variables. The final dataset will have close to 93,900 samples of geophysical logs and 12000 samples of drilling parameters. Approximately 20% will be kept hidden final blind test of the algorithm.

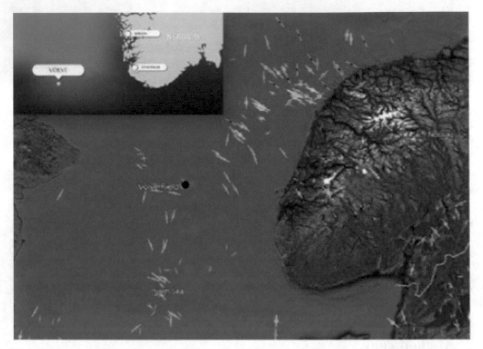

Fig. 1. Location of Volve Field on Google maps. The Volve field located in the Norwegian North Sea

2.2 Data Pre-processing

The dataset collected from any field is usually available for individual sections acquired using different tools and technologies. The well logs have outliers in the form of tool noise, poor borehole measurements etc. The proposed workflow intends to evaluate approaches like Z-Score, Density-Based Spatial Clustering of Applications (DBSCAN), K-means, Hierarchical clustering, SVM and Isolation Forests to identify anomalies and remove them from the main datasets. The focus will be on the key features which go as an input to 1D geomechanical modelling including compressional slowness, shear slowness and bulk density log. The missing values caused by the removal of outliers can be replaced with predictive ML or neural network-based approaches like multilinear regression, ANN etc. However, considering the time constraints the missing values will not be imputed.

2.3 Data Modelling

Data-preprocessing showed multiple outlier detection techniques. These techniques fall under stage-2 of the study are discussed in the subsections (Fig. 2). The research study includes two more stages that required unsupervised clustering and predictive ML algorithms. In stage-3 of the study, the pre-processed drilling parameters of 6 well datasets were divided into the optimum number of clusters using K-means and Hierarchical unsupervised ML algorithms. The concept of silhouette score, elbow curve and dendrograms was utilized to identify the optimal number of clusters. These clusters were later analyzed further to see associated characteristics and their relation to ROP. The last stage of the work attempted to predict the lithology using drilling parameters. Two ensemble algorithms were tried to relate different drilling parameters with geophysical logs like a gamma-ray or neutron density or bulk density log.

Fig. 2. Outline of data processing in this study

2.4 Model Evaluation

The capabilities and robustness of outlier detection techniques will be evaluated using accuracy metrics and by performing visual quality checks by comparing the manually cleaned data with the algorithm cleaned data. To check the clustering tendency of the geophysical logs and drilling data, Hopkins's test is performed. The Hopkins test examines whether data points differ significantly from uniformly distributed data in the multidimensional space. The final clusters obtained using the drilling data as input to K-means and Hierarchical clustering algorithm were evaluated using crossplots and boxplots. The resulting clusters were reclassified into high performance (HP) and low performance (LP) groups based on their relation to ROP.

The relationship between the geophysical logs and drilling parameters must be accurate to determine the drilled rock type. Both independent and dependent variables are continuous numerical values and hence coefficient of determination (R2) metric and visual quality check by overlaying predicted and actual numerical variables is expected to give us a good understanding of the robustness of the model. Literature review suggests that the other researchers have also used similar metrics to evaluate on well log dataset. In the addition to the final rock classification problem accuracy, precision, Roc Auc Score and F1-Score metrics were utilized. These model evaluation metrics are briefly explained in the results and discussion.

3 Results and Discussion

The dataset used in this research project belongs to the Volve field located in the Norwegian North Sea and was released by Equinor to promote research and insights. The datasets included geophysical logs and drilling parameters of 6 different wells.

3.1 Interpretation Based on EDA

It is evident that GR has a decent negative correlation with ROP. This observation is in line with the conventional belief of the drillers. GR also has a good correlation with MD, TORQ_N and Zones. Other geophysical logs like NPHI and RHOB features have a decent correlation with the same set of drilling parameters.

3.2 Outlier Identification in Geophysical Logs

The geophysical logs collected or acquired in any field are usually available for individual sections acquired using different tools and technologies. The data points present in each feature include outliers coming from tool noise, poor borehole measurements etc. and before proceeding to the modelling stages, the geoscientist invests nearly 4–5 h to clean the data and prepare a composite data file devoid of these outliers. The ML-driven workflow evaluated different approaches like Z-Score, DBSCAN, Isolation Forests, K-Means and Hierarchical Clustering to identify anomalies and remove them from the main datasets. The focus was on the key features like bulk density and compressional slowness which go as an input to 1D geomechanical modelling. The Hopkins statistic which is used to understand the clustering tendency of the dataset was derived. The Hopkins statistic estimate on geophysical logs has a value of greater than 0.95 suggesting that the dataset has a high tendency to cluster. The sections below summarize the results obtained from different methods.

DBSCAN

DBSCAN is an unsupervised clustering algorithm that is seen to work for nonconvex clusters and noise. The value of the "eps" which defines the distance between two points is observed to vary between 0.3 to 0.5. The other hyperparameter i.e., the minimum number of neighbours is 100. The resulting clusters are highlighted in Fig. 3. The shown cross plot has compressional slowness variation with bulk density. The snapshot on the left is manually cleaned by a geoscientist and highlighted values (in red) are outliers. The snapshot on the right is showing the final clusters obtained from the DBSCAN algorithm. It is very clear that DBSCAN is selecting partially correct outliers displayed in the cluster "−1". However, this cluster has corrected data points that should not be removed. A similar observation was made on the other well datasets (Fig. 3).

SVM

SVM (Support Vector Machine) belongs to the category of supervised algorithm which is used for both clustering and regression. Literature review suggests that this approach is more suitable for non-linear separation problems. The value of the "nu" is key hyperparameter which is defined as the upper bound on the fraction of training errors the lower bound of the fraction of support vectors. The "nu" value is observed to vary between 0.2 to 0.4 for optimum results. The other hyperparameters were kept to default. The resulting clusters are displayed in Fig. 4 showing a cross plot between compressional slowness and bulk density. The snapshot on the left is manually cleaned by a geoscientist and highlighted values (in blue) are outliers. The snapshot on the right is showing the final clusters obtained from the OneClass-SVM algorithm. It is very clear that the SVM

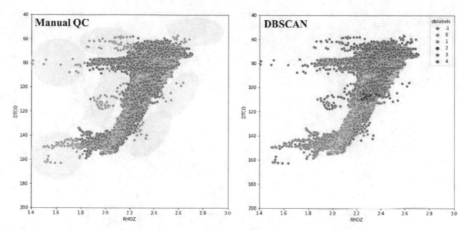

Fig. 3. Crossplot showing the variation of compressional slowness with bulk density in well -1. The snapshot on the left is manually cleaned by a geoscientist and highlighted values in red are outliers. The snapshot on the right is showing the final clusters obtained from the DBSCAN algorithm.

approach is selecting a few outliers in the cluster "−1". Most of the data points in this group are correct (Fig. 4).

Z-Score

Z-score is a simple algorithm that is used to clip extreme values lying far away from the mean value. The cut-off point in this method is chosen as 3.5 which removes some of the outliers. The Z-score approach is selecting some of the correct outliers. Some of the outliers closer to the existing cluster were not spotted by the algorithm. Similar observations on the performance of the Z-score method were made on the datasets prepared for other wells (Fig. 5).

K-Mean

K-means is an example of an unsupervised algorithm that divides the data into user-defined k clusters. It groups the data points based on its distance from centroids. It is expected that the outliers will be clubbed in different clusters and hence this algorithm is used. From the elbow curves, an optimum number of clusters were seen to be in the range of 5 to 6. Using a k value of 5, clusters were formed. The K-means approach is unable to identify correct outliers in comparison to the other methods.

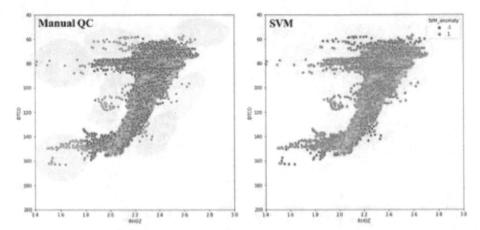

Fig. 4. Crossplot showing the variation of compressional slowness with bulk density in well -1. The snapshot on the left is manually cleaned by a geoscientist and highlighted values in red are outliers. The snapshot on the right is showing the final clusters obtained from the OneClass-SVM algorithm.

Hierarchical Clustering

Hierarchical Clustering (HC) is also an unsupervised algorithm that uses the hierarchy of clusters in the form of a tree called a dendrogram. In this study, an agglomerative approach with a complete linkage method is used. The preparation of a dendrogram is time-consuming especially for larger datasets. The dendrogram helped to identify the optimal clusters which were found to be 5 to 8. The HC approach is selecting most of the correct outliers with minor inclusion of correct data (Fig. 5). This method is time-consuming and might not be suitable for larger datasets.

Isolation Forests

Isolation Forests (IF) belongs to the category of unsupervised algorithm which uses decision trees in the background. The algorithm follows a basic principle that the outliers are few and different. The depth of the trees can be related to the presence of outliers in the features. Usually, trees that travel deeper are less likely to have outliers. The key hyperparameter is the contamination value which is observed to vary between 0.03 to 0.05 for optimum results (Fig. 5).

3.3   Clustering of Drilling Parameters

One of the objectives was to understand and identify interesting patterns in data such as high performing and low performing drilling groups. The plan was to utilize K-means and hierarchical unsupervised ML algorithms to divide the drilling parameters of 6 wells. The estimated Hopkins statistic for drilling parameters is above 0.95 indicating a high clustering tendency. The final dataset has Flowrate, Mud Weight, ROP, SPP, TORQ, TRPM, WOBM, MSE, Zones features with 12740 samples each.

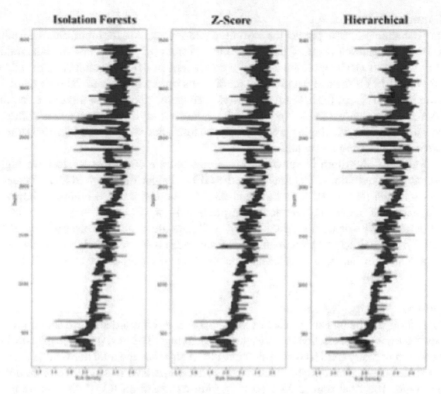

Fig. 5. Line plot showing the comparison of original bulk density curve in red with the cleaned curve in red. The plot compares the result of Isolation Forrest, Z-Score and Hierarchical clustering method.

3.4 Prediction of Rock Facies

As per the plan, a relationship between the drilling parameters recorded by MLU and geophysical logs like a gamma-ray and neutron porosity log was attempted. This relationship was tried with ensemble methods like random forests and XGboost [21]. The robust random forests and XGboost algorithm were expected to capture the diversity and complex relationship present between drilling parameters and geophysical well logs. These algorithms were trained on 5 wells dataset consisting of 9 drilling-related features having 7183 samples each and different geophysical logs as the target variable. The algorithm was then tested on a dataset having 3101 samples and further evaluated on a blind well (hidden dataset) having 2456 samples of each feature.

Gamma Ray Evaluation

The first attempt was to predict the most used geophysical log for lithofacies determination which is gamma-ray (GR) using the random forest (RF) algorithm. The final trained random forests (RF) algorithm showed a coefficient of determination (R2) of 0.82 on train and 0.79 on the test dataset. The predictions on the blind-well data are relatively lower with an R2 of 0.22. From the analysis, it seems that the RF algorithm requires more training datasets to be exposed all the variations in drilling parameters with rock type. The features with higher predictability are identified as mud weight, zones, torque, rotation of the string and depth.

XGBoost algorithm is known to be more accurate and expected to have higher accuracy of predictions. The final trained XGBoost algorithm showed a coefficient of determination (R2) of 0.98 on train and 0.82 on the test dataset. Figure 6 shows the comparison of actual (in blue dots) and predicted GR (in orange dots) for train and test datasets. The predictions on the blind-well data are relatively lower with an R2 of 0.29. The higher accuracy on the test and train datasets is most likely as the algorithm is optimized for the exposed datasets. The XGBoost can predict the trend of GR better than RF (Fig. 6).

Neutron-Porosity Evaluation

Considering the better performance of XGBoost over RF, it was decided to continue using only XGBoost for predictions of other geophysical logs. The next in line geophysical log is neutron porosity (NPHI) for predictions. NPHI signifies the total porosity of the rock and shales usually have higher total porosity in comparison to sandstone or carbonate rock type. The final trained XGBoost algorithm showed an R2 of 0.88 on train and 0.78 on the test dataset. The predictions on the blind-well data are relatively lower with an R2 of 0.23. From the analysis, it seems that the XGBoost algorithm requires more training samples capturing all the variations in drilling parameters with NPHI. The higher accuracy on the test and train datasets is most likely as the algorithm is optimized for the exposed datasets. The key features include mud weight used to drill the rock, depth, rotation of the string and ROP.

Bulk-Density Evaluation

The other geophysical log is bulk density (RHOB) for predictions. RHOB signifies the total density of the rock, and it can also give some indication of the rock type being drilled. The final trained XGBoost algorithm showed an R2 of 0.89 on train and 0.79 on the test dataset. The predictions on the blind-well data are relatively lower with an R2 of 0.31. The higher accuracy on the test and train datasets is most likely as the algorithm is optimized for the exposed datasets. The key features include mud weight used to drill the rock, depth, and zones (Fig. 7).

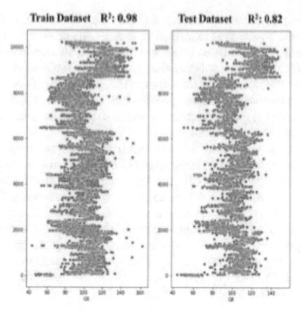

Fig. 6. Performance of XGBoost algorithm to predict gamma ray (GR dots in orange) on train and test dataset.

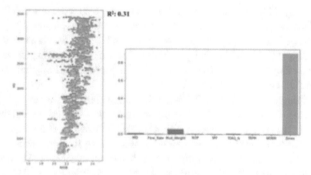

Fig. 7. Performance of XGBoost algorithm to predict bulk-density (RHOB dots in orange) on blind well data. Predictions can capture the original RHOB trend shown in blue dots. The importance of different features is shown in the bar graph

4 Conclusions

The results and conclusions drawn from research work demonstrate applications of supervised and unsupervised ML techniques in the oil and gas sector. The idea of using clustering algorithms to identify anomalies and high-performance groups is unique. The results establish that a combined Z-score and Isolation Forrest algorithms are suitable for outlier identification and removal from bulk-density and sonic logs. The proposed workflow reduces the time in the data preparation stage of the geomechanical modelling by approximately 50% time.

The other two objectives are associated with the effective utilization of drilling parameters. Firstly, a process is explained to classify the drilling data based on their drilling performance. These hidden trends and ranges of features will be of great importance to the drilling engineers and operation geologists to optimize the drillability and lowering of operating costs (OPEX). Lastly, the utility of Random Forest and XGBoost algorithms is demonstrated to predict geophysical logs and rock facies using drilling parameters. This is quite crucial information to the stakeholders to recognize downhole rock types being drilled in near-real-time.

The results and conclusions of this research work are encouraging. However, there are certain areas of research in all three stages of the executed workflow which can be explored as an extension of this study. The first recommendation is related to the consistency of unit systems, missing values, and data acquisition systems. The presence of having varied unit systems and corrections due to different MLUs impacts the performance of the algorithm. It is also found that there is scope to explore other methods like extended isolation forests and modified hyperparameters to enhance the performance of automatic outlier detection techniques. An attempt can be made to improve the accuracy of predictive algorithms by using more wells in the basin which might help the algorithms capture possible variance in the area with rock type.

References

1. Universitas Gadjah Mada Well log analysis for reservoir characterization (2019). https://wiki.aapg.org/Well_log_analysis_for_reservoir_characterization. Accessed 10 Oct 2022
2. Brown, N., Roubíčková, A., Lampaki, I., MacGregor, L., Ellis, M. Newton, P.V. de.: Machine learning on Crays to optimize petrophysical workflows in oil and gas exploration. Concurrency Comput. Pract. Experience, **32**(20), e5655 (2020). https://onlinelibrary.wiley.com/doi/full/10.1002/cpe.5655. Accessed 10 Oct 2022
3. Talreja, R., Bahuguna, S., Kumar, R., Zacharia, J., Kundan, A. Kalpande, V.: Geomechanics insights for successful well delivery in complex kutch - Saurashtra Offshore Region. In: Proceedings of the IADC/SPE Asia Pacific Drilling Technology Conference, APDT (2021)
4. Threadgold, P.: Some Problems And Uncertainties In Log Interpretation. The Log Analyst, pp. 132 (1972). https://onepetro.org/petrophysics/article-abstract/171541/Some-Problems-And-Uncertainties-In-Log?redirectedFrom=fulltext. Accessed 8 Oct 2022
5. Gardner, G.H.F., Gardner, L.W. Gregory, A.R.: Formation velocity and density - the diagnostic basics for stratigraphic traps. Geophysics, **39**(6), 770–780 (1974). https://library.seg.org/doi/10.1190/1.1440465. Accessed 9 Oct 2022
6. Iwuoha, S.C., Pedersen, P.K., Clarkson, C.R. Gates, I.D.: A working method for estimating dynamic shear velocity in the montney formation. MethodsX, **61**, 1876–1893 (2019). https://www.sciencedirect.com/science/article/pii/S221501611930216X?via%3Dihub. Accessed 15 Oct 2022
7. Ester, M., Kriegel, H.-P., Sander, J. Xu, X.: A density-based algorithm for discovering clusters in large spatial databases with noise. In: Proceedings of the 2nd International Conference on Knowledge Discovery and Data Mining. AAAI, pp.226–231 (1996). www.aaai.org. Accessed 3 Oct 2022
8. Liu, F.T., Ting, K.M. Zhou, Z.H.: Isolation forest. In: Proceedings - IEEE International Conference on Data Mining, ICDM, pp. 413–422 (2008)

9. Momeni, M., Hosseini, S., Ridha, S., Laruccia, M.B. Liu, X.: An optimum drill bit selection technique using artificial neural networks and genetic algorithms to increase the rate of penetration. J. Eng. Sci. Technol. **125**, 361–372 (2018). https://www.semanticscholar.org/paper/An-optimum-drill-bit-selection-technique-using-and-Momeni-Hosseini/dfebafd3fa4cec51234860b48ee0cc2f2d1cb7e8. Accessed 9 Oct 2022

10. Abbas, A.K., Assi, A.H., Abbas, H., Almubarak, H. Al Saba, M.: Drill bit selection optimization based on rate of penetration: application of artificial neural networks and genetic algorithms. In: Society of Petroleum Engineers - Abu Dhabi International Petroleum Exhibition and Conference 2019. Abu Dhabi, OnePetro (2019). https://onepetro.org/SPEADIP/proceedings-abstract/19ADIP/3-19ADIP/D032S207R002/216855. Accessed 9 Sep 2022

11. Tewari, S., Dwivedi, U.D. Biswas, S.: A novel application of ensemble methods with data resampling techniques for drill bit selection in the oil and gas industry. Energies, **14**(2), 432 (2021). https://www.mdpi.com/1996-1073/14/2/432/htm. Accessed 10 Oct 2022

12. Daneeva, Y., Glebova, A., Daneev, O. Zvonova, E.: Digital transformation of oil and gas companies: energy transition. Atlantis Press, pp.199–205 (2020). https://www.atlantis-press.com/proceedings/rudeck-20/125942684. Accessed 15 Sep 2022

13. Mohammadpoor, M., Torabi, F.: Big data analytics in oil and gas industry: an emerging trend. Petroleum **64**, 321–328 (2020)

14. Oberwinkler, C. Stundner, M.: From real time data to production optimization. In: Proceedings of the SPE Asia Pacific Conference on Integrated Modelling for Asset Management. Kuala Lumpur, OnePetro, pp.91–104 (2004). https://onepetro.org/SPEAPCIMAM/proceedings-abstract/04APCIMAM/All-04APCIMAM/SPE-87008-MS/72201. Accessed 4 Jan 2022

15. Babu, V.: How to Remove Outliers in Python. Kanoki (2020). https://kanoki.org/2020/04/23/how-to-remove-outliers-in-python/. Accessed 4 Jan 2022

16. Snyder, J., Scott, S., Kassim, R.: Self-adjusting anomaly detection model for well operation and production in real-time. In: Society of Petroleum Engineers - SPE Oklahoma City Oil and Gas Symposium 2019, OKOG 2019. OnePetro (2019). http://onepetro.org/SPEOKOG/proceedings-pdf/19OKOG/1-19OKOG/D011S002R005/1171102/spe-195234-ms.pdf. Accessed 7 Oct 2022

17. Fuad, I.I.M., Demon, M.F.N. Husni, H.: Automated real time anomaly detection model for operation and production data at scale. In: Society of Petroleum Engineers - Abu Dhabi International Petroleum Exhibition and Conference 2020, ADIP 2020. OnePetro (2020). http://onepetro.org/SPEADIP/proceedings-pdf/20ADIP/1-20ADIP/D012S116R179/2382153/spe-203194-ms.pdf. Accessed 7 Oct 2022

18. Khalaf, M., Hussain, A.J., Al-Jumeily, D., Fergus, P., Al Kafri, A.S.: A data science methodology based on machine learning algorithms for flood severity prediction. In: 2018 IEEE Congress on Evolutionary Computation, CEC 2018 - Proceedings (2018). 8477904

19. Hussain, A.J., Al-Jumeily, D., Al-Askar, H., Radi, N.: Regularized dynamic self-organized neural network inspired by the immune algorithm for financial time series prediction. Neurocomputing **188**, 23–30 (2016)

20. Montanez, C.A.C., Fergus, P., Hussain, A., Hind, J., Radi, N.: Machine learning approaches for the prediction of obesity using publicly available genetic profiles. In: Proceedings of the International Joint Conference on Neural Networks, 2017, pp. 2743–2750, (2017). 7966194

21. Mohamed, A.H.H.M., Tawfik, H., Norton, L., Al-Jumeily, D.: e-HTAM: A technology acceptance model for electronic health. In: 2011 International Conference on Innovations in Information Technology, IIT 2011, pp. 134–138 (2011). 5893804

22. Alloghani, M., Aljaaf, A., Hussain, A., Al-Jumeily, D., Khalaf, M.: Implementation of machine learning algorithms to create diabetic patient re-admission profiles. BMC Med. Inform. Decis. Mak. **19**, 253 (2019)
23. Keight, R., Aljaaf, A. J., Al-Jumeily, D., Hussain, A. J., Özge, A., Mallucci, C.: An intelligent systems approach to primary headache diagnosis. In: Huang, D.-S., Jo, K.-H., Figueroa-García, J. C. (eds.) ICIC 2017 LNCS, vol. 10362, pp. 61–72. Springer, Cham (2017). https://doi.org/10.1007/978-3-319-63312-1_6

Deep Learning-Based Approach for Classifying the Severity of Metal Corrosion Using Sem Images

Saranga Veeramangal Hebbar[1], Basheera M. Mahmmod[2], Iznora Aini Zolkifly[3], Abdulaziz Al-Nahari[4], and Sadiq H. Abdulhussain[2]([envelope])

[1] Faculty of Engineering, Liverpool John Moores University, Liverpool, UK
[2] Department of Computer Engineering, University of Baghdad, Baghdad, Iraq
sadiqhabeeb@coeng.uobaghdad.edu.iq
[3] Faculty of Business and Technology, UNITAR International University, Petaling Jaya, Malaysia
[4] UNITAR Graduate School, UNITAR International University, Selangor, Malaysia

Abstract. Metals are used everywhere by humans and used extensively in the field of industrial structures, factories, pipelines for Oil and Gas, Railways etc. As the corrosion effect on the metals can lead to severe repercussions on the system, it is of higher importance to identify the corrosion on the metal surface and take required precautionary actions. There have been multiple approaches that are being employed to identify the corrosion in the initial stages and take appropriate actions, such as physical examinations, inspecting through cameras and CCTVs. But it becomes impossible to have such solutions for biomaterials or metals that gets into continuous exposure of some medium. This study proposes a more objective and automatic way of examining the metal surfaces and classifying the corrosion intensity through convolutional neural networks using the images from scanning electron microscope. In this study, different convolutional neural network (CNN) models of custom and transfer learning methods are explored to analyze the images taken from scanning electron microscope (SEM) to learn and predict the severity of the metal corrosion. The images are taken from electron microscope of the Magnesium and Steel metal surfaces that are corroded with different levels of severity. The images are classified into 3 categories: low-medium-high. Also, the study explores the various options to generate good quantities of images for the limited images that are available such as Super-Resolution Generative Adversarial Network (SRGAN) which helps in making the model to learn the minute surface level textures analyzed and ensure a model built with higher degree of accuracy. The best accuracy results were found for the of the ResNet50 model which are 88% and 94% on Steel and Mg images.

Keywords: Deep learning · Metal corrosion · SEM images

1 Introduction

Metals are used everywhere by humans. It is used extensively in the field of industrial structures, factories, pipelines for Oil and Gas, Railways etc. Metals and alloys are also

© The Author(s), under exclusive license to Springer Nature Singapore Pte Ltd. 2023
Y. B. Wah et al. (Eds.): DaSET 2022, LNDECT 165, pp. 403–418, 2023.
https://doi.org/10.1007/978-981-99-0741-0_29

an important material in the bio-medical field like equipment for surgery, materials for bio implants. Steel is used heavily in the construction of buildings, roads, flyovers, and bridges. Hence humans are heavily dependent on metals due to their inherent strength and quality. But unfortunately, metals are always prone to get oxidized when exposed to the atmosphere, or some specific liquid medium and causing rust or corrosion. The metal when gets exposed to the atmosphere, the oxygen in the air oxidizes metal and converting into hydroxides [1]. Similarly, when the biomaterials are implanted inside the body, the physiological medium like blood, enzymes and other chemicals gets into continuous exposure with the implanted materials and this can cause oxidation of the metals. When a metal gets corroded, the metal which is having strength as its inherent property, gets weakened. When the corrosion of the metal increases, it becomes a weak point in the structure where the entire structure can breakdown, thus causing severe loss and pain. The breakdown of industrial structures can lead to severe damage and cause a potential risk to the entire region around the structure. Corrosions on the bridges and flyovers can lead to life threatening to the people driving on the bridges and cause severe loss to life. Similarly, any corrosion in the biomaterials can lead to severe health issues to the patient and even can lead to death of the person [2]. Metal corrosion can be identified in different methods. Simplest way is to do inspection technique where the metal surfaces are captured through a camera and inspected for any visible corrosions on the surfaces [3]. Though this is a cost-effective technique, it would not be a good fool proof approach as it will not be possible to have the camera capturing every metal surface as well as there would be corrosion happening on the parts where the images cannot be captured. When it comes to biomaterials, it will not be possible to use metallic alloys that are not tested properly for its behaviour on physiological medium. Hence the corrosion identification approach from the images taken from scanning electron microscopy of the metal surface would be able to identify the intensity of the corrosion and there are already multiple classical approaches that are being followed to analyse the surface. But this has a lot of dependency on the Research analyst who is analysing the intensity of corrosion. The classical approach can evaluate and predict the intensity only on the surface that is used for taking the images from the microscope, the generalizability or extending the prediction to larger metal surface would not be possible. This drawback creates a dependency with the research analyst skills to carefully select the right areas of the metal surface to evaluate the corrosion on metal surface.

This is where the deep learning models would have a higher edge on the prediction of intensity of corrosion. Deep learning models can give attention to the texture of the metal surfaces in detail whereas human vision can give attention to the shape of the object. Due to this behaviour of the deep learning models, it would be possible for the DL models to predict more accurately and extend the prediction to the surface which the model is not trained for. In this study, deep learning models are built using electron microscopic images that are classified based on the severity of the corrosion on the metal surfaces. The experiment was conducted using Magnesium and Steel alloy metal coupons immersed in the physiological medium for a continuous duration and capture the images from the electron microscope at different time intervals. The images are classified based on the corrosion intensity as low, medium, and high. By experimenting with different layers and parameters, the different CNN models are built using these images, and one with highest accuracy and least loss would be selected.

2 Related Work

Metal corrosion is a critical challenge in various industries where metallic materials are used. The corrosion in the field of industries and factories would be disastrous and identification of the corrosion at the early stages would ensure such risks are mitigated and safety measures are taken. The corrosion in the biomaterials would be life threatening to the implanted body and no risk be taken in this field in terms of the corrosion of the biomaterials. The tendency of the metallic body to get corroded on a physiological medium need to be analysed before it has been identified for implanting on the body.

The authors in [4] have explored the ways to use the deep learning techniques to perform the autonomous marine vessel inspection to identify the corrosion of the metal surfaces. The authors have explored ten different architectures from traditional classification to object detection to instance segmentation for the purpose. All these architectures were provided with the input images of ballast of the marine vessels having various severity of the corrosion. ResNet50 implementation to classify the severity of the corrosion intensity provided the best possible solution for the segmentation. For localizing the defect, the Mark-RCNN outperformed the object detection.

In [5], the authors have compared the classical computer vision and deep learning techniques for automatic metal corrosion detection. The classical approach used the OpenCV libraries to analyse the images and identify the corrosion based on the colour of the metal surface by making usage of HSV format and applying some transformations on the image to enhance the colour saturation levels. In comparison to this, the deep learning method is applied with Berkeley Vision and Learning Center (BVLC) framework. The framework was further fine-tuned with a good number of images of corroded surfaces which resulted in the overall accuracy of the deep learning model above 88%.

A computer-vision based approach have been proposed to detect corrosion in water, oil, and gas pipelines [1]. The authors have developed a custom Convolution Neural Network (CNN) model on the large number of CCTV images of the corroded pipelines were classified for its severity of corrosion (no-low-medium-high corrosion) with very high classification accuracy (98.8%). Along with that, the proposed localization algorithm would identify the corroded locations in the provided images with high precision. This process ensured there is no need for manual inspection and evaluation techniques for identifying the metal corrosions.

The classical approaches to measure the efficiency of inhibitors are done using electrochemical measurements on the samples and capturing the microscopic images before and after the electrochemical process on the metal for evaluating the intensity of corrosion [6]. But this approach would be considering only the limited surface area under the microscope, only if the care is taken to capture the microscopic images from the right areas of the lamina would be able to provide good insight into the efficiency. But the deep learning models explained on the paper would be able to extend the examination to the entire surface area of the lamina and identify the inhibition behaviour. The images data provided to the CNN attempts to automatically differentiate between standard, unprotected and protected states of the surface that can identify the inhibition behaviour [7].

Stoean et al. in [7] used deep learning models on the microscopic images of corroded copper and steel laminas and analysed the impact of inhibitor solutions in the process of corrosion to measure efficiency. The author has used steel and copper metal coupons for analysing the impact of inhibitor solutions in reducing the corrosion level on these metals. The microscopic images of these metal surfaces are taken on various time frames with and without the contact of inhibitor solution and the images of these surfaces are studied. In this research the number of images taken were very low in number and hence augmentation technique [8] is used to increase the dataset size.

It is found that there has been limited work on applying the deep learning techniques on the microscopic images of the corroded samples to build and predict the severity of the metal corrosion. The deep learning models, unlike the classical approaches that can examine the limited surface that is under the microscope, would be able to extend the examination of the entire metal surface for the identification of severity of corrosion. The usage of microscopic images for the analysis would be able to get the microstructural features on the corroded metal surface which could help in identifying the corrosion on a much earlier stage and on the metal surfaces that cannot be identified using the regular images.

3   Methodology

The image classification based on the severity of the metal corrosion images are done using the various CNN architectures. In the current scenario, the number of training images available for the purpose of research has been a smaller number of images. To ensure developing a CNN model with stable accuracy, effort for creating larger images set by various techniques is performed. The images available for the purpose of research have images of the metal surface with various levels of magnification. Using these images, newer set of images are built using deep learning technique that can expand the magnification on the area where the higher magnification is not available. This would help in increasing the dataset and getting the images with much higher pixel resolution. Various custom models with different Convolution, normalization, pooling, and dropout layers are built for extracting the different features from the training dataset. Along with custom models, pre-trained models are also used to fine tune the model for the current dataset by using transfer learning. At the end, different models that are built for classifying the images are compared against each other for its accuracy, precision, recall, and F1 Score and the one with the best score is selected [9].

3.1   Data Acquisition

Magnesium alloy (AZ31 alloy) coupons and SS316 coupons were used for the experiments. Magnesium alloy coupons and SS316 coupons were polished on grit papers for smooth and mirror finishing. The final size of the coupons was.

• Magnesium coupons – 0.4 cm × 1.4 cm × 0.9 cm

- SS316 coupons – 2 cm × 1 cm × 0.15 cm

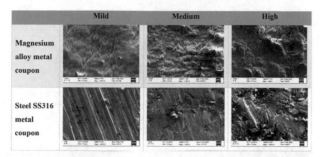

Fig. 1. Samples of Magnesium and Steel metal coupons of different levels of corrosion severity.

Mirror finishing was done so that at the micro-level the surface of the coupons was smooth hence corrosion due to uneven surface structure would be avoided, hence providing precise results. The coupons were then cleansed with acetone to remove all the contaminants, prior to immersion. The coupons were immersed in 40 mL of Saline solution (pH – 7.3–7.4) as simulated physiological solution. The coupons were placed at room temperature for 35 days. Microscopy was performed using Compound Microscope and Scanning Electron Microscope. Microstructure and elemental surface composition of the polished surface and the corroded/fracture sites was observed. For Compound Microscopy, the imaging was performed at 40×. For SEM, imaging was performed at 20 kV under high vacuum with objective aperture 10 μm and 20 μm.

Metal corrosions are studied for both Magnesium and steel, hence there are two distinct classification problems that are investigated within this work. For each task, optical microscopy slides are generated for each of three possible states: Mild, Medium, and High corrosion. In the case of Magnesium, the total amount of available samples is 21:2 with mild corrosion, 11 with medium corrosion, 8 with high corrosion. For steel, there are 16:6 with mild corrosion, 2 with medium corrosion, 8 with high corrosion. Sample images of the 3 categories of corroded samples of both Magnesium and Steel are as shown in Fig. 1.

3.2 Proposed Model Workflow for Custom CNN Model

Here the deep learning (DL) techniques are employed to determine and classify the images. While human vision can pay attention to the shape of an object to identify it, DL computer vision algorithms can focus on their texture instead. The microscopic images can provide very minute details of texture of corroded layer and since deep learning algorithms can focus on the texture of the images in detail, deep convolutional neural networks can be used to detect corrosion on a laboratory environment [7]. Figure 2 shows the data flow diagram of how the classification problem of SEM images are solved using the deep learning technique. Steps are summarized as follows:

1. Pre-processing the images

2. Split the images into train-test group
3. Synthesize and augment the images
4. Fit and validate custom CNN models

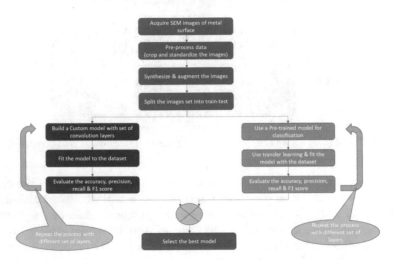

Fig. 2. Data flow diagram of DL approach to Classification problem.

Pre-processing the Images. The deep learning models can be trained well when there are large set of images available for training. Since for the purpose of study, there are limited number of images available, we need to deploy different methods to create larger set of images for the training and validation.

i) The images produced by the SEM are high-definition images having the 1024 × 672 resolution (after removing the label at the bottom of each image). These images have the surface structure of the metal surface in the entire area of the image. Hence these images, if cut into smaller images, will still have the same severity of corrosion on each of these cut images. A deep learning model would be able to learn with the at least input images of 64 × 64 size. Using this, we split the images into smaller images of 256 × 168 non-overlapping patches which can be used for training. This step of creating the non-overlapping patches would be able to create 576 images of 256 × 168 dimensions.

ii) The images that are available for the purpose of training have been captured with different magnification, starting from 200× to 5000×. The higher magnification images are taken from the same metal surface as that of low magnification images. The area covered by 2000× images is 16 times smaller than that of 500× images. Refer to Fig. 3 for the pictorial representation. Using the image reconstruction method, we will generate higher resolution images and with this approach 16 images of 2000x be generated. Using the approach mentioned in this paper [10–12], we would be using the low-resolution images to reconstruct the higher resolution images without further need of electron microscope. Refer to Table 1 for the image's dataset distribution with respect to the magnification. With this approach, we would be able to increase the number of images with higher resolution. By splitting each image with 64×42 dimension and then feeding into super resolution generative adversarial model would generate 9216 non-overlapping patches of 256×168 size.

Fig. 3. Image reconstruction example. Left side of the image has $500\times$ magnification whereas the right-side image has $2000\times$ magnification.

Table 1. Matrix of different severity classes and count of images at different resolution levels.

Severity/Resolution	Low	Medium	High	Total
200×	0	0	1	1
500×	0	4	8	12
1000×	4	6	2	12
1666×	1	0	1	2
2000×	2	3	3	8
5000×	1	0	0	1
Total	**8**	**13**	**15**	**36**

Splitting the Images into Train-Test Group. Splitting of dataset into train and test is necessary for any machine learning or deep learning models to estimate the performance of the algorithm or the model when they are used for prediction on the dataset that was not used for training. The training part of the dataset is used for fitting the model whereas the test part of the dataset would be used for evaluation of the model. The general rule of thumb is to split the dataset into 70-15-15 ratio, 70% of the images would be used for training and 15% of the images to be used for validation and remaining 15% would be used for evaluating the model in the test phase.

Synthesize and Augment the Images. Computer vision deep learning problems are much dependent on the large set of images that can be used for training. To gain sufficient performance of the model, the models would need to be trained with sufficient quality, quantity, and variety of images. We would further be applying the image augmentation technique on these images of 256 × 168 resolutions that would help us to build variety and enough quantity of images. The augmentation process would help in resolving the overfitting problem [12].

The images from SEM are of grayscale in nature, hence the images augmentation would involve the below methods.

- Rotation: the images are rotated clockwise and anti-clockwise for a given angle.
- Flipping: Images are flipped in vertical and horizontal manner.
- Brightness: Increasing and decreasing the brightness of the images.
- Scaling: applying defined level of zoom operation to have the images scaled inwards or outwards.

Fit and Validate Custom CNN Models. Building CNN models would be done keeping in mind the limited number of images available for training. The goal in this step is to build the CNN classifier model to classify the severity of metal corrosion. Deep learning algorithms like CNNs focus on the texture of the image and since the SEM images can provide the texture information in deep detail, the CNN model would be able to predict the severity of the corrosion. Hence, the method assumes that a CNN will be able to classify the severity of the of corroded metal samples. The process would be implemented using python and TensorFlow and the deep learning framework. A CNN model would essentially be having 5 stages [7].

i)   Input layer: Input image of 256 × 168 dimensions would be input-ed to the model
ii) Convolution layers: Series of intermediate NN layers where the kernels or filters of pre-defined size is convolved over the image in the layer. For the experiment we would be having lower units of convolution layers using kernels and depth. Each convolution layer is followed by Rectified Linear Unit (ReLU) as Activation function.
iii) Pooling layer: To down sample the feature detection in the feature maps, this layer is used. Using max of the 2 × 2 pixel values makes it as MaxPooling layer whereas using average value of 2 × 2 pixel values makes it as AvgPooling layer.
iv) Dropout layer: To generalize and resolve the overfitting problem, batch normalization and dropouts are used between each layer. We would be using different combinations between 0.2 to 0.8 for dropouts on the network for the purpose of experiment
v)   Fully connected dense layers: The final layer before the softmax output is to have the fully connected dense layer. The softmax output layer would be having 3 categories of metal corrosion severity classified by the network.

Figure 4 depicts the complete architecture of the custom CNN model that has been designed to generate a CNN classification model.

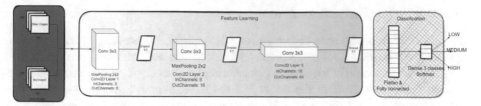

Fig. 4. The CNN layers and parameters for the corrosion analysis of Magnesium and Steel.

3.3   Using Transfer Learning for the Image Classification

Transfer Learning is a method of leveraging state of the art models (pre-trained on huge datasets) for specific use-cases without the hassles of worrying about preparing large datasets or access to the latest GPU setup [13]. We would be using this pre-trained model for classifying the corrosion severity of the SEM images. VGG19 model is input-ed with SEM images dataset after performing augmentation. The weights for the model would be kept same as ImageNet dataset and the layers after the convolution layers are dropped and refitted with the custom fully connected dense layer. The model call is followed with few dense layers and at the end with the softmax output layer with 3 categories of severity of corrosion classified.

To the current scenario, the ResNet50 models are used as the pre-trained model for performing the transfer learning. The Resnet model is designed to perform excellently for image classification problem. In the process of transfer learning, all the layers and the weights in these five convolution layers are kept intact as the task of convolution layers are to extract the various features from the input images and there would be no changes in the current process and the pre-trained process. The weights setup for the entire convolution layers would remain the same and hence no backpropagation be done on these layers to change the weights. With this approach, all the feature selection process of the image would be done in the same way the ResNet-50 models are built. The last layers of fully connected layers are removed from the model and fitted with two new layers with output from each new layer as 256, 128 respectively. The final softmax layer is created to have the model classify the input images into three categories of corrosion severity: Low – Medium – High. The Cross Entropy loss function and Adam optimizer with learning rate as 0.0025. The model is executed for 50 epochs to learn and reduce the loss in each iteration to get a better accuracy model for classifying the images into right categories.

4   Results and Discussion

This section explains in detail about evaluating the different models that are used in the research. The custom CNN model that has three convolution layers are trained and tested on both the Steel and Magnesium images dataset to get the accuracy, precision, recall and F1- Score performance metrics. Similarly, the transfer learning models using ResNet50 is also trained and tested using the dataset and the different performance metrics are evaluated.

4.1   Data Preparation and Analysis

To build a stable predictable deep learning CNN model, the amount of training data be sufficiently large. Also, the CNN model can perform well with images of size 128 × 128 for a 3 channel (RGB) image. Considering the images taken from the scanning electron microscope is from the surface of the metal coupons, the images have the surface captured in the entire area of the image. Hence the images can be split into non-overlapping patches of these images of a definite size. This would help in creating a larger set of images for training and evaluation.

Deep learning models have the capability to generate super resolution images from a low resolution images using Super Resolution Generative Adversarial Network (SRGAN) [10]. The dataset contains images of different magnification of the same metal surface area to validate if the magnification generated by the SRGAN technique is equally good enough for the purpose of building the model.

With this analysis and validation, the SRGAN can reconstruct a low-resolution image with high-resolution with very minimal differences when compared with original high-resolution images. Applying the same technique to the entire set of images, patches of 64 × 42 size to create set of low-resolution images and applying SRGAN to these low-resolution images to generate equivalent higher resolution images.

With this approach, there are a total of 336 + 5376 = 5712 images of Magnesium and 240 + 3840 = 4080 images of Steel metal. For these set of images, the augmentation techniques are applied to generate multiple variations of the generated images.

In all, for Steel there are a total of 9520 High severity images, 2720 medium severity images and 8160 low severity images are for training and testing of the model. Similarly for Magnesium, there are a total of 10880 high severity image, 14960 medium severity images and 2720 low severity images are available. Table 2 has the complete calculation of the entire images generated using various techniques. With these techniques, there is a sufficiently large set of images for the purpose of building a CNN model.

Table 2. Images generated by different techniques and the total count of images.

Metal	Category	Original	SRGAN-HR	Horz-flip	Sharpness	Inverted	Multiple	Total	Train (70%)	(15%) Val	(15%) Test
Steel	High	112	1792	1904	1904	1904	1904	9520	6664	1428	1428
	Medium	32	512	544	544	544	544	2720	1904	408	408
	Low	96	1536	1632	1632	1632	1632	8160	5712	1224	1224
	TOTAL	240	3840	4080	4080	4080	4080	20400	14280	3060	3060
Magnesium	High	128	2048	2176	2176	2176	2176	10880	7616	1632	1632
	Medium	176	2816	2992	2992	2992	2992	14960	10472	2244	2244
	Low	32	512	544	544	544	544	2720	1904	408	408
	Total	336	5376	5712	5712	5712	5712	28560	19992	4284	4284

4.2 Evaluation of Models

The models that are built as described in the previous chapter are trained using the dataset and with different hyper-parameters. The dataset contains images from Magnesium and Steel metal coupons surface. During the development phase, various custom CNN models are built by varying the CNN layers and having different kernel sizes and number of output features from each layer and evaluated for the training and validation dataset accuracy and loss. The one with best training + validation accuracy and minimum loss values are selected to evaluate further with the testing dataset.

Training and Evaluation of the Model for Steel Images. In the first phase, the custom CNN model is trained with Steel training images and validated with validation dataset. The dataset is lightly skewed as the number of images in the medium severity of corrosion images are much lesser compared to the other two categories. The dataset is used for training both custom CNN model and the Resnet50 transfer learning model.

For the custom CNN model training, the model is executed for 100 epochs with the learning rate of 0.01 and weight decay of 0.0001. The training dataset was input-ed to the model with batch of 192 to optimize the GPU utilization and parallel processing. The model could quickly reach 80%+ training accuracy within 20 epochs, and it continued to be above that level until the end of 100 epochs. The same model was able to generate an accuracy of 70%+ validation accuracy consistently with limited movement around 70% mark for entire 100 epochs. The model performed best in the 81st epoch and it was able to provide a training accuracy of 97.73% whereas the validation accuracy of 81.08%. Equivalent loss in this epoch was 0.0867 for training data whereas the loss value was 0.0268 for validation data. The model was mostly consistently generating the validation accuracy of 75%+ in most of the times after 65 epochs. At the end of 100 epochs the model seems to be stabilized at 70–75% validation accuracy and 97–99% training accuracy. Similarly, the validation loss was around 0.03–0.05 and the training loss was hovering around 0.03–0.07. With these observations training of the model with further epochs was not explored and the model that was generated in the 81st epoch with 81.08% validation accuracy and 0.0268 validation loss was picked up as the best model and the weights of this model was saved for performing the prediction on the test dataset.

For the transfer learning model training (**ResNet-50**), the model is executed for 50 epochs with the learning rate of 0.0025. The training dataset was input-ed to the model with batch of 192 to optimize the GPU utilization and parallel processing. The model could quickly reach to 90%+ training accuracy within 10 epochs and crossed 92.5%+ accuracy after 20 epochs, and it continued to be above that level until the end of 50 epochs. The same model was able to generate an accuracy of 89%+ validation accuracy within 3 epochs and an upward movement above 90% after 10 epochs reaching a maximum of 93.9% in the 50th epoch. The model performed best in the 50th epoch, and it was able to provide a training accuracy of 95.7% whereas the validation accuracy of 93.9%. Equivalent loss in this epoch was 0.0011 for training data whereas the loss value was 0.005 for validation data. The model was mostly consistently generating the validation accuracy of 90%+ in most of the times after 20 epochs. At the end of 50 epochs the model seems to be stabilized at 92–94% validation accuracy and 93–96% training accuracy. Similarly, the validation loss was around 0.006–0.007 and the training loss was hovering around 0.001–0.0012. With these observations training of the model with further epochs was not explored and the model that was generated in the 50th epoch with 93.9% validation accuracy and 0.005 validation loss was picked up as the best model and the weights of this model was saved for performing the prediction on the test dataset.

Training and Evaluation of the Model for Magnesium Images. In the second phase, the custom CNN model is trained with Magnesium training images and validated with validation dataset. The dataset is also lightly skewed as the number of images in the low severity of corrosion images are much lesser compared to the other two categories. The dataset is used for training both custom CNN model and the Resnet50 transfer learning model.

In this model training, the model is executed for 100 epochs with the learning rate of 0.01 and weight decay of 0.0001. The training dataset was inputed to the model with batch of 192 to optimize the GPU utilization and parallel processing. The model could quickly reach to 90%+ training accuracy within 20 epochs, and it continued to be above that level until the end of 100 epochs. The same model was able to generate an accuracy of 65%+ validation accuracy consistently with limited movement around 70% mark for entire 100 epochs. The model performed best in the 90th epoch and it was able to provide a training accuracy of 94.94% whereas the validation accuracy of 73.62%. Equivalent loss in this epoch was 0.0952 for training data whereas the loss value was 0.0579 for validation data. The model was mostly consistently generating the validation accuracy of 69%+ in most of the times after 70 epochs. At the end of 100 epochs the model seems to be stabilized at 70–73% validation accuracy and 92–95% training accuracy. Similarly, the validation loss was around 0.08–0.1 and the training loss was hovering around 0.04–0.09. With these observations training of the model with further epochs was not explored and the model that was generated in the 90th epoch with 73.62% validation accuracy and 0.0579 validation loss was picked up as the best model and the weights of this model was saved for performing the prediction on the test dataset.

For the transfer learning model training (**ResNet-50**), the model is executed for 50 epochs with the learning rate of 0.0025. The training dataset was input-ed to the model with batch of 192 to optimize the GPU utilization and parallel processing. The model could quickly reach to 80%+ training accuracy within 10 epochs and crossed 90%+ accuracy after 40 epochs, and it continued to be above that level until the end of 50 epochs. The same model was able to generate an accuracy of 80%+ validation accuracy within 7 epochs and an upward movement above 85% after 30 epochs reaching a maximum of 87.7% in the 44th epoch. The model performed best in the 44th epoch, and it was able to provide a training accuracy of 91.1% whereas the validation accuracy of 87.7%. Equivalent loss in this epoch was 0.0025 for training data whereas the loss value was 0.011 for validation data. The model was mostly consistently generating the validation accuracy of 85%+ in most of the times after 25 epochs. At the end of 50 epochs the model seems to be stabilized at 83–88% validation accuracy and 87–92% training accuracy. Similarly, the validation loss was around 0.011–0.014 and the training loss was hovering around 0.002–0.0025. With these observations training of the model with further epochs was not explored and the model that was generated in the 50th epoch with 93.9% validation accuracy and 0.005 validation loss was picked up as the best model and the weights of this model was saved for performing the prediction on the test dataset.

The final phase of the model building is the prediction on the testing dataset. The performance metrics for evaluating the prediction of the model are done using different parameters such as Accuracy, Precision, Recall, and F-1 Score. These metrics are calculated for each class of the model and the model that can provide best values for the classes for each of these metrics would be selected as the best model for the prediction of the classification of the images.

The performance of the two models that are built for Steel dataset are tested. The test dataset consists of 3060 images that belong to the three categories of corrosion severity. The custom model that generated a training accuracy of 97.73% and validation accuracy of 81.08% is subjected to predict the test dataset and the results are captured and the results for the test dataset are shown in Table 3.

Table 3. Precision, recall, and F1-score for the test dataset of steel for custom CNN model.

	Precision	Recall	F1-score	Support
Low	0.74	0.74	0.74	1224
Medium	0.66	0.32	0.43	408
High	0.69	0.79	0.73	1428
Accuracy			0.71	3060
macro avg	0.7	0.62	0.64	3060
weighted avg	0.71	0.71	0.7	3060

The ResNet50 transfer learning model that generated a training accuracy of 92.5% and validation accuracy of 91.2% is subjected to predict the test dataset and the results are captured and the results for the test dataset are shown in Table 4.

Table 4. Precision, recall, and F1-score for the test dataset of steel for ResNet50 model.

	Precision	Recall	F1-score	Support
Low	0.95	0.96	0.95	1428
Medium	0.92	0.84	0.88	408
High	0.94	0.96	0.95	1224
Accuracy			0.94	3060
macro avg	0.94	0.92	0.93	3060
weighted avg	0.94	0.94	0.94	3060

The performance of the two models that are built for Magnesium dataset are tested. The test dataset consists of 4284 images that belong to the three categories of corrosion severity.

The custom model that generated a training accuracy of 94.94% and validation accuracy of 73.62% is subjected to predict the test dataset and the results are captured and reported in Table 5.

Table 5. Precision, recall, and F1-score for the test dataset of Mg for custom CNN model.

	Precision	Recall	F1-score	Support
Low	0.85	0.64	0.73	408
Medium	0.79	0.88	0.84	2244
High	0.81	0.73	0.77	1632
Accuracy			0.8	4284
macro avg	0.82	0.75	0.78	4284
weighted avg	0.81	0.80	0.80	4284

The ResNet-50 transfer learning model that generated a training accuracy of 92.5% and validation accuracy of 91.2% is subjected to predict the test dataset and the results are reported in Table 6.

Table 6. Precision, recall, and F1-score for the test dataset of Mg for ResNet50 model.

	Precision	Recall	F1-score	Support
Low	0.82	0.86	0.84	408
Medium	0.89	0.90	0.90	2244
High	0.88	0.86	0.87	1632
Accuracy			0.88	4284
macro avg	0.86	0.87	0.87	4284
weighted avg	0.88	0.88	0.88	4284

In all, when comparing the two models for each metal, the ResNet-50 transfer learning model has been able to generate a model that can predict much more accurately with all other metrics with high efficiency. The model was also able to get generated with low training time and lesser GPU need as the weights for the model was re-used from the pre-trained model. Table 7 shows the best model selected for the classification of Steel and Magnesium metal corrosion severity.

Table 7. The performance metrics of the best model for steel and magnesium corrosion severity classification CNN model

	Magnesium			Steel		
	Low	Medium	High	Low	Medium	High
Precision	0.82	0.89	0.88	0.95	0.92	0.94
Recall	0.86	0.90	0.86	0.96	0.84	0.96
F1-Score	0.84	0.90	0.87	0.95	0.88	0.95
Accuracy	0.88			0.94		

5 Conclusion

This study proposed an objective and automatic way of examining the metal surfaces and classifying the corrosion intensity through convolutional neural networks using the images from scanning electron microscope. Different CNN models have been used to analyze the images learn and predict the severity of the metal corrosion. The images have been taken from electron microscope of the Magnesium and Steel metal surfaces that are corroded with different levels of severity. This study explored a various option to generate good quantities of images for the limited images that are available such as SRGAN which helps to learn the minute surface level textures analyzed and ensure a model built with higher degree of accuracy. A comparison was made with the existing works and best accuracy results were found for the of the ResNet50 model which are 88% and 94% on Steel and Mg images.

References

1. Bastian, B.T., Jaspreeth, N., Ranjith, S.K., Jiji, C.V.: Visual inspection and characterization of external corrosion in pipelines using deep neural network. NDT E Int. **107**, 102134 (2019)
2. Eliaz, N.: Corrosion of metallic biomaterials: A review. Materials (Basel) **12**, 407 (2019)
3. Ortiz, A., Bonnin-Pascual, F., Garcia-Fidalgo, E., Company-Corcoles, J.P.: Vision-based corrosion detection assisted by a micro-aerial vehicle in a vessel inspection application. Sensors **16**, 2118 (2016)
4. Andersen, R., Nalpantidis, L., Ravn, O., Boukas, E.: Investigating deep learning architectures towards autonomous inspection for marine classification. In: 2020 IEEE International Symposium on Safety, Security, and Rescue Robotics (SSRR), pp 197–204 (2020)
5. Petricca, L., Moss, T., Figueroa, G., et al.: Corrosion detection using AI: a comparison of standard computer vision techniques and deep learning model. In: Proceedings of the sixth international conference on computer science, engineering and information technology, p. 99 (2016)
6. Samide, A.: A pharmaceutical product as corrosion inhibitor for carbon steel in acidic environments. J. Environ. Sci. Health Part A **48**, 159–165 (2013)
7. Stoean, R., Stoean, C., Samide, A.: Deep learning for metal corrosion control: Can convolutional neural networks measure inhibitor efficiency? In: 2018 20th International Symposium on Symbolic and Numeric Algorithms for Scientific Computing (SYNASC), pp 387–393 (2018)

8. Shorten, C., Khoshgoftaar, T.M.: A survey on image data augmentation for deep learning. J. Big Data **6**, 1–48 (2019)
9. Hameed, I.M., Abdulhussain, S.H., Mahmmod, B.M.: Content-based image retrieval: a review of recent trends. Cogent Eng. **8**, 1927469 (2021). https://doi.org/10.1080/23311916.2021.192 7469
10. Ledig, C., Theis, L., Huszár, F., et al.: Photo-realistic single image super-resolution using a generative adversarial network. In: Proceedings of the IEEE Conference on Computer Vision and Pattern Recognition, pp 4681–4690 (2017)
11. de Haan, K., Ballard, Z.S., Rivenson, Y., et al.: Resolution enhancement in scanning electron microscopy using deep learning. Sci. Rep. **9**, 1–7 (2019)
12. Azuri, I., Rosenhek-Goldian, I., Regev-Rudzki, N., et al.: The role of convolutional neural networks in scanning probe microscopy: a review. Beilstein J. Nanotechnol. **12**, 878–901 (2021)
13. Simonyan, K., Zisserman, A.: Very deep convolutional networks for large-scale image recognition. In: 3rd International Conference on Learn Represent ICLR 2015 - Conference Track Proceedings, pp. 1–14 (2014)

Insurance Risk Prediction Using Machine Learning

Rahul Sahai[1], Ali Al-Ataby[2], Sulaf Assi[3(✉)], Manoj Jayabalan[1], Panagiotis Liatsis[4], Chong Kim Loy[5], Abdullah Al-Hamid[6], Sahar Al-Sudani[7], Maitham Alamran[8], and Hoshang Kolivand[1]

[1] Faculty of Engineering and Technology, Liverpool John Moores University, Liverpool L3 3AF, UK
[2] Department of Electrical Engineering and Electronics, University of Liverpool, Liverpool L69 3GJ, UK
[3] School of Pharmacy and Biomolecular Sciences, Liverpool John Moores University, Liverpool L3 3AF, UK
s.assi@ljmu.ac.uk
[4] Department of Electrical Engineering and Computer Science, Khalifa University, Abu Dhabi, UAE
[5] UNITAR International University, Petaling Jaya, Malaysia
[6] Saudi Ministry of Health, Najran, Saudi Arabia
[7] American University of Iraq- Baghdad, Baghdad, Iraq
[8] Biology Department, College of Science, University of Al-Qadisiyah, Iraq

Abstract. Underwriting decisions by insurance companies make a significant contribution to their profitability. Machine Learning (ML) techniques in underwriting decision making have saved time and improved operational efficiencies. A user-friendly cause-and-effect explanation of model's predictions is useful to stakeholders, financial institutions and regulators. This research performed comparative analysis between tree-based classifiers such as Decision Tree, Random Forest and XGBoost. The study focused on enhancing risk assessment capabilities for life insurance companies using predictive analytics by classifying the insurance risk based on the historical data and propose the appropriate model to assess risk. Its purpose also included incorporating mechanisms that can aid in user friendly interpretation of ML models. Of all the models created as part of this research, the XGBoost classifier performed the best when compared to other classifiers, with an AUC value of 0.86 and F1-score above 0.56 on the validation set. The Random Forest classifier got AUC value of .84 and f1 score of .53 on the validation dataset. The results indicate the importance and advantages of tree-based models. These models i.e., XGBoost, decision tree and random forest are one of the best alternate techniques after the advent and popularity of the new age techniques in the machine learning such as neural networks, deep learning etc. The research also provides an insight on the interpretability of these conventional techniques by way of 'SHAP' or shapley values and 'Feature Importance' or 'Variable Importance'. SHAP was used on complex models such as XGBoost and neural networks whereas Feature Importance is used in supervised learning methods such as Logistic Regression and tree- based models such as Decision Tree and Random Forest. Overall, the study was able to propose XGBoost as the most accurate model for Insurance risk classification and predictions.

© The Author(s), under exclusive license to Springer Nature Singapore Pte Ltd. 2023
Y. B. Wah et al. (Eds.): DaSET 2022, LNDECT 165, pp. 419–433, 2023.
https://doi.org/10.1007/978-981-99-0741-0_30

Keywords: Insurance · Risk prediction · Machine learning · Neural networks · Support vector machine · Random forest

1   Introduction

1.1   Background

Insurance Risk is the term for the risk that Insurance companies want to cover to safeguard their interests. The assessment of this risk is performed by classification of insurance applicants as per their risk profile. This risk assessment applies to all common Insurance types in insurance domain such as Life Insurance, Automobile Insurance, Property and Casualty Insurance, Health and Travel Insurance. Risk assessment is an important process in the insurance industry to classify and categorize applicants. Risk profiles of the primary and joint applicants are thoroughly reviewed by underwriting team and then classified as per risk category. Risk classification is the term used in insurance domain to refer to grouping of customers as per their estimated level of risks, derived from their past data. In the case of life and health insurance, insurers make decisions on claims based on different risk factors such as the age and gender of the applicants, medical condition, lifestyle, occupation, medical history, type of product and amount of insurance [1].

Life insurance firms historically had depended on conventional ways of underwriting such as mortality charts and statistical formulas to estimate life expectancy or mortality to draft rules to govern the underwriting process. These conventional methods take too much time and are cost intensive which results in customers changing their mind and switching to another insurance company. The manual process of detection of claims could result in high costs and inaccuracies. Also, manual process can take much longer turn-around time, which could lead to late-detection of a fraudulent claim. A late detection of the fraud might result in additional costs and possibly losses for the insurance companies. As per statistics published by the Federal Bureau of Investigation (FBI), fraudulent Insurance claims costs are more than eighty billion USD yearly in the United States [2]. Hence, Insurance companies need to make the underwriting process faster, cost-effective and accurate. Predictive analytics have demonstrated the usefulness in simplifying the underwriting process and in making timely and quality decisions [3].

Broadly Life, Travel, Health, Automobile and Property Insurance are some of the main verticals in the insurance industry which cater to most of the needs of insurance applicants across the world [4]. Financial institutions offering insurance services carry a significant amount of risk on themselves [5]. Hence, Insurance companies need to be able to accurately quantify exposures to these risks and to manage these financial risks. These risks for insurance companies can come up from multiple sources and require accurate prediction modelling. One fallout of any errors in the forecasting of risks is underpricing of insurance product and it can lead to financial losses for the insurer in the form of future claims [3]. Some major functional areas which are related to insurance risk and demography related risks are 'Claims Modelling', 'Loss Reserving' and 'Mortality Forecasting' [5].

Insurance underwriting has come a long way from traditional statistics-based methods and actuarial tables to modern day predictive analytics. Modern Machine Learning techniques do help in quick classification of prospective customers based on their level of risk. Insurance companies use machine learning techniques for other benefits such as to run successful marketing campaigns and to determine appropriate insurance premium pricing for insurance products. Some of the process areas in which Insurance companies look for predictive modelling are:

a) Claims modelling, which refers to forecasting of claims which applicants can make in the future. This future cost gets determined by factors such as number of claims (frequency) and the monetary value of claim (claim severity) [9].
b) Mortality risk, which gives Insurance companies an indication of estimated longevity of their customers' lives which could be affected by different factors such as age, occupation, location etc. [11].

Insurance underwriting refers to the process of evaluating the profile and details of an applicant to arrive at the quantum of financial risk for the insurance coverage. This evaluation assists in taking a decision on an insurance application and to arrive at the applicable premium if it is accepted. Insurance underwriting also includes 'Exclusion prediction' which raises the flag for an applicant to be excluded from allowing a specific insurance claim based on factors such as medical history, employment and historical records [10].

Having an ability to understand the causal logic of model's decision is greatly beneficial to Insurance companies to assess the long-term consistency in forecasts and to detect potential causes for model's bias or discriminatory results. This really gives insurance companies an edge when they have capability to perform fact-based validation of models. Hence, not only model accuracy but also the interpretability of predictions are main incentives for transitioning to predictive analytics by Insurance companies [6].

This research analyzes the performance and accuracy of different modelling techniques, for the problem of risk level prediction, such as Decision Tree, Random Forest and Logistic Regression (XGBoost), with parallel focus on interpretability and explainability of the important features or variables responsible for the decisions made by these models. This research will attempt to identify and prioritize features that can determine classification and make predictions.

1.2 Literature Review and Identifying Research Gap

The insurance companies have meaningful reasons to implement Machine Learning algorithms in their processes to protect their long-term and short-term goals and interests such as Risk prediction, customer retention, quicker conversions, profitability, anti-selection and Pricing and Reserving. Insurance companies and Machine Learning researchers face many challenges in the highly complex Insurance domain such as 'data availability', 'data quality' and 'missing values', problem of imbalanced datasets, interpretability of the decisions by the models, lack of model fairness etc.

Pricing 'Reserves' are liabilities for Insurance companies. They reflect an insurance company's contractual commitments with respect to the insurance policies it has issued. Reserves are an insurance company's obligations that it may have to pay in future as claims. Blier-Wong et al. reviewed the literature on pricing and reserving for P&C

insurance and they used neural networks for pricing and reserving [11]. Since actuaries use conventional methods such as Generalized Linear Methods (GLMs) for insurance pricing, the results of the study showed that the transition for insurance companies to generalized additive model (GAMs), gradient boosting machine (GBM) or neural networks is so logical and beneficial.

Pricing of the insurance products is another very important factor that drives profitability of the Insurance company. Premium Pricing, which is also termed as Premium Ratemaking in Insurance parlance, for the insurance products require companies to not to fall for anti-selection. That means they don't want to charge low premium for high-risk applicants and high premium for low-risk applicants. By using different data mining methods for selecting risk factors, some researchers included those factors in GLMs to estimate the insurance premiums. They worked a simple approach called Forward Stepwise Regression and identified the risk factors for claim frequency and claim severity. Their research concluded with results that use of risk factor selection methods provide the insurance companies opportunity to refine the model, by having them remove some of the important risk factors from the model [9–11]. With an intent to prevent adverse selection and generate profits, group of researchers developed insurance premium plans with tree-based learning techniques. They compared the loss ratio for the regression tree, random forest and gradient boosting machines and concluded with results that tree based techniques could bring out various deficiencies in the GLMs benchmark premium plans [12].

In terms of quality of data needed for modelling for risk assessment, there could be high possibility of missing values or information for an insurance applicant's historical data. In such instances, Rusdah & Murfi [13] emphasized that for an accurate analysis and predictions, it is important to manage such missing value problems. However, their research compared the accuracy of an XGBoost model, that was obtained from data with the standard imputation of the missing values, with that of an XGBoost model, run on a dataset that did not go through the imputation process. The results proved that XGBoost can handle missing values without an imputation pre-processing step. The accuracy of the XGBoost model that was obtained from dataset, that did not go through the pre-processing step of imputation of missing values, was found comparable to the accuracy of the XGBoost model that was obtained from data with standard imputation process step [8]. With regards to the typical challenge of an 'Imbalanced Dataset', in the insurance domain, which is more often the case of an insurance dataset since the number of policies registering claims is usually considerably lower than the number of policies not registering any claims. Hence when the dataset is imbalanced, some machine learning techniques do not consider the small class which result in high overall model accuracy. In their study, Hanafy & Ming [14] researched resampling techniques such as over-sampling and under-sampling techniques and concluded using sensitivity that the resampling methods are very effective in handling the imbalanced data [6, 7].

Too many attributes in a dataset are a common phenomenon in the insurance domain since companies try to capture as much information as possible for underwriting purposes so that they can evaluate applications carefully. In their study, Boodhun & Jayabalan [3] reviewed the dataset provided by Prudential Life Insurance, having 128 attributes. They compared two dimensionality reduction techniques i.e., Principal Components Analysis

(PCA), a feature extraction technique, and Correlation-based Feature Selection (CFS), a feature selection technique. The results of the study reflected that almost all the models in the study achieved lower errors in case of CFS when compared to PCA. They proposed customer segmentation for better underwriting process.

By illustrating K-means clustering algorithm-based data mining techniques, Qadadeh & Abdallah [15], observed that the CRM database or customers data of an insurance company brought interesting insights and information using ML and visualization techniques that otherwise will take longer time and with less accuracy by human experts. These insights provide, not only to the underwriters but also to the marketing experts, new perspectives on the customer's demographic and behavioral patterns, by means of effective EDA and visualization, that would help them devise and design cross-marketing campaigns to fulfil customer's interests [1–3]. The survey by Mashrur et al. provided the relationships between different ML learning methods with different applications in Insurance domain. The areas like Claims modelling, Loss Reserving and Mortality modelling can be researched by Regression learning task under Supervised Learning method. Areas such as Insurance pricing can be researched via clustering learning task and Insurance underwriting under dimensionality reduction under Unsupervised Learning method [4].

While the reliability of forecasting methods and accurate predictions are important factors, understanding why a model makes certain forecasts is very crucial for the stakeholders. The technical understanding of the model to derive results is termed as Model Interpretability. The highest accuracy for large datasets is very often accomplished by advanced and complex ensemble or deep learning models but their interpretability always remains a challenge. Various studies have recently proposed methods to help stakeholders understand, in simple terms, the predictions of complex models like XGBoost, for example Shapley additive explanation values (SHAP) and Local Interpretable Model-agnostic Explanation (LIME). The model is refined by reducing the set of features by ranking features according to SHAP values. The SHAP method assigns each feature an importance value which is a sort of rank for that feature in the overall set of features.

For many Machine Learning methods used in Insurance domain, algorithmic bias may cause the insurance pricing models to make forecasts that are, in a sense, discriminatory or unfair to a particular group or section. The bias could arise out of data itself. The survey done by Mehrabi et al. on Bias and Fairness reviewed many real-world instances of Machine Learning algorithms giving suboptimal and discriminatory predictions. Their survey categorized different types of discrimination that may occur and listed different types of bias and fairness definitions [5].

This research attempts to review and address some of the challenges pertaining to insurance risk classification, as indicated in this section, using different tree-based classifiers. This research will also work on the interpretability aspect of the model with the best results and ways to explain individual predictions.

2 Research Methodology

The process of predicting insurance risk is a challenging task for Insurance companies. They rely on traditional methods such as statistical methods and actuarial tables for underwriting purposes. Conventional methods take much longer time to evaluate the risk and can lead to financial exposure for Insurance companies. Insurance datasets tend to be imbalanced datasets. In view of such challenges, Machine Learning techniques and advanced data pre-processing techniques manage these challenges very well and provide the best models for insurance risk forecast process. This methodology section covers information on the dataset and its attributes, pre-processing steps and modelling techniques and evaluation metrics. This methodology also covers information on a balancing technique called 'SMOTE' and dimensionality reduction technique called 'PCA') and interpretability method called 'SHAP'.

2.1 Datasets and Pre-processing

The dataset is the most important aspect of a data science or machine learning research. The dataset being used for this research is of Prudential Life Insurance available at Kaggle.com (Prudential Life Insurance Assessment | Kaggle). This dataset contains 59381 applicant records and 128 features/attributes, which give characteristic details of the applicants. This dataset has all sorts of variables such as nominal, continuous and discrete variables. Table 1 shows the snapshot of the variables/data dictionary in the dataset.

Table 1. Data Dictionary of the dataset.

Variable	Description
ID	A unique identifier associated with an application
Product_Infor_1–7	A set of normalized variables relating to the product applied for
Ins_Age	Normalized age of applicant
Ht	Normalized height of applicant
Wt	Normalized weight of applicant
BMI	Normalized BMI of applicant
Employment_Infor_1–6	A set of normalized variables relating to the employment history of the applicant
InsuredInfor_1–6	A set of normalized variables providing information about the applicant
Insurance_History_1–9	A set of normalized variables relating to the insurance history of the applicant
Family_Hist_1–5	A set of normalized variables relating to the family history of the applicant
Medical_History_1–41	A set of normalized variables relating to the medical history of the applicant
Medical_Keywood_1–48	A set of dummy variables relating to the presence of/absence of a medical keyword being associated with the applicant

(continued)

Table 1. (*continued*)

Variable	Description
Response	This is the target variable, an ordinal variable relating to the final decisions associated with an application

Prudential Life Insurance dataset has many attributes which have considerable amount of missing data. Data pre-processing step will involve analyzing and building a mechanism for suitable imputation method for the dataset. This step will also involve dropping unnecessary columns.

2.2 Transformation/Augmentation

Data transformation is the process of changing the format, structure, or values of data. This step involves building and applying strategies to handle the inconsistencies in the data for model building. Data Augmentation is the process to address the problem of class imbalance in classification tasks. Both these steps will need to be performed before any model building can be done.

Another important step in Data Transformation that could be performed is Discretization. This process involves deriving data intervals out of continuous data. This step makes the data more comprehensible for analysis and visualization.

2.3 Exploratory Data Analysis

Exploratory Data Analysis (EDA) step will be performed for univariate and bivariate analysis of the processed data. This step will also assist in gaining important insights for different features of the dataset such as outliers, correlation between variables etc. These insights are achieved by way of several visualizations' techniques such as charts and graphs which include heat maps, pie and bar charts and histogram etc.

2.4 Balancing Techniques

One of the main challenges that Researchers face in predicting future insurance claims for Insurance companies is the existence of imbalanced datasets. Hence when the data is imbalanced, certain machine learning techniques simply tend to ignore the small class which results in high model accuracy which directly affects the model's performance. To resolve this problem, resampling techniques, such as Over-sampling, under-sampling, hybrid, and the synthetic minority over- sampling technique (SMOTE), can be used. This proposal has shortlisted SMOTE technique for this research. SMOTE doesn't lose data as under sampling does and this approach increases the features available to each class. SMOTE increases the percentage of only the minority cases.

2.5 Modelling Techniques

Primarily three Machine learning algorithms i.e., Decision Tree, Random Forest and XGBoost will be reviewed and built, and their results will be compared.

Decision Tree. A decision tree is a supervised learning approach which is generally used for classification problems. It is a flowchart like structure or a tree like structure in which the internal nodes are the data variables, the branches are the decision rules, and each node is the output. It breaks down a dataset into smaller subsets. It simultaneously develops an associated tree incrementally. In order to build the tree, CART (Classification and Regression Tree) algorithm is used. Since it replicates the human thinking approach while making a decision by way of Yes-No questions, it is very easy to understand and explain.

Random Forest. Random Forests is a tree-based ensemble learning technique which is used for classification and regression problems. The Random Forest classifier builds several decision trees simultaneously and takes predictions from each tree. It bases its final output on the majority vote of predictions of all the trees.

$$g(x) = f_0(x) + f_1(x) + f_2(x) \ldots . + f_n(x) \tag{1}$$

A random forest model is expressed as per Eq. 1 where $g(x)$ is the final model which is sum of all the models. Each model $f(x)$ is a decision tree.

XGBoost. The mathematical expression of XGBoost is as per Eq. 2. The XGBoost supports both regression and classification models and can handle large volumes of complex data with automatic handling of the missing values. It helps in finding the best tree model that works well and prevents overfitting of the data. The objective function (loss function and regularization) at iteration t that we need to minimize is the following:

$$\mathcal{L}^{(t)} = \Sigma_{i=1}^{n} l\left(u_i, \mathbf{0}_i^{(t-1)} + f_t(x_i)\right) + \Omega(f_t) \tag{2}$$

u_i is the real value (label) known from the training dataset. The x in objective function $f(x)$ was the sum of t CART trees and after this it becomes a function of the current tree (step t) only. $f_t(x)$ learns from the residuals of $\hat{u}_i^{(t-1)}$ and is the learner that greedily minimizes an objective function $\mathcal{L}^{(t)}$. .. It is easy to see that the XGBoost objective is a function of functions (i.e. l is a function of CART learners, a sum of the current and previous additive trees).

Interpretability Method. To develop a better understanding regarding the developed model and its predictors variables, SHAP (Shapley additive explanations method) could be used in the models as it can help find the importance or strength of each feature. SHAP will be used because it can provide good explanations for local and global models. In contrast to the existing mechanisms to identify important features in machine learning models, SHAP can help in identifying whether the contribution of each input feature is positive or negative. In particular, the importance of a feature i is defined by the Shapley value in Eq. 3:

$$\Phi_i = \frac{1}{|N|!} \sum_{S \subseteq N \setminus \{i\}} |S|!(|N| - |S| - 1)! [f(S \cup \{i\}) - f(S)] \tag{3}$$

Here, $f(S)$ refers to the output of the ML model to be interpreted using a set of S features, and N is the complete set of all features. The final contribution or Shapley value of feature i (Φ_i) is determined as the average of its contributions across all possible permutations of a feature set.

Required Resources. To carry out the research work, the following software packages, libraries and programming languages are to be used:

a) Anaconda Application
b) Python Programming Language
c) Jupyter Notebook
d) Numpy
e) Pandas
f) Seaborn
g) Scikit-learn
h) Matplotlib
i) Scipy
j) Statsmodel

For the hardware, a machine with the following specifications are required:

a) Windows Operating System
a) x86 64-bit CPU (Intel/AMD architecture)
b) 4 GB RAM
c) 5 GB free disk space

2.6 Performance Evaluation Metrics

Primarily, the following metrics are going to be used for this research. These have been shortlisted basis the existing research that was reviewed and discussed in the Literature Review.

Confusion Matrix. A confusion matrix is a matrix shaped representation used for evaluating performance of binary classification problems. It compares between actual target values and the predicted values provided by model. As shown in Table 2, the rows represent the predicted class, while the columns represent the actual class. In the matrix, True Positive and True Negative represent the quantity of actual positives and negatives which are correctly identified, whereas False Positive and False Negative represent incorrectly predicted positive and negative instances by the model. In insurance parlance, true positive would represent that applicant did not file claim where true negative would mean filing of a claim.

Accuracy. It is defined as the proportion of correctly predicted instances to the proportion of total instances. The efficacy of the accuracy metric is achieved on the balanced datasets. The accuracy is at its best for a model when it is 1.0, whereas it is at worst when

Table 2. Confusion matrix.

	Actual positive	Actual negative
Predictive positive	True Positive (TP)	False Negative (FN)
Predictive negative	False Positive (FP)	True Negative (TN)

it is 0.0. Accuracy is calculated as the total number of two correct predictions (TP + TN) divided by the total number of a dataset (P + N) as shown in the following equation:

$$Accuracy = \frac{TP + TN}{TP + FP + TN + FN} \tag{4}$$

Precision and Recall. The precision is defined as the proportion of correctly predicted positives to the proportion of total positives class (includes those classified correctly and incorrectly) as shown in Eq. 5. This reflects how correctly the class is classified and whether if it belongs to the right class. Recall is used to measure how well the fraction of a positive class becomes correctly classified. It is the ratio of number of correctly predicted positives to the total number of positives as shown in Eq. 6. The higher is the Recall for a model, the more positives are predicted.

$$Precision = \frac{TP}{TP + FP} \tag{5}$$

$$Recall = \frac{TP}{TP + +FN} \tag{6}$$

Sensitivity and Specificity. The sensitivity, also known as 'True Positivity Rate', is the ratio of predicted positive class with the actual positives as shown in Eq. 7. It helps to evaluate a model's ability to predict true positives for each category. Specificity, also known as 'true negative rate' is the ratio of predicted negative class with the actual negatives as shown in Eq. 8. In insurance risk context, sensitivity relates to the ability of a model to predict the occurrence of claims.

$$Sensitivity = \frac{TP}{TP + FN} \tag{7}$$

$$Specificity = \frac{TN}{FP + TN} \tag{8}$$

F1-Score or F-Measure. The F-measure combines precision and recall and calculates their harmonic mean to compute a single score. Its mathematical expression is shown in Eq. 9.

$$F - measure = 2\frac{precision \times recall}{precision + recall} = \frac{2TP}{2TP + FP + FN} \tag{9}$$

Kappa Statistics. In simple terms, Kappa statistics measures how closely the predictions of the classifier are to that of one which predicts as per the frequency of each class.

Kappa statistics measure is good for multi class and imbalanced class problems such as auto insurance claims. It measures how much in agreement are two classifiers who are rating the same quantity. Kappa Statistics is less than or equal to 1. If the value is less than 0 or negative, the classifier is evaluated as not good. Its mathematical expression is shown in Eq. 10.

$$K = \frac{pr(a) - pr(e)}{1 - pr(e)} \tag{10}$$

AUC Measure. The AUC (Area under Curve) is a measure to interpret how perfectly a classifier can distinguish between classes i.e., it measures performance of a model in terms of how much are correct and incorrect classifications. The higher is the value of AUC, the better is the performance of the model in the positive and negative classes. AUC value of greater than 0.5 and less than 1.0 shows that the classifier can distinguish positive class and negative class correctly.

3 Experimental Work

Multiple models are created with 3 classifiers i.e., Decision Tree, Random Forest and Logistic Regression. For each classifier, models are created with class balancing and without the class balancing process (through SMOTE). An additional model is created for each of the previous models (with and without class balancing) with their hyper-parameters tuned.

For data processing, there are 13 features that had data missing. A total of 8 variables, which had missing values greater than 40% were removed from the dataset. The dataset is split into train and test datasets in ratio of 80:20. The response column is removed from the dataset for modelling. For the remaining numerical variables that had missing values, imputation was done using the mean values. As part of pre-processing, dummy variables were created for categorical variables resulting in total of 192 variables for modelling.

For evaluation of the classifiers, metrics such as F1-score, AUC and ROC curve are going to be used. F1-score is a much better metric than accuracy when there is uneven class distribution. If the F1-score is high, both precision and recall are also high. F1-score is considered a measure of combination of precision and recall at a particular threshold value. AUC measures how much a model is capable to measure the separability of the different classes. If the AUC is higher, the model is capable of predicting classes correctly. For e.g., a higher the AUC of a model, the better it is in distinguishing between different species of plants when it is fed with data around measurements of the petals and sepals. A model with measurement of 1 for AUC is very good model whereas AUC of 0.5 or lower is indicative of model not having a good capacity to distinguish between classes. ROC (receiver operating characteristic curve) is a graph that shows the performance of a classifier at all classification thresholds. Classifiers that give curves closer to the top-left corner indicate a better performance. The final evaluation of these classifiers is performed on the hold out dataset which was earmarked for testing at the pre-processing/pre-modelling stage at 20% of overall dataset.

4 Results

Table 3 shows the three classifiers performances when the imbalance is not handled.

Table 3. Classifiers performance (imbalance not handled)

Model	Accuracy	AUC	Recall	Prec	F1	Kappa
Decision Tree	0.4425	0.6617	0.3842	0.4424	0.4424	0.3082
Random Forest	0.5678	0.8461	0.4777	0.5429	0.5369	0.4428
XGBoost	0.5796	0.8579	0.4916	0.5543	0.5570	0.4623

Overall, decision tree model shows a low AUC of 0.66 and F1-score of 0.44, while random forest and XGBoost performed better with 0.85, 0.54, and 0.86, 0.56, respectively, as shown in the above table.

The imbalance was handled using the class balancing technique of SMOTE. Table 4 shows the three classifiers performances when the imbalance is handled. Overall, decision tree model shows a lower AUC and F1-score, while random forest and XGBoost performed better as shown in the table.

Table 4. Classifiers performance (imbalance handled with SMOTE)

Model	Accuracy	AUC	Recall	Prec	F1	Kappa
Decision Tree	0.4332	0.6590	0.3847	0.4409	0.4366	0.3014
Random Forest	0.5512	0.8361	0.4825	0.5202	0.5232	0.4254
XGBoost	0.5768	0.8551	0.4948	0.5521	0.5560	0.4598

Figure 1, Fig. 2 and Fig. 3 shows the ROC curves for three classifiers in the case when the imbalance was handled. From Fig. 1 (decision tree classifier), it can be seen that AUC of any of the individual classes do not exceed 0.75 except for class 8. From Fig. 2 (random forest classifier), it can be seen that there are good AUC scores for classes 3,4,5, 7 and 8. From Fig. 3 (XGBoost classifier), the AUC of classes 3,4 and 8 are good.

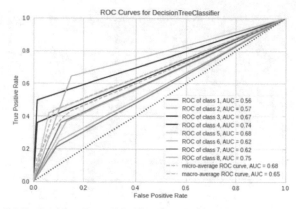

Fig. 1. ROC - Decision Tree Classifier (imbalance handled through SMOTE)

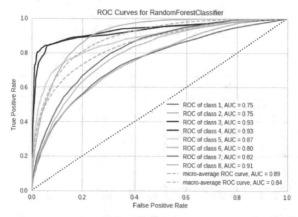

Fig. 2. ROC – Random Forest Classifier (imbalance handled through SMOTE)

Table 5 shows the comparison of performance of different models developed for all the 3 classifiers on the training set. The XGBoost's AOC scores and F1-scores outperform the scores of other classifiers i.e., Decision Tree and Random Forest. There is a marginal effect in the scores when the imbalance through SMOTE is handled and it is marginally better when the imbalance is not handled. The next better scores are of Random Forest.

Table 6 shows the performance of the models on test dataset. Again, the XGBoost's AOC scores and F1-scores are impressive and Random Forest also pretty and not very far behind. Again, there is a marginal effect in the scores when the imbalance through SMOTE is handled and it is marginally better when the imbalance is not handled. The results confirm the fact of superiority of XGBoost over Decision Tree and Random Forest is its ability to give better performance by making use of more trees.

Finally, it was found that the important features that drive insurance claims and which are of practical importance are: BMI, Product_Info_4, Medical_History_15, Ins_Age, Medical_History_23 and Medical_History_4.

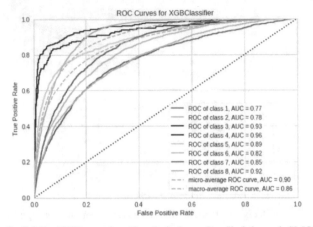

Fig. 3. ROC – XGBoost classifier (imbalance handled through SMOTE)

Table 5. Comparison of performance of models on training set.

	AOC		F1-score	
	Imbalanced	Balanced	Imbalanced	Balanced
Decision tree	0.6617	0.6592	0.4424	0.4366
Random forest	0.8406	0.8361	0.5283	0.5232
XGBoost	0.8579	0.8551	0.5570	0.5560

Table 6. Comparison of performance of models on validation set.

	AOC		F1-score	
	Imbalanced	Balanced	Imbalanced	Balanced
Decision tree	0.6624	0.6576	0.4435	0.4329
Random forest	0.8439	0.8384	0.5355	0.5258
XGBoost	0.8606	0.8562	0.5634	0.5569

5   Discussion and Conclusions

This research was performed on the training dataset of Kaggle's Prudential Life Insurance Dataset. Before the dataset was split into training and test/validation datasets, exploratory data analysis and pre-processing steps were performed. Three different classifiers were used for model building i.e., Decision Tree, Random Forest and XGBoost. These models were applied with SMOTE to handle the class imbalance of the dataset. The models based on XGBoost outperformed models of other classifiers. However, the results showed that SMOTE class balancing technique doesn't necessarily improve the performance metrics in a multi-class dataset. The AUC and F1-score were highest for XGBoost model and the individual classes had highest area under ROC as compared to that of other models. Based on the results, the Insurance risk classification on Prudential Life Insurance dataset proposes XGBoost as the most accurate model, among all three, that can forecast insurance risks.

References

1. Wang, Y.P.: Predictive machine learning for underwriting life and health insurance, pp. , 19–22, October (2021)
2. Dhieb, N., Ghazzai, H., Besbes, H., Massoud, Y.: Extreme gradient boosting machine learning algorithm for safe auto insurance operations. In: 2019 IEEE International Conference on Vehicular Electronics and Safety, ICVES 2019, pp.1–5 (2019)
3. Boodhun, N., Jayabalan, M.: Risk prediction in life insurance industry using supervised learning algorithms. Complex Intell. Syst. **4**(2), 145–154 (2018). https://doi.org/10.1007/s40 747-018-0072-1
4. Rawat, S., Rawat, A., Kumar, D., Sabitha, A.S.: Application of machine learning and data visualization techniques for decision support in the insurance sector. Int. J. Inf. Manag. Data Insights **12**, 100012 (2021)
5. Mashrur, A., Luo, W., Zaidi, N.A., Robles-Kelly, A.: Machine learning for financial risk management: a survey. IEEE Access **8**, 203203–203223 (2020). https://doi.org/10.1109/ACC ESS.2020.3036322
6. Rodríguez-Pérez, R., Bajorath, J.: Interpretation of machine learning models using shapley values: application to compound potency and multi-target activity predictions. J. Comput. Aided Mol. Des. **34**(10), 1013–1026 (2020). https://doi.org/10.1007/s10822-020-00314-0
7. Al-Jumeily, D., Hussain, A., Alghamdi, M., Dobbins, C., Lunn, J.: Educational crowdsourcing to support the learning of computer programming. Res. Pract. Technol. Enhanced Learn. **10**(1), 13 (2015). https://doi.org/10.1186/s41039-015-0011-3
8. Henckaerts, R., Côté, M.P., Antonio, K., Verbelen, R.: Boosting insights in insurance tariff plans with tree-based machine learning methods. North Am. Actuarial J. **25**(2), 1–31 (2020). https://doi.org/10.1080/10920277.2020.1745656
9. Mohamed, A.H.H.M., Tawfik, H., Norton, L., Al-Jumeily, D.: e-HTAM: a technology acceptance model for electronic health. In: 2011 International Conference on Innovations in Information Technology, IIT 2011, pp. 134–138, 5893804 (2011)
10. Alloghani, M., Aljaaf, A., Hussain, A., Al-Jumeily, D., Khalaf, M.: Implementation of machine learning algorithms to create diabetic patient re-admission profiles. BMC Med. Inform. Decis. Mak. **19**, 253 (2019)
11. Keight, R., Aljaaf, A.J., Al-Jumeily, D., Özge, A., Mallucci, A.C.: An intelligent systems approach to primary headache diagnosis. In: Lecture Notes in Computer Science (including subseries Lecture Notes in Artificial Intelligence and Lecture Notes in Bioinformatics) LNCS, vol. 10362, pp. 61–72 (2017)
12. Mehrabi, N., Morstatter, F., Saxena, N., Lerman, K., Galstyan, A.: A survey on bias and fairness in machine learning. ACM Comput. Surv. **54**(6), 1–35 (2021). https://doi.org/10. 1145/3457607
13. Rusdah, D.A., Murfi, H.: XGBoost in handling missing values for life insurance risk prediction. SN Appl. Sci. **2**(8), 1 (2020). https://doi.org/10.1007/s42452-020-3128-y
14. Hanafy, M., Ming, R.: Machine Learning approaches for auto insurance big data. Risks **9**(2), 1–23 (2021)
15. Qadadeh, W., Abdallah, S.: Customers segmentation in the insurance company (TIC) dataset. Procedia Comput. Sci. **144**, 277–290 (2018)

Loan Default Forecasting Using StackNet

Saket Satpute[1], Manoj Jayabalan[1], Hoshang Kolivand[1], Jolnar Assi[2],
Omar A. Aldhaibani[1(✉)], Panagiotis Liatsis[3], Paridah Daud[4], Ali Al-Ataby[5],
Wasiq Khan[1], Ahmed Kaky[6], Sahar Al-Sudani[7], and Mohamed Mahyoub[8]

[1] Faculty of Engineering and Technology, Liverpool John Moores
University, Liverpool L3 3AF, UK
O.A.Aldhaibani@ljmu.ac.uk
[2] Traders Island Ltd., London, UK
[3] Department of Electrical Engineering and Computer Science, Khalifa University, Abu Dhabi,
UAE
[4] UNITAR International University, Petaling Jaya, Malaysia
[5] Department of Electrical Engineering and Electronics,
University of Liverpool, Liverpool L69 3GJ, UK
[6] University of Anbar, Ramadi, Iraq
[7] Singapore Institute of Technology, Singapore, Singapore
[8] eSystem Engineering Society, Liverpool, UK

Abstract. Credit risk analysis is a process used by financial institutions to estimate the creditworthiness of the borrower. Financial institutions do this to protect their revenues against loan default. Peer to peer lending is a growing and popular microfinance tool used by financial institutions to provide their customers with an online platform to match borrowers and lenders. This way of matching lenders with borrowers happens with the use of some algorithm. Unfortunately, no algorithm is foolproof, and where there is a process of loan or lending money there is a risk of loan default. Financial institutions try to solve this problem using various credit risk models. This paper provides a credit risk model which predicts loan default using a stacked ensemble model. The dataset used for this research is of a leading European P2P application called Bondora which is available on Kaggle. Our stacked model will consist of AdaBoost, XGBoost and Random Forest using StackNet framework. StackNet implementation helped in increasing the performance of our credit risk model. The StackNet classifier achieves an AOC-ROC score of 99.4% and accuracy of 96.9% when using 15 features. This proves StackNet classifier demonstrates excellent performance in forecasting default in P2P lending.

Keywords: Credit risk analysis · Financial analysis · Loan · Models · Machine learning

1 Introduction

1.1 Background

Banks and financial institutions are constantly evolving to serve their customer and to increase their business. As a result, the banks and financial institutions increase their

© The Author(s), under exclusive license to Springer Nature Singapore Pte Ltd. 2023
Y. B. Wah et al. (Eds.): DaSET 2022, LNDECT 165, pp. 434–447, 2023.
https://doi.org/10.1007/978-981-99-0741-0_31

avenues from the interest of the loans given to the people. Whenever there is any money lending process there is always a risk of credit default. Banks and Financial Institutions have always tried to reduce this risk using statistical models for predicting credit default risk [1]. As the evolution of technology took place, banking sectors also started to adopt to it through computers and Internet applications. With the introduction of Smart Phone banks were able to serve their customers more efficiently. The statistical model also started to evolve to include new algorithms. Credit is defined as 'a transaction between two parties in which one (the creditor or lender) supplies money, goods, services, or securities in return for a promise of future payment by the other (the debtor or borrower). Such transactions normally include the payment of interest to the lender' [1].

When viewed historically, credit has been used during the early days of civilization even in times of barter transactions system. When carefully used, credit helps in the growth of individual's (borrower's) financial growth and in turn increases economic state of the country. It supports in growth of both household consumption and business investment. It also helps in creating employment and industrial development of the country. Yet, credits are associated with many disadvantages such as default risk. When borrowers cannot meet the obligations on time or the businesses cannot breakeven then the defaults are accumulated and then the financial institutions can add on penalty charges.

This ultimately results in destruction of wealth and business, which also affects the borrowers, credit score and creditworthiness. Creditworthiness of a borrower denotes whether the borrower is worthy of receiving credit. A borrower lacking in creditworthiness is at a risk of defaulting. Establishing borrower's creditworthiness is a part of credit risk analysis. Credit risk analysis can be defined as 'Credit risk refers to the probability of the loss (due to the non-recovery of) emanating from the credit extended because of the non-fulfilment of contractual obligations arising from unwillingness or inability of the counterparty or for any other reason' [2].

High credit risk is associated with high loss. Peer to peer lending or P2P lending is a microfinance tool used by financial institutions to match borrower to lender through some algorithm on a virtual platform [3]. P2P lending offers attractive returns to the lenders. It helps borrowers, as they do not have to go through traditional financial institutions and have less strict credit requirements as well. The borrower and the lender interact through the virtual web platform.

The virtual platform acts as a facilitator between the two parties. P2P platforms offer most types of loans the most common being personal loan and banking loan. P2P lending process have lower turnaround time as compared to traditional financial processes. Borrowers can secure a loan in a few hours' time. Borrowers may also be given lower interest rates than the traditional financial institutions. P2P loans offer more flexibility than the traditional loans.

Despite all these advantage P2P loans cannot avoid the credit risk that apply to traditional loans as well. P2P lending platforms use machine learning (ML) for credit risk analysis as well. Various risk factors and threats are encountered by P2P lending application at various stages and include user registrations, risks, billing, refunds etc.... [4].

In this respect, machine learning (ML) algorithms are used in the banking industry to determine the credit risk default of the borrower. In the research study (Wang et al., 2021) the authors have used different ML approaches to determine if they are suitable for credit risk modeling [5]. They used Random Forest (RF), Logistic Regression (LR) and Decision Tree (DT) for their research. However, other ML techniques can be explored for analysis of credit risk default.

Research (Li, 2019) used LR algorithm with XGBoost ensemble to get better results for credit risk analysis [6]. This research are indications that ensemble models when applied to credit risk modelling are more accurate than traditional single ML algorithms. These current trends show us the potential to explore more ML techniques to create better credit risk analysis models. This study aims to use ML techniques to create a stacked model using StackNet. StackNet helps us in increasing accuracy of the prediction significantly. This study uses a dataset from Bondora, a popular European peer-to-peer lending platform, to conduct credit risk analysis. The Bondora data set contains financial and demographic information about borrowers.

This research aims to provide a model to predict credit risk default in P2P lending applications by evaluating different algorithms related to Random Forest, AdaBoost and XGBoost.

1.2   State of the Art and New Contributions

In the literature, will be firstly introduced basic terms regarding the financial system and P2P banking to get a better understanding of the P2P lending process. This part also explains credit risk analysis and what are the factors associated with it. Afterward, a review of the current market situation in the peer-peer lending market along with the business models used by lending platforms will be presented. This thesis will continue with the review of the current research papers and literature that is mostly concentrated around default prediction of loans and assessment of risks.

Regression models verify whether the size of the loans or demographic, geographical and personal information can be used to explain the overall profitability of loans from an investor's perspective. Previous and current approaches in detecting the credit risk will also be reviewed along with their limitations.

P2P Lending Process
To participate in the process, potential lenders and borrowers must first register on the P2P lending platform and become members. Before authorizing you to borrow or lend money, the P2P lender will do a series of checks on your job, credit history, loan purpose, and so on. Though these checks appear to be comparable to those conducted by a bank, the P2P lending guidelines are far more flexible.

Traditional Banking vs P2P Lending
The main difference between P2P lending and traditional banking lies in their business model as it collects its funds from depositors pay some interest on the amount to the depositors. Traditional banks loan the same collected amount for the lending loan to the borrowers. The borrowers pay back the loan with interest decided by the banks.

The banks build their revenue by paying lower interests on saving deposits and higher interests for the loan payments.

Credit Risk in P2P Lending According to Different P2P Lending Types
For this research CCAF report on alternative finances 2021 has been referenced. According to the report following observations and insights were gathered on the credit risks in different types of P2P lending applications as depicted in Fig. 1.

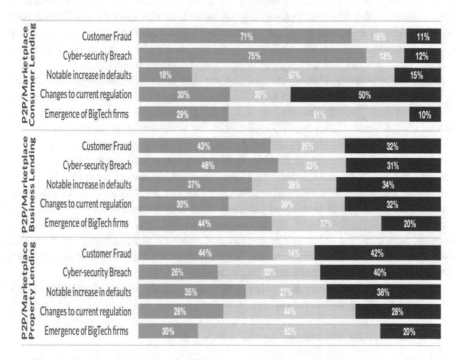

Fig. 1. Credit risk according to different types of P2P lending (Ziegler et al. 2021).

The most common issue with P2P lending applications is the possibility of regulatory changes. These worries were particularly prevalent in platforms that provide services linked to P2P Consumer Lending (50%), where more than half of the users saw this as a high-risk hazard. Furthermore, client fraud is a serious risk for companies that provide P2P Property Lending services (42%). Additionally, more than 34% of platforms that provide services related to P2P Property Lending (38%) and P2P Business Lending (38%) warned of a major increase in defaults (34%).

Customer fraud is common risk in FinTech sector. Customers usually fill incorrected and fake information into the P2P lending application system to create a fake borrower profile. These customers can be identified using rule based and ML based algorithms.

Cyber Security is important when it comes to FinTech companies. As most of the business is carried out on the Internet it makes the applications and the customers data vulnerable to the various cyber-attacks and data thefts. P2P lending applications ensure cyber security by implementing a strong cyber security policy and follow protocols set by the regulatory authorities in the region.

Notable increase in default in P2P lending applications is a result of unforeseen credit risk and information asymmetry in the borrower's information. P2P lending applications use ML algorithms for credit risk analysis of borrowers to determine the credit risk. It would reject the borrowers' loan application for those who are at a high credit risk.

Changes to the current regulation by the regions authorities can harm the P2P lending applications a loss in business. Currently, most of the P2P leading applications are popular because they have emerged as an alternative financial source to traditional bank. Previous sections indicated on of the advantages of P2P lending is of immediateness of its loan lending process. Regulation in P2P lending should be balanced so that both lenders and borrowers feel safe about investing and taking loans from it.

Over half of platforms across all business models in the United States and Canada, the United Kingdom, Europe, and the Asia-Pacific Region (excluding China) thought regulation in their jurisdiction was both adequate and suitable. In the United Kingdom and the United States, more than 70% of platforms believe their regulation is adequate and suitable. It's unclear if this was due to regulators' attempts to build an acceptable regulatory system for alternative finance, or if it was due to the fact that the majority of platforms reporting this data already met regulatory standards and were thus active.

This literature review started with explaining P2P lending process and the difference between P2P lending and traditional banking. The further credit risk analysis and the different factors associated with credit risk analysis were discussed. In this section, a thorough review of the existing approaches and research studies has been done. The limitation of the current approaches has been discussed. Lastly existing data analysis models in credit risk and different existing research have been discussed.

The main contributions of this paper with respect to the above-mentioned state of the art can be summarised as follows:

- To develop a credit risk analysis for P2P lending dataset.
- To propose a layered stacked ensemble model using existing ensemble methods like Random Forest, AdaBoost and XGBoost.
- To evaluate the layered stacked ensemble model using StackNet framework on credit risk analysis performance.

2   Methodology

2.1   Research Methodology

The proposed methodology stacks the ML algorithms efficiently and increases overall accuracy and performance. Ensemble methods in ML learning are techniques, which combine several weak learners to produce an optimal learner. Random forest is an example of ensemble model, which uses bagging ensemble method and decision trees as the weak learner. For a model to be in the ensemble it should guarantee two requirements of acceptability and diversity.

Acceptability is satisfied if a weak learner or model is better than a random guesser. It means that the probability of the learner predicting correctly should be better than 50%.

Diversity is satisfied when the weak learner calculates the prediction independently of the other learners. Some of the popular types of ensemble methods are voting, blending, stacking, boosting, and bagging.

For this research a stacked model, will implemented using Stacknet framework as shown in Fig. 2 below:

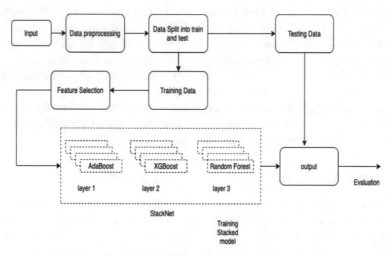

Fig. 2. Breakdown of the methodology.

2.2 Workflow

- Data is loaded and Data pre-processing is applied. Data pre-processing handles missing value treatment, outlier capping and feature engineering.
- The next step is the StackNet modelling where data will be passed through three layers of ML algorithms i.e., AdaBoost, XGBoost and Random Forest. Each layer will compute the output and pass through the next layer.
- The final output prediction will be generated after the last layer has processed the data.

2.3 Modelling Techniques

Ensembles for this research boosting methods such as AdaBoost and XGBoost for analysis along with Random Forest (bagging) [5, 6]. Let us understand the boosting technique. Boosting is a technique in which weak learner are sequentially combined to form one optimal learner. The optimal learner's accuracy is higher than any individual learner. Bagging is an ensemble method where multiple models are fit to different data samples of the training data. The outcome is obtained from combining all the outputs from the models.

Feature Selection Using Recursive Feature Imination (RFE)

- This feature selection strategy works by iteratively removing the least valuable feature. The usefulness of these features are decided by the scores, that are assigned to them by the recursively trained ML models. Initially, the procedure entails training a ML model with the initial set of features and calculating classification performance [7].
- The model's "coef" or "feature importance" characteristics are used to priortise features in RFE. It then iteratively removes a small number of features every loop, eliminating any remaining dependencies and collinearities in the model.
- The model is trained again using the remaining set of features after the least significant feature is eliminated from the current feature set.
- This approach is applied to the remaining features recursively until we achieve a subset with the appropriate number of features. Recursive Feature Elimination reduces the number of features, resulting in a gain in model efficiency.

2.4 StackNet Framework

The deep neural network architecture solution for this research includes MobileNets V3. MobileNets V3 includes input pre-processing layers for rescaling inputs to float tensors of image pixels with values in the $[-1, 1]$ range. This implies no image normalization is needed to feed into the model. Further with MobileNets V3, the input comprises of three input channels – Height, Width and RGB layers with the default shape of (224, 224, 3). This requires the input images are reshaped to (224, 224, 3) specification.

StackNet is a framework used to implement a stacking ensemble. StackNet is implemented in Java. StackNet resembles a feedforward neural network. In feed forward neural network when input is given to a perceptron it is passed through a transformation method such as sigmoid, relu, tanh etc. Then the output is again fed into the next layer. Similarly, StackNet assumes the method can take any form of a ML algorithm. StackNet combines various layers to increase accuracy of the prediction.

StackNet model has the following two different running modes.

- Normal mode
  This is like a standard feedforward neural network; the Normal stacking mode implies that each layer utilises the predictions of the direct prior one.
- Restacking mode

The Restacking mode implies that layer nth uses the activations of previous layers' $n-1$ neurons as well as the activations of all preceding levels' $(n-2, n-3 \ldots 1st)$ neurons (including the input layer). This is based on the idea that the powerful algorithm can retrieve new information from the rescanning of the initial data that could not be obtained in initial runs. It is based on the approach that convergence must occur within a single model iteration (forward training).

K-fold cross validation training: In neural networks during model training data is split into two samples (Fig. 3). One is used for training and other is used for testing with predictions. This is done to prevent model overfitting. The data is split into two parts, but it causes increase in the bias as the data is processed in each successive layer. To solve this problem, the stackNet uses k-fold cross validation.

1. Transformation functions may be any supervised algorithms.
2. Each neuron's outputs can be logically passed into subsequent layers. Classifiers, regressors, and any estimator that provides an output are examples of algorithms.
3. The stackNet layer generates output as a prediction corresponding to the number of unique categories.
4. K-fold cross-validation is performed in following way:

    a. Data is shuffled randomly.
    b. Data is split into k groups.
    c. For each unique group:

        i. Use the group as test data.
        ii. Other groups are used as training data.
        iii. Fit the model on training data and evaluate it on the test data.
        iv. Evaluation score is retained, and the model is discarded.

    d. Model performance is evaluated on the sample of model evaluation scores.
    e. Rerun the algorithm again on the whole training data as it will be used to score the external test set later.

5. StackNet allows algorithms within the same layer to run asynchronously, but to process the layer $n + 1$ all operations and outputs in layer n must be completed.

AdaBoost

The AdaBoost algorithm is an improved boosting algorithm in which the predictions of weak learners are combined into a weighted average, becoming the final output of the boosted classification model. Adaptive boosting also tweaks the weak learners when they misclassify objects so that the next learner will correctly classify it.

A set of training data $(X1, Y1), (X2, Y2),\ldots, (Xn, Yn)$, where every "Xi" as the input belongs to some domain or instance space X, and all labels are in a finite set $\{1, 2,\ldots N\}$ where K is the number of classes, and each of "Yi" as the output is qualitative. Start with the unweighted training sample to build a classifier such as a classification tree by AdaBoost.

AdaBoost algorithm:

1. Take input as a sequence of N labeled examples $<(x1, y1),\ldots.. (xn, yn)>$, distribution D over the N examples.
2. Weak leaning algorithm WeakLearn integer T specifying Number of iterations
3. Initialize the weight vector $wi1 = D(i)$ for $i = 1 \ldots\ldots, N$.
4. Do for $t = 1, 2, \ldots.. T$
5. Output the hypothesis

The weight of a training data point will increase (boosted) if that training data point misclassified; and then a second classifier will be built using the new weights, but they are not equal anymore. So the weight of misclassified training data will boost; This process will continue and maybe build more than 1000 classifiers might be built in this way

XGBoost

A type of gradient boosting method known as XGBoost or extreme gradient boosting is one that continuously corrects the loss or residue of all the previous learners when a new learner is added. The final prediction is a sum of the predictions from multiple learners. The XGBoost algorithm has high computational speed and gives high model performance [6, 12, 13].

Extreme Gradient Boosting is comprised of sequence of decision trees utilizing gradient descent algorithm in order to minimize the errors of weak estimators in which the objective function consists of training loss and regularization term, indicates the loss function.

The final weights gained by training the model have become smooth using this additional regularization. In this algorithm, all the trees are trained once at a time improving the performance of the algorithm in terms of its run time. Every loss function at step t can be optimized by taking the first and the second order gradient statistics. Accordingly, the objective function for the new tree in the general setting.

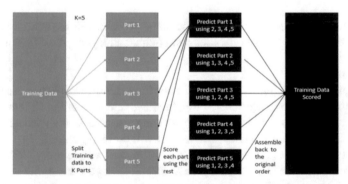

Fig. 3. K-fold cross validation (KazNova, Sept 2018).

Random Forest

Random forest is a special ensemble model which consist only of decision trees. Random forest is constructed by using many decision trees which are independent to each other. Random features are assigned to trees and the outcome is the class which was predicted by most of the trees individually in the random forest [8–11].

1. First create n-tree bootstrap samples from the data.
2. For each of the bootstrap samples, create an unpruned classification or regression tree, with the conditions: at each node, instead of choosing best split from all possible

features, take random sample m of the features and choose the best split from among those variables.
3. Take majority votes for classification and predict the new data by aggregation of the output predictions of the n-tree trees.

During the training phase the error rate estimation is done by:

1. At every iteration of the bootstrap, predict the output that is not in the sample using the tree created with that sample. This is also called as out of bag data (OOB).
2. By aggregating the OOB predictions error rate is calculated. This error rated is called OOB estimate of error rate.

3   Results and Discussion

3.1   Exploratory Data Analysis

Exploratory Analysis is a type of analysis where features from the dataset are explored with the help of different plots and graphs. For the features in the dataset in this study, exploratory analysis was performed to get an idea of the data distribution and other observations and inference which could be explored through graphs and plots.

Education
Analysis done on the dataset found that most of the loan applicant had higher, secondary, or vocational education. Applicants with basic or primary education have been granted a smaller number of loans. It can be inferred that the higher your education the more chances are that your loan application is accepted (Fig. 4).

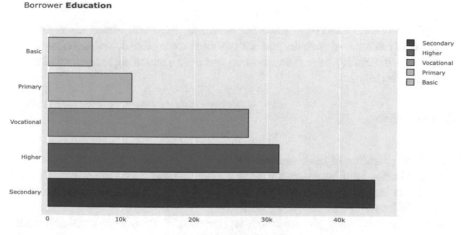

Fig. 4. Education bar chart

Figure 4 further shows that applicants with vocational education have repaid almost all their loans. Similarly, the applicants having higher education have all most all of their loans in default.

Rating
The loan rating is given by Bondora's rating model. We can see that the maximum of loans 21.5% are rated E and 18.8% are rated F. This can be inferred as the maximum loans are considered as bad loans. The good loans are rated as A, AA, B and C are around 30%. So, most of the loan application accepted are bad loans. It can also be observed that loans having AA, A and B rating and defaulted are misclassified and those that were rated E, F and HR and repaid are also misclassified by the Bondora's Rating model (Fig. 5).

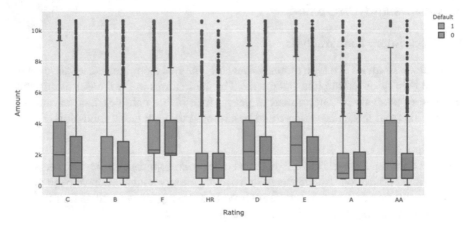

Fig. 5. Rating bar plots

Exploratory data analysis is done for getting insights and observations from the dataset. A fair number of features and various plots were plotted, and the observations were presented. Finally, the feature selection was conducted using the recursive feature elimination technique.

3.2   Evaluation

SHAP Interpretation of the Results
SHAP or shapley additive explanations is a model agnostic technique that is used to interpret and explain predictions. The stackNet model results have been interpreted using the following SHAP analysis graphs.

SHAP summary plots give us a high-level view of feature importance and what is driving it. The plot is made up of many dots each dot. Each dot determines which feature its representing, colour shows whether it was high or low for that row and the horizontal location states whether the effect of that value caused higher or lower prediction. In

Fig. 6, the stackNet model summary plot has been plotted using the SHAP values. It can be observed that Principal Payments Made and Amount of loan are the most important features in determining the Default of loan. It can also be determined from the plot that lower value of Principle Payments Made, and Higher values of Loan Amount contribute significantly to the credit default of loan.

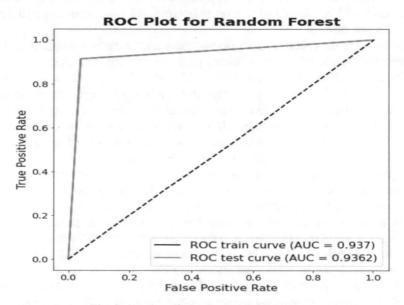

Fig. 6. Random Forest final model ROC curve

The evaluation and interpretation of the stackNet model has been discussed and presented. First the hyperparameter tuning process for the single models was presented and the evaluation was show by comparing the initial and final models. After finding the best models, the models were stacked using the stackNet framework to create the stackNet model. The stackNet model was compared with the single model performance using the classification metrics of Accuracy, Precision, Recall, F1score, AUC score and AUC ROC curve. The stackNet model performed better than all individual models. Moreover, the results of the stackNet model were evaluated using the shap summary and decision plots. Feature importance and decision process of the stackNet model was explained through these plots.

4 Conclusions

In this research, credit risk analysis of P2P lending application was implemented. A meta-modelling framework for stacking multiple ML algorithms called stackNet was applied to a real-world dataset provided by Bondora, a European P2P lending platform. For the final model, a total of 15 most relevant features were used. The final credit risk model was able to classify default with good accuracy of 96.9%.

A layered stacket model using stackNet was implemented. The layers consisted of Random Forest, AdaBoost with LR and XgBoost algorithms. StackNet is a meta-modelling framework that employs Wolpert's layered generalisation at several layers to increase prediction accuracy. It simplifies the process to stack and combine different ML models together. It uses Kfold training internally to reduce overfitting on the given data. Also because of Kfold training it also can be applied to less sample data and train a high performing model. Stacking the ML models improves the overall learning of the model along with its efficiency.

According to the results of the analysis, the stackNet method outperformed other ensemble methods in P2P lending default prediction on Bondora dataset when measured by classification performance. The stackNet classifier was able to achieve 99.4% AUC-ROC score. This research also found that the best performing ensemble models individually evaluated over the same dataset could not achieve better performance than the combination of the same ensemble algorithms when stacked together in the stackNet model. The stackNet model allowed us to combine the strengths of the multiple ML methods to performance to get better results on the dataset.

As part of our future work, there are some limitations that should be acknowledged. This study has been conducted only on dataset of Bondora platform. The credit risk analysis of different P2P platform will differ according to the dataset and application. Hence it is recommended that this research be repeated on the other platforms. The other limitations that are encountered is due to the scope of the study. This study focussed on combining three ML algorithms (XGBoost, AdaBoost with LR, and RF) through stack-Net framework. StackNet framework allows any type of ML algorithm to be stacked. So in the future studies multiple types of ML algorithms could be used to perform this research to increase the model performance, which were not part of this study.

Another aspect that could be explored further is with respect to the external circumstances for example the Covid-19 pandemics has caused economic crisis in many countries it could be beneficial to combine Covid-19 data to the Bondora dataset and explore how did it impact the default rate in the applications. Considering these factors, this work should motivate others to perform more research on improving default probability prediction.

References

1. Mann, R.J.: Explaining the pattern of secured credit. In: The Creation and Interpretation of Commercial Law, pp. 347–405. Routledge (2022)
2. Joseph, C.: Advanced Credit Risk Analysis and Management. John Wiley & Sons (2013)
3. Mezei, J., Byanjankar, A., Heikkilä, M.: Credit risk evaluation in peer-to-peer lending with linguistic data transformation and supervised learning (2018)
4. Suryono, R.R., Budi, I., Purwandari, B.: Detection of fintech P2P lending issues in Indonesia. Heliyon 7(4), e06782 (2021)
5. Wang, T., Zhao, S., Zhu, G., Zheng, H.: A machine learning-based early warning system for systemic banking crises. Appl. Econ. 53(26), 2974–2992 (2021)
6. Li, Y.: Credit risk prediction based on machine learning methods. In: 2019 14th International Conference on Computer Science & Education (ICCSE), pp. 1011–1013. IEEE, 2019, August
7. Samb, M.L., Camara, F., Ndiaye, S., Slimani, Y., Esseghir, M.A.: A novel RFE-SVM-based feature selection approach for classification. Int. J. Adv. Sci. Technol. 43(1), 27–36 (2012)

8. Bagga, S., Goyal, A., Gupta, N., Goyal, A.: Credit card fraud detection using pipeling and ensemble learning. Procedia Comput. Sci. **173**, 104–112 (2020)
9. Khalaf, M., et al.: A data science methodology based on machine learning algorithms for flood severity prediction. In: 2018 IEEE Congress on Evolutionary Computation (CEC), pp. 1–8. IEEE, July 2018
10. Hussain, A.J., Al-Jumeily, D., Al-Askar, H., Radi, N.: Regularized dynamic self-organized neural network inspired by the immune algorithm for financial time series prediction. Neurocomputing **188**, 23–30 (2016)
11. Montañez, C.A.C., et al.: Machine learning approaches for the prediction of obesity using publicly available genetic profiles. In: 2017 International Joint Conference on Neural Networks (IJCNN), pp. 2743–2750. IEEE, May 2017
12. Mohamed, A.H.H., Tawfik, H., Norton, L., Al-Jumeily, D.: e-HTAM: a technology acceptance model for electronic health. In: 2011 International Conference on Innovations in Information Technology, pp. 134–138. IEEE, April 2011
13. Alloghani, M., et al.: Implementation of machine learning algorithms to create diabetic patient re-admission profiles. BMC Med. Inform. Decis. Mak. **19**(9), 1–16 (2019)
14. Keight, R., Aljaaf, A.J., Al-Jumeily, D., Hussain, A.J., Özge, A., Mallucci, C.: An intelligent systems approach to primary headache diagnosis. In: Huang, D.S., Jo, K.H., Figueroa-García, J.C. (eds.) ICIC 2017. LNCS, vol. 10362, pp. 61–72. Springer, Cham (2017). https://doi.org/10.1007/978-3-319-63312-1_6
15. Abdulhussain, S.H., Mahmmod, B.M., Flusser, J., AL-Utaibi, K.A., Sait, S.M.: Fast overlapping block processing algorithm for feature extraction. Symmetry **14**(4), 715 (2022)

Statistical Learning

Neural Network Autoregressive Model for Forecasting Malaysia Under-5 Mortality

Wan Zakiyatussariroh Wan Husin[1(✉)], Aina Nafisya Suhaimi[1],
Nur Shuhaila Meor Zambri[2], Muhammad Azri Aminudin[3], and Nor Azima Ismail[1]

[1] Mathematical Sciences Studies, College of Computing, Informatics and Media, Universiti
Teknologi MARA Cawangan Kelantan, 18500 Machang Kelantan, Malaysia
wanzh@uitm.edu.my
[2] Universiti Teknologi PETRONAS, Seri Inkandar, 32610 Perak, Malaysia
[3] Universiti Teknologi MARA Cawangan Perak, Seri Inkandar, 32610 Perak, Malaysia

Abstract. Under-five mortality is a key point of kid prosperity in which most
countries have discussed on. In 2019, around 5.2 million children died each year
throughout the world. Despite these countries having a high number of deaths,
the lack of data makes it difficult to get accurate estimations. Moreover, this early
childhood mortality is still high and has turned into a huge problem in some
developing countries. Thus, this study aims to study the trend pattern and develop
forecasting models to forecast future trends of under-five mortality in Malaysia by
gender. The yearly under-five mortality rates (U5MR) of 41 years (1980–2020) in
Malaysia were analysed using Neural Network Autoregressive (NNAR). The result
of the NNAR was then compared with the Box-Jenkins Methodology (ARIMA
model) result. It was found that the U5MR in Malaysia fluctuated from year to
year with a slowly decreasing trend pattern for both genders, with males having
a higher rate than females. Moreover, the result from the NNAR model provides
a more accurate forecast compared to the ARIMA model for both genders with
the lowest root mean square error (RMSE) and mean absolute percentage error
(MAPE) value. The future trend increased slightly and the forecast trend for the
male was higher than the female population. The result of this study could become
a reference for other developed and developing countries. It could also become an
indicator for human resource management and health care allocation planning.

Keywords: Autoregressive integrated moving average model · Forecasting ·
Neural network autoregressive · Under-5 mortality · U5MR

1 Introduction

The death of a child before reaching the age of five is known as under-five mortality.
The infant and child mortality rate is one of the most important indicators of a country's
overall development level and one of the most important metrics of child health [1]. As
a result, a high rate of under-five mortality is undesirable because it indicates a country's
declining living conditions. According to available data on under-five mortality rates
(U5MR) in [2], around 5. 2 million children died each year in the world in 2019, and

© The Author(s), under exclusive license to Springer Nature Singapore Pte Ltd. 2023
Y. B. Wah et al. (Eds.): DaSET 2022, LNDECT 165, pp. 451–464, 2023.
https://doi.org/10.1007/978-981-99-0741-0_32

majority of them died from preventable and treatable causes in impoverished nations, in which infant and child deaths account for the majority of deaths. One of the primary parts of the sustainable development goals (SDGs) is to reduce mortality, with SDG3 focusing specifically on children under the age of five [3]. SDG3 aims to reduce U5MR in all countries to at least 25 deaths per 1000 live births by 2030 [4]. Currently, the reduction of the U5MR in most countries including Malaysia over the last few decades was encouraging, but the question remained whether the progress was comparable across gender. Females have a different mortality age pattern than males, which could be one explanation. At all ages, including before birth, females live longer and have lower death rates than males in almost every country in the world [5]. In addition, Malaysia has made remarkable success in lowering U5MR since the SDG3 objective of 24 per 1,000 live births has been met since 1984 and has now been reduced to 8.6 in 2019 [2]. Even so, U5MR could be reduced to a minimum value since it could reflect the success of new policies and processes. Forecasting U5MR for the following ten years is crucial in order to appropriately establish new policies and procedures [6]. In addition, currently, forecasting U5MR has become a major topic of research in different areas primarily in government planning and insurance firms. Prediction of U5MR is also important for revising the insurance policy. Moreover, forecasting U5MR is also crucial in order to manage health, community service and allowance [7].

Previously, a lot of studies have been done in forecasting U5MR by previous researchers using different forecasting models; i.e., [8] used the Loess regression model, [9] used Seasonal Auto-Regressive Integrated Moving Average (SARIMA) and [10] proposed Spatio-Temporal model. However, most of the previous researchers used the Autoregressive Integrated Moving Average Model (ARIMA) model with Box-Jenkins methodology in forecasting U5MR such as in Beijing, China [11], Ghana [12], Bangladesh [13], Lanzhou, China [14] and Nigeria [15]. However, in the case of Malaysia, researchers were more focused on determinant factors related to under-five mortality such as studies done by [16, 17] and [18]. In Malaysia as well, several studies were conducted on trends and forecasting U5MR such as [19] which proposed a local linear model, and [6] which performed a random walk model under the family of ARIMA model in comparison with a local linear model. As such, it could be concluded that most of the research used ARIMA model to conduct forecasting of U5MR. However, according to [20, 21], ARIMA approximations may not be sufficient for complex nonlinear real-world issues; thus, more sophisticated tools and techniques may be required to take into account the non-linear behaviour. The Neural Network Auto-Regressive (NNAR) model is a type of Artificial Neural Network (ANN) model over other nonlinear statistical models that uses lagged values of time series as input predictors and predicts the series values as output. As mentioned by [22] and [23], one of the key differences between NNAR and ARIMA models is that NNAR does not put any constraints on its parameters to maintain stationarity. In previous research, NNAR models had been applied in some other areas of forecasting by several researchers [22–24]. In addition, [22] stated that the NNAR model provides good results in terms of in-sample and out-of-sample metrics compared to ARIMA model. Therefore, this study is conducted to compare the performance of NNAR with ARIMA model and generate forecast for future trends of U5MR in Malaysia by gender.

2 Material and Methods

2.1 Data Description and Study Area

The study focuses on predicting U5MR in Peninsular Malaysia. Peninsular Malaysia, also known as West Malaysia, is a part of Malaysia. Its totals areas is 132,490 km^2, which is nearly 40% of the total area of the country; the other 60% is East Malaysia. Secondary data on U5MR are provided by Malaysia's Department of Statistics (DOSM). According to the DOSM record, data U5MR pertaining to Peninsular Malaysia are more complete with longer time series compared to those pertaining to overall Malaysia. The data on the U5MR were split according to female and male population. The data utilized in this research spanned from the years of 1980 to 2020, with an emphasis on under-five mortality in Peninsular Malaysia due to the data's availability.

2.2 Autoregressive Integrated Moving Average Model

Autoregressive Integrated Moving Average (ARIMA) through Box and Jenkins methodology is the most prominent approach for time series forecasting. ARIMA model is a combination of Autoregressive (AR) and a Moving Average (MA) model. In the ARIMA model, AR (p) refers to p lag error terms. Equation 1 is the general form of the ordered AR process where p denotes the number of the lag term of the series, in which it is U5MR series of this study [25]. According to [26], the AR(p) model can be written as:

$$y_t = \mu + \phi_1 y_{t-1} + \phi_2 y_{t-2} + \ldots + \phi_p y_{t-p} + \varepsilon_t \tag{1}$$

where y_t is the U5MR, μ is constant terms, $\phi_j (j = 1, 2, \ldots, p)$ is the estimated parameters, y_{t-p} is the pth order of the lagged U5MR and ε_t is the error term which is assumed *iid* with mean zero and variance, σ_ε^2. The Moving Average (MA) model connects current time series estimations to random errors that occur in the previous period rather than the values of the actual series. The MA(q) model is written as

$$y_t = \mu - \theta_1 \varepsilon_{t-1} - \theta_2 \varepsilon_{t-2} - \ldots - \theta_q \varepsilon_{t-q} + \varepsilon_t \tag{2}$$

where μ is the mean about which the series fluctuates, $\theta's$ are the estimated moving average parameters, and $\varepsilon'_{t-q}s$ are the error terms ($q = 1, 2, 3, \ldots$) that are assumed to be independently distributed over time.

Next, ARMA model is simply the combination between AR(p) and MA(q) models. ARMA is predicated on the assumption of stationarity and a series that do not involve differencing. Reference [26] stated ARMA(p,q) model as

$$y_t = \mu + \phi_1 y_{t-1} + \phi_2 y_{t-2} + \ldots + \phi_p y_{t-p} - \theta_1 \varepsilon_{t-1} - \theta_2 \varepsilon_{t-2} - \ldots - \theta_q \varepsilon_{t-q} + \varepsilon_t \tag{3}$$

where μ is the mean about which the series fluctuates, $\phi's$ are the estimated AR parameters, $\theta's$ are the estimated MA parameters and $\varepsilon'_{t-q}s$ are the error terms ($q = 1, 2, 3, \ldots$) assumed to be independently distributed over time. The classical ARMA models assume the time series is stationary, that is, the mean and variance of the series are essentially constant through time. When stationarity assumption is not meet, the data series need to be differenced and the model becomes an ARIMA model with the I is integrated that represents the level of differencing. The forecasting equation could be expressed as ARIMA(p, d, q) where 'd' is the number of nonseasonal differences needed for stationarity. Specifically, p and q denoted the order of the AR and MA models, respectively. One of the special cases of the ARIMA (p, d, q) model is ARIMA (0, 1, 0) with and without constant (μ) value which is known as random walk model. The random walk is a basic of ARIMA for a non-stationary series when there is no AR and MA component. A random walk model is written as

$$y_t = y_{t-1} + \varepsilon_t \tag{4}$$

where y_t is the U5MR series, y_{t-1} is the observation in the previous time period and ε_t is white noise term. If the random walk fluctuates around a constant mean, it is referred to as a random walk with drift in which μ is a constant value or drift. Then, (4) can be written as

$$y_t = y_{t-1} + \mu + \varepsilon_t \tag{5}$$

The development of the ARIMA model through the Box-Jenkins methodology consists of five iterative steps. The process begins by investigating the trend pattern of the series using a simple time plot, Autocorrelation function (ACF) plot and Partial Autocorrelation function (PACF) plot. At this stage, the stationarity assumption is checked and the level of trend differencing (d) is identified. The second step is transforming the series using differencing approach in order to achieve time series stationarity. At this stage, test of unit root (stationarity test) using Augmented Dickey Fuller (ADF) is also conducted. Subsequently, historical data are used to identify the possible ARIMA model by identifying the order of AR (p) and order of MA (q) using ACF and PACF of the stationary series. Then, in the third step, the data are used to estimate model parameters that have been identified in the previous step. Next, adequacy of each of the model is verified by Ljung-Box statistics that is useful for testing the randomness of residuals by testing the null hypothesis of the residuals which is white noise. Finally, based on their adequate predictions, the most appropriate model is chosen based on the smaller values of root mean square error (RMSE) and mean absolute percentage error (MAPE).

2.3 Neural Network Autoregressive Model

Neural Network Autoregressive (NNAR) model is a forecasting model that is parametric and non-linear. It is a type of Artificial Neural Network (ANN) that employs delay time series data as inputs to a neural network to analyse non-linear interactions between input regressors and responses [27]. The model is appropriate with time series data since the lagged values of the time series can be used as inputs to a neural network. It is just like lagged values which are used in a linear AR model. The methodology of the NNAR model applies the concept of neural networks (NN). The NN is a statistical model that were commonly used in machine learning. It is superior in estimating functions based on a large number of training sets, and it consists of a network of numerous nodes that work in parallel, as shown in Fig. 1. Reference [24] found that the NN can describe complex non-linear relationships without making any assumptions about the underlying relationship. Each node is associated with an activation function that converts the node's input into an output that serves as an input for nodes in following layers. Each node multiplies the input signal by a weight w_{ij}, which is a feature of the link between nodes i and j of adjacent layers, and then adds the weighted input signals together [24].

As shown in Fig. 1, each hidden layer node executes a single 'sigmoid' transformation on its input. As each layer of nodes receives input from the preceding level, this NN is a multilayer feed-forward neural network, as illustrated Fig. 1. The outputs of one layer's nodes would be sent into the next layer as inputs. The inputs to each node are combined using a weighted linear combination. A nonlinear function modifies the result before it becomes output. For example, the inputs into hidden neuron j in Fig. 1 are combined linearly to allow:

$$z_j = g(\Sigma_i y_i w_{ij} - \beta_j) \tag{6}$$

where z_j is the output of the jth node which is hidden neuron, y_i is the ith input, and β_j is parameter of the jth node. In the hidden layer, (6) is modified using a nonlinear function $g(\ )$ to provide the input for the next layer.

$$g(x) = \frac{1}{1 + e^{-x}} \tag{7}$$

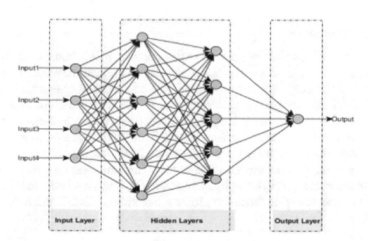

Fig. 1. Neural network structure

The NN learns to control the weights associated with each link by utilizing a learning method throughout the training phase. All neural network models use the back-propagation algorithm, which is the most well-known and commonly used learning method. This algorithm instructs the network to give the best possible response to each received input. Then it adjusts the weights by comparing the obtained response with the target output as to minimize the error.

The step in NNAR model is started by determining the order of AR for the U5MR series. The AR order indicates the number of prior values in which the current value of the time series is based. The NN is trained in the second stage with a training set that takes the order of AR into account. The number of input nodes is determined by the AR order, and hence the inputs to the NN are previous lagged observations. The NN model's output is the expected values. Because there is no theoretical basis for selection, the number of hidden nodes is frequently determined through trial and error or experimentation [24]. To avoid the issue of overfitting, the number of iterations should be carefully chosen. The relationship between the model output (y_t) and the inputs $(y_{t-1}; ...; y_{t-p})$ has the following mathematical representation:

$$y_t = w_0 + \Sigma_{j=1}^k w_j \cdot g\left(w_{0,j} + \Sigma_{i=1}^p w_{i,j}y_{t-i}\right) + \varepsilon_t \tag{8}$$

where, w_{ij} $(i = 0, 1, 2, ..., p, j = 1, 2, ..., k)$ and w_j $(j = 0, 1, 2, ..., h)$ are model parameters or connection weights; p is number of input nodes; and k is number of hidden nodes. A sigmoid function is used as the hidden layer transfer function that is shown in (7). A NNAR model is used to enter the lagged values of a time series. The hidden layer has p lagged inputs and k nodes, as shown by the notation NNAR (p, k) where p is lagged inputs and k is nodes in hidden layers [23]. Following [28], this study only considers feed-forward networks with one hidden layer. A NNAR (p, k) model is a neural network with the last p observations $(y_{t-1}, y_{t-2}, ... y_{t-p})$ used as inputs for forecasting the output y_t and with k neuran in the hidden layer. NNAR $(p, 0)$ is equivalent to an ARIMA $(p, 0, 0)$ model but without the restrictions on the parameters to ensure stationarity.

2.4 Model Evaluation

The data series is divided into two parts. The first component is in-sample estimation and the second is out-sample assessment, which are used to evaluate the model's forecasting performance. Following [29], this study used 80% of the data for in-sample estimation and 20% of the data for out-sample evaluation. Hence, 9 observations were used for out-sample evaluation and the remaining 32 observations were in-sample estimation. The data of under-five mortality for the period 1980 until 2011 were used for all models' in-sample estimation while the data from 2012 until 2020 were used for out-sample evaluation. The estimated model was then evaluated by comparing the forecast performance of different estimated models using RMSE and MAPE. The model with the lowest error was used for forecasting. Referring to [26], the function of RMSE and MAPE can be written as:

$$RMSE = \sqrt{\frac{\Sigma_t^n e_t^2}{n}} \tag{9}$$

where $e_t = y_t - \widehat{y_t}$, with y_t as the actual observation at the point and $\hat{y}$ is the fitted value at time t.

$$MAPE = \Sigma_{t=1}^n \frac{\left|\left(\frac{e_t}{y_t}\right) * 100\right|}{n} \tag{10}$$

where n denotes effective data points and $\left|\left(\frac{e_t}{y_t}\right) * 100\right|$ is defined as an absolute percentage error calculated on the fitted values for a particular forecasting method.

2.5 Forecasting

Forecasting was conducted for 10-years ahead, from 2021 to 2030 using the best model with the least error. By 2030, all nations aim to reduce total under-five mortality to at least 25 per 1,000 live births, as specified in Goal 3 of the SDGs [2]. Therefore, it is critical for Malaysia to anticipate in forecasting under-five mortality over the next ten years so that new policies and strategies may be developed.

3 Results

This section presents the results of developing ARIMA and NNAR models for U5MR in Malaysia. Data from 1980 until 2011 are used for in-sample estimation while the data for the remaining years of 2012 until 2020 are used for out-sample evaluation. Then, the best model is used in forecasting U5MR for the next 10 years which is from 2021 until 2030. The process of model development begins by investigating the trend pattern of the U5MR in Malaysia from 1980 to 2020. Figure 2 depicts the time plots that are used to indicate the historical trend of U5MR in Peninsular Malaysia according to gender which changes over time. The U5MR for the male population is higher than the female population from 1980 until 2020. The trend decreased consistently from 1980 to 1996 for both genders. The male population's U5MR decreased from 7.52 to 2.56 while the mortality dropped from 6.24 to 2.13 for the female population. However, the rate for both gender massively rose in 1997 and 1998 due to the economic crisis that happened during those years. In 1998, food prices increased by 9% which affected the health status as buying expenditure increased. However, many programs were raised in improving economic position and importantly, health status [30]. This resulted in the declining of U5MR starting from the year 1999 for both gender populations and showed a uniform pattern in the years ahead until 2011. In 2012, there was a slight increment and the trend fluctuated until 2016. There was a slight increase in the year 2017; however, the rate started to decrease in 2018 and the years ahead.

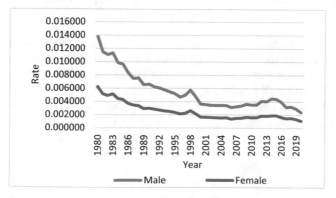

Fig. 2. Trend for peninsular Malaysia U5MR from 1980 to 2020

The U5MR for male and female population of in-sample data were used to examine the data in the first stage. Figure 2 obviously shows that the U5MR data were non-stationary over time. Besides, as seen in trend analysis, a trend existed that indicates it is a non-stationary series. Furthermore, as shown in Fig. 3, the values of the ACF and PACF confirmed that the U5MR series is non-stationary. Figure 4 displays the trend analysis of the first difference of U5MR series for both male and female population. The series seemed to show the upward pattern toward the on-going data in both figures. Then, ACF and PACF plots were used in further analysis to validate the data's stationarity for both population as demonstrated in Fig. 5. There was no obvious trend in the ACF and PACF for different series of both population. The series of U5MR for both population were stationary. Then, the unit root ADF test was conducted for both population and the calculated t-statistic for the male and female population was -7.545 and -6.502 respectively, with a probability value for both population of <0.0001 which is smaller than the 5% level of significance. As a result, the U5MR series for both male and female are confirmed to be stationary.

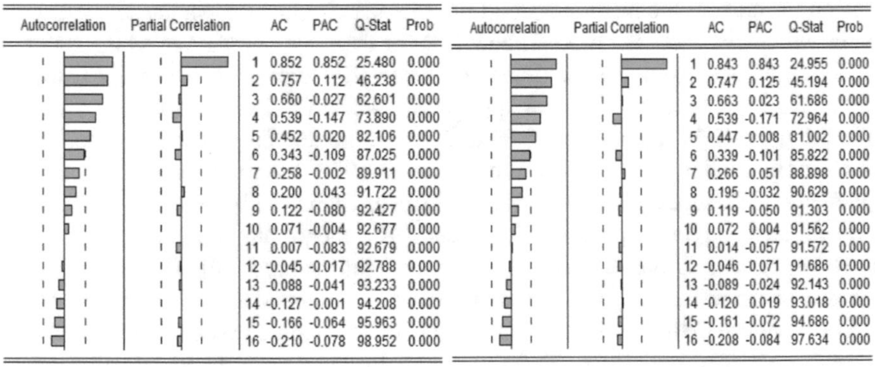

Autocorrelation	Partial Correlation		AC	PAC	Q-Stat	Prob	Autocorrelation	Partial Correlation		AC	PAC	Q-Stat	Prob
		1	0.852	0.852	25.480	0.000			1	0.843	0.843	24.955	0.000
		2	0.757	0.112	46.238	0.000			2	0.747	0.125	45.194	0.000
		3	0.660	-0.027	62.601	0.000			3	0.663	0.023	61.686	0.000
		4	0.539	-0.147	73.890	0.000			4	0.539	-0.171	72.964	0.000
		5	0.452	0.020	82.106	0.000			5	0.447	-0.008	81.002	0.000
		6	0.343	-0.109	87.025	0.000			6	0.339	-0.101	85.822	0.000
		7	0.258	-0.002	89.911	0.000			7	0.266	0.051	88.898	0.000
		8	0.200	0.043	91.722	0.000			8	0.195	-0.032	90.629	0.000
		9	0.122	-0.080	92.427	0.000			9	0.119	-0.050	91.303	0.000
		10	0.071	-0.004	92.677	0.000			10	0.072	0.004	91.562	0.000
		11	0.007	-0.083	92.679	0.000			11	0.014	-0.057	91.572	0.000
		12	-0.045	-0.017	92.788	0.000			12	-0.046	-0.071	91.686	0.000
		13	-0.088	-0.041	93.233	0.000			13	-0.089	-0.024	92.143	0.000
		14	-0.127	-0.001	94.208	0.000			14	-0.120	0.019	93.018	0.000
		15	-0.166	-0.064	95.963	0.000			15	-0.161	-0.072	94.686	0.000
		16	-0.210	-0.078	98.952	0.000			16	-0.208	-0.084	97.634	0.000

a) Male   b) Female

Fig. 3. ACF and PACF for original series of U5MR

Then, the possible ARIMA models as in Table 1 are identified using ACF and PACF plot in Fig. 5. The model comparison for both genders in finding the best ARIMA model that fits the sample was conducted as shown in Table 1. In addition, the Ljung-Box test was also conducted for all possible ARIMA models and it was found that errors for all the models were white noise at 5% level of significance. Based on the values of RMSE and MAPE in Table 1, it shows that the best fit for both male and female population of U5MR comes from ARIMA (0,1,0) model which is often known as a random walk model.

Next, the NNAR model was developed for the U5MR in Malaysia for both populations. According to Table 2, the NNAR models for male and female have been learned to predict the data within the U5MR series. By using the nnetar function in R software, NNAR (1,1) was identified and generated for both male and female population as the best model. The nnetar function in the forecast package for R fits a NN model to a time

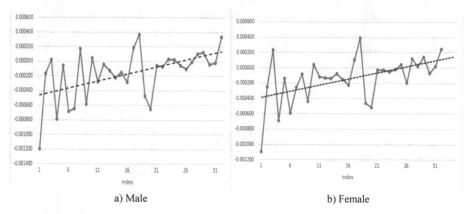

a) Male   b) Female

Fig. 4. Trend analysis after first differencing of U5MR

Autocorrelation	Partial Correlation		AC	PAC	Q-Stat	Prob	Autocorrelation	Partial Correlation		AC	PAC	Q-Stat	Prob
		1	-0.016	-0.016	0.0084	0.927			1	0.002	0.002	0.0002	0.989
		2	-0.002	-0.003	0.0087	0.996			2	-0.092	-0.092	0.2996	0.861
		3	0.257	0.257	2.4169	0.490			3	0.225	0.228	2.1533	0.541
		4	-0.110	-0.109	2.8749	0.579			4	0.051	0.039	2.2507	0.690
		5	0.341	0.370	7.4516	0.189			5	0.239	0.297	4.4923	0.481
		6	0.121	0.041	8.0474	0.235			6	0.007	-0.048	4.4940	0.610
		7	-0.025	0.072	8.0745	0.326			7	0.091	0.159	4.8447	0.679
		8	0.059	-0.153	8.2298	0.411			8	0.070	-0.084	5.0613	0.751
		9	-0.110	-0.084	8.7942	0.456			9	-0.150	-0.144	6.1021	0.730
		10	0.089	-0.039	9.1821	0.515			10	0.029	-0.120	6.1418	0.803
		11	-0.100	-0.200	9.6917	0.558			11	0.026	-0.027	6.1775	0.861
		12	-0.041	-0.002	9.7837	0.635			12	-0.062	-0.079	6.3838	0.896
		13	0.115	0.090	10.538	0.649			13	0.022	0.065	6.4111	0.930
		14	-0.028	0.183	10.586	0.718			14	0.007	0.080	6.4142	0.955
		15	0.074	0.100	10.940	0.757			15	0.143	0.265	7.7199	0.935
		16	-0.058	0.011	11.170	0.799			16	-0.123	-0.106	8.7542	0.923

a) Male   b) Female

Fig. 5. ACF and PACF after first differencing of U5MR

series. The topology 1-1-1 for both genders means that there is one lag and one hidden layer.

U5MR for the female population has lower values for RMSE and MAPE compared to the male population. The coefficient of determination (R^2) obtained for both female and male population was 0.730199 and 0.741670, respectively. R^2 implies that 73.02% of the series variance is described by U5MR for male input while the remaining 26.98% of the variance could not be attributed to these inputs. Meanwhile, for the female population, 74.17% of the series variance is described and the remaining 25.83% of the variance could not be attributed to these inputs. Finally, the performance of the ARIMA model, which is represented by the random walk with drift model and NNAR model, was evaluated based on their predicted (in-sample estimation) and forecast ability (out-sample evaluation). Table 3 shows the comparison estimation and evaluation between Random Walk with Drift and NNAR model using RMSE and MAPE values and it shows that both RMSE and

Table 1. Summary of in-sample evaluation of ARIMA models

Statistics	ARIMA (0,1,0)	ARIMA (1,1,0)		ARIMA (0,1,1)	ARIMA (1,1,1)
Male					
Q-value	15.767	15.573		15.577	15.580
p-value	0.609	0.554		0.554	0.483
Decision (5% sig. Level)	The errors are white noise	The errors are white noise		The errors are white noise	The errors are white noise
RMSE	**0.000338**	0.000344		0.000344	0.000350
MAPE	**8.520452**	8.526277		8.526252	8.526114
Female					
Q-value	12.476	13.746	13.747		13.737
p-value	0.711	0.685	0.685		0.746
Decision (5% sig. Level)	The errors are white noise	The errors are white noise	The errors are white noise		The errors are white noise
RMSE	**0.000293**	0.000298	0.000298		0.000293
MAPE	**7.537731**	8.339387	8.338742		8.342657

MAPE values for the NNAR model were smaller than Random Walk with Drift. Hence, the NNAR model is selected as the best model for forecasting U5MR in Malaysia.

Figure 6 shows the pattern of actual U5MR data from 1980 to 2020 for both male and female population as well as the future trend of U5MR from 2021 to 2030. The pattern of future forecast trends for both populations shows a slight increase. However, the future trend for the male population is expected to increase higher than the female population. In the year 2021, the forecast value for male is 0.000739 and 0.000983 for the year 2030. Besides, the forecast values for female are 0.000592 and 0.000763 in the year 2021 and 2030 respectively as shown in Table 4.

Table 2. Topology and goodness of fit measures for NNAR models

Variable	Model	Topology	RMSE	MAPE	R2
Male	NNAR (1,1)	1-1-1	0.000265	7.204815	0.730199
Female	NNAR (1,1)	1-1-1	0.000204	6.559554	0.741670

Table 3. In-Sample and out-sample evaluation for Random Walk with Drift and NNAR model

		In-sample		Out-sample	
		Random Walk with Drift	NNAR	Random Walk with Drift	NNAR
Male	MAPE	8.520452	**5.863241**	11.86651	**11.82579**
	RMSE	0.000338	**0.000259**	0.000329	**0.000285**
Female	MAPE	7.537731	**5.722626**	12.80801	**9.442303**
	RMSE	0.000293	**0.000214**	0.000271	**0.000165**

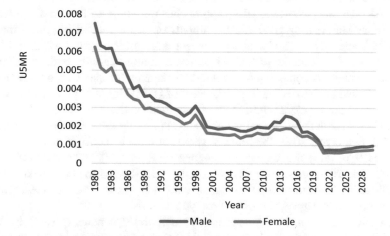

Fig. 6. Trend of actual and forecast values for U5MR of male and female population

Table 4. Forecast value U5MR of male and female population

Year	Male	Female
2021	0.000739	0.000592
2022	0.000761	0.000614
2023	0.000752	0.000603
2024	0.000762	0.000601
2025	0.000811	0.000643
2026	0.000839	0.000665
2027	0.000899	0.000706
2028	0.000935	0.000721
2029	0.000936	0.000735
2030	0.000983	0.000763

4   Conclusion

This study aims to forecast the future trend in Malaysia's U5MR by gender. The analysis begins by studying the trend pattern of U5MR in Malaysia by gender from 1980 to 2020. The trend and irregular components are discovered to have an impact on the series of U5MR for both male and female population. From 1980 to 1996, the U5MR for both male and female population shows a decreasing pattern and there was a random shock of the irregular component in U5MR data with rates dramatically increasing in 1997 and 1998 due to the economic crisis. However, from 2000 to 2011, the U5MR trend followed a consistent pattern. The trend for U5MR had a slight increase from 2012 to 2015, then a decline from 2016 until 2020. This situation shows that the U5MR series has an unstable pattern. ARIMA and NNAR modeling approaches were then used to predict the U5MR in Peninsular Malaysia by gender. The comparison of the forecasting performance of the ARIMA and NNAR model using forecast error measures of RMSE and MAPE shows that the NNAR model provides a more accurate forecast in both in-sample and out-sample evaluation compared to the ARIMA model with specification random walk with drift. Based on the 10-step ahead forecast of U5MR for both population of gender generated from the year 2021 until 2030, it is concluded that the future U5MR for both male and female population in Peninsular Malaysia slightly increases from the year 2021 until 2030. In the year 2021, the forecast value for male is 0.000739 and 0.000983 for the year 2030. Besides, the forecast values for female are 0.000592 and 0.000763 in the years 2021 and 2030 respectively. Based on the U5MR trend for Peninsular Malaysia from 1980 to 2020, the rate for both genders massively rose in 1997 and 1998 due to the economic crisis that happened during those years.

Therefore, it is recommended for future researchers to conduct research by considering other models, which are able to predict U5MRs more accurately and capture their patterns more specifically in order to investigate the possibility of the random shock of the crisis affecting the trend of U5MR in Malaysia. In addition, Malaysia has made considerable progress in reducing U5MR. However, the issue concerned is whether the improvement is uniformly distributed throughout the country. Thus, future researchers may consider a spatio-temporal forecasting model because the model is able to produce spatially out-of-sample forecasts.

Acknowledgment. The authors would like to thank Universiti Teknologi MARA (UiTM) for the facilities provided to conduct this study and the DOSM for providing the data on U5MR. This research project was supported by the Ministry of Higher Education Malaysia through the Fundamental Research Grant Scheme (FRGS/1/2019/STG06/UITM/02/4).

References

1. Worku, M.G., Teshale, A.B., Tesema, G.A.: Determinants of under-five mortality in the high mortality regions of Ethiopia: Mixed-effect logistic regression analysis. Arch. Public Health **79**(1), 1–9 (2021)
2. World Health Organization, Children: Improving survival and well-being (2020). https://www.who.int/news-room/fact-sheets/detail/children-reducing-mortality. Accesses 10 Oct 2022

3. Costa, J.C., da Silva, I.C.M., Victora, C.G.: Gender bias in under-five mortality in low/middle-income countries. BMJ Glob. Health **2**(2), e000350 (2017)
4. Van Malderen, C., Amouzou, A., Barros, A.J., Masquelier, B., Van Oyen, H., Speybroeck, N.: Socioeconomic factors contributing to under-five mortality in sub-Saharan Africa: a decomposition analysis. BMC Pub. Health **19**(1), 1–19 (2019)
5. Pascariu, M.D., Canudas-Romo, V., Vaupel, J.W.: The double-gap life expectancy forecasting model. Insur. Math. Econ. **78**, 339–350 (2018)
6. Husin, W.Z.W., Ramli, R.Z., Muzaffar, A.N., Abd, N.N., F., Rahmat S. N. E.: Trend analysis and forecasting models for U5MR in Malaysia. PalArch's J. Archaeol. Egypt/Egyptol. **17**(10), 875–889 (2020)
7. Kontis, V., Bennett, J.E., Mathers, C.D., Li, G., Foreman, K., Ezzati, M.: Future life expectancy in 35 industrialised countries: projections with a Bayesian model ensemble. Lancet **389**(10076), 1323–1335 (2017)
8. Murray, C.J., Laakso, T., Shibuya, K., Hill, K., Lopez, A.D.: "Can we achieve millennium development goal 4? New analysis of country trends and forecasts of under-5 mortality to 2015. Lancet **370**(9592), 1040–1054 (2007)
9. Rostami, M., Jalilian, A., Hamzeh, B., Laghaei, Z.: Modeling and forecasting of the under-five mortality rate in Kermanshah province in Iran: a time series analysis. Epidemiol. Health. **37**, e2015003 (2015)
10. Aheto, J.M.K., Taylor, B.M., Keegan, T.J., Diggle, P.J.: Modelling and forecasting spatio-temporal variation in the risk of chronic malnutrition among under-five children in Ghana. Spat. Spatio-Temporal Epidemiol. **21**, 37–46 (2017)
11. Cao, H., et al.: Trend analysis of mortality rates and causes of death in children under 5 years old in Beijing, China from 1992 to 2015 and forecast of mortality into the future: an entire population-based epidemiological study. BMJ Open **7**(9), 1–11 (2017)
12. Bosson-Amedenu, S.: Nonseasonal ARIMA modeling and forecasting of malaria cases in children under five in Edum Banso Sub-district of Ghana. Asian Res. J. Math. **4**(3), 1–11 (2017)
13. Rajia, S., Sabiruzzaman, M., Islam, M.K., Hossain, M.G., Lestrel, P.E.: Trends and future of maternal and child health in Bangladesh. PLoS ONE **14**(3), 1–17 (2019)
14. Liang, W., et al.: Mortality analysis and model prediction of children under 5 years old in a city of Northwest China. Ann. Epidemiol. Pub. Health **3**(1), 1–7 (2020)
15. Eke, D.O., Ewere, F.: Modeling and forecasting under-five mortality rate in Nigeria using auto-regressive integrated moving average approach. Earthline J. Math. Sci. **4**(2), 347–360 (2020)
16. Wahab, M.A.A., Jamadon, N.K., Mohmood, A., Syahir, A.: River pollution relationship to the national health indicated by under-five child mortality rate: a case study in Malaysia. Bioremediat. Sci. Technol. Res. **3**(1), 20–26 (2015)
17. Aziz, F.A.A., et al.: Prevalence of and factors associated with diarrhoeal diseases among children under five in Malaysia: a cross-sectional study 2016. BMC Pub. Health **18**(1), 1–8 (2018)
18. Waziri, S.I., Nor, N.M., Hook, L.S., Hassan, A.: Access to safe drinking water, good sanitation, occurrence of under-five mortality and standard of living in developing countries: system GMM approach. Jurnal Ekonomi Malays. **52**(2), 279–289 (2018)
19. Abd Nasir, N.F., Muzaffar, A.N., Rahmat, S.N.E., Husin, W.Z.W., Abidin, N.S.Z.: Forecasting Malaysia under-5 mortality using state space model. J. Phys. Conf. Ser. **1496**(1), 012001 (2020)
20. Khashei, M., Bijari, M.: A novel hybridization of artificial neural networks and ARIMA models for time series forecasting. Appl. Soft Comput. J. **11**(2), 2664–2675 (2011)
21. Zhang, G., Patuwo, B.E., Hu, M.Y.: Forecasting with artificial neural networks: the state of the art. Int. J. Forecast. **14**(1), 35–62 (1998)

22. Maleki, A., Nasseri, S., Aminabad, M.S., Hadi, M.: Comparison of ARIMA and NNAR models for forecasting water treatment plant's influent characteristics. KSCE J. Civ. Eng. 22(9), 3233–3245 (2018)
23. Thoplan, R.: Simple v/s sophisticated methods of forecasting for Mauritius monthly tourist arrival data. Int. J. Stat. Appl. 4(5), 217–223 (2014)
24. Sena, D., Nagwani, N.K.: A neural network auto regression model to forecast per capita disposable income. ARPN J. Eng. Appl. Sci. 11(22), 13123–13128 (2016)
25. Shetty, C.: Time series models (2020). https://towardsdatascience.com/time-series-models-d9266f8ac7b0. Accessed 1 Oct 2021
26. Lazim, M.A.: Introductory Business Forecasting. A practical approach 3rd Edition (2013)
27. Ahmar, A.S., Boj, E.: Application of neural network time series (NNAR) and ARIMA to forecast infection fatality rate (IFR) of Covid-19 in Brazil. JOIV: Int. J. Inf. Vis. 5(1), 8–10 (2021)
28. Hyndman, R. J., Athanasopoulos, G.: Forecasting: principles and practice. OTexts. (2018)
29. Yaffee, R., McGee, M.: Time series analysis and forecasting with applications of SAS and SPSS. Int. J. Forecast. 17(2), 301–302 (2000)
30. Ramesh, M.: Economic crisis and its social impacts. Lessons from the 1997 Asian Economic Crisis, pp. 79–99 (2009)

Robustness of Support Vector Regression and Random Forest Models: A Simulation Study

Supriadi Hia[1,2], Heri Kuswanto[1(✉)], and Dedy Dwi Prastyo[1]

[1] Institut Teknologi Sepuluh Nopember, Surabaya, Indonesia
supriadi.hia@bps.go.id, {heri_k,dedy-dp}@statistika.its.ac.id
[2] Badan Pusat Statistik, Jakarta, Indonesia

Abstract. Classical statistics are usually based on parametric models, where the performance depends heavily on assumptions and is not robust in the presence of outliers in the data. Due to the COVID-19 pandemic, our daily lives have changed significantly, including slowing economic growth. These extreme changes can manifest as an outlier in time series studies and adversely affect the results of data analysis. Many classical methods of official statistics are prone to outliers. In this work, we evaluate machine learning methods: Support Vector Regression (SVR) and Random Forest (RF) and compare it with ARIMA to determine the robustness through simulation studies. Robustness is measured by the sensitivity of the SVR and Random Forest hyperparameter and the model's error in the presence of outliers. Simulations show that more outliers lead to higher RMSE values, and conversely, more samples lead to lower RMSE values. The type of outliers significantly impacts the RMSE value of the ARIMA model, where additional outliers (AO) have a worse impact than temporary change (TC). Consecutive outliers produce a smaller RMSE mean than non-consecutive outliers. Based on the sensitivity of hyperparameters, SVR and Random Forest models are relatively robust to the presence of outliers in the data. Based on the simulation results of 100 iterations, we find that SVR is more robust than ARIMA and Random Forest in modeling time series data with outliers.

Keywords: Outlier · Random forest · Robustness · Support vector regression

1 Introduction

Classical statistics are usually based on parametric models (Ronchetti 2006). Many classical statistical and econometric methods are notoriously unrobust, as their results depend heavily on precise stochastic assumptions and the nature of some observations in the sample. Robustness refers to the strength of statistical models, tests, and procedures according to the specific conditions of statistical analysis achieved. The basic idea of robust statistics is to stabilize a statistical procedure against small changes in data or models so that large changes do not seriously damage the procedure (Gather and Davies 2004).

Outlier data are often found in actual data, and the analytical results are not robust. Outliers in time series data are different from outliers in cross-sectional data. In time

© The Author(s), under exclusive license to Springer Nature Singapore Pte Ltd. 2023
Y. B. Wah et al. (Eds.): DaSET 2022, LNDECT 165, pp. 465–479, 2023.
https://doi.org/10.1007/978-981-99-0741-0_33

series, the order of data is essential as it relates to time. The COVID-19 pandemic has drastically changed our daily lives, including lifestyles and work. Economic growth is slowing, and these extreme changes can be detected as outliers in time series studies. Many studies have found that outliers influence the results of data analysis. There are no fixed rules for outliers, but they should be treated with caution. According to Yung et al. (2018), many classical methods of official statistics are prone to outliers. Traditional time series models such as ARIMA are unsuitable for modeling data with outliers because they can violate normality assumptions. Therefore, this study uses machine learning approaches.

Machine learning statistical models have become increasingly established over the last few decades and have become serious competitors to classical statistical models in the field of forecasting (Ahmed et al. 2010). Support Vector Regression (SVR) and Random Forest (RF) are supervised learnings suitable for classification and regression problems (Khan et al. 2021). These two methods were subsequently developed and can be used for time series analysis. SVR and RF are relatively new machine learning techniques in time series forecasting and have yet to be widely adopted. Both methods are computationally fast (Baba et al. 2015). Several researchers have used SVR for time series data. For example, Hong (2009) developed the SVR model for forecasting electrical loads with the immune algorithm (IA) to determine parameters (σ, C, ε); Pai et al. (2010) developed the seasonal SVR (SSVR) model for seasonal time series forecasting problems; Purnama and Setianingsih (2020) applies SVR in forecasting data on the number of aircraft passengers because the data has a nonlinear data pattern; Priliani et al. (2018) implements the Weight Attribute Particle Swarm Optimization (WAPSO) optimization technique to obtain optimal SVR parameters, etc. On the other hand, several researchers have used RF to predict time series data. For example, Dudek (2015) proposes using a random forest model for short-term electricity load forecasting; Mueller (2020) forecasts ticket sales for Major League Baseball matches and identifies important predictors through RF; Alanis (2022) found that using RF in estimating private companies' equity beta gave better results with a smaller average error than comparable company analysis (CCA).

This study aims to evaluate the robustness of SVR and RF models based on simulation studies and compare it with one of the time series analysis techniques, ARIMA. The simulated data follow an empirical model; a basic model derived from the ARIMA model from Indonesian economic growth data (year-on-year). Robustness is measured by the sensitivity of the hyperparameter and error of models to the presence of outliers in data.

2 Literature Review

2.1 Outliers in Time Series Data

According to (Sapankevych and Sankar 2009), a time series is a sequence of historical measurements y_t observations of a variable Y over the same time interval. There are several well-known types of outliers in time series. Suppose a series of Y_t has an outlier (additive outlier (AO), innovational outlier (IO), level shift (LS), or temporary change (TC)) at $t = t_1$ where $I_t(t_1) = 1$ at $t = t_1$ and 0 others. An ARIMA-based time series model Y_t with outliers can be written (Chen and Liu 1993) as follows:

$$Y_t = Z_t + \omega L(B)I_t(t_1).\tag{1}$$

Y_t is the time series containing outliers, Z_t follows the ARIMA model as in (7) and (8), and ω is the initial effect when $t = t_1$. The criteria for the value of $L(B)$ are as follows:

$$AO : L(B) = 1\tag{2}$$

$$IO : L(B) = \frac{\theta(B)}{(1 - B)^d\phi(B)}\tag{3}$$

$$LS : L(B) = \frac{1}{(1 - B)}\tag{4}$$

$$TC : L(B) = \frac{1}{(1 - \delta B)}\tag{5}$$

where B is the backshift operator.

Detecting outliers in a series begins with decomposition of time series data into trend components T_t, seasonal S_t and remainder R_t, as in the following equation (Hyndman 2021):

$$Y_t = T_t + S_t + R_t\tag{6}$$

Then, remove the seasonal and trend components and find the outliers in the remaining series R_t. Y_t is detected as an outlier at point t if $R_t < Q_1 - 3 \times IQR$ or $R_t > Q_3 + 3 \times IQR$, where Q_1 and Q_3 are the quartiles 1 and 3 of R_t and IQR is the interquartile. We can clean up outliers by replacing detected outliers with linearly interpolated values using neighboring observations, and this process is repeated (Hyndman 2021).

2.2 ARIMA Model

Time series can be modeled using ARIMA (p, d, q), where the ARIMA model can be written as follows (Chen and Liu 1993):

$$\phi(B)(1 - B)^d Z_t = \theta(B)a_t, \ t = 1, 2, \ldots, n,\tag{7}$$

where n is the number of observations, d is a non-negative integer, B is the backshift operator, ϕ is the autoregressive component, θ is the moving average component, and $a_t \sim IIDN(0, \sigma_a^2)$. Hyndman and Khandakar (2008) developed an automatic forecasting method. One of them is based on ARIMA model, called automatic ARIMA. Automatic ARIMA tries various combinations of ARIMA models until it reaches the smallest AIC value.

2.3 Support Vector Regression for Time Series

Support Vector Machines (SVMs) are widely used in machine learning cases such as pattern recognition, object classification, time series forecasting, and regression analysis (Sapankevych and Sankar 2009). In the case of time series, given a time series data set x(t), where t is a series of n discrete samples. Regression analysis provides predictive functions for linear and nonlinear regression applications (Sapankevych and Sankar 2009):

$$\text{linear}: f(x) = (w \cdot x) + b. \tag{8}$$

$$\text{nonlinear}: f(x) = (w \cdot \phi(x)) + b. \tag{9}$$

where $\phi(x)$ is referred to as the kernel function.

Then find the optimal weight w and threshold b, and a criterion for determining the optimal weight. The average "flatness" of the weights can be measured by minimizing the Euclidean norm, $\|w\|^2$, and the empirical risk function as the error produced by the estimation process. The goal is to minimize the normal risk $R_{reg}(f)$, where f is a function of x(t) defined as (Sapankevych and Sankar 2009)

$$R_{reg}(f) = R_{emp}(f) + \frac{\lambda}{2}\|w\|^2. \tag{10}$$

The scale factor λ is commonly called the regularization constant, and this term is often referred to as "capacity control". Its function is to reduce "overfitting" of the data and minimize the effects of poor generalization. Empirical risk is defined as:

$$R_{emp}(f) = \frac{1}{N}\Sigma_{i=0}^{N-1}L(x(i),\gamma(i), f(x(i), w)), \tag{11}$$

where i is the index for discrete time series t = {0, 1, 2, . . . , N − 1}, $\gamma(i)$ is the actual data (training data) of the prediction sought and $L(\cdot)$ is the "loss function" or "cost functions".

In order to solve the optimal weights and minimize the regular risk, a quadratic programming problem is formed (using an ε-insensitive loss function), where there are parameters such as C and ε which are usually calculated empirically. In detail, the problem of optimal weight and minimum regular risk can be found in Sapankevych and Sankar (2009). An approximation of the function f(x) is given as the optimal weighted sum multiplied by the dot product between the data points:

$$f(x) = \Sigma_{i=1}^{N}(\alpha_i - \alpha_i^*)\langle x, x(i)\rangle + b. \tag{12}$$

Data point with a non-zero Lagrangian multiplier α that is on or outside the e-tube defined as support vectors.

To perform nonlinear regression on the SVR, we need to map the input space x(i) to a high-dimensional feature space $\varphi(x(i))$. Considering that the SVR solution relies on the inner product of the input data, a kernel function that satisfies the Mercer condition can be written as

$$k(x, x') = \langle \phi(x), \phi(x')\rangle. \tag{13}$$

Some kernel functions (required to generate kernel functions) satisfy Mercer constraints such as Gaussian, polynomial and hyperbolic tangent.

2.4 Random Forest for Time Series

Goehry proposed Random Forests for time series data on 2020 in the journal "Random Forest for Time-Dependent Processes" (Goehry 2020). The idea is to replace the bootstrap method used in the standard Random Forest with a moving block bootstrap, thus preserving dependencies over time (original data structure). Suppose we have a random sequence $(X_t, Y_t)_{t \in Z} \in R^p \times R$ such that

$$Y_t = f(X_t) + \epsilon_t, \tag{14}$$

ϵ_t is an error where $[\epsilon_t | X_t] = 0$. The purpose of Random Forest is to estimate the regression function $\forall x \in R^p, f(x) = E[Y_t | X_t = x]$. In the context of statistics, we only observe the training sample $D_n = ((X_1, Y_1), \ldots, (X_n, Y_n))$ which is used to construct the Random Forest estimator which is denoted by $\hat{f}_n$.

The Random Forest algorithm for time series is as follows (Goehry 2020) and (Goehry *et al.* 2021):

1. Input training data: $((X_1, Y_1), \ldots, (X_n, Y_n))$.
2. Specify the parameters $M, \alpha_n, m_{try}, \tau_n, l_n$. M is the number of trees, α_n is the number of observations for each tree, where $\alpha_n \in \{1, \ldots, n\}$; m_{try} a is the size of the variable previously selected for splitting where $m_{try} \in \{1, \ldots, p\}$; τ_n is a threshold or number of leaves in each tree, where the splitting process will stop when the number of leaves cannot exceed the parameters τ_n, $\tau_n \in \{1, \ldots, \alpha_n\}$; and l_n is the block size/block length.
3. Stopping criteria: the variance in a node is zero or the number of observations in a node below the threshold τ_n.
4. Construct the *j*th tree with the following procedure:

   a. Drawing $\alpha_n \leq n$ observations using one of the block bootstrap variants with parameter l_n.
   b. Repeat the following steps recursively at each node until the stopping criteria are met:

   - For each node, randomly select the m_{try} variable.
   - Select the best split using the variance criterion among the previously selected variables.
   - Cut according to the best split selected.

5. Repeat Step 4 M times.
6. The output for the new observation x is the average of the predictions M given by the tree for x.

In the case of regression, the estimator from Random Forest is as follows (Goehry, 2020):

$$\hat{f}_{M,n}(x; \Theta_1, \ldots, \Theta_M; D_n) = \frac{1}{M} \sum_{j=1}^{M} \hat{f}_n(x, \Theta_j, D_{n,}), \tag{15}$$

where $\hat{f}_n\left(x, \Theta_j, D_n\right)$ is the predicted value at point x for the j th random tree, $\Theta_1, \ldots, \Theta_M$ are independent random variables and identically distributed. Θ_j characterizes the j th Random Forest tree in terms of split variables, cutpoints at each node, and terminal node values (Hastie *et al.*, 2009).

2.5 Selection of the Best Method

There are several criteria in determining the best method, one of which is the root mean square error (RMSE). RMSE can be written as follows:

$$\text{RMSE} = \sqrt{\frac{1}{n}\sum_{i=1}^n \left(Y_t - \hat{Y}_t\right)^2}, \tag{16}$$

where n is the length of the series, Y_t is the actual value at time t and $\hat{Y}_t$ is the predicted value at time t.

3 Methodology

3.1 Data and Data Source

The study uses the Indonesian economic growth data series (y-on-y) obtained from the official website of the Badan Pusat Statistik (https://www.bps.go.id) as empirical data. The period used is Q1 2001 to Q4 2021. Simulation data are generated from his ARIMA model (p, d, q) based on the results of economic growth modeling.

3.2 Analysis Steps

The design of simulation studies considers several factors, such as the length of the series, the type of outliers, the number of outliers, and the the nature of the outliers in the series (consecutive or not consecutive) according to the nature of the outliers that occur in the macroeconomic indicator series, i.e., economic growth. This study's simulation designs included 20 simulations, as shown in Table 1. The analytical steps for the simulation study are as follows:

1. Data exploration (Indonesia's economic growth)
2. Identifying outliers in the data as described in Sect. 2.
3. Eliminating outliers and imputing missing data with linear interpolation values using neighbor observations.
4. Perform ARIMA modeling on the data in step 3 with automatic ARIMA.
5. Generate time series data that follows the ARIMA order obtained from step 4, with the following procedure:

   a. Generate normal distribution error with mean 0 and variance σ^2.
   b. Generate initial value for lag in model.
   c. Generating Z_t as much as n + 50 according to the model generated using a_t, ARIMA parameter and lag value. Z_t is a series that does not contain outliers.

6. Adding outliers to Z_t according to the scenario in Table 1 to obtain Y_t, where Y_t is a series containing outliers. Outlier placement is done randomly on Z_t where $t > 50$.

7. Next, create a data frame containing Y_t and lag of Y_t according to the ARIMA model by automatic ARIMA in step 4. Input variables are restricted to AR lag only. Next, get the last n observations from the data frame to eliminate the initial effect on the first 50 data.

8. Perform automatic ARIMA, SVR, and Random Forest modeling on Y_t. The input variable for SVR and Random Forest follows the lag obtained in Step 4. Hyperparameter settings are made on the SVR and Random Forest to get the optimum parameters. For SVR, this study took epsilon values (ε) from 0 to 1.05 with multiples of 0.01; cost value (C) from 1 to 105 with multiples of 1; and the value of gamma (γ) from 0 to 10.5 in multiples of 0.1. For Random Forest, this study takes the value of the number of trees (M) from 50 to 1000 with multiples of 25; m_{try} value is 1; node size value from 2 to 5; block size value from 2 to 12. The optimum parameter is obtained from the model that produces the smallest error.

9. Calculating RMSE from ARIMA, SVR and Random Forest Models.

10. Repeat steps 5–9 100 times.

11. Compare ARIMA, SVR, and Random Forest models based on the average RMSE values generated in step 9.

Table 1. Simmulation design

Sim	n	Num of outliers	Outlier types	Nature of outliers	Location of outliers
1	50	1	AO	-	$T = 25$
2	200	1	AO	-	$T = 100$
3	50	1	TC	-	$T = 25$
4	200	1	TC	-	$T = 100$
5	50	2	AO	NC	$T = 17, 33$
6	200	2	AO	NC	$T = 67, 133$
7	50	2	TC	NC	$T = 17, 33$
8	200	2	TC	NC	$T = 67, 133$
9	50	2	AO, TC	NC	$T = 17, 33$
10	200	2	AO, TC	NC	$T = 67, 133$
11	50	2	AO, TC	C	$T = 25, 26$

(continued)

Table 1. (*continued*)

Sim	n	Num of outliers	Outlier types	Nature of outliers	Location of outliers
12	200	2	AO, TC	C	T = 100,101
13	50	4	AO	NC	T = 10,20,30,40
14	200	4	AO	NC	T = 40,80,120,160
15	50	4	TC	NC	T = 10,20,30,40
16	200	4	TC	NC	T = 40,80,120,160
17	50	4	AO, TC, AO, TC	NC	T = 10,20,30,40
18	200	4	AO, TC, AO, TC	NC	T = 40,80,120,160
19	50	4	AO, TC, AO, TC	C	T = 15,16,35,36
20	200	4	AO, TC, AO, TC	C	T = 90,91,110,111

Notes: AO = Additive Outlier, TC = Temporary Change, C = Consecutive, NC = Non-Consecutive

4 Discussion

4.1 Data Exploration

In this study, economic growth uses the growth rate (year-over-year) from the first quarter of 2001 to the fourth quarter of 2021. Therefore, the base year should be adjusted before further analysis. The base year adjustment begins by setting the 2000 constant-price GDP value equal to the 2010 constant-price GDP value. Next, recalculate the GDP growth rate (year-on-year) to produce the series shown in Fig. 1.

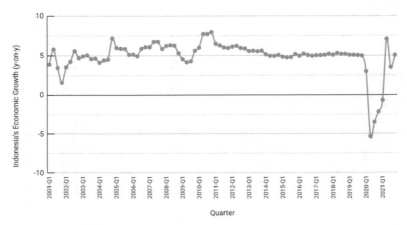

Fig. 1. Indonesia's economic growth y-on-y (January 2000 to February 2022).

Based on the results of outlier detection, there are two types of outliers in economic growth data (y-on-y): additive outliers (AO) and temporary changes (TC). AO occurred in Q2 2001 (5.77%), Q4 2004 (7.16%), Q1 2020 (2.97%), and Q2 2021 (7.07%). On the other hand, TC occurred in Q4 2001 (1.56%) and Q2 2020 (-5.32%).

4.2 Input for SVR and Random Forest Models

The input variables for the SVR and Random Forest models are based on the ARIMA model of the Indonesian economic growth by first replacing outliers with linear interpolation. The new series clean from outliers is modeled with automatic ARIMA methods, and the best model produced is ARIMA (1,0,2). Mathematically, ARIMA (1,0,2) can be written as

$$Z_t = 5, 1895 + 0, 7029 Z_{t-1} + a_t + 0, 2375 a_{t-1} + 0, 3781 a_{t-2}. \tag{17}$$

According to this model, Y_{t-1} will be the input for SVR and Random Forest.

4.3 Forest Models Generating Data Containing Outliers (Y_t)

The process of generating the Y_t data begins by generating a_t with $\mu = 0$ and $\sigma^2 = 0, 2087$. From ARIMA (1,0,2), we get $\phi_1 = 0, 7029$, $\theta_1 = 0, 2375$, $\theta_2 = 0, 3781$ and intercept $\theta_0 = 5, 1895$. From the obtained parameters, we generate $Z_t = \mu + 0, 7029 Z_{t-1} + 0, 2375 a_{t-1} + 0, 3781 a_{t-1} + a_t$ by first setting the value of $Z_{t-1} = 0$ and $Z_{t-2} = 0$, then iterate from $t = 1, \ldots, n + 50$ so that it will get the Z_t series with data length n + 50. For this simulation, we specify $\mu = 5$.

Next, add the outlier value in the Z_t series at a predetermined t point. The outlier on Z_t is added at the point t > 50 by Z_t plus the mean of Z_t. Furthermore, the first 50 data will be discarded to eliminate the effect of the initial value so that the Y_t series is obtained (Table 1).

4.4 Modeling with ARIMA, SVR and Random Forest

After getting Y_t, we create a new record containing Y_t and Y_{t-1}, then take the last n observations from the new data to eliminate the effect of the initial value on the first 50 data. Furthermore, the new data will be used for modeling ARIMA, SVR, and Random Forest. Then model Y_t with ARIMA, SVR, and Random Forest, replicate 100 times, calculate each RMSE for each iteration and calculate the average. The average RMSE values sof 100 replicates for each model are shown in Table 2.

4.5 Modeling with ARIMA, SVR and Random Forest

The number of outliers greatly impacts the accuracy of the results. The more outliers in time series, the larger the RMSE value. This applies to all three models and outlier types. In the ARIMA model, the average RMSE value with one outlier is 0.7473; the number of outliers two is 0.8857, and the number of outliers four is 1.1271. In the SVR model, the average RMSE value with one outlier is 0.6957; the number of outliers 2 is

Table 2. Average RMSE values of ARIMA, SVR, and random forest models

Sim	ARIMA	SVR	RF
1	0.93	0.79	1.04
2	0.64	0.58	0.71
3	0.83	0.82	1.04
4	0.58	0.59	0.71
5	1.20	1.04	1.32
6	0.76	0.68	0.82
7	1.10	1.08	1.36
8	0.68	0.68	0.81
9	1.20	1.11	1.45
10	0.73	0.70	0.87
11	0.84	0.81	1.08
12	0.58	0.59	0.72
13	1.58	1.41	1.74
14	0.95	0.84	1.01
15	1.42	1.41	1.74
16	0.85	0.84	1.00
17	1.55	1.47	1.91
18	0.92	0.86	1.07
19	1.07	1.04	1.34
20	0.68	0.69	0.84

0.8354, and the number of outliers 4 is 1.0710. In the Random Forest model, the average RMSE value with one outlier is 0.8747; the number of outliers 2 is 0.0546, and the number of outliers 4 is 1.3314. Therefore, outliers harm ARIMA, SVR, and Random Forest prediction results, and the impact increases as the number of outliers increases.

Adding a data series can reduce the effect of outliers. In the ARIMA model, the average RMSE value with a sample of 50 is 1.1722, while for samples of 200 is 0.7370. In the SVR model, the average RMSE value with a sample of 50 is 1.0980, while for samples of 200 is 0.7054. In the Random Forest model, the average RMSE value with a sample of 50 is 1.4019, while for samples of 200 is 0.8568. The decrease in the average RMSE value applies to any outlier. Thus, regardless of the type of outlier, increasing the number of samples for the three models (ARIMA, SVR, and Random Forest) will result in a lower average RMSE.

In the ARIMA model, the type of outlier is exceptionally influential on the RMSE value. Outlier type AO gives a worse impact than TC. AO generally produces an average RMSE value of 1.0119, followed by a combination of AO and TC of 0.9456, then TC of 0.9092. More specifically, when the number of outliers in the series is 2 and 4, the combination of AO and TC produces a lower RMSE value than the AO and TC types.

In the SVR model, the combination of AO and TC produces an average RMSE of 0.9103, followed by TC of 0.9013, then AO of 0.8907. When the outliers in the series are 1 and 2, TC has a worse impact than AO and the combination (AO and TC). Meanwhile, when there are four outliers, AO has a worse impact than TC. In the Random Forest model, the combination of AO and TC produces an average RMSE of 1.1603, followed by TC of 1.1112, then AO of 1.1062. Similar to SVR, when the outliers in the series are 1 and 2, TC has a worse impact than AO and a combination (AO and TC). Meanwhile, when there are four outliers, AO has a worse impact than TC. Therefore, we conclude that the type of outlier affects the ARIMA model; however, it does not affect SVR and Random Forest models.

Consecutive outliers produce a smaller RMSE mean than non-consecutive outliers. This can occur because consecutive or adjacent outliers can combine to form one type of outlier. This reduces the number of outliers and, thus, the average RMSE value will be lower. In the ARIMA model, consecutive outliers produce an average RMSE of 1.0779, while non-consecutive outliers are 0.7917. The consecutive outlier in SVR model produces an average RMSE of 1.0092, while the non-consecutive outlier is 0.7853. In the Random Forest model, consecutive outliers produce an average RMSE of 1.2591, while non-consecutive outliers are 0.9948.

From 20 simulations, each run with 100 iterations, the average RMSE value of SVR is lower than that of Radom Forest (Fig. 2). From this, we can conclude that modeling time series data with outliers using SVR is superior to modeling using ARIMA and Random Forest.

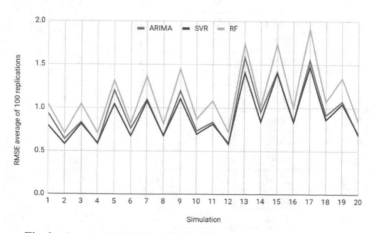

Fig. 2. Average RMSE of ARIMA, SVR and random forest models

4.6 Sensitivity of Hyperparameter

Table 3 shows the variation in parameter values for each simulation design is very high. For example, in simulation design 1, the optimal epsilon value is 0.35 with a standard deviation of 0.27. The value of 0.27 is relatively large because the multiple of the epsilon value is very small, i.e., 0.01. The same applies to parameter cost and gamma.

Among the 20 simulation designs in Table 3, the epsilon values range from 0.22 to 0.41, with an average of 0.32. The costs range from 19.11 to 38.29 with an average of 20, and gamma values range from 0.35 to 3.16 with an average of 1.48. Each simulation design differs in terms of the number of outliers, number of samples, type of outliers, and nature of outliers (consecutive or non-consecutive). The standard deviation of epsilon between simulation designs is 0.06, a cost of 4.34, and a gamma of 0.84. Variation in parameter values between iterations is relatively larger than the variation between simulation designs (Table 3). Therefore, the sensitivity of the parameter values of SVR models tends to be influenced more by the randomness of the generated data than by the design of the simulation itself (outliers). In other words, the SVR model is relatively robust to outliers.

Table 3. Mean and standard deviation of SVR hyperparameter

Sim	Average of hyperparameter			Standard deviation of hyperparameter		
	Epsilon	Cost	Gamma	Epsilon	Cost	Gamma
1	0.35	26.27	2.11	0.27	30.02	2.94
2	0.37	26.27	1.26	0.22	31.66	1.75
3	0.23	33.35	0.59	0.17	34.55	1.75
4	0.33	38.29	0.35	0.20	36.81	0.76
5	0.32	23.58	2.50	0.20	32.83	2.77
6	0.36	31.18	1.43	0.21	32.44	1.66
7	0.29	35.90	1.00	0.21	34.83	1.84
8	0.25	28.92	0.38	0.18	31.56	0.66
9	0.39	27.63	2.69	0.26	33.71	3.21
10	0.39	26.17	1.82	0.25	32.87	2.74
11	0.22	29.60	1.02	0.16	36.79	1.28
12	0.27	28.72	0.50	0.18	33.83	0.62
13	0.31	30.51	3.16	0.19	34.87	3.24
14	0.36	27.14	1.74	0.23	33.41	2.10
15	0.33	26.88	1.34	0.20	33.49	2.35
16	0.24	23.64	0.59	0.17	27.18	1.10
17	0.41	19.11	2.00	0.25	28.55	2.70

(*continued*)

Table 3. (*continued*)

Sim	Average of hyperparameter			Standard deviation of hyperparameter		
	Epsilon	Cost	Gamma	Epsilon	Cost	Gamma
18	0.38	27.63	2.54	0.24	33.52	3.17
19	0.31	31.38	1.87	0.19	35.27	2.78
20	0.33	26.08	0.75	0.24	31.01	1.08

Table 4 shows that the standard deviation for each parameter tends to be very high. For example, simulation 1 has a mean M value of 101 and a standard deviation of 106.24; a value of 106.24 is relatively high. The only input variable used is 1, which is no need to discuss. The node size standard deviation of 1.33 and block size of 2.93 is relatively small values.

Table 4. Mean and standard deviation of random forest hyperparameter

Sim	Average of hyperparameter			Std dev of hyperparameter		
	M	Node size	Block size	M	Node size	Block size
1	101.00	3.91	4.55	106.24	1.33	2.93
2	148.00	4.81	5.24	155.40	0.46	3.24
3	116.50	3.66	3.15	106.44	1.45	2.32
4	146.00	4.84	4.57	150.95	0.53	3.28
5	103.00	4.03	4.82	93.70	1.27	3.14
6	119.00	4.68	5.55	114.27	0.71	3.49
7	114.25	3.49	3.74	97.33	1.48	2.55
8	144.75	4.74	4.47	144.00	0.58	3.17
9	104.50	4.10	3.28	108.62	1.18	1.94
10	107.25	4.62	4.48	87.13	0.78	3.03
11	128.25	3.20	2.57	119.75	1.48	1.63
12	191.50	4.75	2.95	179.51	0.67	1.83
13	108.25	4.10	4.47	72.04	1.31	3.16
14	101.75	4.64	4.91	70.64	0.70	3.04
15	107.00	3.91	4.26	82.65	1.30	3.21
16	109.75	4.55	4.84	111.68	0.83	3.02
17	109.75	4.09	4.10	99.49	1.25	2.94
18	115.75	4.64	4.64	98.45	0.75	2.80

(*continued*)

Table 4. (*continued*)

Sim	Average of hyperparameter			Std dev of hyperparameter		
	M	Node size	Block size	M	Node size	Block size
19	116.75	3.80	3.60	94.22	1.41	2.59
20	122.75	4.41	4.17	110.10	1.06	2.69

From the 20 simulations in Table 4, the M values range from 101 to 191.5, averaging 120.79. Node sizes range from 3.2 to 4.84, with an average of 4.25. The block size ranges from 2.57 to 5.55, with an average of 4.22. The standard deviation of M between simulation plans is 22.04, node size 0.49, and block size 0.78. Variation in parameter values between iterations tends to be greater than the variation between simulation designs (Table 4). Similar to SVR models, the sensitivity of the parameter values of Random Forest models tends to be influenced more by the randomness of the generated data than by the simulation design itself (outliers). In other words, the Random Forest model is relatively robust against outliers.

5   Conclusion

Robustness simulation studies of SVR and Random Forest models on time series data with outliers reveal that: The more outliers in the data, the larger the RMSE value. This applies to all models used and to any type of outliers such as AO, TC, or a combination of both; (ii) adding more samples lowers the average RMSE value for any model (ARIMA, SVR, Random Forest) and any type of outlier (AO, TC, Combination of AO and TC); (iii) The type of outliers has a large impact on the RMSE value of the ARIMA model, where AO has a worse impact than TC. On the other hand, for SVR and Random Forest, the outlier type has no significant effect; (iv) Consecutive outliers produce smaller RMSE means than non-consecutive outliers. This happens because consecutive outliers can combine to form only one type of outlier. This reduces the number of outliers in the series so that the average RMSE value will be lower; (v) Based on the sensitivity of hyperparameter values, the SVR and Random Forest models are quite robust against the presence of outliers in the data. However, from the simulations run with 100 iterations, the average RMSE values of SVR are lower than ARIMA and Radom Forest models. Therefore, we can conclude that the SVR model is more robust than Random Forest when modeling time series data with outliers.

References

Ahmed, N.K., Atiya, A.F., Gayar, N.E., El-Shishiny, H.: An empirical comparison of machine learning models for time series forecasting. Economet. Rev. **29**(5–6), 594–621 (2010)

Alanis, E.: Forecasting betas with random forests. Appl. Econ. Lett. **29**(12), 1134–1138 (2022)

Baba, H., Takahara, J.I., Yamashita, F., Hashida, M.: Modeling and prediction of solvent effect on human skin permeability using support vector regression and random forest. Pharm. Res. **32**(11), 3604–3617 (2015)

Chen, C., Liu, L.M.: Forecasting time series with outliers. J. Forecast. **12**(1), 13–35 (1993)

Dudek, G.: Short-term load forecasting using random forests. In: Dudek, G. (ed.) Intelligent Systems' 2014. Advances in Intelligent Systems and Computing, vol. 323, pp. 821–828. Springer, Cham (2015).https://doi.org/10.1007/978-3-319-11310-4_71

Gather, U., Davies, P.L.: Robust Statistics (No. 2004, 20). Papers (2004)

Goehry, B.: Random forests for time-dependent processes. ESAIM: Probab. Stat. **24**, 801–826 (2020)

Goehry, B., Yan, H., Goude, Y., Massart, P., Poggi, J.M.: Random forests for time series (2021)

Hastie, T., Tibshirani, R., Friedman, J.H., Friedman, J.H.: The Elements of Statistical Learning: Data Mining, Inference, and Prediction, vol. 2, pp. 1–758. Springer, New York (2009). https://doi.org/10.1007/978-0-387-21606-5

Hong, W.C.: Electric load forecasting by support vector model. Appl. Math. Model. **33**(5), 2444–2454 (2009)

Hyndman: Detecting time series outliers (2021). https://robjhyndman.com/hyndsight/tsoutliers/. Accessed 8 Aug 2022

Hyndman, R.J., Khandakar, Y.: Automatic time series forecasting: the forecast package for R. J. Stat. Softw. **27**, 1–22 (2008)

Khan, W., Crockett, K., O'Shea, J., Hussain, A., Khan, B.M.: Deception in the eyes of deceiver: a computer vision and machine learning based automated deception detection. Expert Syst. Appl. **169**, 114341 (2021)

Mueller, S.Q.: Pre-and within-season attendance forecasting in major league baseball: a random forest approach. Appl. Econ. **52**(41), 4512–4528 (2020)

Pai, P.F., Lin, K.P., Lin, C.S., Chang, P.T.: Time series forecasting by a seasonal support vector regression model. Expert Syst. Appl. **37**(6), 4261–4265 (2010)

Priliani, E.M., Putra, A.T., Muslim, M.A.: Forecasting inflation rate using support vector regression (SVR) based weight attribute particle swarm optimization (WAPSO). Scientific Journal of Informatics **5**(2), 118–127 (2018)

Purnama, D.I., Setianingsih, S.: Support vector regression (SVR) model for forecasting number of passengers on domestic flights at Sultan Hasanudin airport Makassar. Jurnal Matematika, Statistika dan Komputasi **16**(3), 391–403 (2020)

Ronchetti, E.M.: The historical development of robust statistics. In: Proceedings of the 7th International Conference on Teaching Statistics (ICOTS-7), pp. 2–7, July 2006

Sapankevych, N.I., Sankar, R.: Time series prediction using support vector machines: a survey. IEEE Comput. Intell. Mag. **4**(2), 24–38 (2009)

Yung, W., et al.: The use of machine learning in official statistics. UNECE Machine Learning Team report (2018)

The Impact of Restricting Community Activities on COVID-19 Transmission: A Case Study in Sumatra Island, Indonesia

Abdullah Sonhaji[1], Sapto W. Indratno[1(✉)], Kurnia Novita Sari[1], Adi Pancoro[2], Ernawati Arifin Giri-Rachman[2], Udjianna S. Pasaribu[1], and Susi Setiyowati[1]

[1] Statistics Research Group, Program Study of Mathematics, Faculty of Mathematics and Natural Science, Institut Teknologi Bandung, Bandung, Indonesia
saptowi@itb.ac.id

[2] Genetics and Molecular Biotechnology Group, School of Life Science and Technology, Institut Teknologi Bandung, Bandung, Indonesia

Abstract. Sumatra Island is the third largest island with the second largest population in Indonesia which has the following eight provinces: Aceh, North Sumatra, West Sumatra, Riau, Jambi, South Sumatra, Bengkulu and Lampung. The connectivity of these eight provinces in the economic field is very strong. This encourages high mobility between these provinces. During this Covid-19 pandemic, the high mobility between provinces affects the level of spread of Covid-19 on the island of Sumatra. The central government ordered local governments to implement a community activity restriction program called PPKM. In this article, a study is conducted on the impact of the PKKM program on the spread of Covid 19 on the island of Sumatra, Indonesia. The spread of Covid-19 is modeled using the Susceptible-Infected-Recovered-Death (SIRD) model which considers the mobility factor of the population. The model parameters were estimated using Approximate Bayesian Computation (ABC). The results of the study using this model show that the application of PKKM in several provinces in Sumatra can reduce the level of spread of COVID-19.

Keywords: Approximate Bayesian Computation (ABC) · COVID-19 · Origin-destination matrix · Spatial SIRD Model · Restrictions community activities

1 Introduction

On December 31, 2019 in Wuhan, China, it was reported that a group of people associated with the Huanan Seafood Wholesale Market had symptoms of "pneumonia of unknown origin". A few days later, WHO released this disease related to the Novel Coronavirus, which was later called COVID-19. The disease is related to the bat-derived acute respiratory syndrome (SARS)-like coronavirus (bat-SL-covzc45 and bat-SL-covzxc21) (with an 88% similarity) but far from SARS-Cov (79% similarity) and MERS-Cov (approximately 50% similarity). These potentially related components in COVID-19 are given

© The Author(s), under exclusive license to Springer Nature Singapore Pte Ltd. 2023
Y. B. Wah et al. (Eds.): DaSET 2022, LNDECT 165, pp. 480–493, 2023.
https://doi.org/10.1007/978-981-99-0741-0_34

by Yang et al. in 2020 [1]. In the same year Ren et al. [2] also investigated this case regarding the spread of the virus in humans and animals. With the rapid increase in cases of Covid-19, WHO declared the outbreak a global pandemic. The spread and increase in COVID-19 cases has spread among countries including Indonesia.

The first cases reported in Indonesia occurred in early March, where two people aged 31 and 64 were infected with COVID-19. On April 18, 2021, it was reported that there was an increase in new cases in Indonesia, with a total of 1,604,348 cases. Of these cases, 1,455,065 patients recovered (90.70%) and 43,424 patients died (2.71%). Currently, the graph of COVID-19 cases in Indonesia still shows an upward trend every day [3]. With the increasing number of Covid-19 cases in Indonesia, the Indonesian government's efforts to reduce the number of COVID-19 transmissions continue to be carried out. One of the ways the Indonesian government suppresses the transmission of Covid-19 is by implementing the 3M health protocol program, namely the use of masks, washing hands, and keeping a distance. Of course, the application of 3M is adapted to the culture of each province. The objectives of the 3M program include practicing social distancing to prevent close contact and improving hygiene by washing hands regularly with soap or 62–71% ethanol, which can reduce the infectivity of the spread of the virus.

Information on behavior and public knowledge regarding the prevention and transmission of Covid-19 was discussed by Sari et al.[4], Zhong [5], Elachola et.al [6] and Taghrir et.al [7]. Zhang [9] in 2020 discussed that the media can be used as a source of socialization to ask people to stay at home, not to use public transportation, always wear masks, keep a distance from crowds, and not visit areas with a high number of COVID-19 infections.

The mobilization of the Indonesian people with high connectivity has caused the pandemic to continue to spread rapidly to all regions in Indonesia. So the factor of population mobility between provinces in Indonesia is an important component that must be considered. The transmission of COVID-19 is controlled by limiting the movement of the population through social distancing interventions to reduce the number of contacts. By limiting the mobility of the population Nouvellet et al. [10] shows a pattern of sharp declines in Covid-19 deaths. This shows that there is a relationship between mobility and transmission of Covid-19.

In this article, normal population mobility between provinces (before the pandemic) refers to population mobility data between provinces in 2018. Here connectivity between provinces is described as a directed graph with side weights describing the intensity of population mobility between provinces connected by that side. As an archipelagic country, Indonesia has a natural sea boundary that separates the islands in Indonesia. This can be used as an advantage in regulating the mobilization of people between islands. However, for the internal mobility of large islands such as Sumatra, it is more difficult to regulate population mobility due to the large number of land routes that can be used by residents to communicate between provinces on the island of Sumatra.

The impact of COVID-19 is a challenge for the Indonesian government to formulate preventive policies and actions in prevention, handling, control, and recovery in various sectors. Therefore, it is necessary to estimate the number of infected cases and the possibility of new cases to assist and direct the government in making policies in the health sector in particular and the economy in general. The government's policy to reduce mobilization is to impose restrictions on community activities.

Given that the culture of each province in Indonesia is different, in analyzing the impact of the PPKM program, it is necessary to consider a spatial distribution model that can capture the behavior of each of these provinces. Spatial disease transmission studies have been carried out by many researchers, as mentioned in articles [11–17]. In this article, to obtain information for each province, the Spatial SIRD model is used which considers the following local parameters: contact rate, death rate, recovery rate and proportion of displaced population. There are 136 parameters in this SIRD dynamic model that must be estimated. The process of estimating these parameters using the classical method or the maximum likelihood approach is difficult. Therefore, the Approximate Bayesian Computation (ABC) method [24, 25] was used. In principle, this method uses the MCMC approach in generating the expected parameters.

The SIRD model used will provide local parameters for each province whose values can vary. It is through these parameters that an analysis of the success of the PPKM program is carried out. This article contributes to the spatial analysis method of COVID-19 transmission by considering the mobility of the population between provinces in Indonesia. It is hoped that this COVID-19 spatial transmission approach can be used for preventive measures against COVID-19 in every province in Indonesia.

Sumatra has an area of 473,481 km^2 and is the third largest of the islands in Indonesia. Mainland Sumatra is generally dominated by the Barisan mountain range that stretches along the island. This makes people mobilize more often using road trips, which causes the movement to become more intense. Therefore, it is very interesting to conduct research related to the increase in infected cases on the island of Sumatra. The framework was applied to the eight largest provinces on the island of Sumatra with the transmission of COVID-19 from data on the increase in cases of infection, recovery, and death. In addition, it is also proposed to analyze the data completely by using a spatial SIRD model that can handle several distributions and contact modes.

The structure of this journal is as follows. We provide the spatial SIRD framework given in Sect. 2, with data sources, detailing the formation of the Original Destination Matrix (O-DM), including the ABC algorithm in Sect. 3. In Sect. 4 the analysis of the spatial SIRD model is given. In addition, this section also provides results and discussions related to the implementation of the prediction model for the number of infected cases before and after PPKM. This article closes with the conclusion in Sect. 5.

2   Methods

2.1   Data Sources and SIRD Model

Data from the Ministry of Health of the Republic of Indonesia consists of a series of daily reported cases of COVID-19 from March 2nd, 2020 to the present, grouped by province. The data set report on 9,288,988 people had been examined by PCR test with 7,684,640 people were proven negative. In addition, there are 1,604,348 people confirmed cases of COVID-19, 1,455,065 recovered, and 43,424 deaths in 34 provinces and 510 districts/cities in Indonesia [3]. Data from the Indonesian Ministry of Health for 2020 to date are spatially collected on the island of Sumatra over certain time intervals.

Sumatra Island contributes 9.91% to the national. In addition, data related to the size and mobilization of Indonesia's population in each province is also needed [19–21]. The need for mobilization data from province A to province B, inversely, is used to create the Origin-Destination Matrix (O-DM) in the modeling stage. There are several studies that discuss the model of the spread of the disease by considering the location factor. Bjornstad (2007) conducted a study on disease transmission spatially and measured the time of outbreak [11]. Brockmann and Helbing conducted follow-up research in 2013 on the spread of epidemics. The global spread of the epidemic is due in part to a dynamic, network-driven process. Effective distance reliably predicts disease arrival time, so mobilization is an important factor in this modeling. The model used is a complex spatiotemporal pattern. This can be reduced to a very simple homogeneous wave propagation pattern, if conventional geographic distances are replaced by probabilistically motivated effective distances [12]. Further research was applied to MERS in 2016 regarding the use of openly accessible data including the airline transportation network to parameterize a hazard based risk prediction model. The hazard was assumed to follow an inverse function of the effective distance (i.e., the minimum effective length of a path from origin to destination), which was calculated from the airline transportation data [13]. The mobile phone data to quantify seasonal travel and directional asymmetries were applied in this study on January 1st, 2020, and equal to zero thereafter. Besides that, it is quantifying mobility patterns from mobile phone data [14]. In a recent study included the zoonotic force factor of infection in Wuhan during the baseline scenario before market closure and hence international case exportation occurred according to a non-homogeneous process [15]. Sonhaji et al. (2021) [23] discussed a spatial-SIR model which resampling method used for estimating the model parameters.

In this article a modified SIRD Model by considering the mobility factor is given as follows:

$$
\begin{cases}
\dfrac{dS_i(t)}{dt} = \left(-\beta_i \dfrac{I_i(t)}{N_i} - \dfrac{\alpha_i \sum_{\ell \in \tau_i} \beta_\ell m_{i,\ell} \frac{I_\ell(t)}{N_\ell}}{N_i + \sum_{\ell \in \tau_i} m_{i,\ell}} \right) S_i(t) + \eta_i R_i(t) \\[4mm]
\dfrac{dI_i(t)}{dt} = \left(\beta_i \dfrac{I_i(t)}{N_i} + \dfrac{\alpha_i \sum_{\ell \in \tau_i} \beta_\ell m_{i,\ell} \frac{I_\ell(t)}{N_\ell}}{N_i + \sum_{\ell \in \tau_i} m_{i,\ell}} \right) S_i(t) - \left(\dfrac{\gamma_i + \delta_i}{2} \right) I_i(t) \\[4mm]
\dfrac{dR_i(t)}{dt} = \left(\dfrac{\gamma_i}{2} \right) I_i(t) - \eta_i R_i(t) \\[4mm]
\dfrac{dD_i(t)}{dt} = \left(\dfrac{\delta_i}{2} \right) I_i(t),
\end{cases}
\tag{1}
$$

where:

$S_i(t)$: the number of susceptible individuals in province i who can be infected. at day t.

$I_i(t)$: the number of individuals in province i who have been infected (from susceptible) and able to transmit the virus at day t.

$R_i(t)$: number of individuals in province i who have been recovered and death (from infected) at day t.

$D_i(t)$: number of individuals in province i who has been death (from infected) at day t.

β_i: the individual contact rate in province i.

δ_i: the death rate in province i.
γ_i: the recovery rate in province i.
α_i: the percentage of incoming migrant using public transportation in province i.
$m_{i,\ell}$: the number of individuals migrate from province ℓ to province i.
τ_ℓ: the neighboring association from province ℓ.
N_i: the population in province i.

In the modified model, the people mobility factor is depicted as the variable $m_{i,\ell}$. Meanwhile medic readiness in a province is represented by parameters γ_i and δ_i which picture the survival probability of infected people in province i. In this case, a uniform prior distributions $Unif(a_i, b_i)$ is used for both parameter γ_i and δ_i, with a_i and b_i represent hyper parameters which are related to medical readiness in province i.

2.2 Origin-Destination Matrix

The modeling stage that estimates the distribution pattern of human mobility from one area to another, this process is commonly called a mobility distribution model. This form is a two-dimensional matrix that provides information on the amount of mobility between regions.

In the context of SIRD modeling, positive case estimation is very important in determining pandemic management policies and strategies. Therefore, positive case estimation involving O-DM is very important in SIRD modeling. O-DMs are made between provinces in Indonesia by using distance records from Google map information and mobility data in 2018 from the Central Bureau of Statistics Republic of Indonesia, Badan Pusat Statistik (BPS), both land, sea and air transportation. Furthermore, the distance between provinces is used as a determinant in making proportions / weights. The farther the distance from a province, the smaller weight is given, and vice versa. [19–22]. The patterns of mobility in the transportation system are often described in terms of the flow of mobility (vehicles, passengers, goods) moving from the origin to the destination during a certain period of time.

The data obtained from BPS 2018 report show only the total number of people migrate from one province to other province. There is no information on the mode of transportation used by the people. There are several obstacles in making the land transportation O-DM, including (i) the absence of data related to the origin and destination, (ii) the absence of information regarding the number of passengers traveling by land transportation from one province to another, (iii) there are various modes of land transportation, namely motorized vehicles, trains and commuter lines, (iv) mobility is only recorded in one province.

To overcome the constraints (i) and (ii) on land transportation, the distance calculation is carried out using Google maps as a reference in determining the proportion / weight. We assume that the farther a province from the targeted province is, the smaller proportion of people enter the targeted province from the province. To compute the proportion of people entering the targeted province A from a province, we do the following way:

a.   Let $\{d(A, C_j)\}_j$ be the collection of Euclidean distance between the targeted province A and the province $C_j, j = 1, 2, \ldots, N$. Here we only take provinces which have land transportation to the targeted province A, suppose there is N such province. Define $L = \max\{d(A, C_j), j = 1, 2, \ldots, N\}$ and the distance ratio $P_{A,C_j} = \frac{d(A,C_j)}{L}, j = 1, 2, \ldots, N$.

b. Calculate the distance $d(A, C_j) = 1 - \log(P_{A,C_j}), j = 1, 2, \ldots, N$.
c. Calculate the proportion of people entering the targeted province A as follows:
$$P_{A,C_j} = \frac{d(A,C_j)}{\sum_{k=1}^{N} d(A,C_j)}.$$

Information on the number of residents has been obtained in units of people per year. To overcome (iv), it is necessary to multiply the proportion of distance with population mobility within the province, so that population mobility between provinces can be obtained using land transportation. Always be reminded that the units of this BPS data information are a year, so it is necessary to equalize the units in days. Therefore, the population mobility data between provinces is divided by 365 to obtain daily data unit. Thus, the mobility of land and air transportation passengers between provinces is obtained, with units of people per day.

3 Approximate Bayesian Computation (ABC) Method

Based on [24], the unknown model parameters are treated as random variables and as fixed quantities. It is used to quantify the uncertainty of the estimates in a coherent and probabilistic manner. The first step in the Bayesian paradigm is to establish the prior distribution that assumes an unknown parameters $\{\alpha, \beta, \gamma, \delta, \eta\} = \theta$. Subsequently this prior information is updated in the light of experimental data $\{S, I, R, D\} = \Omega$ produced by model (1) using Bayes theorem by multiplying it with the likelihood function $\pi(\Omega|\theta)$ and renormalizing, thus the next process is obtained posterior distribution:

$$\pi(\theta|\Omega) = \frac{\pi(\Omega|\theta)}{\int_{\theta} \pi(\Omega|\theta)\pi(\theta)d\theta} \propto \pi(\Omega|\theta)\pi(\theta)$$

The experimental data are insufficient to write down a tractable likelihood. Nevertheless, ABC is possible to simulate from the model to perform inference without having to compute the likelihood. One of the difficulties when fitting models to COVID-19 outbreak data is that the infection process is unobserved. Observation data along with its posterior distribution will be difficult to observe, especially in data on infections involving displacement of population mobilization. Therefore, the ABC algorithm is suitable for estimating the parameters of the COVID-19 pandemic model based on available observational data in Indonesia.

The ABC method can be used even without the likelihood function and highly efficient [25, 26]. The Best algorithm selected based on the efficiency of time and a high degree of accuracy [26]. This method is widely applied to the modeling of infectious diseases. In fact, infectious disease data have missing or incomplete information due to the partially observed nature of the pandemic. To overcome this, the ABC algorithm is the easiest method to do [25, 26]. The following algorithms can be applied to other cases as well [26].

Approximate Bayesian Computation, ABC-SMC-MNN Algorithms steps
1. Set the number of generations, the number of observation, and the number of nearest neighbours $(T, N,$ and $M,$ respectively)
2. Set the prior distribution $P(\theta)$
3. Set the initial generation $t = 1$
4. Set the tolerance value as much as $G,$ so that $\varepsilon_1 < \varepsilon_2 < \cdots < \varepsilon_G$
5. Set n the number of observations
6. Set the initial indicator $i = 1$
7. If $t > 1,$ then $\exists M$ as the nearest neighbour of $\theta_{g-1}^{(i)}$
8. Calculate empirical covariance matices $\Sigma\big(\theta_{g-1}^{(i)}, M\big), \forall i = 1, \cdots, N$ For a case $M = N,$ then the empirical covariance matrice $\Sigma\big(\theta_{g-1}^{(i)}\big) = \Sigma(\theta_{g-1})$
9. If $g = 1,$ sample θ^{**} from the prior distribution $P(\theta)$ If $g > 1,$ sample θ^* from the previous generation $\{\theta_{g-1}\}$ with weights $\{w_{g-1}\},$ with $\theta^{**} \sim N(\theta^*, \Sigma(\theta^*, M))$
10. If $P(\theta^{**}) = 0,$ return to step 5
11. Generate n data set D_j^{**} from the SIRD model (1) using θ^{**}
12. Calculate $\hat{P}(D \vert D^{**}) = \left(\frac{1}{n}\right)\sum_{j=1}^n \mathbb{I}\big(d\big(D, D_j^{**}\big) \leq \varepsilon_g\big)$
13. If $\hat{P}(D \vert D^{**}) = 0,$ return to step 5
14. Set $\theta_g^{(i)} = \theta^{**}$
15. Calculated corresponding weight of the accepted observation $-i$ $$w_g^{(i)} = \begin{cases} \hat{P}(D\vert D^{**})P(\theta^{**}), & if\ g = 1 \\ \dfrac{\hat{P}(D\vert D^{**})P(\theta^{**})}{\sum_{j=1}^N w_{g-1}^{(j)} \mathcal{N}\left(\theta_g^{(i)} \big\vert \theta_g^{(j)}, \Sigma(\theta_g^{(i)}, M)\right)}, & if\ g > 1 \end{cases}$$
16. If $i < N,$ increment $i = i + 1$ and go to step 5
17. Normalise the weights so that $\sum_{i=1}^N w_g^i = 1$
18. If $g < G,$ set $t = t + 1$ and go to step 8

4   Results and Discussion

In the results of this time, the impact of PPKM inference will be examined in several provinces. This activity starts from January 11th 2021, until now. Therefore, the observation data carried out were one week before PPKM (4th-10th January 2021) and three months – one week after the PPKM was implemented (11st-18th April 2021). From the results of observations and analysis, it will be seen the impact of PPKM on pandemic transmission. The observation limitation will focus on Sumatra Island, because in [23] the SIR-Spatial analysis has been carried out in Java Island. Sumatra Island has eight major provinces including Aceh, North Sumatra, West Sumatra, Riau, Jambi, Bengkulu, South Sumatra, and Lampung.

The first step is to build O-DM for the eight provinces. The matrix used is the mobility of the Sumatran population using land and sea transportation as in Table 1. The gray row shows O-DM for air transportation while the other in land transportation. The table was obtained by collecting secondary data from BPS regarding the number

Table 1. Origin -Destination Matrix of the island of Sumatra (the white bold: land transportation, the grey bold: air transportation)

D \ O	Aceh	N.Sum	W. Sum	Riau	Jmb	Bkl	S.Sum	Lpg
Aceh	0	2,433	2,089	2,098	1,983	1,937	1,922	1,867
	0	963	0	0	0	0	0	0
N. Sum	4,813	0	4,403	4,462	4,101	3,966	3,943	3,836
	520	0	454	677	41	0	296	0
W. Sum	1,834	1,930	0	2,308	2,171	2,049	2,003	1,832
	0	477	0	55	39	28	36	0
Riau	2,381	2,640	2,988	0	2,752	2,556	2,569	2,454
	0	685	58	0	30	0	42	0
Jmb	2,493	2,640	3160	3,094	0	3,083	3,425	3,078
	0	32	26	26	0	5	106	22
Bkl	513	537	627	604	665	0	693	628
	0	0	36	0	19	0	89	0
S. Sum	3,182	3,346	3,853	3,817	4,541	4,400	0	4,319
	0	280	33	32	112	87	0	130
Lpg	1,380	1,438	1,624	1,606	1,834	1,779	1,941	0
	0	0	0	0	0	0	117	0

of passengers mobilizing per 2018. The main data source is obtained in the form of a table which is then transformed into a matrix that is sized according to the number of provinces to be analyzed. Retrieval of data for air transportation is already in a matrix form and does not have any problems. However, for land transportation data, there are three types of vehicles, motorized vehicles and trains, which only apply on the island of Sumatra. Motorized vehicles consist of cars and motorcycles, with three categories of cars, goods cars, passenger cars, and buses. Trains also consist of two categories, freight and passenger trains. For transportation using this rail, it already has a passenger unit.

The information obtained from BPS is only the distribution of mobility in the form of vehicle units a year. There are seven types of transportation and the most are freight cars 24%. In this case, freight cars are not counted in mobility because there is no movement of people from one region to another. The largest contribution came from passenger cars, motorcycles and buses. The largest number of land transportation types are passenger cars and motorcycles (23%). O-DM requires passenger or mobile person information. Thus, the unit information obtained from the 2018 annual report on Land Transportation Statistics is multiplied by the maximum capacity of transportation to carry the number of people.

Information on the number of residents has been obtained in units of people a year. It is just that this mobility information can only be observed in one province, this is an obstacle that arises in land transportation. Other data input needed is related to the daily report of COVID-19 on the island of Sumatra, in particular, which contains $\Omega = \{S, I, R, D\}$ a uniform distribution of each parameter. The parameters used were asumied to follow $\beta_i \sim$ $Unif[0.02, 0.1]$, $\frac{1}{\gamma_i} \sim Unif[14, 20]$, $\alpha_i \sim Unif[0.3, 0.98]$, $\delta_i \sim Unif[0.04, 0.09]$, $\frac{1}{\eta_i} \sim$ $Unif[1020, 1070]$.

4.1 Prediction of the Number of Infected Cases Previous to PPKM

The analysis before the PPKM was carried out on January 4–10, 2021, exactly a week before the PPKM activities began. The analysis was carried out based on local parameters estimated from the Spatial SIRD Model using the ABC method. This ABC method will provide the best parameters in representing the observation data. The measure of goodness of the parameter used is the Sum of Square Error (SSE). Parameters that provide simulation data that are closest to the observed data will give the smallest SSE.

Before PPKM started, there were five provinces that had value $R_0 > 1$ (Table 2), meaning that further countermeasures were needed, such as social restrictions and quarantine to prevent the spread of pathogens. The five provinces are Aceh, West Sumatra, Riau, Jambi, and Lampung. One of the supporting factors for the high rate R_0 is that population mobilization is still high, it can be seen from the value $\alpha \geq 0.64$, except that West Sumatra and Lampung have a low level of mobilization, in the range of 0.391 and 0.359, respectively. The same thing happens with the probability of contracting the suspect described by the value $\beta > 0.02$. Outbreaks also occur as early as mid-2021.

Table 2. Comparison of parameter estimates before and after PPKM was implemented(Previous to; Afterwards)

Provinces	$R_0(person)$	$\frac{1}{\gamma}(day)$	$\delta(\%)$	$\frac{1}{\eta}(day)$	$\alpha(\%)$	$\beta\left(\frac{1}{person}\right)$
Aceh	(1.200;0.567)	(18;18)	(0.078;0.066)	(1,067;1,025)	(0.710; < 0.40)	(0.067;0.032)
N. Sumatera	(0.966;1.035)	(18;19)	(0.052;0.063)	(1,050;1,047)	(0.780;0.413)	(0.054;0.054)
W. Sumatera	(1.219;0.991)	(14;17)	(0.078;0.059)	(1,055;1,042)	(0.391; > 0.80)	(0.087;0.060)
Riau	(1.700;1.672)	(17;16)	(0.063;0.059)	(1,049;1,059)	(0.976;0.477)	(0.100;0.103)
Jambi	(1.000;0.946)	(15;17)	(0.056;0.049)	(1,054;1,026)	(0.639;0.653)	(0.067;0.056)
S.Sumatera	(0.943;0.639)	(17;17)	(0.045;0.066)	(1,065;1,028)	(0.853; > 0.80)	(0.056;0.038)
Bengkulu	(0.558;1.810)	(19;18)	(0.065;0.072)	(1,052;1,054)	(0.670;0.723)	(0.029;0.102)
Lampung	(1.323;0.750)	(17;18)	(0.070;0.065)	(1,070;1,050)	(0.359; > 0.80)	(0.080;0.043)

4.2   Prediction of the Number of Infected Cases Afterward PPKM

Analysis after PPKM starts from April 11–18, 2021 or after PPKM lasts for three months – one week. There was a change after the implementation of PPKM. It is seen that there is a decrease in the value of R_0 for several provinces. Based on the results obtained before the implementation of the PPKM program, there were five provinces that had an R_0 > 1 value and after the implementation of the PPKM program there were only three provinces that still had an R_0 > 1 value. The three provinces are North Sumatra, Bengkulu, and Riau. So that these three provinces must make improvements in implementing this PPKM program. It can be seen that the three provinces experienced a significant increase in the value of R_0, namely for North Sumatra and Bengkulu by 7.14% and 224.37%, respectively. Meanwhile in Bengkulu, despite a 1.65% decrease in the R_0 value, the R_0 value is still above 1. This is the main focus of the regional government to again tighten PPKM or create new programs to emphasize the increase in the number of COVID-19.

The increase is in line with increasing parameters for North Sumatra and Bengkulu has increased for four until five parameters. Meanwhile, other provinces experienced a decline of more than 51.13% although three provinces declined with the basic reproduction rate still above one.

An interesting phenomenon occurred in Aceh, Lampung, and South Sumatra which experienced R_0 significant decline. This shows that the PPKM program in these provinces is very efficient and must be maintained. One of the methods taken by the South Sumatra provincial government is to carry out a persuasive approach to the sub-district task force team to carry out PPKM to hamlet or urban village areas. Its application from the initial stage is in the form of education for the community. The health team also conducts a tracing process of close patient contact in order to minimize transmission. The goal is that all levels of society follow what their role models instruct. This is one way to reduce COVID-19 transmission.

The recovery process has an increase in time period by an average of one day or a maximum increase of 17.86%. While the death rate has mostly decreased, this has reduced the number of deaths that occur due to COVID-19 cases. This decline is not in line with North Sumatra, South Sumatra, and Bengkulu which experienced increases δ. The η is robust parameter in modeling this time, which means that how much the change does not have much impact on the estimation results obtained. So that how long it takes to get reinfected does not really affect the increase in infections that will occur. This is because the parameter value is very small. The mobilization factor becomes an obstacle that makes cases increase again. Almost all provinces experienced an increase in the percentage of mobility by $\alpha > 2.19\%$, except for Aceh, North Sumatra, and Riau which experienced a decrease of respectively by $> 90.00\%$, 47.05%, and 51.13%. Therefore, policies to reduce mobilization really help emphasize COVID-19 transmission.

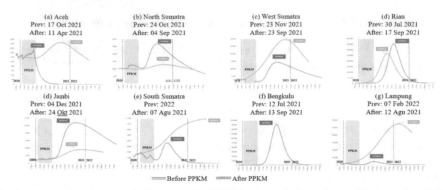

Fig. 1. Comparison between before and after PPKM was carried out along with the outbreak date

Another change that occurred because of this PPKM caused the peak of the outbreak to occur in 2021 (Fig. 1) using observation data from April 11st-18th, where most provinces experienced a decrease in the number of patients infected with COVID-19 by more than 27.70%. The estimation results will change depending on the observed data. It could be that after April 18th, certain provinces experienced a significant increase in infections, making the outbreak date faster and the number of infected cases increasing. Therefore, this modeling is devoted to observe the impact of PPKM on the increase in COVID-19 cases in Sumatra with an observation period of one week before PPKM was implemented and three months – one week after PPKM.

The phenomenon that occurred in Aceh caused a shift in the outbreak value in the same year 2021. What is surprising is that the number of infected cases has decreased along with the implementation of the PPKM program. Many people in Aceh obeying the regulations and a bit of getting bored with the imposition of these social restrictions. What about conditions in North Sumatra? The same thing happened to a shift in the value of the outbreak, but it was still in the same year. Here it is clear that the number of infected cases in the outbreak period was higher than before the PPKM took place. However, the rate of decline after PPKM has been slow but sure.

Meanwhile, the reverse situation from Aceh occurred in West Sumatra. Sumatran people tend to obey the rules more, and this could be due to elements of customary leaders who have a strong role in controlling their communities well. The outbreak date has shifted to a month earlier with the increase in infected cases decreasing significantly. Further observations, namely the province of Riau, show that after PPKM increased cases compared to before this government program was implemented. Even the addition can reach up to 32.96%.

The incidence in Jambi is almost the same as in North Sumatra, with the number of infected cases after PPKM has increased compared to before. The Jambi government needs to take firm action against people who do not comply with these regulations. The date of the outbreak in South Sumatra has indeed shifted forward to 2021, but it appears that there will be two local peaks after the PPKM has started. Bengkulu is in the spotlight for the provincial government to fix PPKM-related regulations, after the PPKM lasted for three months-one week, increasing the number of infected cases. There needs to be an evaluation related to this program for the government of Bengkulu Province. The opposite happened in Lampung, this program is running well and can even reduce the incidence of infected cases by 91.77%.

5 Conclusion

In this journal, the spatial factor of the SIRD model has been developed which allows it to be applied to the COVID-19 pandemic data that is collected spatially for analysis over time. This model was run to see the pattern of additional infected cases before and after the PPKM was implemented.

As stated in the introduction, there are two main objectives in this work, namely to see the changes in mobilization that have occurred and to assess the effect of the PPKM on the increase of infected cases. Obviously, the second objective was achieved given that most provinces experienced postponement of the infection outbreak date. For the first objective, the results here show that most provinces experienced increased mobilization during the PPKM. According to the spatial model SIRD causes a decrease in the level of transmission of COVID-19 in Sumatra Island with the implementation of PPKM than before the implementation of this program. This could happen because Sumatra has a strong culture and customs, so that PPKM does not have a significant effect. Especially related to maintaining distance and wearing a mask.

Spatial SIRD modeling through the ABC method can be done well and yields satisfactory results in the application of Restrictions Community Activities that the local government carries out. Government policies related to PPKM, especially in Sumatra Island, were very effective of observed from the reduced value R_0 in the provinces of Aceh, West Sumatra, Riau, Jambi, South Sumatra, and Lampung. Meanwhile, for North Sumatra and Bengkulu, monitoring and increasing the level of compliance again must be carried out. However, from the aspect of the increase in infected cases, Bengkulu experienced a significant decrease. In addition, the level of mobilization for all Provinces in Sumatra needs to be minimized to reduce COVID-19 transmission.

References

1. Yang, J., et al.: Prevalence of comorbidities in the Novel Wuhan Coronavirus (COVID-19) infection: a systematic review and meta-analysis. Int. J. Infect. Dis. **94**, 91–95 (2020). https://doi.org/10.1016/j.ijid.2020.03.017
2. Ren, L.L., et al.: Identification of a novel coronavirus causing severe pneumonia in human: a descriptive study. Chin. Med. J. **133**, 1015–1024 (2020). https://doi.org/10.1097/CM9.0000000000000722
3. Ministry of Health Republic on Indonesia. Dashboard Kasus Covid-19 di Indonesia (2021)
4. Sari, D.K., Amelia, R., Dharmajaya, R., Sari, L.M., Fitri, N.K.: Positive correlation between general public knowledge and attitudes regarding COVID-19 Outbreak 1 month after first cases reported in Indonesia. J. Commun. Health. **46**, 182–189 (2021). https://doi.org/10.21203/rs.3.rs-35827/v1
5. Zhong, B.L., et al.: Knowledge, attitudes, and practices towards COVID-19 among Chinese residents during the rapid rise period of the COVID-19 outbreak: A quick online cross-sectional survey. Int. J. Biol. Sci. **16**(10), 1745–1752 (2020). https://doi.org/10.7150/ijbs.45221

6. Elachola, H., Ebrahim, S.H., Gozzer, E.: COVID-19: facemask use prevalence in international airports in Asia, Europe and the Americas. Travel Med. Infect. Dis. **35**, 101637 (2020). https://doi.org/10.1016/j.tmaid.2020.101637

7. Taghrir, M.H., Borazjani, R., Shiraly, R.: COVID-19 and Iranian medical students: a survey on their related-knowledge, preventive behaviors and risk perception. Arch. Iran. Med. **23**(4), 249–254 (2020). https://doi.org/10.34172/aim.2020.06

8. Meo, S.A., Al-Khlaiwi, T., Usmani, A.M., Meo, A.S., Klonoff, D.C., Hoang, T.D.: Biological and epidemiological trends in the prevalence and mortality due to outbreaks of novel coronavirus COVID-19. J. King Saud Univ. **32**(4), 2495–2499 (2020). https://doi.org/10.1016/j.jksus.2020.04.004

9. Zhang, M., et al.: Knowledge, attitude, and practice regarding COVID-19 among healthcare workers in Henan. China. J. Hospital Infect. **105**(2), 183–187 (2020). https://doi.org/10.1016/j.jhin.2020.04.012

10. Nouvellet, P., et al.: Reduction in mobility and COVID-19 transmission. Nature Commun. **12**, 1090 (2021). https://doi.org/10.1038/s41467-021-21358-2

11. Bjørnstad, O.N., Grenfell, B.T.: Hazards, spatial transmission and timing of outbreaks in epidemic metapopulations, environmental and ecological statistics 2007. Environ. Ecol. Stat. **15**, 265–277 (2008). https://doi.org/10.1007/s10651-007-0059-3

12. Brockmann, D., Helbing, D.: The hidden geometry of complex, network-driven contagion phenomena. Science **342**, 1337–1342 (2013). https://doi.org/10.1126/science.1245200

13. Wesolowski, A., et al.: Multinational patterns of seasonal asymmetry in human movement influence infectious disease dynamics. Nature Commun. **8**, 2069 (2017)

14. Wu, J.T., Leung, K., Leung, M.D.G.M.: Nowcasting and forecasting the potential domestic and international spread of the 2019-nCoV outbreak originating in Wuhan, China: a modeling study. Lancet **395**(10225), 689–697 (2020). https://doi.org/10.1016/S0140-6736(20)30260-9

15. Metcalf, C.J.E., Munayco, C.V., Chowell, G., Grenfell, B.T., Bjørnstad, O.N.: Rubella metapopulation dynamics and importance of spatial coupling to risk of congenital rubella syndrome in Peru. J. Roy. Soc. Interface **8**, 369–376 (2011). https://doi.org/10.1098/rsif.2010.0320

16. Nah, K., Otsuki, S., Chowell, G., Nishiura, H.: Predicting the international spread of middle east respiratory syndrome (MERS). BMC Infect. Dis. **16**(356), 1–9 (2016)

17. Yang, Z., et al.: Modified SEIR and AI prediction of the epidemics trend of COVID-19 in China under public health interventions. J Thorac. Dis. **12**(3), 165–174 (2020). https://doi.org/10.21037/jtd.2020.02.64

18. William, M.: The History of Sumatra Containing An Account Of The Government, Laws, Customs And Manners Of The Native Inhabitants. 28 September 2005. [EBook #16768]

19. Badan Nasional Penanggulangan Bencana (BNPB) and Gugus Tugas Percepatan Penanganan COVID-19 Indonesia. (2021)

20. Central Bureau of Statistics Republic of Indonesia (BPS). Jumlah Penduduk Indonesia (2019)

21. Central Bureau of Statistics Republic of Indonesia (BPS). Land Transportation Statistics (2019)

22. Central Bureau of Statistics Republic of Indonesia (BPS). Statistik Transportasi Udara (2019)

23. Sonhaji, A., Pasaribu, U.S., Sari, K.N., Husniah, H., Ilmi, N.F.F.: The spatial-SIR model for COVID-19 transmission by resampling method to selecting parameter study case: java island, Indonesia 2021. In: Proceedings of the International Conference on Industrial Engineering and Operations Management Harare, Zimbabwe. 20–22 October 2020

24. Kypraios, T., Neal, P., Prangle, D.: A tutorial introduction to Bayesian inference for stochastic epidemic models using approximate Bayesian computation. Math. Biosci. **287**, 42–53 (2017). https://doi.org/10.1016/j.mbs.2016.07.001

25. Toni, T., Welch, D., Strelkowa, N., Ipsen, A., Stumpf, M.P.H.: Approximate Bayesian computation Scheme for parameter inference and model selection in dynamical systems. J. R. Soc. Interface **6**, 187–202 (2009). https://doi.org/10.1098/rsif.2008.0172
26. Minter, A., Retkute, R.: Approximate Bayesian computation for infectious disease modelling. J. Elsevier B.V. Epidemics **29**, 100368 (2019). ISSN 1755–4365, https://doi.org/10.1016/j.epidem.2019.100368

Predicting Internet Usage for Digital Finance Services: Multitarget Classification Using Vector Generalized Additive Model with SMOTE-NC

Wahyu Wibowo[1]($\boxtimes$), Amri Muhaimin[2], and Shuzlina Abdul-Rahman[3]

[1] Institut Teknologi Sepuluh Nopember, Surabaya, Indonesia
wahyu_w@statistika.its.ac.id
[2] Universitas Pembangunan Nasional Veteran Jawa Timur, Surabaya, Indonesia
[3] University Teknologi MARA, Shah Alam, Malaysia

Abstract. Digital Finance Service has a prominent role in the digital economy. Digital economy can be interpreted as economic and business activities through markets based on digital technology or internet and web technology. Practically, the internet has many purposes not only for entertainment and communication but also for financial services. Therefore, based on demographic characteristics, such as education, occupation, gender, race, age, and place of residence, this study aims to predict internet usage for buying, selling, and banking facilities. This is a classification problem with imbalanced multitarget classification, then the classification method is vector generalized additive model (VGAM). Also, we used Synthetic Minority Over-sampling Technique Nominal-Category (SMOTE-NC) to handle the imbalanced case. The dataset used is derived from the National Socio-Economic Survey (NSES) in 2020. The sample of this research is household members residing in urban districts or villages located in the province of East Java. The result shows that VGAM SMOTE-NC produces a mean geometric accuracy value obtained is 93.1% and can predict the minority class.

Keywords: Classification · Digital finance · Imbalance · Multitarget · Semi-parametric · VGAM

1 Introduction

Currently, advances in information and communication technology have offered many alternative models of interaction between people, not only in the way of communication but also in the way of economic and business activities. In terms of information exchange, various social media platforms accelerate the flow of information in various forms, text, images, audio, and video. Economic activity, which was previously dominant in the form of cash transactions, turned into non-cash transactions. This encourages the emergence of a cashless society, which is a new structure of society that sees money no longer as something that must be in the form of sheets of paper or coins. But more on the nature of money related to its use in transactions. In its development, the cashless society encourages the emergence of the concept of the digital economy.

© The Author(s), under exclusive license to Springer Nature Singapore Pte Ltd. 2023
Y. B. Wah et al. (Eds.): DaSET 2022, LNDECT 165, pp. 494–504, 2023.
https://doi.org/10.1007/978-981-99-0741-0_35

Digital economy can be interpreted as economic and business activities through markets based on digital technology or internet and web technology The activities cover a wide area, including the entertainment business, health, and education services as well as financial and banking services. The digital economy is built based on technological infrastructure (hardware, software, and internet networks), e-business (the process by which organizations carry out computer network-based activities, and e-commerce (online trade transactions). [1] introduced an approach to measuring how well the digital economy practices, these approaches include (i) Information Technology Infrastructure, (ii) E-Commerce, (iii) Firm and Industry Structure, (iv) Demographic and Worker Characteristics, and (v) Price Behavior.

Following the Demographic and Worker Characteristics approach, Badan Pusat Statistik (BPS) conducts a National Socio-Economic survey to identify the use of the internet, including (i) Obtaining information/news; (ii) Doing School/Lecture Assignments; (iii) Sending/Receiving e-mails; (iv) Social Media; (v) Purchase of goods/services; (vi) Sales of Goods/Services; (vii) Entertainment (Downloading/playing games; (viii) Financial facilities (e-banking); (ix) Getting information about goods/services. Research results by [2] show that internet use is mostly for accessing social media and entertainment, while the use for buying, selling and banking facilities is in the lowest 3. In the context of the digital economy, buying, selling, and banking facilities via the internet are a form of active participation to convert cash transactions into non-cash.

Therefore, this study aims to predict internet usage for buying, selling, and banking facilities based on demographic characteristics, such as education, occupation, gender, race, age, and place of residence.

2 Related Works

2.1 Internet Use for Buying and Selling Activities Based on Demographic Characteristics

Demographic characteristics are one of the factors that have an important role in the use of the internet for buying and selling activities, this is evidenced by research conducted by [1] which investigated the impact of socio-demographic factors on the online buying behavior of professionals in Turkey, using the chi-square method to test the hypothesis, the results of the hypothesis concluded that the online buying behavior of professionals is influenced by socio-demographic factors of education level, age, and income [2]. Other studies that can be compared include research conducted by [3] using survey data from the 2002 General Social Survey (GSS02). This study models the impact of demographic factors on consumers' actual online purchases. The results show that a model based on demographic factors alone explains 22.6% of the variance in consumers' overall online shopping behavior. Demographic factors significantly affect age, gender, education level, income, occupation, and marital status. Meanwhile, using the same method, 200 respondents was researched in Medan City, Indonesia, and the results showed that online shopping decisions were strongly influenced by gender, income levels, and online shopping applications [4]. While the research conducted by [5] added a moderating effect to the binary logistic regression model because the influence of beliefs on online buying behavior is moderated by demographic characteristics such as income, education, and generational age. The results of this study show consistency with previous research, namely and as expected, demographic factors of income, level of education, and internet use have a significant and positive relationship with Online Shoppers, while other demographic factors, namely age, have a negative relationship with Online Shoppers.

Furthermore, in this literature review, four studies use regression analysis as a research method, including research conducted by [2] using online survey data on students in the USA showing that Internet knowledge, income, and education level are very strong predictors of Internet purchases among college students. The research explores the use of the internet by telecommunication company employees in the business world [6]. The study findings indicate that statistically significant differences in Internet usability and anxiety were found between different levels of education. The relationship was examined between demographic and motivational variables with e-commerce usage activities in China [7]. The results showed that age and education level had a significant relationship with the activity of using e-commerce. Their research on Gender and E-Commerce, revealed that women in the research sample felt less satisfied emotionally with e-shopping than men [8]. In contrast, men feel greater confidence in shopping with the internet and perceive the internet as a more convenient shopping outlet than women. Factor analysis method with Principal component analysis (PCA) to test whether demographic factors influence risk factors for consumer attitudes towards online shopping in South Africa [9]. The study found that it did not find a moderating effect of gender on the relationship between risk and attitude towards online shopping, while age had a moderating effect on the relationship between product, security, and privacy risk and attitudes towards online shopping. The effect of age and gender were examined using the Bayesian Network Classification method, the results found that spending ability increased with age until the age of 30, stabilized in the early 60s, and then began to decline afterward [10]. Meanwhile, female email users are more likely to be online shoppers than male email users.

2.2   Internet Use for Banking Facilities Based on Demographic Characteristics

In addition to buying and selling activities, today's internet usage is also used in the banking sector. Several studies related to the influence of demographic characteristics on internet use for the banking sector have been carried out, one of which is research conducted by [11] which examines the influence of demographic variables on the use of e-banking in Nigeria using the Pearson Chi-Square-Independence test method. The results showed that the influential variables were the effect of marital status, age, and education level on the use of e-banking. Using the same method, The research revealed that there is a strong correlation between employment status and education level in the use of internet banking services in Ghana [12]. The study showed that mobile banking users differ significantly in the perception of personal and sensory values on the income factor and education level of e-banking users in Turkey through the t-test and variance (ANOVA) analysis method [13]. Meanwhile, [14] applied multiple linear regression methods and the Fisher Z transformation on online banking users in India and their results showed that gender, age, qualifications, experience, occupation, income, and marital status were significant moderating variables. [15] attempted to identify empirically the factors underlying the decision to adopt online banking in Poland through Binomial Logistics Regression. Several demographic characteristics emerged as important determinants of adoption. Internet banking is favored by educated men living in urban areas. In addition, adult customers and minors under the age of 18 are less likely to use virtual banking.

Furthermore, in this literature review, several studies use structural equation modeling as a research method. Conducted partial least squares-structural equations modeling to examine age and gender-dependent variations of consumer intentions and use of mobile banking services, the results show that consumer behavioral intentions are significantly moderated by age through its relationship with facilitating conditions and trust among Lebanese respondents [16]. As for the gender factor, a significant moderating effect was demonstrated in the Lebanese, but not British samples. The factors was examined hinder consumers from using internet banking in Karachi, Pakistan using structural equation modeling (SEM) [17]. The results show that image barriers have a higher impact on Internet banking use, followed by value barriers and risk barriers. The results also showed that men face higher barriers than women. In another study, [18] attempts to measure the impact of demographic variables on various determinants of customer satisfaction in the Indian banking industry. The finding in this study is that age affects customer perceptions of ease of use. Customer perception for the age group (Up to 20 years old) is much lower compared to the other groups. There is a clear difference in perception between the age group (Up to 20 years old) and the age group (Above 50 years old). In addition, the education group (Schooling) has a lower perception of ease of use compared to other groups.

3 Methodology

This section describes the analytical methods related to this research which explained the related dataset until the analysis stage. In the dataset section, it is explained how the data was obtained and explained what variables were used as predictors and responses. Meanwhile, in the analysis stage, the data preprocessing process, and the calculation of the goodness of the model are explained until the conclusion is created.

3.1 Dataset

The dataset used was derived from the National Socio-Economic Survey (NSES) in 2020. The sample of this research is household members residing in urban districts or villages located in the province of East Java. The total data obtained are 65536 observations, with the number of questionnaire instrument variables totaling 40 questions which are then transformed into 40 variables. The data type includes ratio, interval, nominal, and ordinal data scales. This survey was conducted in a cross-sectional method so that the data obtained were not time-series data. The method used was face-to-face interviews for each respondent. From all the variables and observations, only a few research variables and observations fulfill the requirements. This data will be used in the data processing process.

Table 1. Research's variables and description

Variable's name	Data types
Living area (X_1)	Nominal (Urban, or City)
Marital status (X_2)	Nominal (Married or not)
Sex (X_3)	Nominal (Men or woman)
Age (X_4)	Ratio
Last education (X_5)	Ordinal, 7 Categories
Most activity in last 7 days (X_6)	Nominal, 5 Categories
Position in job (X_7)	Nominal, 5 Categories
Cellphone used (X_8)	Nominal (Yes or no)
Type of personal computer in last 3 months (X_9)	Nominal, 3 Categories
Internet banking usage ($X_{10)}$	Nominal (Yes or no)
Tools of internet used (X_{11})	Nominal, 4 Categories
Place of internet used (X_{12})	Nominal, 8 Categories
Bank account ownership (X_{13})	Nominal (Yes or no)
Internet for online payment (Y_1)	Nominal (Yes or no)
Internet for online shop (Y_2)	Nominal (Yes or no)
Internet for internet banking (Y_3)	Nominal (Yes or no)

There are 16 variables in total (responses and predictors). For predictors, there are 13 variables, and three variables for responses. Here is Table 1 for variable description. The number of data is 65536. The data is got from the Economic and Social Survey in Indonesia. From Table 1, there is only one numerical variable, which is Age, the rest is a non-numerical variable with nominal and ordinal types. Based on that Table, this case is binomial classification with multitarget in its responses.

3.2 Generalized Additive Model

In some cases, the correlation between response and predictor variable is not truly linear, sometimes nonlinear. In a generalized linear model, the model is created from a linear relationship between the predictor and response variable. The equation from the model could be $y = x\beta + \varepsilon$, that's the ordinary linear equation called linear regression. It can be used if the relationship between the response and predictor variable is linear.

On the other hand, if the relationship between both is nonlinear, that model can't be used. To accommodate the nonlinear relationship between response and predictor variable, the generalized additive model can be used. This method is a semiparametric method that uses a smoother to handle the nonlinearity relationship in the data. The model is shown in Eq. (1),

$$y = \beta_0 + f_1(x_1) + \cdots + f_d(x_d)$$
$$f_d(x_d) = \beta_d x_d + \sum u_{dk} z_{dk}(x_d) \tag{1}$$

y is the response variable, and it has the predictor with the smoothing function in each variable. The f_i, $1 \leq i \leq d$, are smoothing functions. Thus, the y variable no longer has a linear relationship between x. Usually, the smoothing function that is used is penalized splines, especially in R software. If d is equal to 2 in Eq. 1, then the model becomes,

$$y = \beta_0 + \beta_1 x_1 + \sum u_{1k} z_{1k}(x_1) + \beta_2 x_2 + \sum u_{2k} z_{2k}(x_2) \tag{2}$$

where $z_{1k}(\cdot)$ is an O'Sullivan spline basis over x variable, the number of penalization also applied on the u_{ik}. By Eq. 2, the response variable y will be changing either increasing or decreasing based on the x variable. Moreover, if the response variable is multivariate, the analysis is becoming a vector generalized additive model. Equation (2) changes into Eq. (3)

$$\begin{bmatrix} y_{1i} \\ y_{2i} \end{bmatrix} = \begin{bmatrix} \beta_{10} + \beta_{11} x_{1i} + \beta_{12} x_{2i} + f_{13}(x_{3i}) + f_{14}(x_{4i}) + f_{15}(x_{5i}) \\ \beta_{20} + \beta_{21} x_{1i} + \beta_{22} x_{2i} + f_{23}(x_{3i}) + f_{24}(x_{4i}) + f_{25}(x_{5i}) \end{bmatrix} + \begin{bmatrix} \varepsilon_{1i} \\ \varepsilon_{2i} \end{bmatrix} \tag{3}$$

let's say that there are 5 predictor variables and 2 responses. Then it will contain 5 smoothing functions. Moreover, the error is normally, independently distributed over mean zero and variance $\sum$.

3.3 Analysis Stage

The data that has been obtained is then processed to obtain output information and conclusions. The tool used is the RStudio software. The stages in the process include pre-processing, modeling, model evaluation, and concluding. The following is a description of the analysis stages from the pre-processing stage until the output information is obtained. Mainly, this study splits the analysis stage into 5 parts. The flow process is presented in Fig. 1:

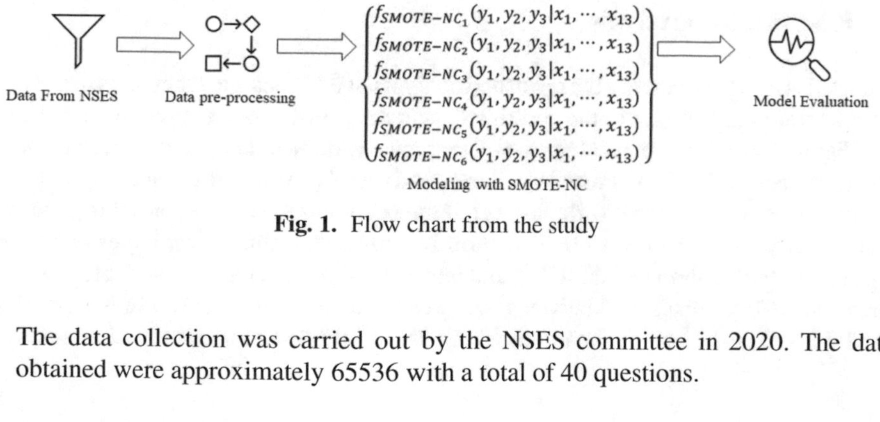

Data From NSES   Data pre-processing

$$\begin{cases} f_{SMOTE-NC_1}(y_1, y_2, y_3 | x_1, \cdots, x_{13}) \\ f_{SMOTE-NC_2}(y_1, y_2, y_3 | x_1, \cdots, x_{13}) \\ f_{SMOTE-NC_3}(y_1, y_2, y_3 | x_1, \cdots, x_{13}) \\ f_{SMOTE-NC_4}(y_1, y_2, y_3 | x_1, \cdots, x_{13}) \\ f_{SMOTE-NC_5}(y_1, y_2, y_3 | x_1, \cdots, x_{13}) \\ f_{SMOTE-NC_6}(y_1, y_2, y_3 | x_1, \cdots, x_{13}) \end{cases}$$

Model Evaluation

Modeling with SMOTE-NC

Fig. 1. Flow chart from the study

1. The data collection was carried out by the NSES committee in 2020. The data obtained were approximately 65536 with a total of 40 questions.

2. The next stage is the data pre-processing stage. At this stage, it is divided into several parts, namely handling missing values, creating a dummy variable, and conducting an observation filter that is adjusted by the survey instrument.

  a. At the stage of handling the missing values, because the data is cross-sectional data, then if the data is of ordinal or nominal type, the mode is used. If the data is of type ratio, then the median or mean is used.
  b. Due to a lot of nominal or ordinal scale data variables with more than 2 classes, the formation of dummy variables is highly recommended. These dummy variables are used up to the modeling stage.
  c. The observation selection is based on the variable called Internet banking usage (X_{10}). If the respondent answers no, then the respondent is not used as the object of this research. So that the final data used in this study was 25477 observations.
  d. Then the stage of splitting the training and testing data is carried out. In this section, we were using an 80% training and 20% testing rule.
  e. We checked the class in the response variables. If there was an imbalance problem, then we will be over-sampling the minority class using SMOTE-NC. Because there are 3 response variables, and we have 6 different order combinations doing SMOTE-NC.

3. There are 2 modeling scenarios, respectively without and with SMOTE-NC. For 2nd model, the ensemble was developed with the majority voting from the predicted values that were being got by each model's combination.
4. The evaluation stage involves calculating the values of accuracy, specificity, and sensitivity. These four things are compared to determine whether the VGAM method works properly with this data.
5. Conclusion drawing is the stage of explaining related to the model produced by the VGAM and VGAM-SMOTE-NC. And to convey VGAM's performance in handling multiclass data classification.

The process analysis was designed to analyze this data. In another case, the process can be re-designed.

4 Result and Analysis

This part is about explaining the result and analysis of the research. It contains descriptive statistics and modelling. For the descriptive statistics, the bar chart is presented in Fig. 2.

Figure 2 shows some bar charts on all response variables. The name of each response could be seen in Table 1. From Fig. 2, respectively (a), (b), and (c) are named by the internet for online payment variable (Y_1), if the person used internet for online payment, than Y_1 is yes, otherwise is no. This term is applied for other Y variables, which are internet for online shop variable (Y_2), and internet for internet banking variable (Y_3). The chart looks very imbalanced between the "Yes" and "No" categories. For this problem, we did SMOTE-NC to the data to tackle it. The SMOTE-NC needs only one response

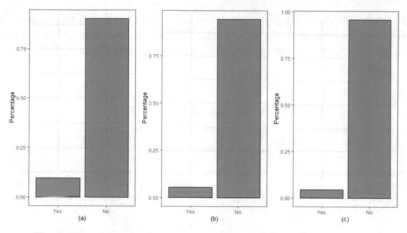

Fig. 2. The proportion between yes and no class in response variables.

Table 2. Order Combinations for SMOTE-NC

No	Order
1	$Y_1 \prec Y_2 \prec Y_3$
2	$Y_1 \prec Y_3 \prec Y_2$
3	$Y_2 \prec Y_1 \prec Y_3$
4	$Y_2 \prec Y_3 \prec Y_1$
5	$Y_3 \prec Y_1 \prec Y_2$
6	$Y_3 \prec Y_2 \prec Y_1$

variable to run it. Thus, we create 6 combinations to do SMOTE-NC. Table 2 presents the following combinations:

For the first combination, when the model did the SMOTE-NC for y_1, then the rest of the response variables became predictor variables. After the SMOTE-NC has been done, then the model was created, in this phase, we have a model that was created from the first combination. It repeated until all combinations were run. Thus, we have six models to create an ensemble model with a majority vote from each model.

Modelling is done by dividing the data into training and testing. Because the data used have imbalanced characteristics, the distribution of training testing also pays attention to the proportions in each class. So that the distribution of training-testing is carried out using the stratified sampling method. The purpose of stratified sampling is to maintain the proportion of minority classes in the testing data and avoid unexpected things such as the testing data does not contain minority classes. So, the amount of training and testing data used in this study is 80% training data and 20% training data. So that the number of training data is 20382 observations, and the testing data is 5095 observations.

The model used is a full linear model, the response variables are assumed to be vectors and analysed simultaneously with predictor variables. Here is the simple model for the current case, $y_j = \beta_{j0} + \beta_{j1}x_{jk} + \varepsilon_j$, where y_j is the vector of response variables, β_j is the vector of parameters, x_j is the vector of predictor variables with $k = 1, 2, 3,...,$ 12, and ε is the error for each response in the vector, where $j = 1, 2, 3$. The result represents in Table 3.

Table 3. Metric result between models

Score	VGAM			VGAM SMOTE-NC		
	Y1	Y2	Y3	Y1	Y2	Y3
Accuracy	0.904	0.946	0.956	0.893	0.946	0.955
Sensitivity	0.993	0.998	0.997	0.966	0.993	0.991
Specificity	0.075	0.058	0.071	0.213	0.140	0.152
Accuracy GMEAN	0.935			0.931		
Sensitivity GMEAN	0.996			0.983		
Specificity GMEAN	0.067			0.165		

Based on Table 3, overall, the VGAM method can predict people's behaviour in using the internet for economic activities well. If the geometric mean value with the formula $\left(\Pi_i^n x_i\right)^{\frac{1}{n}}$ is calculated for accuracy, it is 0.935. This means that the VGAM model can predict response variables simultaneously with an accuracy of 93.52%. Meanwhile, the specificity and sensitivity values explain the model in predicting each class. Overall, based on the model, it can predict the minority class or customers who use the internet for economic activities on each variable with an average specificity level of 0.067, while the model has an average level of sensitivity in predicting the majority class of 0.996. It is a very imbalanced result. On the other hand, with the VGAM SMOTE-NC result, the specificity is rising. But there are slightly decreasing in accuracy and sensitivity. Yet the VGAM SMOTE – NC produces better results. That model can classify the minority case better than the VGAM only.

5   Conclusion and Future Works

Analysis related to the behaviour of the internet user community towards economic activity is rarely carried out. Especially concerning demographic data. Plus, the variables analysed are multi-variable or similar to multivariate analysis. In this study, the response variables used were three and all of them were binary. The proportion between classes on the response variable is not balanced thus stratified sampling technique was applied. The method used was VGAM which is a semi-parametric method. After modelling using VGAM with 80% training data rules and 20% testing data, we produced a model that can predict multitarget classification, with a sample from the household of digital use in finance. VGAM performance in handling the case of unbalanced data is not good. Thus we did SMOTE-NC to the model and the result is better, especially in the specificity score. For further research, it is possible to use the dimension reduction method to optimize the feature that is used in the VGAM model.

Acknowledgments. The authors would like to thank the Ministry of Research, Technology, and Higher Education of the Republic of Indonesia for supporting this research through the Priority Fundamental Research Grant of Institut Teknologi Sepuluh Nopember with the contract number 935/PKS/ITS/2021.

References

1. Akman, I., Rehan, M.: Online purchase behavior among professionals: a socio-demographic perspective for Turkey. Econ. Res. - Ekonomska istraživanja **27**(1), 689–699 (2014). https://doi.org/10.1080/1331677X.2014.975921
2. Case, T., Burns, O.M., Dick, G.: Drivers of On-Line Purchasing Among U.S. University Students. In: AMCIS 2001 Proceedings, vol. 169 (2001). http://aisel.aisnet.org/amcis2001/169
3. Naseri, M.B., Elliott, G.: Role of demographics, social connectedness and prior internet experience in adoption of online shopping: applications for direct marketing. J. Target. Meas. Anal. Mark. **19**, 69–84 (2011). https://doi.org/10.1057/JT.2011.9
4. Lubis, A.: Evaluating the customer preferences of online shopping: demographic factors and online shop application issue. Acad. Strateg. Manage. J. **17**, 1 (2018)
5. Rodgers, S., Harris, M.: Gender and e-commerce: an exploratory study. J. Advert. Res. **43**, 322–329 (2003). https://doi.org/10.1017/S0021849903030307
6. Zhang, Y.: Age, gender, and Internet attitudes among employees in the business world. Comput. Hum. Behav. **21**, 1–10 (2005). https://doi.org/10.1016/j.chb.2004.02.006
7. Kooti, F., Lerman, K., Aiello, L., Grbovic, M., Djuric, N., Radosavljevic, V.: Portrait of an online shopper: understanding and predicting consumer behavior. In: Proceedings of the Ninth ACM International Conference on Web Search and Data Mining (2016). https://doi.org/10.1145/2835776.2835831
8. Makhitha, K.M., Ngobeni, K.: The influence of demographic factors on perceived risks affecting attitude towards online shopping. SA J. Inf. Manage. **23**, 9 (2021). https://doi.org/10.4102/SAJIM.V23I1.1283
9. Chong, A.: Mobile commerce usage activities: The roles of demographic and motivation variables. Technol. Forecast. Soc. Chang. **80**, 1350–1359 (2013). https://doi.org/10.1016/J.TECHFORE.2012.12.011
10. Punj, G.N.: Effect of consumer beliefs on online purchase behavior: the influence of demographic characteristics and consumption values. J. Interact. Mark. **25**, 134–144 (2011). https://doi.org/10.1016/J.INTMAR.2011.04.004
11. Izogo, E., Nnaemeka, O.C., Onuoha, O.A., Ezema, K.S.: Impact of demographic variables on consumers' adoption of e-banking in Nigeria: an empirical investigation. Eur. J. Bus. Manage. **4**, 27–39 (2012)
12. Gupta, R., Varma, S.: Impact of demographic variables on factors of customer satisfaction in banking industry using confirmatory factor analysis. Int. J. Electron. Bank. **1**, 283 (2019). https://doi.org/10.1504/IJEBANK.2019.10022902
13. Bk, A.: The impact of customer demographic variables on the adoption and use of internet banking in developing economies. J. Internet Bank. Commer. **20**, 1–30 (2015). https://doi.org/10.4172/1204-5357.1000114

14. Polasik, M., Piotr Wisniewski, T.: Empirical analysis of internet banking adoption in Poland. Int. J. Bank Mark. **27**(1), 32–52 (2009). https://doi.org/10.1108/02652320910928227
15. Merhi, M., Hone, K., Tarhini, A., Ameen, N.: An empirical examination of the moderating role of age and gender in consumer mobile banking use: a cross-national, quantitative study. J. Enterp. Inf. Manage. **34**, 1144–1168 (2020). https://doi.org/10.1108/jeim-03-2020-0092
16. Demirhan, M.: Demographic characteristics and perceived value differences in mobile banking: an empirical study in Turkey.Erciyes Üniversitesi Sosyal Bilimler Enstitüsü Dergisi **48**, 237–263 (2020). https://dergipark.org.tr/en/pub/erusosbilder/issue/55878/696785
17. Arif, I., Aslam, W., Hwang, Y.: Barriers in the adoption of internet banking: a structural equation modeling - neural network approach. Technol. Soc. **61**, 101231 (2020). https://doi.org/10.1016/j.techsoc.2020.101231
18. Chawla, D., Joshi, H.: The moderating effect of demographic variables on mobile banking adoption: an empirical investigation. Glob. Bus. Rev. **19**, S113–S190 (2018). https://doi.org/10.1177/0972150918757883

Text Mining and Classification

Identifying Topic Modeling Technique in Evaluating Textual Datasets

Nik Siti Madihah Nik Mangsor[1], Syerina Azlin Md Nasir[2(✉)],
Shuzlina Abdul Rahman[3], and Zurina Ismail[4]

[1] Department of Computer Science, Faculty of Computer and Mathematical Sciences, Universiti Teknologi MARA Cawangan Kelantan, Kota Bharu, Malaysia
[2] Department of Information Technology, Faculty of Computer and Mathematical Sciences, Universiti Teknologi MARA Cawangan Kelantan, Machang, Malaysia
syerina@uitm.edu.my
[3] Department of Information Systems, Faculty of Computer and Mathematical Sciences, Universiti Teknologi MARA, Shah Alam, Selangor, Malaysia
[4] Faculty of Business Administration, Universiti Teknologi MARA, Shah Alam, Selangor, Malaysia

Abstract. One of the most popular methods of topic modeling is Latent Dirichlet Allocation (LDA). To date, philanthropic corporate social responsibility (PCSR) activities are ad-hoc in nature, where assistance is provided more to basic needs with very little attention to activities that can contribute to eradicating poverty. Based on previous related literature, it is found that there is no proper categorization and documentation of PCSR-related activities. Therefore, this research is aimed to identify the most suitable LDA approaches for categorizing PCSR activities. The analysis involved five-years data from the annual reports of 19 CSR-award winning companies in Malaysia. For this study, three LDA techniques were considered and compared namely Variational Bayes Inference, Gibbs Sampling and Expectation Maximization. Then, performance measurement was carried out using coherence value and pyLDAvis technique. As a result, the study showed that the LDA Expectation Maximization method is the best topic modelling technique for clustering PCSR documents. Furthermore, this approach can estimate parameters in probabilistic models when dealing with partial, noisy or missing data. The findings offer an insight to be considered by companies in strategizing the CSR activities, particularly philanthropic responsibility in ensuring optimum impact to innovatively support the society.

Keywords: Topic modeling · Corporate social responsibility · CSR · PCSR

1 Introduction

To solve environmental concerns and create sustainable growth for businesses, the use of corporate social responsibility (CSR) as a firm strategy has become an imperative necessary tool in today's dynamic commercial arena. Several academics have proposed that CSR implemented by organisations may promote sustainability as a developing

© The Author(s), under exclusive license to Springer Nature Singapore Pte Ltd. 2023
Y. B. Wah et al. (Eds.): DaSET 2022, LNDECT 165, pp. 507–521, 2023.
https://doi.org/10.1007/978-981-99-0741-0_36

company strategy in difficult competitive working conditions, long-term development, and stability [1–4]. To effectively manage CSR activities, specifically philanthropic CSR (PCSR) activities, identifying categories and types of philanthropic activities is crucial.

Understanding the primary themes of a document collection is a critical challenge in today's information age [5]. As a result, more efficient methods and tools for recognizing and analyzing content in large documents such as news, articles, and books are required [6]. Natural language processing (NLP) is a field of computer science that combines the power of computational linguistics, computer science, and artificial intelligence to enable machines to understand, analyse, and synthesize the meaning of natural human speech [7]. Keyword extraction is the most essential work in various domains, including information retrieval, text mining, and NLP applications, namely topic recognition and tracking [6].

In this research, we concentrated on topic modelling technique as a technique for locating clusters of words (themes) in a text corpus. Since it is difficult to identify themes manually, which is inefficient due to the enormous number of data, topic modelling techniques have been devised for text mining [8]. Various topic modelling techniques can automatically extract topics from short text and standard long-text [9–11].

Therefore, this study is aimed to further investigate the issue by conducting experimentation of several LDA topic modelling techniques to help in categorizing PCSR activities. The data used in this study were taken from 5 years annual report of 19 CSR-Award winning companies in Malaysia 2020.

The remainder of this paper is organized in the following structure. Section 2 describes the related work. We explain our research methodology in Sect. 3. The results and discussion are explained in Sect. 4. Finally, Sect. 5 describes the conclusion of our research.

2 Related Works

Topic modelling is a sophisticated text mining method that may disclose the hidden integrated structure in textual material and then provide vital support for researchers and practitioners in the large area of decision making [12]. Topic modelling is an efficient approach of assessing a large number of documents. Topic modelling is used to uncover the hidden structure of a document set [13]. The topic modelling methodology can be thought of as a method for discovering a collection of words, i.e., the subject of a series of documents conveying group information. It is also a type of text mining to find repeated word patterns in text documents. Because it serves more functions than a clustering or classification strategy, a topic model is regarded as a powerful tool. Objects can be modelled as latent topics that reflect the meaning of a collection of documents [13].

The most popular topic modeling algorithms that contribute every sphere of text analysis in multiple domains include Latent semantic analysis (LSA), Probabilistic Latent semantic analysis (PLSA), and Latent Dirichlet Allocation (LDA) [14]. Landauer and Dumais [9] proposed latent semantic analysis technique. They used Singular value decomposition (SVD) that has been applied to many different areas including information retrieval, natural language processing and modeling of human language knowledge [15]. PLSA is a method that was developed after the LSA method to address some of the flaws discovered in LSA. It was first introduced in 1999 by Jan Puzicha and Thomas Hofmann. PLSA is an automated document indexing method based on a statistical latent class model for factor analysis of count data. Using a generative model, it also seeks to improve the probabilistic use of LSA. PLSA's major purpose is to identify and differentiate between distinct contexts of word usage without the use of a dictionary or thesaurus [16]. The purpose of LDA model is to upgrade the existing technique of capturing the exchangeability of both words and documents using PLSA and LSA models [17]. LDA is one of the most widely used techniques for analysing data. LDA extends PLSA by using Dirichlet priors for document-specific subject mixtures, resulting in previously unnoticed documents. LDA has been a huge success in text mining due to its ability to be generalised and extensible [18, 19].

LDA is built on the premise that each text document is made up of a large number of subjects, each of which comprises a big number of words. [20–22]. LDA can be used to identify and characterize hidden topic hierarchies within text document collections [23, 24]. The [25] method is based on a semantic analysis of 'big data software engineering' (BDSE) job advertisements using LDA, a probabilistic topic model for detecting latent semantic patterns (themes). The core knowledge and abilities for BDSE based on LDA themes are discovered as a result.

Overall, LDA topic modelling has shown promise in communication studies. [26] used LDA, which provides methods and measures for estimating and increasing validity and dependability. [27] used latent Dirichlet allocation (LDA) to extract meaningful and actionable knowledge from the UGC of Airbnb users in New York City in order to inform researchers and industry practitioners about the themes of interest, the relative proportionate importance, and the relationship between distinct themes of interest. Another study used latent Dirichlet allocation to identify 30 areas of interest among 266,544 English ratings of hotels from 16 countries on TripAdvisor [28]. Another study used LDA to examine 104,161 English reviews of Korean lodging in South Korea and revealed differences in the incidence of 14 subjects across different types and regions of lodging. [16] showed that themes differ according to lodging type and context.

3   Methodology

The proposed method consists of a few steps which are dataset collection, text preprocessing and topic modelling. The algorithms employed in this study is LDA, which has been tested using the Python software, Jupyter Notebook. A brief description about the proposed method is presented in the subsections.

3.1   Data Collection

This experiment uses a dataset gathered from the 5-year annual reports of 19 CSR-award winning companies in Malaysia 2020, available on the data stream. The annual report is a detailed description of the activities of a corporation during the previous year. Firms released the CSR reports in order to share their initiatives and outcomes on social responsibility. The "Sustainability Section" is the focus of each annual report that represents a detailed description of company CSR reports. Then, the selected paragraphs of textual data based on PCSR activities of each annual report are extracted and converted into Excel format. The documents are collated and summarized for the cleaning process. The list of 19 CSR-Award winning companies is shown in Table 1.

Table 1. List of 19 CSR-award winning companies

CSR-award winning companies			
1	Tenaga Nasional Berhad (TNB)	11	Celcom Axiata Berhad
2	Kenangan Investment Bank	12	KPMG
3	Kumpulan Perangsang Selangor (KPS)	13	Bank Rakyat
4	Kwantas Corporation Berhad	14	Pharmaniaga
5	Mah Sing Group Berhad	15	7-Eleven
6	Malaysia Airports Holdings Berhad	16	Great Eastern
7	Serba Dinamik Holdings Berhad	17	RHB Bank
8	Kuala Lumpur Kepong Berhad (KLK)	18	Sunway Berhad
9	Hektar Property Services	19	ABM Fujiya Berhad
10	Heineken Malaysia Bhd		

3.2 Text Pre-processing

Before the dataset is effectively analyzed and mined, corpus documents must be noise-free. The phrase "corpus pre-processing" refers to the practice of deleting duplicate and less informative phrases from a corpus in order to establish a clean corpus. Data cleaning is performed by removing typographical errors or validating and correcting values against a known list of entities. The pre-processing involves several processes to prepare the text data for qualitative analysis which are as follows:

1) Normalization: The data should be normalized or standardized to bring all of the variables into proportion with one another. Non-numeric qualitative data should be converted to numeric quantitative data.
2) Tokenization: This stage divides the given text into smaller sections called sentences, and the phrases into smaller portions called tokens. White space or line breaks separate tokens.
3) Stop words removal: Stop words are words that are not relevant to the desired analysis. Stop words, such as a, and are or do, are eliminated as they are usually seen in the papers and do not provide any useful information.
4) Stemming: Stemming is the process of reducing words to their original root. The aim of stemming is to limit the diversity in text data by transforming words into their common form. For example, "contributes", "contributing" and "contributed" are converted to "contribute".

5) Building corpus: Each document is represented in the corpus by a sequence of pairings. The first digit of the pair conveys that the numeric ID relates to a word, while the second digit expresses how frequently that word occurs.

3.3 Topic Modeling

Previous research discovered that topic modelling can provide a global view of the topics and how they differ from one another, as well as a complete evaluation of the phrases most closely associated with each specific issue. Topic modelling has recently become more available and is now used in conjunction with document clusters to improve clustering accuracy. The goal of topic modelling is to find a set of topics in a collection of text documents; each topic is defined as a distribution across a set of words. This is accomplished through the use of statistical modelling approaches [29, 30].

LDA is one of the most often used approaches. LDA enhances PLSA by using Dirichlet priors for document-specific subject mixtures, resulting in previously unnoticed documents. Because of its strong generality and flexibility, LDA has been a huge success in text mining [31–33]. LDA is built on the premise that each text document is made up of a large number of subjects, each of which comprises a big number of words. LDA only requires text documents and the predicted number of topics as input.

The LDA technique is employed in this study, which includes 95 texts, by randomly assigning a subject to each word. Based on the word distribution of all subjects, two frequencies can be computed: topic-document and word-topic. The frequency counts collected during initialization (topic frequency) and a Dirichlet-generated multinomial distribution over subjects for each document are used to determine the popularity of each topic in the document. Meanwhile, the frequency counts collected during initialization (word frequency) and a Dirichlet-generated multinomial distribution across words for each topic are used to determine the popularity of the word in each topic. The word is then reassigned to the topic with the highest conditional probability. Finally, repeat the process until the word to subject assignments become meaningful and stable.

The followings are three LDA approaches considered in this study to identify the topic of PCSR activities:

Variational Bayes Inference: One alternative to MCMC is Variational Bayes (VB) Inference [34]. VB inference is much faster than MCMC since it depends on rapid optimization algorithms. The goal is to identify the family member with the least Kullback-Leibler (KL) divergence from the intended posterior distribution. As indicated by the vast number of software libraries that enable it, such as Stan [34] and Edward [35], variational inference is becoming increasingly popular in machine learning.

Gibbs Sampling: It is a technique for sampling conditional distributions of variables on a regular basis until the distribution over states converges to the true distribution in the long run (Srihari, S. Gibbs Sampling). It also includes a basic and widely applicable Markov Chain Monte Carlo (MCMC) technique that samples each random variable of a probabilistic graphical model one by one (PGM). For many graphical models (e.g., mixture models, LDA), Gibbs Sampling is relatively simple to derive and has low compute and memory needs. Gibbs Sampling is used when the joint distribution is unknown or difficult to sample directly. Gibbs Sampling is used in order to achieve the distributions and in the study to produce an effective description of product features or qualities-based opinions [23, 36].

Expectation Maximization: Expectation Maximization (EM) is a method for conducting maximum likelihood estimation in the presence of latent variables [30]. The purpose of EM is to use the dataset's available observable data to estimate the missing data of the latent variables [31]. The use of EM can be advantageous in detecting the values of latent variables and filling in missing data during a sampling. For a long time, the expectation-maximization (EM) approach has been used to calculate missing values and address the incomplete clustering problem, and it provides an efficient technique for parameter estimation for incomplete data [32]. The original LDA proposes using the EM estimate to identify and maximise the log likelihood of the data, according to [25]. The EM algorithm is a general approach for estimating maximum likelihood when the data is "incomplete" or the probability function includes latent variables [26].

3.4   LDA Visualization

The topic model output is visualised in Python using Gensim and the pyLDAvis package. pyLDAvis is a web-based interactive topic model visualisation that is based on LDAvis [30]. We may comprehend the LDA model by exploring the link between topic and terms using pyLDAvis. PyLDAvis contains two panels: a distribution map for each topic and an intensity graph for the most frequently occurring phrases in the corpus. The three LDA techniques are then compared to determine which approach is best for recognising PCSR activities. Topic coherence measures is used to automatically detect the coherence of a topic. The core premise is based on the linguistic distributional hypothesis, which asserts that words with similar meanings tend to reoccur in analogous contexts. Coherent topics are ones in which all or the majority of the words, such as the top N words of the topic, are linked. The computational problem is to create a measure that correlates well with human topic ranking data, such as word and topic incursion test results. As a result, a metric that correlates well with human topic ranking data is a favourable indicator of topic interpretability.

   To compare different LDA techniques, this study used the CV coherence metric for topic coherence calculations. Human interpretability improves with higher coherence scores. CV is comprised of four components: (i) data segmentation into word pairs; (ii) calculation of word or word pair probabilities; (iii) calculation of a confirmation measure that quantifies how strongly one word set supports another word set; and (iv) aggregation of individual confirmation measures into an overall coherence score [33]. [34] states that the coherence score is utilised to determine the optimal number of topics to extract using LDA. It is used to assess how well the subjects have been retrieved. The coherence value for any topic ranges between 0 and 1. The greater the value, the more compact the topic, which is preferable [37]. Since there is seemingly no scikit-learn equivalent to Gensim's Coherence Model, the study uses pyLDAvis technique to compare between LDA Gibbs Sampling and LDA Expectation Maximization algorithm.

4 Experiment Results

This section presents the findings from the experiments conducted in this study.

4.1 Result of Variational Bayes Inference

The experiment first starts with the implementation of Variational Bayes Inference LDA algorithm in Python gensim model as shown in Fig. 1 below:

```
ᴴ  # Build LDA model                                              #Variational Bayes
   lda_model = gensim.models.ldamodel.LdaModel(corpus=corpus,
                                               id2word=id2word,
                                               num_topics=11,
                                               random_state=42,
                                               update_every = 1,
                                               passes = 100,
                                               alpha ='auto',
                                               per_word_topics=True
                                               )
```

Fig. 1. LDA variational bayes inference sample code

The LDA model is built where each topic is a combination of keywords and each keyword contributes a certain weightage to the topic. The keywords for each topic and the weightage (importance) of each keyword from Variational Bayes Inferences LDA approach can be seen using lda_model.print_topics() as shown in Fig. 2 below:

```
[(0,
  '0.009*"campaign" + 0.008*"event" + 0.007*"family" + 0.006*"employee" + '
  '0.006*"management" + 0.005*"held" + 0.005*"awareness" + '
  '0.005*"central_square" + 0.005*"child" + 0.004*"health"'),
 (1,
  '0.014*"student" + 0.012*"school" + 0.007*"child" + 0.007*"education" + '
  '0.006*"development" + 0.005*"initiative" + 0.005*"project" + 0.005*"family" '
  '+ 0.005*"training" + 0.004*"home"'),
 (2,
  '0.018*"staff" + 0.015*"employee" + 0.010*"safety" + 0.008*"training" + '
  '0.008*"organized" + 0.007*"various" + 0.007*"health" + 0.006*"activity" + '
  '0.006*"support" + 0.005*"campaign"'),
 (3,
  '0.012*"employee" + 0.007*"education" + 0.007*"student" + '
  '0.007*"development" + 0.007*"fund" + 0.007*"initiative" + 0.006*"child" + '
  '0.006*"contribution" + 0.006*"organised" + 0.005*"activity"'),
 (4,
  '0.024*"child" + 0.012*"underprivileged" + 0.009*"childrencare" + '
  '0.009*"raised" + 0.008*"school" + 0.007*"support" + 0.007*"health" + '
  '0.007*"organised" + 0.007*"live" + 0.007*"insurance"'),
 (5,
  '0.015*"school" + 0.012*"airport" + 0.009*"employee" + 0.008*"local" + '
  '0.008*"student" + 0.007*"initiative" + 0.006*"development" + '
  '0.005*"training" + 0.005*"english" + 0.004*"well"'),
```

Fig. 2. Top 10 keywords of each topics

Topic 0 is represented as 0.009 "campaign" + 0.008 "event" + 0.007 "family" + 0.006 "employee" + 0.006 "management" + 0.005 "held" + 0.005 "awareness" + 0.005 "central_square" + 0.005 "child" + 0.004 "health". It means the top 10 keywords that contribute to this topic are: 'campaign', 'event', 'family'... and so on and the weight of 'campaign' on topic 0 is 0.009. The weights reflect how important a keyword is to that topic.

According to [30], it is impossible to understand it simply by looking at a combination of words and numbers like the one above. Visualization is one of the most effective ways to understand data. As a result, PyLDAvis is used as another method to interpret the topics in a topic model, as depicted in Fig. 3 below:

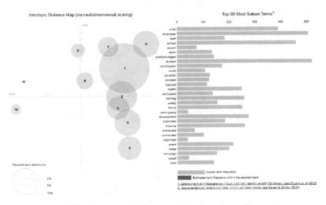

Fig. 3. Top 10 keywords of each topics

Each bubble indicates a different topic. The larger the bubble, the greater the percentage of documents in the corpus of that topic. The total frequency of each term in the corpus is represented by blue bars. If no topic is chosen, blue bars representing the most commonly used terms are displayed. The number of times a certain phrase is generated by a given topic is indicated by red bars. Based on Fig. 4 below, there are around 510 words for 'student', and this term is used approximately 320 times inside topic 1. The word with the longest red bar is the one that appears the most frequently in the paper related to that topic.

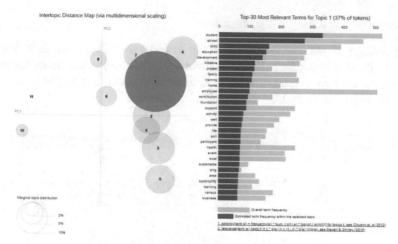

Fig. 4. pyLDAvis- red bars visualization

The more apart the bubbles are, the more unlikely they are. For example, it is difficult to distinguish between subjects 2 and 6, whereas it is much easy to distinguish between topics 2 and 5. Furthermore, rather than being grouped in one quadrant, a good topic model includes large non-overlapping bubbles distributed around the chart [31].

4.2   Result of Gibbs Sampling

The keywords for each topic and the weightage (importance) of each keyword from Gibbs Sampling LDA approach is determined using lda_model.print_topics() as shown in Fig. 5.

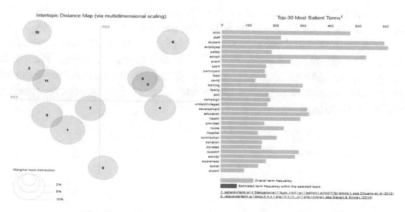

Fig. 5. pyLDAvis visualization of gibbs sampling

4.3   Result of Expectation Maximization

The expectation-maximization (EM) algorithm is used in the scikit-learn library's version of LDA to iteratively update parameter estimates [32]. The Python Sklearn approach is then used to create the Expectation Maximization LDA algorithm. Figure 6 depicts the words per subject from the LDA Sklearn implementation.

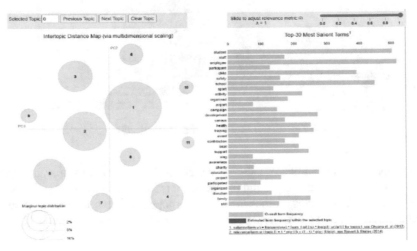

Fig. 6. pyLDAvis visualization of expectation maximization

4.4   Comparison of LDA Algorithm

The comparison between those three LDA algorithms is summarized in Table 2. According to the table analysis, LDA Gibbs Sampling method outperforms LDA Variational Bayes algorithm in terms of coherence value, implying that LDA Gibbs sampling provides higher human interpretability than Variational Bayes algorithm. The pyLDAvis approach generates bubbles according to the topic model, with a good topic model having reasonably large and non-overlapping bubbles dispersed around the chart rather than clustered in one quadrant.

Table 2. Comparison between LDA approaches

LDA	COHERENCE VALUE	pyLDAvis TECHNIQUE
Variational Bayes Inference Algorithm	0.33826	
Gibbs Sampling Algorithm	0.40415	
Expectation Maximization Algorithm	No coherence value	

As a result, bubbles generated by LDA Expectation Maximization have larger, non-overlapping bubbles than bubbles generated by LDA Gibbs Sampling. That is, the LDA Expectation maximisation algorithm outperforms the LDA Gibbs sampling technique. To conclude, the LDA Expectation Maximization method is found to be the best topic modelling technique for identifying PCSR activity groups. Table 3 displays the top 10 most frequent words from LDA Expectation Maximization.

Table 3. Top 10 most frequent words

Topic 1	Topic 2	Topic 3	Topic 4	Topic 5	Topic 6	Topic 7	Topic 8	Topic 9	Topic 10	Topic 11
Staff	Child	Child	Event	Participant	Student	Staff	Employee	Child	Employee	Employee
Various	Employee	Development	Child	Student	School	Employee	School	Family	Student	Organised
Employee	Airport	Education	Campaign	Sport	Contribution	Safety	Student	Health	Education	Eleven
Local	Support	Student	Management	School	Education	Organized	Training	Underprivileged	Local	Charity
Care	School	School	Family	Player	Initiative	Training	Development	Support	Safety	Participated
Charity	Project	Employee	Held	Badminton	Home	Health	Child	School	School	Campaign
Awareness	Activity	Local	Central_square	Orang_asal	Scholarship	Various	Health	Home	Training	Activity
Activity	Student	Project	Program	Development	Area	Activity	Skill	Donated	Support	Child
Participate	Various	Member	Awareness	Activity	Family	Workplace	Initiative	Employee	Development	Distributed
Child	Foundation	Activity	Donation	Organised	Business	Awareness	Activity	Life	Initiative	Home

5 Conclusion

This study is aimed to categorise PCSR activities based on the annual reports of selected CSR enterprises in Malaysia. The findings show that LDA Expectation Maximization is the best topic modelling for this study. This further confirms that Expectation Maximisation approach can estimate parameters in probabilistic models using partial, noisy, or missing data. One of the most typical estimate challenges is estimating the mean of a signal which is noise. Many parameter estimation challenges are found to be more difficult because direct access to the data required to estimate the parameters is either impossible or incomplete. When an outcome is the consequence of the aggregation of simpler outcomes, or when outcomes are clumped together as in a binning or histogram process, such difficulties arise. Data dropouts or clustering could also occur, resulting in an unknown number of underlying data points. The Expectation Maximization approach is excellent for such issues because it gives maximum-likelihood parameter estimates when an underlying distribution has a many-to-one mapping to the distribution governing the observation [36]. Furthermore, the findings could assist in conducting further research to cluster PCSR activities which aid firms and NGOs in successfully managing fund distribution based on the identified clusters. It is important to have an accurate prediction of allocation based on the identified clusters because such knowledge may become a guide to companies and NGOs in allocating the resources into those areas and populations where they are mostly needed. Thus, it could increase the efficiency of fund distribution through integrated activities among companies, avoid wasted resources and safeguard the benefits of the beneficiaries.

References

1. Baumgartner, R.J.: Managing corporate sustainability and CSR: a conceptual framework combining values, strategies and instruments contributing to sustainable development. Corp. Soc. Responsibil. Environ. Manage. 21(5), 258–271 (2014)
2. Flammer, C., Luo, J.: Corporate social responsibility as an employee governance tool: Evidence from a quasi-experiment. Strateg. Manag. J. 38(2), 163–183 (2017)
3. Alrubaiee, L.S., Aladwan, S., Joma, M.H.A., Idris, W.M., Khater, S.: Relationship between corporate social responsibility and marketing performance: the mediating effect of customer value and corporate image. Int. Bus. Res. 10(2), 104–123 (2017)
4. Derevianko, O.: Reputation stability vs anti-crisis sustainability: under what circumstances will innovations, media activities and CSR be in higher demand? Oeconomia Copernicana 10(3), 511–536 (2019)
5. Churchill, R., Singh, L.: The evolution of topic modeling. ACM Comput. Surv. (CSUR) (2021)
6. Albalawi, R., Yeap, T.H., Benyoucef, M.: Using topic modeling methods for short-text data: a comparative analysis. Front. Artif. Intell. 3, 42 (2020)
7. Hannigan, T.R., et al.: Topic modeling in management research: rendering new theory from textual data. Acad. Manag. Ann. 13(2), 586–632 (2019)
8. Reisenbichler, M., Reutterer, T.: Topic modeling in marketing: recent advances and research opportunities. J. Bus. Econ. 89(3), 327–356 (2019)
9. Landauer, T.K., Dumais, S.T.: A solution to Plato's problem: the latent semantic analysis theory of acquisition, induction, and representation of knowledge. Psychol. Rev. 104(2), 211 (1997)

10. Qiang, J., Qian, Z., Li, Y., Yuan, Y., Wu, X.: Short text topic modeling techniques, applications, and performance: a survey. IEEE Trans. Knowl. Data Eng. **34**(3), 1427–1445 (2020)
11. Sbalchiero, S., Eder, M.: Topic modeling, long texts and the best number of topics. Some Problems and solutions. Qual. Quant. **54**(4), 1095–1108 (2020). https://doi.org/10.1007/s11 135-020-00976-w
12. Cheng, X., et al.: Topic modelling of ecology, environment and poverty nexus: an integrated framework. Agr. Ecosyst. Environ. **267**, 1–14 (2018)
13. Barde, B.V., Bainwad, A.M.: An overview of topic modeling methods and tools. In: 2017 International Conference on Intelligent Computing and Control Systems (ICICCS), pp. 745–750. IEEE (2017)
14. Kherwa, P., Bansal, P.: Topic modeling: a comprehensive review. EAI Endors. Trans. Scalable Inf. Syst. **7**(24) (2019)
15. Kherwa, P., Bansal, P.: Latent semantic analysis: an approach to understand semantic of text. In: International Conference on Current Trends in Computer, Electrical, Electronics and Communication (CTCEEC), pp. 870–874 (2017)
16. Sutherland, I., Sim, Y., Lee, S.K., Byun, J., Kiatkawsin, K.: Topic modeling of online accommodation reviews via latent Dirichlet allocation. Sustainability **12**(5), 1821 (2020)
17. Sutherland, I., Kiatkawsin, K.: Determinants of guest experience in Airbnb: a topic modeling approach using LDA. Sustainability **12**(8), 3402 (2020)
18. Gurcan, F., Cagiltay, N.E.: Big data software engineering: analysis of knowledge domains and skill sets using LDA-based topic modeling. IEEE Access **7**, 82541–82552 (2019)
19. Nallapati, R.M., Ahmed, A., Xing, E. P., Cohen, W.W.: Joint latent topic models for text and citations. In: Proceedings of the 14th ACM SIGKDD International Conference on Knowledge Discovery and Data Mining, pp. 542–550 (2008)
20. Alghamdi, R., Alfalqi, K.: A survey of topic modeling in text mining. Int. J. Adv. Comput. Sci. Appl. (IJACSA) **6**(1) (2015)
21. Bastani, K., Namavari, H., Shaffer, J.: Latent Dirichlet allocation (LDA) for topic modeling of the CFPB consumer complaints. Expert Syst. Appl. **127**, 256–271 (2019)
22. Blei, D.M.: Probabilistic topic models. Commun. ACM **55**(4), 77–84 (2012)
23. Maier, D., et al.: Applying LDA topic modeling in communication research: toward a valid and reliable methodology. Commun. Methods Meas. **12**(2–3), 93–118 (2018)
24. Guo, Y., Barnes, S.J., Jia, Q.: Mining meaning from online ratings and reviews: tourist satisfaction analysis using latent Dirichlet allocation. Tour. Manage. **59**, 467–483 (2017)
25. He, S., Shin, H.S., Tsourdos, A.: Distributed multiple model joint probabilistic data association with Gibbs sampling-aided implementation. Inf. Fusion **64**, 20–31 (2020)
26. Zhai, Z., Liu, B., Xu, H., Jia, P.: Constrained LDA for grouping product features in opinion mining. In: Huang, J.Z., Cao, L., Srivastava, J. (eds.) PAKDD 2011. LNCS (LNAI), vol. 6634, pp. 448–459. Springer, Heidelberg (2011). https://doi.org/10.1007/978-3-642-20841-6_37
27. Zeng, J., Liu, Z.Q., Cao, X.Q.: Fast online EM for big topic modeling. IEEE Trans. Knowl. Data Eng. **28**(3), 675–688 (2015)
28. CHIRAG676. Complete Guide to Expectation-Maximization Algorithm, 21 May 2021. https://www.analyticsvidhya.com/blog/2021/05/complete-guide-to-expectation-max imization-algorithm/
29. Debortoli, S., Müller, O., Junglas, I., Vom Brocke, J.: Text mining for information systems researchers: an annotated topic modeling tutorial. Commun. Assoc. Inf. Syst. **39**(1), 7 (2016)
30. Fourment, M., Darling, A.E.: Evaluating probabilistic programming and fast variational Bayesian inference in phylogenetics. PeerJ **7**, e8272 (2019)
31. Tran, D., Kucukelbir, A., Dieng, A.B., Rudolph, M., Liang, D., Blei, D.M.: Edward: a library for probabilistic modeling, inference, and criticism (2016). http://arxiv.org/abs/1610.09787

32. Hidayatullah, A.F., Pembrani, E.C., Kurniawan, W., Akbar, G., Pranata, R.: Twitter topic modeling on football news. In: 2018 3rd International Conference on Computer and Communication Systems (ICCCS), pp. 467–471. IEEE (2018)
33. Prabhakaran, S.: Topic modeling with gensim (Python). Machine learning plus, 26 March 2018. https://www.machinelearningplus.com/nlp/topic-modeling-gensim-python/
34. Millen, G.: Python machine learning, 26 December 2019. https://millengustavo.github.io/python-ml/
35. Syed, S., Spruit, M.: Full-text or abstract? Examining topic coherence scores using latent Dirichlet allocation. In: 2017 IEEE International Conference on Data Science and Advanced Analytics (DSAA), pp. 165–174. IEEE (2017)
36. Tamizharasan, M., Shahana, R.S., Subathra, P.: Topic modeling-based approach for word prediction using automata. J. Crit. Rev. **7**(7), 744–749 (2020)
37. Sharma, I., Sharma, H.: Document clustering: how to measure quality of clusters in absence of ground truth (2018). http://www.ijcst.com/vol9/issue2/6-iti-sharma.pdf
38. Yin, J., Zhang, Y., Gao, L.: Accelerating distributed expectation-maximization algorithms with frequent updates. J. Parallel Distrib. Comput. **111**, 65–75 (2018)

Y-X-Y Encoding for Identifying Types of Sentence Similarity

Thanaporn Jinnovart and Chidchanok Lursinsap[✉]

Chulalongkorn University, Bangkok 10330, Thailand
lchidcha@chula.ac.th

Abstract. Determining the semantic similarity of any two arbitrary sentences requires two steps, i.e. sentence encoding and semantic similarity measure. The most important step is to encode a set of sentences into a set of equal-length vectors for similarity measure in forms of classification. Two practical encoding schemes had been proposed, statistical-based direct encoding and pretrained encoding. The first approach lacks considering word correlation and the dimension of encoded vector is very large. For the second approach, it requires an extra training time prior to the classification process. This study compromises the previous approaches by considering shallow neural networks for encoding sentences and classifying entailment relations between two sentences. A set of *y-x-y* encoder models is proposed where *y* can be greater or less than *x* depending on given dataset. Neither encoder models nor their corresponding classifiers are built upon big and complex structure, and hence is suitable for carrying out such task. The encoding scheme is tested with SICK 2014 dataset [1], specially designed for neutral, entailment, and contradiction sentence pairs. Comparison results (neutral 97.1%, entailment 91.1%, contradiction 94.6%) support the possibility of the proposed scheme to sentence similarity measure.

Keywords: Encoder · Classifier · SemEval · Sentence similarity

1 Introduction

The ability to measure similarity between sentences is one of the fundamental traits of a viable Natural Language Processing (NLP) system. Although not explicitly mentioned in general, a system with such capability may lead to several useful applications such as question-answering (QA) system, semantic search, paraphrase detection, and many more. For example, one may choose to query with a sentence, an information would then be retrieved by comparing the input sentence with the system's existing sets of sentences, given that the existing ones can be mapped to at least one answer. These applications require the steps of sentence similarity measure. Generally, determining the similarity of two sentences in terms of semantic or other aspects requires two essential steps, sentence encoding and similarity measure by classification. The most important step is the sentence encoding steps.

Typically, there are two approaches, statistical-based direct encoding and pretrained encoding, to encode a sentence and transform it into a vector of fixed length prior to the

© The Author(s), under exclusive license to Springer Nature Singapore Pte Ltd. 2023
Y. B. Wah et al. (Eds.): DaSET 2022, LNDECT 165, pp. 522–534, 2023.
https://doi.org/10.1007/978-981-99-0741-0_37

classification step. One of classical encoding schemes is by representing each word at any location in the sentence as a bit string, where a bit is set to 1 if a particular word exists. This approach is simple but the dimension of the encoded vector is very large, perhaps larger than the number of training patterns. Statistical-based encoding is also widely applied. The method measures the frequency of word occurrence in forms of n-gram [2]. Another popular encoding schemes recently applied is applying long-short term memory (LSTM) [3] and BERT [4].

With the advancement in NLP field, giving a huge credit to the introduction of the Transformers model [5] and, in particular BERT model, many downstream NLP tasks have been implemented with satisfactory results. Although specially designed as a Neural Machine Translation, as for the Transformers model, and a prediction for next word, as for BERT model, many downstream tasks including measuring sentence similarity also benefit from such models. Multiple variants of pretrained models have been developed and are made public for the community [6, 7]. Many, if not all, of which are of at least hundreds and thousands of embedding dimensions. Each of which may or may not be interpretable [8]. However, they are capable of producing rich text representations.

Furthermore, it is common to extract the last layer of a BERT model via mean-pooling in order to obtain a sentence embedding which can be used as one of the two inputs to a cosine similarity measurement [4]. The results would indicate the degree of similarity between zero and one; the former implying no relationship between the two sentence representations. Hence, two input sentences are not related in any way, while the latter implies a strong relationship between two sentences. Using such methods can be one solution to measuring sentence similarity. However, if how the embedding models are concerned, i.e. the pretrained language models, it is remarkable that masked language models are not specifically trained for the task. Therefore, training on large-scale data with heavy computations and resources may be unnecessary.

Although these approaches are rather practical, they have some points to be improved. The direct and statistical-based encoding schemes do not consider the positional relation of words in a sentence. When words in a sentence are permuted, the resulted sentence obviously indicates different meaning or meaningless. The pretrained encoding scheme must spend an extra training time to obtain a new encoded vector. However, the time to reach convergence of training step is uncontrollable. This study compromises these aspects to improve encoding schemes by transforming a word into a vector of appropriate attributes and computing the covariance of all word vectors in a sentence to capture the positional relation. The encoding step can be achieved in $O(n^3)$, where n is the number of word vectors in the sentence. The encoding scheme is combined with classification steps. The experiment was tested with SICK [1] datasets to identity whether a sentence pair belongs to neutral, entailment, or contradiction classes.

The rest of this paper is organized as follows. Section 2 summaries the related work. Section 3 explains the proposed method. Section 4 provides the result and discussion. Section 5 concludes the paper.

2 Related Work

The goal of word embedding is not merely to map words into random numerical representations but also to capture meaningful information, usually in vector space. Some common word embeddings are Word2Vec [9], GloVe [2], ELMo [10] and FastText [11].

Word2Vec [9] is widely known and used due to its ability to capture semantic relations between words. The famous vector ("King") – vector ("Man") + vector ("Woman") is closest to vector ("Queen") example has undoubtedly led to several attempts to exploit the high-quality word representation at sentence-level [12]. However, direct summations and subtractions do not necessarily lead to high quality sentence representations because a sentence does not merely contain a set of words but also a set of relations between words. The latter is an information which word2vec [9] and GloVe [2] embeddings lack.

BERT's bidirectional training strategy allows its [CLS] token to hold information on the input sentence with a vector of size n, where n is also the size of vectors in other positions of the sentence [4]. This means that if there were m words to the input, hence the output, there would be m vectors of size n, including the [CLS] token at the beginning of the text sequence. BERT uses two training strategies: Masked LM and Next Sentence Prediction. The former is used for pre-training a model and the latter is used for fine-tuning. At fine-tuning, this is a task-specific learning and therefore can be adjusted to desired downstream tasks. Regardless of the downstream tasks, BERT requires pre-training on about 3,300M words. It is mentioned as critical to use a document-level corpus to extract long contiguous sequences.

Upon our experiments on BERT, despite giving satisfiable results, every input sequence is required to be of same length. This means that short sentences would inevitably require padding and so consumes a portion of unnecessary information during computation. Although might not usually be the case for sentence similarity task, sentences that are too long would be truncated. Further handling options would then come into play, such as, using only the first 512 tokens, and more expensively classify and combine all input subtexts back together. Moreover, it is reported that the masked language model converges slower than sequential approaches because only 15% of words are predicted in each batch [4].

The idea of knowledge distillation has been imposed on the original BERT-base model of a total of 110M parameters [7]. The goal of this is to obtain smaller models with competitive performance. This can be seen in TinyBERT [6]. The idea comprises a teacher, which is a state-of-the-art original BERT model and a student, which is a much smaller model of similar architecture. The goal of the student is to replicate the behaviour of the teacher. In its predecessor DistillBERT [7], the student is trained to replicate the teacher based on the output distribution. The weight of the student is initialized using the teacher model. This means that not only knowledge transfer is forced to become less efficient but also their internal dimensions must be the same. TinyBERT uses learnable projection matrices in embedding and hidden loss functions to enable element-wise comparison between internal representations of student and teacher whose dimensions are different. TinyBERT follows two-step distillation very similar to training a general BERT on general data like its teacher at embedding and Transformer layers. At fine-tuning step, the teacher becomes a task-specific model to train the student. Despite being $7.5\times$ smaller and $9.4\times$ faster with 96% of the BERT teacher, the student still contains over millions of parameters in total [6]. TinyBERT may also benefit from data augmentation for low resource tasks which includes sentence similarity.

This study avoids heavy general pre-training of millions of parameters on a large-scale dataset and the attempt to impose a constraint to the length of input sentence, where unnecessary padding and truncation were to be neglected.

3 Proposed Method

There are few most important issues concerning the identification of types of sentence similarity. The first issue concerns the set of words and their meanings directly indicating each type of similarity without deploying any complex computation such as neural learning. For example, for a given sentence pair, if there exists some phrase such as *there is no* appearing in one sentence but not in the other sentence, then this pair can be identified as *contradiction* type. The second issue concerns how to extract the appropriate features of the sentence pair for the use of classification in forms of feature vectors. This issue is rather difficult since each sentence has different number of words. Several encoding schemes have been proposed, such as Word2Vec [9], bag-of-word [13], mixture of BERT [4] and LSTM [3]. These schemes are rather powerful but their space and time complexities are also high. In this paper, a shallow network encoding scheme is proposed to achieve lower space and time complexities as well as maintaining the compatible accuracy.

The framework of the proposed method in this paper is illustrated in Fig. 1. The framework consists of five main stages. The first stage is detecting synonymous verbs as well as phrasal verbs in both sentences. This is to prepare for contradiction detection in the next stage and for word code translation in later stages. The second stage is to filter all contradiction pairs of sentences by detecting the opposite words in both sentences. The sentences in *neutral* and *entailment* classes are classified by employing the rest of the processes. If contradiction is not detected, the synonymous words are assigned word codes in a numeric form. The third process is to assign the word codes to all other non-synonymous words, also in a numeric form. Any non-content words can still be discarded at the stage. The word vectors from each of the two sentences are sequentially combined as an input to the fourth stage. A set of encoders is trained in order to obtain the encoded sentence from the middle layer of the encoder model. In the last stage, a set of classifiers are trained subsequent to each encoder. The final output is a classification between *neutral* and *entailment* types. The details are in the following sections.

3.1 Preprocessing

In prior to performing all of the five stages, both sentences given at the beginning are preprocessed. This involves lemmatizing and removing stop words. The following stages are included in the five stages. Content words, which are verbs, are searched through a list to find and generalize any synonyms and phrasal verbs. Synonyms are handled by translating corresponding words to same word codes. A phrasal verb is also generalized by assigning a word code to one of the words in the compound and neglecting the other words in the compound, which would eventually be discarded during word code translation.

3.2 Filtering Contradiction Class

There are scenarios that two sentences are in contradiction class. The first scenario focuses on the set of opposite words. Two sentences contradict each other if one of them has a word or phrase and the other sentence has the opposite word or phrase to the first sentence. For example, sentence 1 has this set of words: *"The brown horse is **near** red barrel at the rodeo"* and sentence 2 has this set of words: *"The brown horse is **far from** red barrel at the rodeo"*. The opposite words of this case are **near** and **far from**. Both words describe the distance of the *horse* with respect to the **red barrel**. It is important to note that the proposed method not only looks for opposite words but also the context of which they are describing. In other words, if a pair of opposite words exist but are referring to different objects, this won't be counted as contradiction because it is possible that the sentence pair is irrelevant, or neutral. The second scenario is the exception of the first scenario. There are some implicitly opposite phrases appearing in only one sentence. These phrases imply the contradiction of the sentence pair. Consider the following sentence pair:

sentence 1: *"**There is no** biker jumping in the air"*.
sentence 2: *"A lone biker is jumping in the air"*.

Some of these implicitly opposite words and phrases are **no, none, no one, nobody, there is no, there are not, there have no**. These are also known as negative markers. The negative marker in sentence 1 of the case above may be considered as one-sided because only one sentence contains such marker. In the sentence pair above, a contradiction exists because the negative marker is modifying *biker* in one sentence, whereas there is none for *biker* in the other sentence.

3.3 Encoder Models

A word in this paper has two morphological forms. For the first form, a word refers to a morpheme or a compound word such as **man** and **football**. For the second form, a word refers to a group of words orderly arranged in the form of a phrase such as **get together** and **have a good relationship**. The part of speech of any word in both forms may be noun, pronoun, verb, adjective, adverb, preposition, conjunction, and interjection. In the experiment, it is found that part-of-speech has the minimum impact on classification accuracy. Two words may have either the same or opposite semantics. For example, **turn back** has the same semantics as **change direction**, but **agree** has opposite semantics to **refuse** and **argue**. Words with the same semantics must be assigned the same word code.

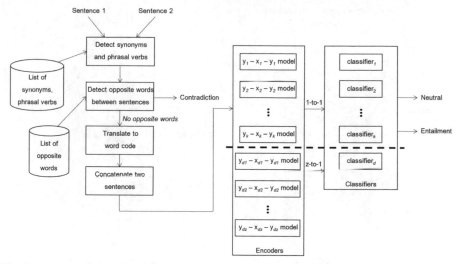

Fig. 1. Proposed framework. Contradiction sentence pairs are filtered out using a database of opposite words at an early stage. Encoders and classifiers are specially trained for binary classification between neutral and entailment sentence pairs. There are k y-x-y models and classifiers plus a dedicated set d to handle those accumulated samples whose amount alone are insufficient for training.

ALGORITHM 1: TRAINING ENCODING AND CLASSIFYING ALGORITHM

Input: a set of preprocessed sentence pairs $S^{(i)} = \left\{ w_1^{(i)}, \cdots, w_m^{(i)} \right\}, S^{(j)} = \left\{ w_1^{(j)}, \cdots, w_n^{(j)} \right\}$.

Output: a set of encoders and classifiers

1 Assign word code to each $w_x^{(i)}$ and $w_y^{(j)}$ to obtain $S^{(i)} = \left\{ c_1^{(i)}, \cdots, c_m^{(i)} \right\}$ and $S^{(j)} = \left\{ c_1^{(j)}, \cdots, c_n^{(j)} \right\}$.

2 Concatenate $S^{(i)}$ and $S^{(j)}$ to obtain $I = \left\{ S^{(i)}, S^{(j)} \right\} = \left\{ c_1^{(i)}, \cdots, c_m^{(i)}, c_1^{(j)}, \cdots, c_n^{(j)} \right\}$.

3 For each y from I's with sufficient samples, vary x on training $y - x - y$ model to obtain separate optimal encoding sizes.

4 For each y from I's with few samples, vary x on training $y - x - y$ model to obtain one common optimal encoding size.

5 For each separate optimal encoding size, train separate classifiers.

6 For common optional encoding size, train one classifier.

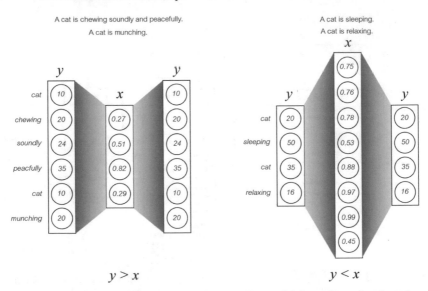

Fig. 2. *y-x-y* encoder model. The structure of an encoder model depends on the given data. The encoded layer of size *x* is adjusted to be either greater or less than the input size *y*. Two pairs of sentences are demonstrated after going through the preprocessing stage which involves removing stop words and detecting synonyms.

A sentence composes of a set of words. The number of words in each sentence is different. To classify a set of sentences, each sentence must be transformed into a set of vectors of equal size first by encoding all the words in the sentence. An encoder model is employed to achieve this. Let $\mathbf{S}^{(i)} = \left\{ w_1^{(i)}, \cdots, w_m^{(i)} \right\}$ be sentence *i* consisting of a set of words $w_x^{(i)}$. Each word is represented by its corresponding word code $c_x^{(i)}$ in numeric form. Thus, the elements in a sentence become a set of word vectors, $\mathbf{S}^{(i)} = \left\{ c_1^{(i)}, \cdots, c_m^{(i)} \right\}$. The input to the encoder is a concatenation of the two sentence pairs; $\mathbf{I} = \left\{ \mathbf{S}^{(i)}, \mathbf{S}^{(j)} \right\} = \left\{ c_1^{(i)}, \cdots, c_m^{(i)}, c_1^{(j)}, \cdots, c_n^{(j)} \right\}$ where *m* is the number of words in sentence *i* and *n* being that of sentence *j*. The relation among these words is captured by a set of simple encoder models, Fig. 2.

ALGORITHM 2: SENTENCE PAIR CLASSIFICATION ALGORITHM

Input: preprocessed sentences $\mathbf{S}^{(i)} = \left\{ w_1^{(i)}, \cdots, w_m^{(i)} \right\}, \mathbf{S}^{(j)} = \left\{ w_1^{(j)}, \cdots, w_n^{(j)} \right\}$.

Output: sentence pair type (*neutral, contradiction, or entailment*)

1 If found opposite words and one-sided negative marker, return *contradiction*.

2 Assign word code to each $w_x^{(i)}$ and $w_y^{(j)}$ to obtain $\mathbf{S}^{(i)} = \left\{ c_1^{(i)}, \cdots, c_m^{(i)} \right\}$ and $\mathbf{S}^{(j)} = \left\{ c_1^{(j)}, \cdots, c_n^{(j)} \right\}$.

3 Concatenate $\mathbf{S}^{(i)}$ and $\mathbf{S}^{(j)}$ to obtain $\mathbf{I} = \left\{ \mathbf{S}^{(i)}, \ \mathbf{S}^{(j)} \right\} = \left\{ c_1^{(i)}, \cdots, c_m^{(i)}, c_1^{(j)}, \cdots, c_n^{(j)} \right\}$.

4 Encode $\mathbf{I}$ using optimal encoding size.

5 Classify using corresponding classifier, return *neutral* or *entailment*.

6 For each separate optimal encoding size, train on separate classifiers.

7 For common optimal encoding size, train on one dedicated classifier.

An encoder model, Fig. 2, is configured in an *y-x-y* manner where *y* is the length of the concatenated preprocessed sentence pair and *x* is the desired encoding size. The input to the model is of one dimension of arbitrary length, as well as the encoding size. A set of encoder models is used because each encoder model is suited for one length of input only. The goal is to find the optimal encoding size for each *y* available in the dataset. Therefore, for each *y*, *x* varies, and vice versa. It is possible for *y* to be greater than *x*, and vice versa. Ideally, the optimal encoding size for all *y* s would be the same which would allow the use of only one classifier for all encoders. In practice, this is unlikely to happen and therefore each classifier maps to one encoder. The overall process of training encoders and classifiers is summarized in Algorithm 1.

Ideally, the number of encoders, and hence classifiers, should be the total unique length of the concatenated preprocessed sentence pairs. This might not be possible, in practice, if a dataset contains very few samples of certain lengths as training the classifier would not be feasible which could lead to unreliable classification. To handle such samples, these samples are separately encoded to same length and accumulated. They are then passed into a classifier for training and testing at once. This method is believed to unveil characteristics of different lengths of sentence pairs. For example, a sentence pair concatenated to length 12 could give an optimal encoding size at 25 whereas length 11 could give size 250. This could be interpreted as two sentences of total preprocessed length of 11 generally contain overlapping words, while length 12 does not, that the encoding dimension is 10 times higher. All classifiers are identical. The network consists of two fully connected layers with sigmoid activation functions and a softmax layer. All classifiers are optimized using stochastic gradient descent with learning rate 0.001.

4   Results and Discussion

Our results show that contradiction samples were successfully separated from neutral and entailment by 92% in training set. Since there are only 666 samples of contradiction out of 4501 training data, SICK's testing data is also evaluated in which contains additional contradiction samples of 720 pairs and the results confirms successful separation by 92.8%. This proves our method of detecting contradiction pairs to be effective. Our proposed method attempted to detect negative markers and antonyms between sentence pairs and then confirmed whether those cues were referring to the same subject, or object, in both sentences.

All contradiction sentence pairs are filtered out and the remaining samples are fed into encoder models and classifiers. The remaining samples contain both neutral and entailment sentence pairs. We may argue that our method does not need any heavy pretrained models as the number of parameters does not exceed 1M. This makes our proposed system lightweight and easy to be used across platforms, and environments.

An attempt to classify among the three types of sentence pairs in [12] can be seen as an example of using pretrained word embeddings with high dimensionality. Each word is referred to a unique GloVe vector and the baseline sentence embedding model was the sums of all word embeddings in each sentence. Despite the sentence model being 100 dimensions for each premise and each hypothesis, their word embeddings were generally of 300 dimensions. When compared to our sentence embedding scheme in which each word only contains one unique value, the use of pretrained model becomes comparably large. Although, in the actual sentence embeddings for our proposed method, encoding sizes may vary dramatically up to 1000, the encoder model arguably consumes much lower resources at training time.

Handling multi-word expressions can be seen in sentences with phrasal verb expressions where groups of words are detected and converted to single verbs. In many times, these single verbs led to synonym detection between sentence pairs. This, in turn, results in same word code assignment.

We agreed with [14] that there are some samples not correctly labeled according to their actual semantics. For example, sentences between "The **turtle** is following the **fish**." and "The **fish** is following the **turtle**." are labeled as contradiction. We argued that this sentence pair is instead neutral. For a contradiction pair, it should explicitly be the **turtle** not following the **fish**. Another example that showed our disagreement is "A group of children is playing in the house and **there is no** man standing in the background." and "A group of kids is playing in a yard and an old man is standing in the background.". This is labelled as a neutral type in the SICK dataset. It is clear that "**there is no man standing**" and "**an old man is standing**" imply contradiction. There were a couple of these doubtful labels that result in incorrect classifications.

One error we observed in our program is when opposite words are detected but they are seen as not referring to the same entity. For example, "A **blond** child…" and "A child with **dark hair**…" both refer to a **child** with different hair colour. Our program would see that **blond** refers to **child** and **dark** refers to **hair**. **Child** in the second sentence would have no descriptive features like the one in the first sentence. Nevertheless, this issue did not have much effect on the overall contradiction accuracy as this type of sentences is subsidiary to the general characteristics of the overall sentences.

According to [1], SICK dataset focused on generic semantic knowledge and semantic compositionality. In our work, it can be said that we captured the former in the process of finding opposite words and the latter in handling multiword scenarios, such as, phrasal verbs. We chose to evaluate on entailment relation as the performance should better reflect a system's understanding of computational semantics at a more general level. In other words, a system is to be able to indicate the type of similarity or relatedness. Although we achieved over 90% accuracy for contradiction type, the proportion of contradiction to the other two types are relatively low in all training, trial and test sets, and mostly involve explicit negations [15]. This is evident in Table 2 where Illinois-LH [15] with negation only has higher accuracy, 86.4, than combining all its features, 77.0. Therefore, we may consider evaluating our system in other datasets that have more diverse contradiction sentence pairs.

From SemEval 2014 Task 1, Illinois-LH is the top of category and therefore is chosen for comparison. Another state-of-the-art method, to the best of our knowledge, is NeuralLog. This method incorporates a BERT variant which may be considered as a BERT-like model.

The descriptions of SemEval 2014 Task 1's Chance, Majority, Probability and Overlap baselines in Table 1 can be found in [1]. Our system's performance is higher than Chance, Majority and Probability baselines because the predictions from those three baselines were built upon randomness and so sentence semantics was irrelevant.

Meanwhile, our system evaluated on sentence semantics. Although Overlap baseline accounted for words in sentences, it included stop words which our system did not because we viewed stop words as non-content words which would only lead to inaccurate evaluation.

Table 1. Performance on the SICK test set.

	Accuracy
Chance baseline	33.3
Majority baseline	56.7
Probability baseline	41.8
Overlap baseline	56.2
Illinois-LH [15]	84.5
NeuralLog [14]	90.3
- without neural-based	71.4
- without logic-based	74.7
Our results	95.2

NeuralLog [14] uses a joint logic-based and neural-based method to perform Natural Language Inference. With accuracy drops to 71.4 and 74.7 from removing either neural-based or logic-based module, respectively, we may deduce that our performance is higher. We agree with [14] that there are times that handcrafting knowledge relations is inevitably required. In our work, this can be seen in building up a list of opposite words and altering the list affects the decision on classifying relation types. Nevertheless, in the case of finding which entity a modifier refers to, we only consider a modifier and its corresponding entity whose position is to the right. We take an example of *red rose* from [14]. Here, *rose* is to the right of *red* and therefore we know that the *rose* is *red*. If a sentence is "...*a rose which is red*...", our system might ignore the modifier *red* because there is no possible entity to the right of it. We believe that our system may benefit from [14]'s neural syntactic variation inference module in this area. However, this may be out of scope for this work as we aim to avoid using models with large number of parameters such as a Transformer model.

Table 2. Performance on the SICK trial set.

	Accuracy		
	Neutral	Entailment	Contradiction
Overlap baseline	77.3	44.8	0.0
Illinois-LH [15]	86.5	83.3	77.0
- negation	85.4	0.0	86.4
Our results	97.1	91.1	94.6

In an ideal situation where opposite words and synonyms are given, our proposed method works. The method of acquiring such resources is secondary in this work which could be the primary focus in another work. In essence, we are proving that handling input sentence pairs of different lengths by not truncating or padding, like other models that do in order to handle variable sized inputs, is possible. We have proven to be able to capture the important information by concatenating sentence vectors which is suited for the similarity task at lower dimensions.

Furthermore, a set of encoders and classifiers were trained on neutral and entailment types. Note that the network structures for all classifiers are identical as shown in Fig. 3. The accuracy is accounted at 97.14% for neutral type, and 91.09% for entailment type. For samples of certain lengths whose amount did not exceed 100, they are considered few and are accumulated at encoding size 750 as it gives the highest accuracy out of the most frequent optimal encoding sizes for those containing substantial number of samples. Additionally, a classifier is dedicated to this group. As show in Table 3, optimal encoding size x varies without any correlation with the concatenated sentence pair length y. For length 12, the encoding size is 1000, whereas length 11 is optimal at encoding size 750. A more dramatic variation is at length 10 where the optimal encoding size is 75. Although optimal encoding sizes for different encoders may be of hundreds, no encoder model contains up to 1M parameters. This is also true for the configurations of all classifiers used. The training time is of seconds and therefore meets our goal in avoiding use of heavy computational resources and large-scale dataset.

Table 3. Sentence pair length vs Optimal encoding size.

y	x
8	25
9	250
10	75
11	750
12	1000
13	250
14	750
15	250
16	250
17	50
18	1000

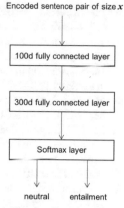

Encoded sentence pair of size x

Fig. 3. One classifier.

5 Conclusion

In this paper, a lightweight encoding scheme, in terms of low dimensionality, is proposed. It involves filtering out contradiction sentence pairs, detecting synonyms and phrasal verbs, assigning word codes, and applying a set of y-x-y encoders and classifiers. With over 90% accuracy for all, our encoding method is able to capture the semantics through the sentence pair vector representations. Pretrained word embedding models, which require training on large-scale datasets, are not involved in the process of distinguishing between the three types of sentence pairs in the SICK dataset. Only shallow, at most two fully connected layers, neural networks were used in the whole process. No complex network structure is used. Moreover, the accuracy can be improved if a word is attributed to additional appropriate features such as types of verb (action, stative, etc.), and types of noun (abstract, collective, etc.) in addition to word codes. A further study on what should be a good set of features representing a word must be made.

References

1. Marco, M., et al.: SemEval-2014 task 1: evaluation of compositional distributional semantic models on full sentences through semantic relatedness and textual entailment. In: Proceedings of the SemEval (2014)
2. Pennington, J., Socher, R., Manning, C.D.: GloVe: global vectors for word representation. In: Proceedings of the 2014 Conference on Empirical Methods in Natural Language Processing (EMNLP) (2014)
3. Hochreiter, S., Schmidhuber, J.: Long short-term memory. Neural Comput. **9**(8), 1735–1780 (1997)
4. Devlin, J., et al.: BERT: pre-training of deep bidirectional transformers for language understanding. Association for Computational Linguistics, Minneapolis, Minnesota (2019)
5. Vaswani, A., et al.: Attention is all you need. Adv. Neural Inf. Process. Syst. **30** (2017)

6. Jiao, X., et al.: TinyBERT: distilling BERT for natural language understanding. arXiv preprint arXiv:1909.10351 (2019)
7. Sanh, V, et al.: DistilBERT, a distilled version of BERT: smaller, faster, cheaper and lighter. arXiv preprint arXiv:1910.01108 (2019)
8. Li, B., et al.: On the sentence embeddings from pre-trained language models. arXiv preprint arXiv:2011.05864 (2020)
9. Mikolov, T., et al.: Efficient estimation of word representations in vector space. arXiv preprint arXiv:1301.3781 (2013)
10. Peters, M.E., et al.: Deep contextualized word representations. Association for Computational Linguistics, New Orleans, Louisiana (2018)
11. Mikolov, T., et al.: Advances in pre-training distributed word representations. arXiv preprint arXiv:1712.09405 (2017)
12. Bowman, S.R., et al.: A large annotated corpus for learning natural language inference. Association for Computational Linguistics, Portugal (2015)
13. Harris, Z.S.: Distributional structure. Word **10**(2–3), 146–162 (1954)
14. Chen, Z., Gao, Q., Moss, L.S.: NeuralLog: natural language inference with joint neural and logical reasoning. Association for Computational Linguistics (2021)
15. Lai, A., Hockenmaier, J.: Illinois-LH: a denotational and distributional approach to semantics. Association for Computational Linguistics, Dublin, Ireland (2014)

Evaluation of Extractive and Abstract Methods in Text Summarization

Ranjita Kumari Biswal Lenka[1], Thomas Coombs[2], Sulaf Assi[3], Manoj Jayabalan[1], Jamila Mustafina[4], Panagiotis Liatsis[5], Abdullah Al-Hamid[6], Sahar Al-Sudani[7], Noor Lees Ismail[8], and Dhiya Al-Jumeily OBE[1(✉)]

[1] Faculty of Engineering, Liverpool John Moores University, Liverpool L3 3AF, UK
d.aljumeily@ljmu.ac.uk
[2] University Hospital Dorset, Bournemouth BH7 7DW, UK
[3] School of Pharmacy and Bimolecular Science, Liverpool John Moores University, Liverpool, UK
[4] Kazan Federal University, Kazan, Russia
[5] Department of Electrical Engineering and Computer Science, Khalifa University, Abu Dhabi, UAE
[6] Saudi Ministry of Health, Najran, Saudi Arabia
[7] American University of Iraq-Baghdad, Baghdad, Iraq
[8] Faculty of Business and Technology, UNITAR International University, 47301 Petaling Jaya, Selangor, Malaysia

Abstract. Text summarization has become very essential tool to record important points and has been used by several websites and applications to lessen length, difficulty, and to preserve the vital information of the original file. The requirement on well-organized and useful text summarization of the website content, news feed and other kinds of legal documents with judgments and predilection is the demand of the present requirement. Hence several attempts have been made to automate the summarizing process. The recent development and state of the art models in natural language processing demonstrated outstanding results in text summarization, however major focus of these analysis was on large dataset with large parameters. This study's primary purpose is to evaluate the performance of ensemble abstractive and extractive models on text summarization. Combined core of BERT and PEGASUS models' output were applied to LexRank model on News Summary dataset to evaluate the performance through ROUGE metric. The results showed the performance of combined and ensemble model is better than individual performance.

Keywords: Text summarization · Natural language processing · Large datasets · BERT · PEGASUS · Evaluation

1 Introduction

On rapid growing of mobile user, social network, big data, online information and cost-effective access to internet service, the information in the digital world is growing exponentially and makes it difficult for user to efficient information consumption. Text summarization has become very essential tool to record important points and has been used

© The Author(s), under exclusive license to Springer Nature Singapore Pte Ltd. 2023
Y. B. Wah et al. (Eds.): DaSET 2022, LNDECT 165, pp. 535–546, 2023.
https://doi.org/10.1007/978-981-99-0741-0_38

by several websites and applications to lessen length, difficulty, and to preserve the vital information of the original file.

The requirement on well-organized and useful text summarization of the website content, news feed and other kinds of legal documents with judgments and predilection is the demand of the present requirement. Hence several attempts have been made to automate the summarizing process. It is well known fact that human brains have the limitation to process many online documents in an efficient manner to verify authentic and useful data. Hence the requirement of condensed form of information or summarization in an automatic, authentic, and efficient process is inevitable to quickly analyse the content in the ocean of information [1]. Summarization helps the user to get the shorter version of the content having original input.

Automatic summarizers have been source of income by the abstracting providers to professionally increase their revenue on text summarization of official documents in an optimum time limit [2]. The technique, where a computer program shortens longer texts and generates summaries to pass the intended message, is defined as Automatic text summarization and reported as general constraints in machine learning and natural language processing (NLP) [3]. Text summarization has been explored extensively and different summarization techniques also have been developed. Most popular and generic methods of automatic text summarization are abstractive and extractive. Extractive summarization engrosses the selection of key phrases and sentences from the source document to generate the new summary without changing the original sentence. It engages ranking the concerned of phrases to choose only those most relevant to the meaning of the input text document [4, 5]. Whereas abstractive summarization derives human understandable new phrases and sentences to capture the meaning of the source text.

The advancement and accelerative development of various models of text summarization in 2 NLP can be briefly described under three major breakthroughs in Deep Learning and Neural Network models. Thereafter the work of text summarization was dominated by sate of art models based on the combination of the three techniques as mentioned below:

a. Seq2seq model: One of the important developments in NLP based on encoder and decoder [6].
b. Attention Mechanism: Excellent progress both in NLP as well as DL [7].
c. Transformer: It possess both encoder/decoder like Seq2seq and self/multi head attention, initially proposed by member of Goole team [1].

Recent state of the art abstractive summarization techniques targeted on innovative self-supervised objectives in pretraining the model [8–10]. Transformer-based encoder-decoder architecture has also been demonstrated as summarization method within the contribution from attention mechanism and can nicely capture the syntactical and contextual information in the linked words in documents [1].

The work on techniques, models, and metrics of evaluation of both abstractive and extractive method have been reported extensively, discussed in Sect. 2 or review section, however most sate of art techniques on text summarization has received much attention of researcher. The Pegasus pre-trained model one of the abstractive method has been reported to perform outstanding on large dataset, state of art results on diverse summarization datasets [13], BERT [12] and other transformer based encoder decoder models along with self-supervised pre-training have revealed to be a powerful framework for getting SOTA performance when fine-tuned on a larger array of datasets in abstractive method [1], similarly STRASS [14] in extractive method enhance the correlation of meaningful and truthiness of summery to original text [13].

Although prior literature reported several extractive and abstractive summarization models, a comprehensive evaluation of these models is further required. Moreover, the analysis and standardization on extractive and abstractive text summarization models on the text summaries are yet to be performed. Hence, the aim of this study is to fill the gaps by formulating a detailed assessment of the output of recent state of the art extractive and abstractive text summarization models in most popular dataset on news summary.

The analysis and evaluation of both extractive and abstractive methods in automatic text summarization is still an open issue, there is no concurrence among different researchers on various approach, need further measurement and evaluation. The development of more targeted and reliable summaries may lead to a more consistent evaluation and better convergence on most accepted metrics and automatic evaluation methods. However, the study of automatic summarization and its analysis is still an upcoming research area with many challenges. Recent state of the art techniques of text summarization have established more reliable to human based evaluation.

The results of transformer-based models like BERT and PEGASUS were reported to be outstanding on both abstractive and extractive methods and even mixed models of the abstractive and extractive methods. The problem statement may be mentioned as "dose the performance get more accurate and better if both the state-of-the-art model such as BERT and PEGASUS ensemble?" The performance of ensemble model in terms of metric like ROUGE score and performance of individual models using BERT and PEGASUS are same or different? Detailed analyses of ROUGE score to check the performance of advanced methods in on news sources are the tenet of this research.

The requirement of reliable and accurate text summarization is growing day to day on dependency over the content in internet and cloud. Manifold raise in research and excellent models on text summarization from 2017 onwards. Most of the models have tested over large sentences and large database. Still, there is further scope to utilize the existing models to fine tune parameter, architectural changes for better outcome. The application specific database like news summery, medical, official documents and legal papers has not been explored extensively. Most of the models have been tested on Wikipedia/ CNN news/Amazon food reviews and internet resources. Since summaries tend to be more oriented towards specific needs or type of text document, it is necessary to tune existing evaluation methods accordingly. However, because of the importance of the research target in automatic summarization, a series of proposals have been introduced to partially or fully automate the evaluation. Therefore, the scope of the study is to examine each SOTA model results on news summery separately and to compare the result on ensemble model of two SOTA models.

In this new digital world, where plenty of tremendous information is available on the Internet and overloaded on daily at each interval of minute to second, mechanism to provide concise, fast and reliable information on input text is demand of the time. It is almost impossible sometimes for normal user or human beings to manually extract the summary of large documents of text. There are plenty of text materials available on the Internet. So, there is a time-consuming exercise for searching for relevant documents from the number of documents available, and getting relevant information from it. It is imperative that text summarization is very much essential to address the said two problems. The different dimensions of text summarization can be generally categorized based on its input type (single or multi document), purpose (generic, domain specific, or query-based) and output type (extractive or abstractive). There are existing models of summarization problem, viz., abstractive versus extractive, single versus multi-document, and syntactic versus semantic. Each has own merits and performance. Therefore, it is pertinent to evaluate the most common method of summarization to address of digitized world.

The study focuses to evaluate both methods as mentioned above using standard metrics in data science and reports the result on academic purpose. The main aim of this research is to evaluate the performance of extractive and abstractive methods in text summarization through automatic evaluation metric using Rouge and BELU.

The objectives of this study are:

- To review published literature on results of both extractive and abstractive methods in text summarization 4
- To analyse limits and performance of extractive and abstractive methods in text summarization
- To examine the performance of PEGASUS and BERT on News summery
- To examine the performance of ensemble model consists of both PEGASUS and BERT
- To evaluate the results automatic evaluation metric is used such as ROUGE-1, ROUGE-2 and ROUGE-L

2 Methodology

The aim of this work is to contribute to the field of automatic text summarization with evaluating the text summarization methods and with the proposed new method in this study. The domain of NLP has recently undergone rapid and substantial progress, driven by the success of sequence to sequence (seq2seq) modeling in NLP. However, the research target has more on developing architectures suitable for long/multiple documents, large database and has neglected short and single documents. Therefore, we evaluated the state of the art model like PEGASUS, BERT and other conventional model like Page Rank LSA and Text Rank in news summery with the help of ROUGE scores like precision, recall and F-score.

The three methods (BERT, PEGASUS and ROUGE) applied in this study were first pre-processed by removing unwanted characters, special characters, spaces, transferring all words to lowercase characters.

Steps involved include:

- Removing the News office information if any;

- Splitting the line using white space tokens or tokenize on white space;
- Normalizing the case to lowercase;
- Removing all punctuation characters from each token;
- Removing any words that have non-alphabetic characters;
- Putting this all together, new function named clean text that takes a sequence of list of lines of text and returns a list of clean lines of text;
- Saving and load the data for further processing tails.

2.1 BERT

BERT is an open-source machine learning framework for natural language processing (NLP) introduced by Google AI team stands for Bidirectional Encoder Representations from Transformers [15]. In brief, BERT uses dynamic bidirectional encoders, transformer and attention. BERT was introduced in two sizes, BERT 22 (Base) and BERT (Large). The BASE model was used to measure the performance of the architecture comparable to other architecture and the LARGE model demonstrated state of the art results.

Pre-training of BERT was conducted in two steps:

1. masked language modelling (MLM): In contrast to conventional language modeling (LM), which aims at predicting the next word given the sequence of previous words, masked language modeling is the task of predicting a percentage of input tokens which are randomly masked. The MLM training objective was chosen over the traditional LM objective because of the bi-directionality of BERT. However, the final output of MLM use the conventional LM optimizer function such as softmax. The demonstrated the final inputs of hidden vectors in connection to that particular masked token are provided to an output of softmax.
2. next sentence prediction (NSP): t is a binary classification technique used to understand the relationship between two sentences when comes as pair to input of the model. It predicts the subsequent sequence in relation to first sentence in the input text. Generally, it is not possible by conventional language modeling. This prediction task can be easily generated from any single language text data. Particularly during selection of the sentences, A and B for each pre-training example, 50% of the time B is the actual next sentence that follows A (labeled as Is Next), and the other 50% of the time it is a random sentence from the corpus (labeled as Not Next). The last hidden vector corresponding to the [CLS] token is input to an output softmax over the two possible predictions.

Moreover, downstream of pre-trained BERT models was carried out in two steps being:

1. fine tuning: Self attention mechanism in the Transformer makes the BERT Fine tuning much simpler to model many downstream tasks, either on single text or text pairs or by swapping out the suitable inputs and outputs. BERT uses the self-attention mechanism to unify these two stages, as encoding a concatenated text pair with self-attention effectively includes bidirectional cross attention between two

sentences. One only needs to plug in the task-specific inputs and outputs into BERT and fine-tune all the parameters end-to-end for a few epochs for each task. In the input, sentence A and sentence B from pre-training are equivalent to, sentence pairs in paraphrasing, hypothesis-premise pairs in entailment, question-passage pairs in question answering and a degenerate text-∅ pair in text classification or sequence tagging.

2. feature-based steps: BERT use feature-based approach to represent word extracted from the pre-trained model and serve as inputs to other task specific architectures. It was noticed that all the task may not be recognized by transformer encoder, therefore task specific feature approach are suitable for downstream. There are several ways of extracting contextual word embeddings from BERT output but selection of approach can be better and enhanced if correlated to task. The work was explained in a study of named entity recognition (NER) by Devlin et al. (2018) for the task of where they applied a feature-based approach by extracting the activations from one or more layers without fine tuning any parameters of BERT on the task, and then used word context embeddings as inputs to a randomly assigned 2 layer 768-dimensional BiLSTM before the classification layer. The outcome showed that concatenating the last four hidden layers as the contextual word embeddings led to the best F1 scores for that specific task.

2.2 PEGASUS

PEGASUS is the short form of Pre-training with Extracted Gap-Sentences for Abstractive Summarization. It was introduced by Google AI team in 2020 and developed by Zhang et al. 2020 [11]. The main difference between BERT and PEGASUS is the choice of pre-training modules. Together GSG and MLM were applied concurrently to for the pre-training objectives. Initially, there were three sentences. First sentence was masked with [MASK1] and used as target generation text (GSG). The other two sentences remain in the input, but some tokens are arbitrarily masked by [MASK2] (MLM).

GAP-SENTENCE GENERATION (GSG): IT selects and masks whole sentences from documents. Those masked sentences would then be concatenated to form pseudo summaries. It may calculate the gap sentences ratio which computes the number of selected gap sentences over the total number of sentences in the document.

The commonly used three strategies to select gap sentences (without replacement) are given below:

1. Random
2. Lead
3. Principal

Principal strategy engrosses selecting top-m scored sentences based on their importance. The importance is measured by the ROUGE-1 score between the original sentence and the rest of the document.

There are four variants to the principal strategy:

1. Sentences are scored independently

2. Sentences are scored sequentially, maximizing ROUGE-1 score between selected sentences and remaining sentences.
3. Computing ROUGE-1 score by considering n-grams as a set (a. unique n-gram)

Masked Language Model (MLM): In MLM, words from the sentences are randomly masked and use other words from the sequence to predict these masked words. The GSG task can be construed as a document-level MLM and is derived from this very concept.

2.3 LexRank

Introduced by Erkan et al. 2004, LexRank, a modified version of Page Rank algorithm was used for automated text summarization [16]. The sentences of the documents may be viewed as a network of sentences that are related to each other. Many of the words may be similar to words of other sentences; few may not have any link. In TexRank method, for word similarity, it assumes all weights to be unit weights and calculate ranks like a typical PageRank execution, but in LexRank, based on the degrees of cosine similarity between words and phrases, the centrality of the sentences to allocate weights. Steps involved in LexRank Algorithm include: (1) Data input to the model, (2) word embedding, (3) intra-sentence cosine similarity, (4) adjacency matrix, (5) connectivity matrix, (6) eigenvector centrality, and (7) output of the model.

2.4 Rouge

The Recall Oriented Understudy for Gisting Evaluation (ROUGE) was introduced by Lin et al. 2004, still as on today, ROUGE, is the most widespread summarization evaluation metric [17]. It works by comparing an automatically produced summary with a set of reference summaries (typically human-produced).

2.5 Implementation

In this study, the dataset related to news summary available in Kaggle [18] was considered for analysis and downloaded from kaggle site. The data set composed of 4,515 examples and contains Author_name, Headlines, Url of Article, Short text, Complete Article. It was mentioned in the website that the information on news article was gathered in short and only scraped the news articles from Hindu, Indian times and Guardian. Time period ranges from February to August 2017 were included.

The dataset was divided into csv files as follows: (1) *news_summary.csv*: news_summary.csv file contained the columns author', 'date', 'headlines', 'read_more', 'text', 'ctext For this study, 'ctext' column data is used as input for text summarization, (2) *news_summary_more.csv*: news_summary_more csv files contains two columns as 'headlines', 'summary'. News_summary_more files data was used for most of the summarization evaluation.

2.6 Dataset Preparation

The data was pre-processed by removing unwanted columns, transferring all words to lowercase characters and removing null values, HTML tags, parenthesis, punctuations, special characters, and stop-words. Steps involved in dataset preparation are listed in Table 1.

Table 1. Steps involved in dataset preparation.

Step number	Step
1	Removing the white space characters if any
2	Splitting the line to tokens
3	Removing all punctuation characters from each token
4	Removing any words that have non-alphabetic characters
5	Normalizing the case to lowercase
6	Putting this all together, new function named 'clean text' that takes a sequence of list of lines of text and returns a list of clean lines of text
7	Saving and load the data for further processing

2.7 Model Implementation

The pre-training models of BERT and PEGASUS used in the experiments was "BERT (bert_extractive_summarizer)" and "PEGASUS-Paraphrase" respectively. Both models were pre-trained for abstractive summary tasks. After cleaning the data set, the subsets of news summery fed separately to each model and to ensemble core model (BERT + PEGASUS) in two steps.

The procedure of data analysis was adopted and summarized as follows:

(1) The input of news summery data frame (800 subsets) fed to each BERT, PEGAUS and LexRank separately and the output RUGE score recorded.

(2) The input subset of news summery split into two group such as Group 'A' and Group 'B', Group 'A' consists of 400 subsets fed to BERT, Group 'B' consists rest 400 subset fed to PEGASU, then summery recorded. The output of both Group 'A' and Group 'B' combined to form input of LexRank.

To evaluate the experiment results, we used ROUGE (Recall-Oriented Understudy for Gisting Evaluation) metric proposed by Lin (2004) [17], which compares a generated summary against a reference summary. Specifically, we used ROUGE-1, ROUGE-2, and ROUGE-L. ROUGE-1 computes the overlap of unigram, ROUGE-2 computes the overlap of bigrams, and ROUGE-L computes the longest common subsequence's that focus on sentence-level structure and identify longest co-occurring in sequence ngrams.

3   Results and Discussion

The two experiments (one with individual model and other with ensemble or core consists of BERT plus PEGASUS) as discussed in the method sections are evaluated, analysed and interpreted. The core models were tested by the divide and conquer method of input text. The input of news text was functionally divided by 50% of total news summery data frame (800 examples). To compare, marking was made for first 50% of input text as – 'A' text, rest 50% of input text as –'B' text. The input of core model was sequentially divided into two patterns. Both the output compared individually to standard models considered here for this study and both compared with ROUGE scores.

The proposed ensemble model output compared with the existing models in this section. The results of BERT, PEGASUS and LexRank Model on the News Summery are displayed. All the important variants of ROUGE used in this study such as ROUGE-1, ROUGE-2 and ROUGE-L. Comparison made on the metrics for all individual performance using the three score. We have compared the proposed model with others based on these three variants.

The results in the Table 2 listed in tabular form that on the News summery dataset, PEGASUS performed better and shown the best result out of all the algorithms on ROUGE-1 metrics, and PEGASUS delivers the best performance for ROUGE-2 and ROUGE-L metrics out of all the compared algorithms. If we compare the overall average of F-score, then PEGASUS has the best F-score for all 15% improvement than BERT, 70% better than LexRank, and BERT has the second-best average F-score.

Table 2. List of ROUGE metrics for BERT, PEGASUS and LexRank on News Summary.

Model	ROUGE-1			ROUGE-2			ROUGE-3		
	Precision	Recall	F-score	Precision	Recall	F-score	Precision	Recall	F-score
BERT	0.53	0.67	0.63	0.29	0.46	0.29	0.41	0.73	0.49
PEGASUS	0.69	0.72	0.52	0.39	0.35	0.30	0.56	0.79	0.59
LexRank	0.16	0.65	0.25	0.07	0.36	0.11	0.11	0.46	0.18

Our study reported similar result to published literature of previous studies. The results to compare the summarization output indicators of 3 transformer-based model including BERT (BART model based on BERT) and PEGASUS reported by Goodwin et al., in 2020 that PEGASUS(FL) provides better performance [19], in line with our study results. The study also supported the claim and results of study reported by Zhang et al. in 2020 during his experiment of PEGASUS model [11, 19].

Here a visual representation of the above-gathered data, which will analyse the performance of different algorithms in Fig. 1. The precision of ROUGE-1 score of PEGASUS was highest near to 80%, whereas the precision of LexRank was less than 20% in ROUGE-1.

The results in Table 3 listed in tabular form that on the News summery dataset for all the models including proposed model. The output of the proposed model indicated improvement in comparison to existing models. The f-score of ROUGR-L improved

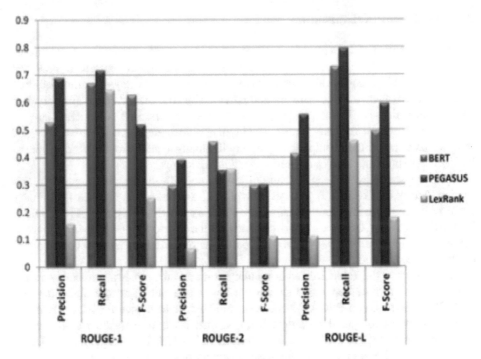

Fig. 1. Visual representation of ROUGE Score in existing models.

around 5%, however it has not made any justifiable difference compared to PEGASUS. The transformer-based model including PEGASUS and BERT demonstrated dramatic performance with respect to conventional LexRank model. In our study, Figs. 1 and 2 explained how both pre-trained models with performed well, the results are in accordance with similar studies [12, 20, 21].

Table 3. ROUGE score of all models

Model	ROUGE-1			ROUGE-2			ROUGE-3		
	Precision	Recall	F-score	Precision	Recall	F-score	Precision	Recall	F-score
Proposed model	0.62	0.73	0.62	0.41	0.46	0.40	0.61	0.70	0.63
BERT	0.53	0.67	0.63	0.29	0.46	0.29	0.41	0.73	0.49
PEGASUS	0.69	0.72	0.52	0.39	0.35	0.30	0.56	0.79	0.59
LexRank	0.16	0.65	0.25	0.07	0.36	0.11	0.11	0.46	0.18

In order to check, the performance of proposed model over the existing models, The ROUGE-L F-score value of our experiment on proposed model when statistically compared with student's t test, the p value was 0.2760 ($p < 0.05$), this indicate the results of the comparison is statistically not significant.

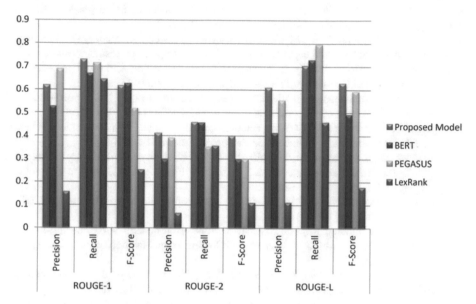

Fig. 2. Visual representation of all the existing Models including proposed one.

4   Conclusion

In this study, the news summery data in Kaggle for automatic text summarization evaluated over the in Zhang Ustry standard models. Attempt was made to compare the ensemble average of BERT output and PEGASUS output with standard models; however, the results indicate better performance but did not show any significant difference ($P < 0.05$). It was proposed and evaluated the model on standard datasets which demonstrated results comparable to the state-of-the-art models without access to any linguistic information. The results demonstrate the superior performance of our proposed model over the existing models. Due to time limitation, more databases on different models could not be analyzed. Attempt may be made by other researcher to improve upon this aspect in the future.

References

1. Vaswani, A., et al.: Attention is all you need. In: Advances in Neural Information Processing Systems, vol. 30 (2017)
2. See, A., Liu, P.J., Manning, C.D.: Get to the point: summarization with pointer-generator networks. arXiv preprint arXiv:1704.04368 (2017)

3. Aksenov, D., Julián, M., Peter, B., Robert, S., Leonhard, H., Georg, R.: Abstractive text summarization based on language model conditioning and locality modeling. arXiv arXiv: 2003.13027 (2020)
4. Radford, A., Wu, J., Child, R., Luan, D , Amodei, D., Sutskever, I.: Language models are unsupervised multitask learners. OpenAI Blog **1**(8), 9 (2019)
5. Fabbri, A.R., Kryściński, W., McCann, B., Xiong, C., Socher, R., Radev, D.: Summeval: re-evaluating summarization evaluation. Trans. Assoc. Comput. Linguist. **9**, 391–409 (2021)
6. Sutskever, I., Vinyals, O., Le, Q.V.: Sequence to sequence learning with neural networks. In: Advances in Neural Information Processing Systems, vol. 27 (2014)
7. Cho, K., Van Merriënboer, B., Bahdanau, D., Bengio, Y.: On the properties of neural machine translation: encoder-decoder approaches. arXiv preprint arXiv:1409.1259 (2014)
8. Agarwal, A., Lavie, A.: METEOR: an automatic metric for MT evaluation with high levels of correlation with human judgments. In: Proceedings of WMT-08 (2007)
9. Tjandra, A., Sakti, S., Nakamura, S.: Multi-scale alignment and contextual history for attention mechanism in sequence-to-sequence model. In: 2018 IEEE Spoken Language Technology Workshop (SLT), pp. 648–655. IEEE (2018)
10. Cohan, A., Goharian, N.: Revisiting summarization evaluation for scientific articles. arXiv preprint arXiv:1604.00400 (2016)
11. Zhang, J., Zhao, Y., Saleh, M., Liu, P.: PEGASUS: pre-training with extracted gap-sentences for abstractive summarization. In: International Conference on Machine Learning, pp. 11328–11339. PMLR (2020)
12. Devlin, J., Chang, M.W., Lee, K., Toutanova, K.: BERT: pre-training of deep bidirectional transformers for language understanding. arXiv preprint arXiv:1810.04805 (2018)
13. Celikyilmaz, A., Bosselut, A., He, X., Choi, Y.: Deep communicating agents for abstractive summarization. arXiv preprint arXiv:1803.10357 (2018)
14. Bouscarrat, L., Bonnefoy, A., Peel, T., Pereira, C.: STRASS: a light and effective method for extractive summarization based on sentence embeddings. arXiv preprint arXiv:1907.07323 (2019)
15. Hao, Y., Dong, L., Wei, F., Xu, K.: Visualizing and understanding the effectiveness of BERT. arXiv preprint arXiv:1908.05620 (2019)
16. Erkan, G., Radev, D.R.: LexRank: graph-based lexical centrality as salience in text summarization. J. Artif. Intell. Res. **22**, 457–479 (2004)
17. Lin, C.Y.: Rouge: a package for automatic evaluation of summaries. In: Text Summarization Branches Out, pp. 74–81 (2004)
18. Kaggle (2022). https://www.kaggle.com/
19. Goodwin, T.R., Savery, M.E., Demner-Fushman, D.: Flight of the PEGASUS? Comparing transformers on few-shot and zero-shot multi-document abstractive summarization. In: Proceedings of COLING. International Conference on Computational Linguistics, vol. 2020, p. 5640. NIH Public Access (2020)
20. Lewis, M., et al.: BART: denoising sequence-to-sequence pre-training for natural language generation, translation, and comprehension. arXiv preprint arXiv:1910.13461 (2019)
21. Liu, Y.: Fine-tune BERT for extractive summarization. arXiv preprint arXiv:1903.10318 (2019)

Correction to: Well Log Data Preparation and Effective Utilization of Drilling Parameters Using Data Science Based Approaches

Rahul Talreja, Thomas Coombs, Sulaf Assi, Noor Azma Ismail,
Manoj Jayabalan, Panagiotis Liatsis, Mohamed Mahyoub,
Abdullah Al-Hamid, and Hoshang Kolivand

Correction to:
**Chapter "Well Log Data Preparation and Effective Utilization
of Drilling Parameters Using Data Science Based Approaches"
in: Y. B. Wah et al. (Eds.):** *Data Science and Emerging
Technologies***, LNDECT 165,
https://doi.org/10.1007/978-981-99-0741-0_28**

Inadvertently, the first author's name was published with a type as Rhul instead of Rahul. The name has now been corrected.

The updated original version of this chapter can be found at
https://doi.org/10.1007/978-981-99-0741-0_28

© The Author(s), under exclusive license to Springer Nature Singapore Pte Ltd. 2023
Y. B. Wah et al. (Eds.): DaSET 2022, LNDECT 165, p. C1, 2023.
https://doi.org/10.1007/978-981-99-0741-0_39

Author Index

A

Ab Hamid, Siti Haslini   77, 92
Abdul Aziz, Nik Amnah Shahidah   77
Abdulhussain, Sadiq H.   403
Abdullah Sani, Nur Hurriyatul Huda   52
Abdullah, Azizi   293
Abdullah, Mohammad Nasir   329
Abdullah, Zul Hilmi   293
Abdul Rahim, Fiza   131
Abdul-Rahman, Shuzlina   494, 507
Aditsania, Annisa   282
Adnan, Noradilah   237
Ahmad, Azlin   222
Alamran, Maitham   357, 419
Al-Ataby, Ali   419, 434
Alatrany, Abbas   209
Aldhaibani, Omar A.   434
Alghenaim, Mohammed Fahad   131
Al-Hamid, Abdullah   209, 357, 373, 388,
         419, 535
Al-Jumeily OBE, Dhiya   37, 117, 209, 343,
         357, 373, 535
Alkawsi, Gamal   131
AL-Khassawneh, Yazan Alaya   3
Almajali, Sufyan   305
Almatarneh, Sattam   3
Al-Muni, Kdasy   37
Al-Nahari, Abdulaziz   168, 222, 357, 403
Al-Rimy, Bander Ali   168
Al-Sudani, Sahar   419, 434, 535
Aminudin, Muhammad Azri   451
Arifa, Irfanul   282
Ariffin, Ahmad Najmi   237
Assaggaf, Abdulrahman Mohammed Aqel
         168
Assi, Jolnar   343, 434
Assi, Sulaf   37, 209, 357, 373, 388, 419, 535

B

Bade, Abdullah   67
Bakar, Nur Azaliah Abu   131

C

Chakraborty, Snigdha   357
Coombs, Thomas   388, 535
Cuzzocrea, Alfredo   187

D

Daud, Paridah   67, 103, 195, 434

F

Fithriasari, Kartika   266

G

Giri-Rachman, Ernawati Arifin   480
Govindarajulu, Shatheesh Kumar   37

H

Hanandeh, Essam Said   3
Harper, Matthew   117, 343, 373
Hebbar, Saranga Veeramangal   403
Hia, Supriadi   465
Husin, Wan Zakiyatussariroh Wan   451
Hussein, Mohamad Fairul   195

I

Indratno, Sapto W.   480
Ismail, Noor Azma   222, 373, 388
Ismail, Noor Lees   117, 168, 195, 535
Ismail, Nor Azima   451
Ismail, Shaharudin   293
Ismail, Waidah   293
Ismail, Zurina   507

J

Jain, Aditi   117
Jamilat, Veronica S.   52
Jayabalan, Manoj   37, 117, 209, 388, 419,
         434, 535
Jinnovart, Thanaporn   522

© The Editor(s) (if applicable) and The Author(s), under exclusive license
to Springer Nature Singapore Pte Ltd. 2023
Y. B. Wah et al. (Eds.): DaSET 2022, LNDECT 165, pp. 547–548, 2023.
https://doi.org/10.1007/978-981-99-0741-0

K

Kaky, Ahmed   209, 434
Khan, Wasiq   117, 434
Kolivand, Hoshang   388, 419, 434
Kumar, Arav   316
Kurniawan, Isman   282
Kusuma, Purba Daru   18
Kuswanto, Heri   465

L

Lenka, Ranjita Kumari Biswal   535
Liatsis, Panagiotis   37, 117, 209, 388, 419, 434, 535
Lolla, Ravikanth   343
Loy, Chong Kim   343, 419
Lunn, Jan   343, 357
Lursinsap, Chidchanok   522

M

Mahmmod, Basheera M.   403
Mahyoub, Mohamed   388, 434
Mangsor, Nik Siti Madihah Nik   507
Mansor Noordin, Nur Aziha   52
Mohamad, Normaiza   37, 103, 195
Mohamad, Nur Maisarah   92
Mohd Aris, Fatin Ezzati   237
Muhaimin, Amri   494
Murugan, Saravanan   209
Musa, Omar   195
Mustafina, Jamila   37, 117, 209, 343, 357, 373, 535
Mustapa, Mazliana   237
Mustapha, Muhammad Firdaus   77, 92

N

Nasir, Syerina Azlin Md   507
Nasrallah, Gabriel   316
Nassiri Abrishamchi, Mohammad Ali   154

P

Pancoro, Adi   480
Pasaribu, Udjianna S.   480
Prastyo, Dedy Dwi   465

R

Rahim, Fiza Abdul   131
Rifki, Kevin Agung Fernanda   266
Rosdi, Ros Ameera   253

S

Sahai, Rahul   419
Saleh, Abdulrazak Yahya   253
Sapri, Nik Nur Fatin Fatihah   329
Sari, Kurnia Novita   480
Satpute, Saket   434
Seng, N. G. Meng   222
Setianingsih, Casi   18
Setiyowati, Susi   480
Sonhaji, Abdullah   480
Suhaimi, Aina Nafisya   451

T

Talreja, Rahul   388
Tengku Jalal, Tengku Noradilah   52

V

Vanduhe, Vanye Zira   131
Verma, Aniket   373

W

Wan Jaafar, Wan Ahmad Ridhuan   237
Wan Yaacob, Wan Fairos   329
Watson, Megan   37
Wibowo, Wahyu   494
Wilson, Megan   357
Wulandari, Rizka Shinta   18

Y

Yao, Danny Ngo Lung   67, 209
Yap, Bee Wah   329
Yousif, Maitham G.   209, 373
Yudho, Sasongko   52

Z

Zainal, Anazida   154
Zakaria, Lailatul Qadri   293
Zambri, Nur Shuhaila Meor   451
Zolkifly, Iznora Aini   67, 403
Zulkifli, Mohamad Khairul Naim   103

Printed in the United States
by Baker & Taylor Publisher Services